Foundation Engineering

FRONTISPIECE

Karl Terzaghi (1883–1963)

Founder and guiding spirit of soil mechanics, outstanding engineering geologist, and preeminent foundation engineer. He was the first to make a comprehensive investigation of the engineering properties of soils; he created or adapted most of the theoretical concepts needed for understanding and predicting the behavior of masses of soil; and he devised the principal techniques for applying scientific methods to the design and construction of foundations and earth structures.

Foundation Engineering

SECOND EDITION

RALPH B. PECK
Professor of Foundation Engineering
University of Illinois at Urbana–Champaign

WALTER E. HANSON
Consulting Engineer and Senior Partner
Hanson Engineers Incorporated, Springfield, Illinois

THOMAS H. THORNBURN
Professor of Civil Engineering
University of Illinois at Urbana–Champaign

JOHN WILEY & SONS

New York • Chichester • Brisbane • Toronto • Singapore

Copyright © 1953, 1974, by John Wiley & Sons, Inc.

All rights reserved. Published simultaneously in Canada.

Reproduction or translation of any part of this work beyond that permitted by Sections 107 or 108 of the 1976 United States Copyright Act without the permission of the copyright owner is unlawful. Requests for permission or further information should be addressed to the Permissions Department, John Wiley & Sons, Inc.

Library of Congress Cataloging in Publication Data:

Peck, Ralph Brazelton.
    Foundation engineering.

    Bibliography: p.
    1. Foundations. I. Hanson, Walter Edmund, 1916–   joint author. II. Thornburn, Thomas Hampton, joint author. III. Title.

TA775.P4 1974   624'.15   73-9877

ISBN 0-471-67585-7

Preface

The general organization and purpose of this edition do not differ from the first edition. To make the book more useful to instructors who wish to combine the teaching of soil mechanics and foundation engineering, the treatment of stress-deformation characteristics of soils has been expanded, and elementary discussions of flow nets, earth pressures, and stability of slopes have been added.

The scope of the book has been enlarged to include discussions of foundations on rock, on swelling soils, and on fill. Also a new chapter (Chapter 27) has been added on flexible earth-retaining structures such as braced and tieback walls and anchored bulkheads. The subject of pile-driving dynamics has been modernized and its implications brought to a practical level. These and other revisions, including the addition of several design plates, increase the usefulness of the book for practicing engineers.

References have been added so that the reader can consult sources. The suggested readings at the end of each chapter have several functions, depending on the subject. Some references provide additional details or background, but most citations include applications of the principles or procedures discussed in the text. All are annotated.

Ralph B. Peck
Walter E. Hanson
Thomas H. Thornburn

June 1973

Acknowledgments

The comments by E. J. Daily and H. O. Ireland on the manuscript of the first edition helped to establish the pattern of the book; they are still reflected in the present volume. As in the first edition, much information has been drawn from *Soil Mechanics in Engineering Practice*, by K. Terzaghi and Ralph B. Peck.

The sections dealing with rock and rock foundations have had the benefit of discussions with Don U. Deere and A. J. Hendron, Jr. M. T. Davisson's assistance was invaluable in developing the treatment of pile-driving dynamics and pile foundations in general. The section on pier foundations was revised extensively on the basis of many helpful comments by T. R. Maynard.

Various sections and design plates were reviewed by the partners and associates of Hanson Engineers; their comments and encouragement contributed substantially to the development of the present edition. The design plates in Part D that incorporate current practice in reinforced concrete were reviewed, together with the relevant text, by Narbey Khachaturian. The drafting of all the design plates was done by Rodney A. Huffman.

The ever-changing manuscripts were typed with care and skill by Mrs. Jane Dowding, Mrs. Grete Carlson, and Mrs. Claudie Daniels.

We thank all these people.

R.B.P.
W.E.H.
T.H.T.

Foreword to First Edition

In a broad sense, foundation engineering is the art of selecting, designing, and constructing the elements that transfer the weight of a structure to the underlying soil or rock. In practice, however, the actual construction is usually not carried out by the organization responsible for the design; the role of the engineer is generally considered to consist only of the selection of the type of foundation, the design of the substructure, and the supervision of construction. It is the purpose of this book to provide the basic information necessary to carry out this role.

The art of foundation engineering had its origins in antiquity. It developed with the accumulation of experience but without the aid of science until, in about 1920, it had reached a considerable degree of refinement. Yet, occasional inexplicable failures indicated that the limitations of the empirical procedures were not properly understood.

In the early 1920's there began a concerted scientific effort to determine the physical laws responsible for the behavior of the subsurface materials from which foundations derive their support. The new field of endeavor, known as soil mechanics, attracted and still holds the attention of many workers. It has provided new techniques for selecting the appropriate types of foundation under a given set of conditions and for predicting the performance of the completed substructure. It has by no means decreased the importance of the accumulated experience of the ages, but it has defined the limits within which the traditional techniques are applicable and has provided new techniques suitable under the circumstances in which the traditional procedures are not valid.

In recent years the power of science has become increasingly apparent, and there has been a tendency to discount the importance of the vast store of knowledge acquired during past generations by trial and error. This attitude has been reflected in many engineering schools by the replacement of courses in foundation engineering by others in soil mechanics, and by the prevalent opinion that detailed training in soil mechanics must precede and may even eliminate the need for training in foundation engineering.

In reality, soil mechanics is only one of the bodies of knowledge upon which the foundation engineer may draw. If studied to the exclusion of the other aspects of the art, it leads to the erroneous and dangerous impression that all problems in foundation engineering are susceptible of direct scientific solution. Unfortunately, the vagaries of nature and the demands of economy combine to eliminate this possibility.

The purpose of this book is to introduce the undergraduate student to the field of foundation engineering and to provide him with the ability to investigate and evaluate subsurface conditions, select the most suitable types of foundation for a given site, judge the performance of each type in service, and design the structural elements of the type finally selected. Where soil mechanics contributes to these purposes it is introduced, but it is nowhere presented merely for its own sake. Experience in the classroom has indicated that this approach develops in the student an excellent comprehension of the successful practice of foundation engineering.

Because much of the information concerning the characteristics of soils is common to the fields of foundation, highway, railway, and airport engineering, the curricula of several engineering schools contain a basic course in the properties of soils. To meet the needs of such a course, the first part of this book has been expanded beyond the limits adequate for a discussion of the foundations of structures. Appropriate sections for omission by students concerned only with the latter subject will be obvious to the experienced instructor.

RALPH B. PECK
WALTER E. HANSON
THOMAS H. THORNBURN

Contents

SYMBOLS   xvii

PART A. PROPERTIES OF SUBSURFACE MATERIALS

Chapter 1. Identification and Classification of Soils and Rocks   3

1.1.	Definition of Soil and Rock	3
1.2.	Purpose of Identification and Classification	3
1.3.	Description and Identification of Soils	4
1.4.	Index Properties of Soils	7
1.5.	Soil Grain Properties	8
1.6.	Weight-Volume Relationships of Soil Aggregate	11
1.7.	Structure and Consistency of Soil Aggregate	18
1.8.	Soil-Classification Systems	24
1.9.	Description and Classification of Rocks	30

Chapter 2. Hydraulic Properties of Soil and Rock   39

2.1.	Introduction	39
2.2.	Permeability of Soil	39
2.3.	Permeability of Rock	44
2.4.	Effective and Porewater Pressures	44
2.5.	Soil Moisture, Drainage, and Frost Action	47
2.6.	Seepage and Flow Nets	51

Chapter 3. Consolidation Characteristics of Soils   59

3.1.	Significance of Stress–Strain Characteristics of Earth Materials	59
3.2.	Consolidation Tests on Remolded Clays	59
3.3.	Consolidation Characteristics of Normally Loaded Deposits	61
3.4.	Computation of Settlement	62
3.5.	Consolidation Characteristics of Preloaded Deposits	63
3.6.	Consolidation Characteristics of Sensitive Clays	64
3.7.	Consolidation Characteristics of Residual Soils	65
3.8.	Consolidation Characteristics of Collapsible Soils	65
3.9.	Consolidation Characteristics of Sands	66
3.10.	Evaluation of Compressibility in Practice	66

3.11. Swelling Clays and Clay-Shales 67
3.12. Rate of Consolidation 68

Chapter 4. Stress-Deformation-Strength Characteristics of Soil and Rock   81

4.1. Behavior of Soils Under Complex States of Stress 81
4.2. Behavior in Shear of Idealized Granular Mass 81
4.3. Triaxial Tests and Mohr's Circle of Stress 83
4.4. Stress–Strain Relations for Dry Sands and Gravels 85
4.5. Mohr's Rupture Diagram 86
4.6. Shearing Strength of Dry Sands and Gravels 87
4.7. Influence of Water in Voids 88
4.8. Behavior of Fine-Grained Soils 90
4.9. Shearing Resistance of Unsaturated Soils 94
4.10. Effects of Repetitive Loads and Time 95
4.11. Selection of Test Procedures for Determining Shear Strength of Soils in Practice 96
4.12. Strength and Deformability of Rock 99

Chapter 5. Techniques of Subsurface Investigation   103

5.1. Methods of Exploration 103
5.2. Exploratory Borings 103
5.3. Sampling 106
5.4. Direct Measurements of Consistency and Relative Density 112
5.5. Miscellaneous Methods of Soil Exploration 116
5.6. Record of Field Exploration 120

Chapter 6. Character of Natural Deposits   125

6.1. Origin of Natural Deposits 125
6.2. Deposits Associated with Glaciation 129
6.3. Windblown Deposits 140
6.4. River and Continental Deposits 144
6.5. Organic and Shore Deposits 147
6.6. Unweathered Bedrock 149
6.7. Weathered Rock and Residual Soil 153

Chapter 7. Program of Subsurface Exploration   163

7.1. Development of Exploratory Program 163

PART B. TYPES OF FOUNDATIONS AND METHODS OF CONSTRUCTION

Chapter 8. Excavating and Bracing   169

8.1. Introduction 169
8.2. Open Excavations with Unsupported Slopes 169

8.3.	Sheeting and Bracing for Shallow Excavations	170
8.4.	Sheeting and Bracing for Deep Excavations	170
8.5.	Movements Associated with Excavation	173

Chapter 9. Drainage and Stabilization — 177

9.1.	Introduction	177
9.2.	Ditches and Sumps	177
9.3.	Well Points	178
9.4.	Deep-Well Pumps	180
9.5.	Sand Drains	180
9.6.	Miscellaneous Methods of Drainage and Stabilization	181

Chapter 10. Footing and Raft Foundations — 185

10.1.	Types of Footings	185
10.2.	Historical Development	185
10.3.	General Considerations	186
10.4.	Allowable Soil Pressures	186
10.5.	Combined Footings	187
10.6.	Raft Foundations	188
10.7.	Drainage, Waterproofing, and Dampproofing	189

Chapter 11. Foundations on Compacted Fill — 193

11.1.	Historical Development	193
11.2.	Design Considerations	193
11.3.	Deep-Seated Settlement	194
11.4.	Placement and Compaction of Fill	195
11.5.	Control of Compaction	198
11.6.	Proportioning and Details of Foundation Elements	199

Chapter 12. Pile Foundations — 203

12.1.	Function of Piles	203
12.2.	Types of Piles	203
12.3.	Installation of Piles	209
12.4.	Action of Piles Under Downward Loads	212
12.5.	Dynamics of Pile Driving	216
12.6.	Choice of Type of Pile	225
12.7.	Lateral and Upward Loads on Pile Foundations	225
12.8.	Negative Skin Friction	226

Chapter 13. Pier Foundations — 229

13.1.	Definitions	229
13.2.	Methods of Construction	229
13.3.	Drilled Piers	235

Chapter 14. Pier Shafts, Retaining Walls, and Abutments 245

- 14.1. Pier Shafts 245
- 14.2. Retaining Walls 246
- 14.3. Abutments 247

Chapter 15. Shoring and Underpinning 251

- 15.1. Shoring 251
- 15.2. Underpinning 251

Chapter 16. Damage Due to Construction Operations 255

- 16.1. Settlement Due to Excavation 255
- 16.2. Settlement Due to Vibrations 257
- 16.3. Settlement Due to Lowering the Water Table 257
- 16.4. Displacement Due to Pile Driving 258
- 16.5. Importance of Field Observations for Control of Construction Operations 258
- 16.6. Influence of Construction Procedures on Design 259

PART C. SELECTION OF FOUNDATION TYPE AND BASIS FOR DESIGN

Chapter 17. Factors Determining Type of Foundation 263

- 17.1. Steps in Choosing Type of Foundation 263
- 17.2. Bearing Capacity and Settlement 264
- 17.3. Design Loads 264

Chapter 18. Foundations on Clay and Plastic Silt 269

- 18.1. Significant Characteristics of Deposits of Clay and Plastic Silt 269
- 18.2. Footings on Clay 270
- 18.3. Rafts on Clay 276
- 18.4. Piers on Clay 277
- 18.5. Piles on Clay 281
- 18.6. Settlement of Foundations Underlain by Clay 286
- 18.7. Excavation in Clay 297
- 18.8. Lateral Displacements Due to Vertical Loads on Clay 301

Chapter 19. Foundations on Sand and Nonplastic Silt 307

- 19.1. Significant Characteristics of Sand and Silt Deposits 307
- 19.2. Footings on Sand 307
- 19.3. Rafts on Sand 318
- 19.4. Piers on Sand 321
- 19.5. Piles in Sand 322
- 19.6. Excavation in Sand 325
- 19.7. Effect of Vibrations 327

Chapter 20. Foundations on Collapsing and Swelling Soils — 333

- 20.1. General Considerations — 333
- 20.2. Foundations on Collapsing Soils — 334
- 20.3. Foundations on Swelling Soils — 337

Chapter 21. Foundations on Nonuniform Soils — 349

- 21.1. Introduction — 349
- 21.2. Soft or Loose Strata Overlying Firm Strata — 350
- 21.3. Dense or Stiff Layer Overlying Soft Deposit — 350
- 21.4. Alternating Soft and Stiff Layers — 352
- 21.5. Irregular Deposits — 353
- 21.6. Excavation and Stability of Slopes in Nonuniform Soils — 354

Chapter 22. Foundations on Rock — 361

- 22.1. Basis for Design — 361
- 22.2. Foundations on Unweathered Rock — 361
- 22.3. Treatment of Rock Defects — 364
- 22.4. Foundations on Weathered Rock — 367
- 22.5. Excavation in Rock — 369

PART D. DESIGN OF FOUNDATIONS AND EARTH-RETAINING STRUCTURES

Chapter 23. Individual Column and Wall Footings — 375

- 23.1. Basis for Design Procedures — 375
- 23.2. Critical Sections — 376
- 23.3. Placing of Reinforcement — 377
- 23.4. Depth of Spread Footings — 378
- 23.5. Procedure for Design and Use of Minimum-Depth Curves — 378
- 23.6. Isolated Column Footings on Piles — 379

Chapter 24. Footings Subjected to Moment — 385

- 24.1. Introduction — 385
- 24.2. Resultant Within Middle Third — 386
- 24.3. Resultant Outside Middle Third — 387
- 24.4. Moment About Both Axes — 391
- 24.5. Footings Having Unsymmetrical Shapes — 392
- 24.6. Moment on Pile Footings — 392
- 24.7. Piles Subjected to Tension — 394

Chapter 25. Combined Footings and Rafts — 401

- 25.1. Purpose of Combined Footings — 401
- 25.2. Combined Footings of Rectangular and Trapezoidal Shapes — 401

25.3.	Cantilever Footings	402
25.4.	Choice of Column Loads	403
25.5.	Structural Design of Combined Footings	403
25.6.	Basis for Design of Raft Foundations	405

Chapter 26. Retaining Walls and Abutments 415

26.1.	Introduction	415
26.2.	Proportions of a Cantilever Retaining Wall	415
26.3.	Summary of Forces Acting on Retaining Walls	416
26.4.	Earth Pressure	417
26.5.	Vertical Pressure Against Base	425
26.6.	Forces Resisting Sliding	426
26.7.	Summary of Procedure for Design of Cantilever Retaining Wall	427
26.8.	Pile-Supported Retaining Walls	428
26.9.	Abutments	437

Chapter 27. Flexible Earth-Retaining Structures 447

27.1.	Behavior of Flexible Earth-Retaining Structures	447
27.2.	Anchored Bulkheads	448
27.3.	Braced Cuts	456
27.4.	Tieback Bracing System	466

Bibliography 473

Index 487

Symbols

The dimensions of the quantities in the following list are the ones generally used in the text and in practice. The foundation engineer deals with soil mechanics, in which both metric and English units are common, as well as with several branches of structural engineering, in which English units are ordinarily used. This situation leads to an assortment of dimensions that is undesirable from an academic point of view. With wider adoption of *SI* units this assortment will be greatly reduced, but the International System is still so unfamiliar in foundation practice in North America that its use would not be justifiable in the present edition.

Conversion between English and metric systems is facilitated by the following relationship:

$$1 \text{ ton/sq ft} = 2 \text{ kips/sq ft} \approx 1 \text{ kg/sq cm} \approx 34 \text{ ft of water}$$
$$= 15 \text{ lb/sq in.}$$

in which the ton is the short ton of 2000 lb. Other useful conversion factors are 1 lb = 454 g, and 1 ft = 30.5 cm.

If no dimension is added to a symbol, the symbol indicates a pure number.

A (cm² or ft²) = area
A (cm) = spacing of electrodes, resistivity survey
A_b (cm²) = cross-sectional area of reinforcing bar
A_p (lb or lb/ft) = anchor pull
A_r (%) = area ratio of sampling spoon
A_s (in.²) = area of tensile steel (concrete design)
a (in.) = width of column; depth of compressive-stress block (concrete design)
a (cm² or ft²) = area (standpipe)
a_v (cm²/g) = coefficient of compressibility

B (ft) = width of footing or foundation; diameter of base of pier
B' (ft) = adjusted width of rectangular footing, for use in curves of Fig. 23.2
b (in.) = width of beam (concrete design)

C (any dimension) = constant; coefficient
C (lb) = total compressive force (concrete design)

C_c = compression index
C_N = correction factor for N-values in standard penetration test
C_r (%) = inside clearance ratio of sampling tube; correction for rate effect
C_t = secondary compression index
C_u = uniformity coefficient = D_{60}/D_{10}
C_w = groundwater correction factor
C_z = coefficient of curvature = $D_{30}^2/D_{10}D_{60}$
C.g. = center of gravity (design plates)
c (tons/ft²) = cohesion; intercept of rupture line on vertical axis of Mohr's rupture diagram; undrained shear strength
c (ft) = distance from neutral axis to extreme fiber (concrete design)
c (ft/sec) = velocity of stress wave in pile
c_a (lb/ft²) = ultimate skin friction on pier
c_b (lb/ft²) = undrained shear strength below base of excavation
c_v (cm²/sec) = coefficient of consolidation

D (ft) = depth; depth of embedment
D_e (in.) = external diameter of sampling spoon
D_f (ft) = depth of foundation (Fig. 18.1b)
D_i (in.) = internal diameter of cutting shoe
D_{it} (in.) = inside diameter of sampling tube
D_P (mm) = diameter of grain corresponding to percentage P on grain-size curve
D_w (ft) = depth to groundwater level
D_{10} (mm) = effective grain size (expressed in centimeters in equations)
DL (tons) = dead load (design plates)
d (ft) = distance; distance from given pile to center of gravity of pile group; depth from top of slab or beam to center of tension reinforcement (concrete design); diameter of pile or pier
d_b (ft) = diameter of bell
d_w (ft) = depth of water

E (tons/ft²) = modulus of elasticity
E (volt) = potential
E (ft lb) = energy of pile hammer per blow
E_i (tons/ft²) = initial tangent modulus
e = void ratio
e_0 = void ratio in situ
e_{max} = void ratio in loosest state
e_{min} = void ratio in densest state
Δe = change in void ratio
e (ft) = eccentricity

F = factor of safety

F (lb) = force
F (%) = percentage passing No. 200 sieve (in group index)
F_y (lb/in.^{2}) = specified yield strength of structural steel
f (lb/in.^{2}) = maximum fiber stress
f_c' (lb/in.^{2}) = specified compressive strength of concrete
f_s (lb/in.^{2}) = allowable stress in reinforcing steel
f_y (lb/in.^{2}) = specified yield strength of reinforcing steel

G = specific gravity of solid constituents
g (cm/sec^2) = acceleration of gravity

H (lb) = total horizontal load
ΣH (lb) = resultant horizontal load
H (ft or cm) = distance; thickness of stratum or fill, except when used in connection with theory of consolidation. In this event, H = half-thickness of layer drained top and bottom
H (ft) = height of fall of ram of pile driver
H_c (ft) = critical height of slope
H_s (ft) = reduced height of solid matter; equivalent height of surcharge; height of uniform surcharge equivalent to wheel loads behind abutments
ΔH (ft or cm) = position head (hydraulics)
h (ft or cm) = hydraulic head; height; vertical distance
Δh (ft or cm) = equipotential drop
h_c (cm) = height of capillary rise
h_{cc} (cm) = height of complete saturation by capillary rise

I (ft^4) = moment of inertia
I (amp) = current
I_d = density index of cohesionless soil
I_L (%) = liquidity index
I_P (%) = plasticity index
i = hydraulic gradient
i_c = critical hydraulic gradient
i_p (g/cm^3 or lb/ft^3) = pressure gradient (hydraulics)

K = earth-pressure coefficient
K (lb/ft) = spring stiffness, pile analysis
k = coefficient; ratio of least width of column or pedestal to width of footing
k (cm/sec) = coefficient of permeability
k (lb/in.^{3} or tons/ft^3) = modulus of subgrade reaction
k_A = coefficient of active earth pressure
k_h (lb/ft^2/ft) = coefficient of horizontal component of earth pressure
k_P = coefficient of passive earth pressure

k_v (lb/ft²/ft) = coefficient of vertical component of earth pressure
k_0 = coefficient of earth pressure at rest

L (ft or cm) = length; length of footing
LL (tons) = live load (design plates)
l (ft or cm) = length
l_d (in.) = development length of reinforcement (concrete design)
l (ft or cm) = length of arc

M (ft lb) = total moment
ΣM (ft lb) = resultant total moment
M_u (ft lb) = ultimate design moment (reinforced concrete)
m (in./ft) = batter of pile (horizontal/vertical)
m_v (cm²/g) = coefficient of volume compressibility

N = number of blows per foot in standard penetration test
N (lb) = normal component of force
N_c = bearing-capacity factor
N_{cq} = bearing capacity factor
N_d = number of equipotential drops (flow net)
N_f = number of flow channels (flow net)
N_q = bearing-capacity factor
N_s = stability factor
N_γ = bearing-capacity factor
$N_{\gamma q}$ = bearing-capacity factor
n = porosity; number of piles in cluster
n_b = number of bars in central width equal to short dimension of footing
n_d = depth ratio (stability of slopes)
n_t = total number of bars required by moment at critical section of footing
n_1 = number of piles in row

P (%) = per cent of grains smaller than given size
P (lb) = total load
P (ft) = perimeter of group of piles
P_A (lb/ft) = total active earth pressure per lineal foot of wall
P_h (lb/ft) = horizontal component of total active earth pressure per lineal foot of wall
P_P (lb/ft) = total passive earth pressure per lineal foot of wall
P_u (lb) = ultimate pile resistance; ultimate column load (concrete design)
P_v (lb/ft) = vertical component of total active earth pressure per lineal foot of wall
P_w (lb or lb/ft) = resultant water pressure

p (tons/ft²) = normal stress; total pressure
$\bar{p}$ (tons/ft²) = effective pressure
Δp (tons/ft²) = change in pressure; stress difference
Δp_f (tons/ft²) = stress difference at failure
p_A (tons/ft²) = active earth pressure
p_{cr} (tons/ft²) = critical pressure for crushing of soil structure
p_h (tons/ft²) = effective overburden pressure on horizontal plane; horizontal pressure on vertical plane
p_q (tons/ft²) = earth pressure due to uniform surcharge
p_P (tons/ft²) = passive earth pressure
p_s (tons/ft²) = pressure under which soil is presently in equilibrium
p_v (tons/ft²) = vertical pressure on horizontal plane
p_0 (tons/ft²) = original overburden pressure
p_0' (tons/ft²) = maximum pressure under which soil has previously been consolidated
p_1 (tons/ft²) = major principal stress
p_3 (tons/ft²) = minor principal stress; confining pressure

Q (cm³) = total discharge (hydraulics)
Q (lb) = ultimate pile capacity; point load
q (cm³/sec) = discharge per unit of time (hydraulics)
q (tons/ft²) = load per unit of area; soil pressure beneath footing
q' (lb/ft) = line load
q_a (tons/ft²) = allowable soil pressure
q_b (tons/ft²) = gross soil pressure; contact pressure
q_d (tons/ft²) = net ultimate bearing capacity
q_d' (tons/ft²) = maximum intensity of loading that can be supported by the soil at the base of the footing
q_n (lb/ft²) = net soil pressure for structural design of footing (column load divided by area of footing)
q_p (kg/cm²) = Dutch-cone resistance
q_t (tons/ft²) = toe pressure
q_u (tons/ft²) = unconfined compressive strength

R = ratio of particle sizes (filter requirements)
R (lb) = resultant force; resistance to dynamic penetration of pile
r (cm or ft) = radius; radial distance

S = ratio of long to short side of footing
S (ft) = settlement
S (lb) = total shearing force
S (in.³) = section modulus
S_r (%) = degree of saturation
S_t = sensitivity
s (ft) = spacing of piles in row
s (in.) = penetration of pile under last blow of

hammer; spacing of stirrups (concrete design)
s (tons/ft^2) = shearing resistance per unit of area

T (lb) = tangential component of force; total tensile force (concrete design)
T_s (g/cm) = surface tension of water
T_v = time factor (theory of consolidation)
t (in.) = total thickness of reinforced concrete member
t (sec) = time
t (tons/ft^2) = shearing stress

U (%) = average degree of consolidation
U_z (%) = degree of consolidation at depth z
u (tons/ft^2) = excess hydrostatic pressure
u_a (tons/ft^2) = cell pressure
u_d (tons/ft^2) = porewater pressure associated with stress difference Δp
u_{df} (tons/ft^2) = value of u_d at failure
u_w (tons/ft^2) = porewater pressure

V (cm^3 or ft^3) = total volume
V (lb) = total vertical load
ΣV (lb) = resultant total vertical load
V (lb/ft) = total shear on section of beam or retaining wall
V_g (cm^3 or ft^3) = total volume of gas
V_s (cm^3 or ft^3) = total volume of solids
V_u (lb) = total design shear force (concrete design)
V_v (cm^3 or ft^3) = total volume of voids
V_w (cm^3 or ft^3) = total volume of water
v (cm/sec or ft/sec) = discharge velocity (hydraulics); particle velocity (pile driving)
v_c (lb/in.^{2}) = nominal permissible shear stress for strength design of concrete
v_u (lb/in.^{2}) = total nominal design shear stress (concrete design)

W (g or lb) = total weight
W_c (lb/ft) = weight of concrete in retaining wall
W_H (lb) = weight of ram of pile driver
W_n (lb) = weight of element, pile analysis
W_s (g or lb) = total weight of solids
W_s (lb/ft) = weight of soil above heel of retaining wall
W_w (g or lb) = total weight of water
W.T. = water table (design plates)
w (%) = water content in per cent of dry weight
w_L (%) = liquid limit
w_{opt} (%) = optimum moisture content
w_P (%) = plastic limit

w_S (%) = shrinkage limit

x (ft) = distance

$\bar{x}$ (ft) = distance from reference line to center of gravity

y (ft) = distance

z (cm or ft) = depth

α (deg) = angle; contact angle; angle between given plane and plane on which principal stress acts

α_1 = reduction factor for skin friction on pier

α_2 = reduction factor for skin friction on pile

β (deg) = angle; slope angle of surface

γ (g/cm^3 or lb/ft^3) = unit weight (soil, water, and air)

γ' (g/cm^3 or lb/ft^3) = submerged unit weight

γ_d (g/cm^3 or lb/ft^3) = unit weight of soil if water is entirely replaced by air

γ_{max} (g/cm^3 or lb/ft^3) = maximum dry density for given compaction procedure

γ_p (lb/ft^3) = unit weight of pile material

γ_s (g/cm^3 or lb/ft^3) = unit weight of solid constituents

γ_{sat} (g/cm^3 or lb/ft^3) = unit weight of saturated soil

γ_w (g/cm^3 or lb/ft^3) = unit weight of water

γ_z (g/cm^3 or lb/ft^3) = dry unit weight at zero air voids

δ (deg) = shearing distortion; angle of wall friction

ϵ = strain

μ = Poisson's ratio

ρ (ohm cm) = resistivity

ρ (lb sec^2/ft^4) = mass density of pile material

ρ = ratio of reinforcement area to concrete area

Σ = summation

ϕ (deg) = angle of internal friction; angle of shearing resistance

ϕ_d (deg) = angle of shearing resistance for dry sand; effective stress friction angle

Foundation Engineering

PART A

Properties of Subsurface Materials

The behavior of every foundation depends primarily on the engineering characteristics of the underlying deposits of soil or rock. Therefore, the foundation engineer must be able to distinguish among the various deposits of different character, to identify their principal constituents, and to determine their physical properties. Part A provides the information necessary to accomplish these objectives.

Arthur Casagrande (1902–   )

Professor of Soil Mechanics and Foundation Engineering, Harvard University. He has been responsible for many advances in soil mechanics, particularly in the development of procedures and apparatus for testing soils, and in the techniques for identifying and classifying soil materials. In 1936 he organized the First International Conference on Soil Mechanics and Foundation Engineering. Through the stimulus of this conference, through his extensive consulting practice, and through his outstanding ability as a teacher, he has exerted a powerful influence for the acceptance of soil mechanics in foundation engineering.

PLATE 1.

CHAPTER 1

Identification and Classification of Soils and Rocks

1.1. Definition of Soil and Rock

The terms rock and soil, as used by the civil engineer, imply a clear distinction between two kinds of foundation materials. *Rock* is considered to be a natural aggregate of mineral grains connected by strong and permanent cohesive forces. *Soil*, on the other hand, is regarded as a natural aggregate of mineral grains, with or without organic constituents, that can be separated by gentle mechanical means such as agitation in water. These convenient definitions are generally understood and are used in this book. Nevertheless, in reality there is no sharp distinction between rock and soil. Even the strongest and most rigid rocks may be weakened by the processes of weathering, and some highly indurated soils may exhibit strengths comparable to those of weathered rock.

1.2. Purpose of Identification and Classification

Nature, without benefit of man, has furnished the materials on or within which the engineer founds his structures. The engineer, to prepare his design, must learn what materials are present and what properties they possess. He gains this knowledge partly by reference to geologic and engineering literature, but mainly by extracting, examining, and possibly testing samples he believes will be representative of the materials. He uses the knowledge in combination with other data to visualize the probable state, arrangement, and behavior of the deposits.

In all branches of civil engineering and especially in foundation engineering, experience is a priceless asset. Indeed, the accumulated experience of generations of foundation engineers, including those of our own times, constitutes the essence of the art. Yet, unless the individual experiences of all engineers can be summarized into a body of knowledge that can be assimilated readily, they lose their value to the engineering profession.

In a general way, it has been found that soils, and to a lesser extent rocks, can be classified into groups within each of which the significant engineering properties are somewhat similar. Consequently, proper classification of subsurface materials is an important step in connection with any foundation job, because it provides the first clues to the experiences that may be anticipated during and after construction. The ability to identify and classify rocks and soils properly is, therefore, basic to the analysis of all engineering problems dealing with earth materials.

The detail with which samples are described, tested, and evaluated depends on the type of structure to be built, on considerations of economy, on the nature of the earth materials, and to some extent on the method of sampling. The samples should be described first on the bases of a visual inspection and certain simple tests that can be performed in the field as easily as in the laboratory. Thereupon the material can usually be classified into one of several major groups. Later, simple laboratory tests may be carried out to verify the original classification. Most systems of soil classification utilized by engineers permit an individual with only a limited amount of training to perform a visual classification of soil samples. They also provide for a more precise classification based on numerical values that can easily be determined in the laboratory.

The identification and classification of the products of nature constitute an artificial procedure, because these materials are infinitely varied and do not lend themselves to separation into distinct categories. As a result, various arbitrary systems of classification have been developed, each with certain advantages and disadvantages for a particular purpose. Furthermore, as attempts are made to refine any one system, the system inevitably becomes more complicated and ultimately becomes so cumbersome that it defeats its own purpose.

To avoid this difficulty, it is preferable to make use of relatively simple systems of classification with only a few categories, to one of which a given rock or soil can be assigned. More detailed information concerning the given rock or soil can best be summarized by stating the numerical results, known as *index properties*, of certain physical tests, known as *classification tests*. If the classification tests are properly chosen, soils or rock materials having similar index properties are likely to exhibit similar engineering behavior.

The usefulness of index properties is so great that they are discussed in detail in the following pages. In addition to their value in the correlation of construction experience, they provide a means for checking the correctness of the field identification of a given material. If the material has been improperly identified, the index properties indicate the error and lead to correct classification. Thus, even the beginner with no experience of his own can be assured that his classifications are appropriate, and he can then take advantage of the experiences of other engineers. The techniques for acquiring this knowledge are described in Arts. 1.3 to 1.8.

1.3. Description and Identification of Soils

Principal Types of Soil. The principal terms used by civil engineers to describe soils are gravel, sand, silt, and clay. Most natural soils consist of a mixture of two or more of these constituents, and many contain an admixture of organic material in a partly or fully decomposed state. The mixture is given the name of the constituent that appears to have the most influence on its behavior, and the other constituents are indicated by adjectives. Thus a silty clay has predominantly the properties of a clay but contains a significant amount of silt, and an organic silt is composed primarily of silt-sized mineral matter but contains a significant amount of organic material.

Gravels and sands are known as *coarse-grained* soils, and silts and clays as *fine-grained* soils. The distinction is based on whether the individual particles can be differentiated by the naked eye. The methods of describing coarse-grained soils differ from those appropriate for fine-grained soils; therefore, the procedures are discussed under separate headings.

Coarse-Grained Soil Materials. The coarse-grained soil materials are mineral fragments that may be identified primarily on the basis of particle size.

Particles having a diameter greater than about 5 mm are classified as *gravel*. However, if the diameter exceeds about 200 mm (8 in.) the term *boulder* is usually applied.

If the grains are visible to the naked eye, but are less than about 5 mm in size, the soil is described as a *sand*. This name is usually further modified as *coarse, medium*, or *fine*. The definitions of these terms must be chosen arbitrarily. In the United States the ASTM classification of size limits given in Table 1.1 has been adopted as standard for engineering purposes.

TABLE 1.1 Particle Size Limits of Soil Constituents, ASTM Classification (in Millimeters)

Gravel	Larger than 4.75
Coarse sand	4.75 to 2.00
Medium sand	2.00 to 0.425
Fine sand	0.425 to 0.075
Fines (combined silt and clay)	Smaller than 0.075

A complete verbal description of a coarse-grained soil includes more than an estimate of the quantity of material in each size range. The *gradation, particle shape*, and *mineralogical composition* should also be noted whenever possible. The gradation may be described as *well-graded, fairly well-graded, fairly uniform, uniform*, or *gap-graded*. Well-graded soils contain a good representation of all particle sizes ranging from coarse to fine. The particles of uniform soils are all approximately the same size. Gap-graded soils consist of mixtures of uniform coarse-sized particles and uniform fine-sized particles, with a break in gradation between the two sizes. Any soil not well-graded may be characterized as *poorly graded*.

The shape of the coarse-grained particles in a soil has an influence on the density and stability of the soil deposit. The usual terms describing grain shape are illustrated in Fig. 1.1.

When the coarser particles of the soil are inspected by the naked eye or with a small hand lens, an effort should be made to estimate the degree of weathering. The prevalence of weak rock materials, such as shale and mica, should also be noted since these may influence the durability or compressibility of the deposit.

Fine-Grained Soil Materials. Inorganic silt, which constitutes the coarser portion of the microscopic soil fraction, possesses little or no plasticity or cohesion. The least plastic varieties consisting primarily of very fine rounded quartz grains are called *rock flour*. The most plastic varieties containing an appreciable quantity of flake-shaped particles are called *plastic silt*.

Clay is predominantly an aggregate of microscopic and submicroscopic flake-shaped crystalline minerals. It is characterized by the typical colloidal properties of plasticity, cohesion, and the ability to adsorb ions. These properties are exhibited over a wide range of water content.

The distinction between silt and clay cannot be based on particle size because the significant physical properties of the two materials are related only indirectly to the size of the particles. Furthermore, since both are microscopic, physical properties other than particle size must be used as criteria for field identification.

The *dry strength* provides one basis for distinction. A small briquet of the soil is molded and allowed to dry in the air. It is then broken and a fragment about $\frac{1}{8}$ in. in size is pressed between thumb and forefinger. The effort required to break the fragment provides a basis for describing the strength as *very low, low, medium, high*, or *very high*. A clay fragment can be broken only with great effort, whereas a silt fragment crushes easily.

Since silts are considerably more permeable than clays, the *dilatancy* or *shaking test* may also be used to distinguish between the two materials. In this test a small amount of soil is mixed with water to a very soft consistency in the palm of the hand. The back of the hand is then lightly tapped. If the soil is silty, water rises quickly to its surface and gives it a shiny or glistening appearance. Then if the soil pat is deformed, in some instances by squeezing and in others by stretching, the water flows back into it and leaves the surface with a dull appearance. Usually, the greater the proportion of clay

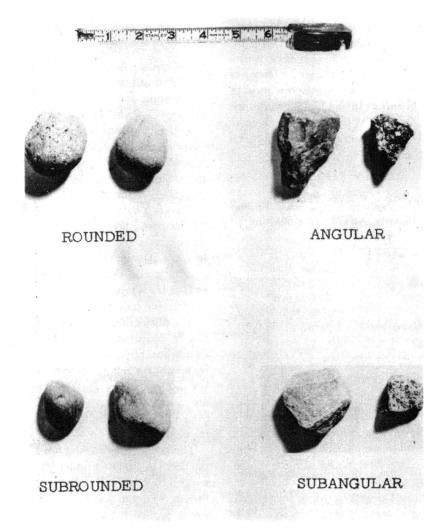

FIGURE 1.1. Typical shapes of coarse particles (after U.S. Bureau of Reclamation, 1963).

in the sample, the slower the reaction to the test. The reaction is described as *rapid*, *slow*, or *none*.

The property of *plasticity* is characteristic of clays and may also be used as the basis for a simple field test. At certain moisture contents a soil that contains appreciable quantities of clay can be deformed and remolded in the hand without disintegration. Thus, if a sample of moist soil can be manipulated between the palms of the hands and fingers and rolled out into a long thin thread, it unquestionably contains a significant amount of clay. As moisture is lost during continued manipulation, the soil approaches a nonplastic condition and becomes crumbly. Just before the crumbly state is reached, a highly plastic clay can be rolled into a long thread, with a diameter of approximately $\frac{1}{8}$ in., which has sufficient strength to support its own weight. A silt, on the other hand, can seldom be rolled into a thread with a diameter as small as $\frac{1}{8}$ in. without severe cracking, and is completely lacking in tensile strength unless small amounts of clay are present. The record of a simple plasticity test should indicate not only whether a plastic thread can be formed, but also the toughness of the thread as it nears the crumbling stage. This condition is described as *weak and friable*, *medium*, or *tough*.

A fourth procedure, known as the *disper-*

persion test, is also useful for distinguishing between silt and clay, and for making a rough estimate of the relative amounts of sand, silt, and clay in a material. A small quantity of the soil is dispersed with water in a glass cylinder or test tube and then allowed to settle. The coarser particles fall out first and the finest particles remain in suspension the longest. Ordinarily, sands settle in 30 to 60 sec. Material of silt size settles in 15 to 60 min, whereas that of clay size remains in suspension for at least several hours and usually for several days unless the particles of clay combine in groups or floccules (see Art. 1.5).

Organic Soil Materials. Very small quantities of organic matter often have a significant influence on the physical properties of soils. Most organic soils are weaker and more compressible than soils having the same mineral composition but lacking in organic matter. The presence of an appreciable quantity of organic material can usually be recognized by the dark gray to black color and the odor of decaying vegetation that it lends to the soil.

Organic silt is a fine-grained, more or less plastic soil containing mineral particles of silt and finely divided particles of organic matter. Shells and visible fragments of partly decayed vegetable matter may also be present.

Organic clay is a clay soil that owes some of its significant physical properties to the presence of finely divided organic matter.

Highly organic soil deposits such as *peat* or *muck* may be distinguished by a dark-brown to black color, by the presence of fibrous particles of vegetable matter in varying states of decay, and by the characteristic organic odor.

Combinations of organic and mineral soil materials are not always easily recognized, particularly if the organic content is small. Nevertheless, the presence of organic matter should always be suspected if the soil has a dark-brown, dark-gray, or black color. If the organic odor cannot be distinguished, it can sometimes be brought out by a slight amount of heat.

A summary of the reactions of the fine-grained inorganic and partly organic soils to the simple field identification tests is given in Table 1.2.

1.4. Index Properties of Soils

In the preceding article, simple methods were described for identifying the major soil components. These methods, however, represent only the first step in the adequate description of soil materials. They must be supplemented by other procedures leading to quantitative results that may be related to the physical properties with which the engineer is directly concerned. As stated in Art. 1.2, the tests required for this purpose are known as classification tests, and the results as the index properties of the soils.

By performing suitable classification tests and determining the corresponding index properties, the engineer acquires the means for describing a given soil accurately with-

Table 1.2 Identification of Fine-Grained Soil Fractions from Manual Tests

Typical Name	Dry Strength	Dilatancy Reaction	Toughness of Plastic Thread	Time to Settle in Dispersion Test
Sandy silt	None to very low	Rapid	Weak to friable	30 sec to 60 min
Silt	Very low to low	Rapid	Weak to friable	15 to 60 min
Clayey silt	Low to medium	Rapid to slow	Medium	15 min to several hours
Sandy clay	Low to high	Slow to none	Medium	30 sec to several hours
Silty clay	Medium to high	Slow to none	Medium	15 min to several hours
Clay	High to very high	None	Tough	Several hours to days
Organic silt	Low to medium	Slow	Weak to friable	15 min to several hours
Organic clay	Medium to very high	None	Tough	Several hours to days

out the use of verbal descriptions that are subject to misunderstanding on account of vague terminology. The development of the ability to think of soils in terms of numerical values of their index properties should be one of the foremost aims of every engineer who deals with foundations.

Index properties may be divided into two general types, soil *grain properties* and soil *aggregate properties*. The soil grain properties are the properties of the individual particles of which the soil is composed, without reference to the manner in which these particles are arranged in a soil deposit. Thus, it is possible to determine the grain properties of any soil sample, whether disturbed or undisturbed. Soil aggregate properties, on the other hand, depend on the structure and arrangement of the particles in the soil mass. Although soil grain properties are commonly used for identification purposes, the engineer should realize that the soil aggregate properties have a greater influence on the engineering behavior of a soil.

1.5 Soil Grain Properties

Size of Grains. The most important grain property of coarse-grained soils is the *particle-size distribution*. This is determined by performing a *mechanical analysis*. The sizes of coarse-grained constituents can be determined by means of a set of sieves. The finest sieve commonly used in the field or in the laboratory is the No. 200 U.S. Standard sieve in which the width of the openings is 0.075 mm. For this reason 0.075 mm has been accepted as the standard boundary between coarse-grained and fine-grained materials.

To determine the particle-size distribution of any soil containing fine-grained material, a wet method of mechanical analysis must be used. All wet methods of analysis are based on Stokes's law, which expresses the velocity at which a spherical particle falls through a fluid medium as a function of the diameter and specific gravity of the particle. A suspension of the soil is first agitated and then allowed to stand at rest. After a given time has elapsed, all particles coarser than a certain size have settled below some plane at an arbitrary depth in the suspension. This size can be computed by means of Stokes's law. The corresponding density of the suspension at the arbitrary depth is a measure of the quantity of soil smaller than the computed size. Thus, by making density measurements at various times, the particle-size distribution can be determined.

For engineering purposes, the density is usually measured by means of a *hydrometer*. The details of the procedure are set forth in ASTM Method D-422. A sample of about 50 g of soil is dispersed in 1 liter of distilled water and poured into a standard sedimentation cylinder. The suspension in the cylinder is then shaken for approximately 1 min and the cylinder set upright on a plane horizontal surface. A special hydrometer of streamlined design is inserted in the suspension, and readings of the density are taken after various intervals of time. It is customary to take readings at 2, 4, 8, 15 min, and so on. The computations based on these readings lead to the particle-size distribution on the assumption that all the particles are spherical. Actually, the finest soil particles are not spherical, but are flake-shaped or needle-shaped. The particle size actually determined, therefore, is the diameter of a sphere that would settle out of suspension at the same rate as the soil particle.

One of the most common sources of error involved in wet mechanical analysis is inadequate dispersion of the fine-grained soil particles. The laboratory technician may believe that he is determining the sizes of the individual particles, whereas in reality he may be measuring the sizes of floccules composed of several particles. To avoid flocculation, a small amount of an electrolyte, known as a *dispersing agent*, is added to the suspension. There is no way to determine by ordinary laboratory tests when ultimate dispersion has been obtained. It is, therefore, sometimes necessary to resort to several different methods of dispersion if there is reason to doubt the validity of the data obtained. It has been found that polyphosphate compounds are generally the most effective dispersing agents. Sodium hexa meta phos-

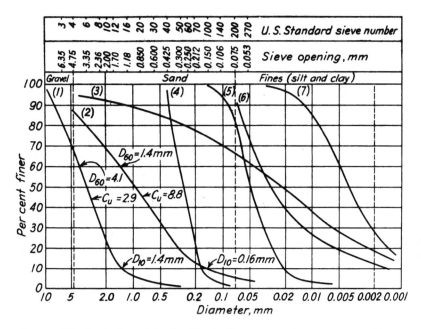

FIGURE 1.2. Typical particle-size distribution curves of natural soils. (1) Pea gravel, Castle Rock, Colo. (2) River gravel, Denver, Colo. (3) Glacial till, Peoria, Ill. (4) Sand, Grenada, Miss. (5) Glacial rock flour, Winchester, Mass. (6) Clayey silt, Smead, Mont. (7) Silty clay, Marathon, Ontario, Can.

phate is the most commonly used, but in some instances trisodium phosphate may produce more complete dispersion.

The use of the electron microscope enables the research investigator to determine the actual size and shape of fine-grained soil particles, but such refinement is neither economical nor practicable in the routine classification of soils.

The results of a mechanical analysis are usually presented in the form of a particle-size distribution curve. The percentage P of material finer than a given size is plotted as the ordinate to a natural scale, and the corresponding particle diameter D_P, in millimeters, is plotted as the abscissa to a logarithmic scale. A plot of this type has the advantage that materials of equal uniformity are represented by curves of identical shape whether the soil is coarse-grained or fine-grained. Moreover, the shape of the curve is indicative of the grading. Uniform soils are represented by nearly vertical lines, and well-graded soils by S-shaped curves that extend across several cycles of the logarithmic scale. Figure 1.2 shows particle-size curves for soils of several types.

The particle-size characteristics of soils can be compared most conveniently by a study of certain significant numerical values derived from the distribution curves. The two most commonly used by engineers are designated as D_{10}, the *effective grain size*, and $C_u = D_{60}/D_{10}$, the *uniformity coefficient*. The effective size is the diameter of the particle corresponding to $P = 10$ per cent on the particle-size plot. Hence, 10 per cent of the particles are finer and 90 per cent are coarser than the effective size (see Fig. 1.2). It is possible to have a gap-graded soil with a large uniformity coefficient which is actually composed of two uniformly graded fractions. The coefficient of curvature, $C_z = (D_{30})^2/(D_{10} \times D_{60})$, is a value that can be used to identify such soils as poorly graded. In well-graded gravels, C_u is greater than 4 and C_z is between 1 and 3. In well-graded sands, C_u is greater than 6 and C_z is between 1 and 3. (See ASTM Designation D-2487, Classification of Soils for Engineering Purposes.)

Mineralogical Composition. The most important grain property of fine-grained soil materials is the mineralogical composition. If the soil particles are smaller than about 0.002 mm, the influence of the force of gravity on each particle is insignificant compared with that of the electrical forces acting at the surface of the particle. A material in which the influence of the surface charges is predominant is said to be in the colloidal state. The colloidal particles of soil consist primarily of *clay minerals* that were derived from rock minerals by weathering, but that have crystal structures differing from those of the parent minerals.

The three most important groups of clay minerals are *smectite*, *illite*, and *kaolinite*. They are all crystalline hydrous aluminosilicates. The results of studies using the electron microscope and X-ray diffraction techniques show that the clay minerals have a lattice structure in which the atoms are arranged in several sheets, similar to the pages of a book. The arrangement and the chemical composition of these sheets determine the type of clay mineral.

The basic building blocks of the clay minerals are the silica tetrahedron and the alumina octahedron. These blocks combine into tetrahedral and octahedral sheets to produce the various types of clays. *Two-layer minerals* have a single tetrahedral sheet joined to a single octahedral sheet to form what is called a 1:1 lattice structure. Kaolinite is a typical two-layer mineral. In *three-layer minerals* a single octahedral sheet

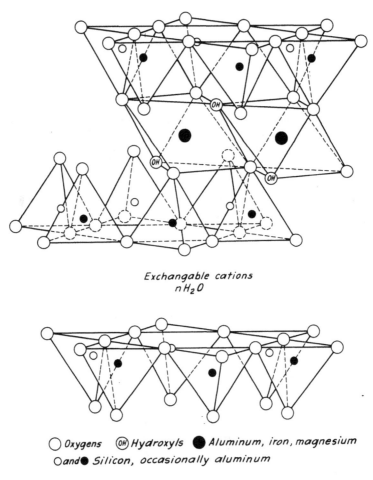

FIGURE 1.3. Diagrammatic sketch of the structure of montmorillonite (after Grim, 1962).

is sandwiched between two tetrahedral sheets to give a 2:1 lattice structure. Figure 1.3 is a diagrammatic sketch of the structure of montmorillonite, one of the smectites, representative of the 2:1 lattice. The structure of illite is similar, but some of the silicon atoms are replaced by aluminum and, in addition, potassium ions are present between the tetrahedral sheets of adjacent crystals.

The differences in the structural configuration of the clay lattices together with variations caused by the substitution of other atoms for silicon and aluminum lead to differences in the intensity of electrical charges that exist on the surfaces of various kinds of clays. These in turn lead to differences in chemical properties, as discussed in Art. 1.7.

1.6 Weight-Volume Relationships of Soil Aggregate

Definitions. The looseness or denseness of a soil sample may be determined quantitatively in the laboratory. The terms porosity, void ratio, and relative density are commonly used to define the density of the sample. Figure 1.4 is a diagram of a soil sample in a sealed container as it would look if the solid, liquid, and gaseous phases could be segregated. The volume of solids is designated by the symbol V_s, the volume of water by V_w, and the volume of gas by V_g. Since the relationship between V_g and V_w usually changes with groundwater conditions as well as under imposed loads, it is convenient to designate all the volume not occupied by solid material as void space, V_v. If the total volume of the sample is designated as V, then the *porosity* is defined by the equation

$$\text{Porosity, } n = V_v/V \qquad 1.1$$

Usually this value is expressed as a percentage. As a soil is compressed, the values of both the numerator and the denominator of the preceding equation change. It is convenient in many of the calculations necessary in settlement computations to refer the void space to an unchanging denominator. For this reason the quantity known as *void ratio* is commonly used. It is defined as

$$\text{Void ratio, } e = V_v/V_s \qquad 1.2$$

One of the most important index properties of fine-grained soils is its *water content, w*. It is defined as

$$\text{Water content, } w(\%) = 100W_w/W_s \qquad 1.3$$

In this equation, W_w is the weight of water and W_s is the weight of oven-dry solid matter. The weight of water is referred to the unchanging quantity W_s rather than to the total weight of the sample. As the temperature of a drying mixture of soil and water is increased, the mixture continues to lose moisture until at rather high temperatures the minerals that constitute the soil break down and lose their water of constitution. For this reason comparisons of water contents are meaningless unless the temperature at which the soil is dried is standardized. The standard oven temperature is 105 to 115°C.

Many soils below the water table and

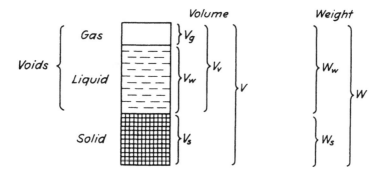

FIGURE 1.4. Diagram of a sample of soil illustrating meaning of symbols used in weight-volume relationships.

some fine-grained soils above it are in a saturated condition. However, the voids of most soils above the water table are filled partly with water and partly with air. Even some submerged soils have a significant air or gas content. The *degree of saturation* is defined as

Degree of saturation,
$$S_r(\%) = 100 V_w/V_v \qquad 1.4$$

Thus, at a degree of saturation of 100 per cent all of the void space is filled with water.

The weight per unit of volume or *unit weight* γ is one of the most important physical properties of a soil. It must be known, for example, before computations of earth pressure or overburden pressure can be made. By definition

$$\text{Unit weight, } \gamma = W/V \qquad 1.5$$

wherein W is the total weight of the soil including the soil moisture and V is the total volume. It is convenient to indicate particular values of unit weight by means of subscripts. If the soil is completely saturated, that is, if $V_g = 0$, its unit weight is designated by γ_{sat}. If the soil is oven-dry, its unit weight is denoted by γ_d, designated as *dry unit weight* or *dry density*, and is defined as

$$\text{Dry unit weight, } \gamma_d = W_s/V \qquad 1.6$$

If the water content is known, the dry density of a moist sample can be computed as

$$\gamma_d = \frac{100 W}{(100 + w)V} = \frac{100\gamma}{100 + w} \qquad 1.6a$$

In studies of the compaction of soils it is sometimes useful to compute the dry unit weight that would be obtained if the volume of a moist sample were decreased by exclusion of the air until the degree of saturation of the sample just reached 100 per cent. This condition is designated as *zero air voids*. The corresponding unit weight,

Dry unit weight at zero air voids,
$$\gamma_z = \frac{W_s}{V_w + V_s} \qquad 1.7$$

In practice, it is often inconvenient to determine the value of γ directly from measurements of the total weight and total volume. It is more commonly determined indirectly by computation based on a knowledge of the *unit weight of the solid constituents* γ_s. This quantity is defined as

Unit weight of solid constituents,
$$\gamma_s = W_s/V_s \qquad 1.8$$

It is often preferable to deal with the *specific gravity of the solid constituents* G, defined as

Specific gravity of solid constituents,
$$G = \gamma_s/\gamma_w \qquad 1.9$$

Table 1.3 Specific Gravity of Most Important Soil Constituents[a]

Gypsum	2.32	Dolomite	2.87
Montmorillonite[b]	2.65–2.80	Aragonite	2.94
Orthoclase	2.56	Biotite	3.0–3.1
Kaolinite	2.6	Augite	3.2–3.4
Illite[b]	2.8	Hornblende	3.2–3.5
Chlorite	2.6–3.0	Limonite	3.8
Quartz	2.66	Hematite, hydrous	4.3±
Talc	2.7	Magnetite	5.17
Calcite	2.72	Hematite	5.2
Muscovite	2.8–2.9		

[a] From E. S. Larsen and H. Berman, *The Microscopic Determination of the Non-Opaque Minerals*, second edition, U. S. Department of the Interior, Bull. 848, Washington, 1934.
[b] From R. E. Olson and G. Mesri (1970). "Mechanisms Controlling the Compressibility of Clays," *ASCE J. Soil Mech.*, 96, No. SM6.

Table 1.4 Porosity, Void Ratio, and Unit Weight of Typical Soils in Natural State

Description	Porosity (n)	Void Ratio (e)	Water Content (w)[a]	Unit Weight			
				g/cu cm		lb/cu ft	
				γ_d[b]	γ_{sat}[c]	γ_d	γ_{sat}
1. Uniform sand, loose	0.46	0.85	32	1.43	1.89	90	118
2. Uniform sand, dense	0.34	0.51	19	1.75	2.09	109	130
3. Mixed-grained sand, loose	0.40	0.67	25	1.59	1.99	99	124
4. Mixed-grained sand, dense	0.30	0.43	16	1.86	2.16	116	135
5. Windblown silt (loess)	0.50	0.99	21	1.36	1.86	85	116
6. Glacial till, very mixed-grained	0.20	0.25	9	2.12	2.32	132	145
7. Soft glacial clay	0.55	1.2	45	1.22	1.77	76	110
8. Stiff glacial clay	0.37	0.6	22	1.70	2.07	106	129
9. Soft slightly organic clay	0.66	1.9	70	0.93	1.58	58	98
10. Soft very organic clay	0.75	3.0	110	0.68	1.43	43	89
11. Soft montmorillonitic clay (calcium bentonite)	0.84	5.2	194	0.43	1.27	27	80

[a] w = water content when saturated, in per cent of dry weight.
[b] γ_d = dry unit weight.
[c] γ_{sat} = saturated unit weight.

where γ_w is the unit weight of water, taken as 1 g/cu cm in the metric system or 62.5 lb/cu ft in the English system. The value of γ_s or G may be determined by test in the laboratory, but it can usually be estimated with sufficient accuracy. For routine computations, the value of G for sands may be taken as 2.65. Tests on a large number of clay soils have indicated that the value of G usually falls in the range from 2.5 to 2.9 with an average value of about 2.7.

Table 1.3 gives the specific gravity of the most important soil constituents. It may be of assistance in estimating the value of G for a soil of known mineral composition.

Typical values of porosity, void ratio, and unit weight of various soils are listed in Table 1.4.

Density of Soil Aggregate. The behavior of any soil is influenced to a considerable extent by its relative looseness or denseness. In this respect, however, a distinction is necessary between coarse-grained cohesionless soils and cohesive materials. In a mass of coarse-grained soil most of the grains touch several others in point-to-point contact and efforts to densify the mass can reduce the void ratio only through rearrangement or crushing of the particles. On the other hand, the densification of fine-grained soil, especially clay, depends on other factors such as cohesion and the presence of water films on the particle surfaces.

The void ratio or porosity of any soil usually does not in itself furnish a direct indication of its behavior under load or during excavation. Of two coarse-grained soils at the same void ratio, one soil may be in a dense state whereas the other may be loose. Thus, the relative density of a coarse-grained material is much more significant than the void ratio alone. The relative density can be expressed numerically by the *density index*, I_d, defined as

$$\text{Density index, } I_d = \frac{e_{max} - e}{e_{max} - e_{min}} \qquad 1.10$$

in which e_{max} is the void ratio of the soil in its loosest state, e is the actual void ratio, and e_{min} is the void ratio in the densest possible state. Hence, $I_d = 1.0$ for a very dense soil and 0 for a very loose soil.

In practice, the relative density of granular soils is usually judged indirectly by penetration or load tests (Arts. 5.4 and 5.5), because direct measurement of the void ratio of a soil in the field is not convenient. However, if e is known, the values of e_{max} and e_{min} can be determined in the laboratory. The loosest state can usually be created by allowing the dust-dry material to fall into a container from a funnel held in such a way that the free fall is about $\frac{1}{2}$ in. If the material is silty, a looser state can sometimes be established by permitting the soil to settle through water. The densest state can usually be achieved by packing the soil into a container by means of a combination of static pressure and vibration or sometimes by "raining" the sand from such a height that the impact of the falling grains compacts the surface layer.

Standard ASTM procedures describe several means of producing e_{min}. Because different procedures lead to different void ratios for various materials, the numerical values of e_{max} and e_{min} cannot always be determined definitively. Consequently, the value of I_d always involves uncertainties and should be accompanied by descriptions of the manner in which e_{max} and e_{min} were ascertained.

For a soil containing appreciable amounts of silt or clay, the density index loses its significance, because the values of e_{max} and e_{min} have no definite meaning. Yet many construction operations deal with such materials. Moreover, the beneficial effects of compacting such soils have been demonstrated by long experience. The need for a method of defining the degree of compaction led in the early 1930s to the development in California of a laboratory compaction test (Proctor, 1933). This test has been refined and standardized by ASTM and AASHO as the *moisture-density relations test* (ASTM Designation D-698 or AASHO Method T-99). Apparatus commonly used is illustrated in Fig. 1.5. It consists of a metal cylinder having a volume of $\frac{1}{30}$ cu ft and an internal diameter of 4.0 in., together with a metal tamper having a weight of 5.5 lb and a circular face of 2-in. diameter. The soil is placed in the cylinder in three approximately equal layers. Each layer is compacted by 25 blows of the tamper falling freely through a distance of 12 in. (An alternative procedure permits the use of a 6.0-in. diameter mold having a volume of 0.075 cu ft; 56 blows of the standard hammer are applied to each of three layers.) After compaction, the soil is struck off level with the top of the cylinder, and the weight

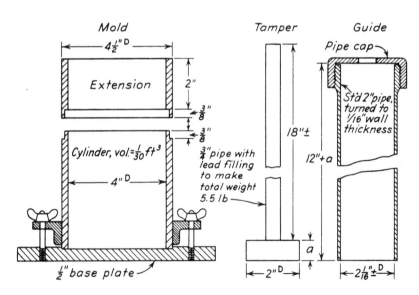

FIGURE 1.5. Apparatus used to determine moisture-density relations of soils (ASTM Method D-698, AASHO Method T-99).

of the soil in the container is determined. A sample is then extracted from the middle of the compacted cylinder of soil for a moisture-content determination.

From the weight and volume of the soil in the container, the unit weight γ of the soil is computed. The measure of compaction, however, is the *dry unit weight* γ_d, the weight per cubic foot of the solid soil constituents in the container. The values of γ_d are determined for a series of samples of the soil, each of which has a different initial or placement water content. Ordinarily the first determinations are made with the soil in a fairly dry state; successive determinations are made on increasingly wetter soils until the weight of moist soil that can be packed into the mold has reached a maximum and starts to decrease.

The procedure described above, widely known as the *Standard Proctor* test, was developed to duplicate in the laboratory, as nearly as possible, the results that could be obtained by equipment commonly used in the 1930s for compaction of soils in the field. Since that time, field compaction equipment has improved to the point where it is possible to produce higher dry unit weights by field compaction than by the Standard Proctor procedure. The greater compaction is often required in the construction of airfields and high dams. For this reason, other moisture-density relations tests have been adopted in connection with higher compactive efforts. The most common of these, sometimes referred to as the *modified AASHO* test, but more properly designated as ASTM Method D–1557 or AASHO Method T–180, can also be made with the same mold shown in Fig. 1.5. The face of the tamper has the same dimensions, but the weight of the tamper is increased to 10 lb and the height of fall to 18 in. Furthermore, the soil is compacted by 25 blows on each of five layers instead of three. (The 6.0-in. diameter mold may be used as an alternative; the number of blows per layer is then increased to 56.)

The results of the tests are represented by moisture-density curves in which γ_d for each determination is plotted against the corresponding value of the placement moisture content w. The ordinate of the peak of the curve is designated the *maximum dry density* γ_{max}, or 100 *per cent compaction*, and the abscissa the *optimum water content* or *optimum moisture content* w_{opt}. Figure 1.6 shows the compaction curves obtained on a glacial till by the two different methods of compaction. The relation between dry unit weight γ_z at zero air voids and placement water content w is also shown. Since the line representing this relation corresponds to 100 per cent saturation, it must lie to the right of all points on any moisture-density curve for the soil.

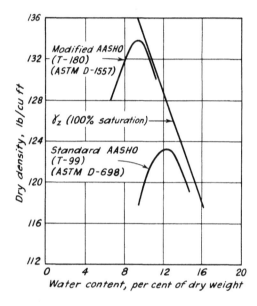

FIGURE 1.6. Moisture-density relations for a glacial till obtained using two different compaction efforts.

The two ASTM procedures for performing compaction tests lead, as illustrated by Fig. 1.6, to two different moisture-density relations for the same soil. Similarly, still different curves would be obtained in the field, depending on such variables as type, weight, and number of passes of compaction equipment, or thickness of layers being compacted. Hence, such terms as optimum moisture content or 100 per cent compaction do not represent unique properties of a particular soil, but depend also on the compaction procedure. For this reason, the pro-

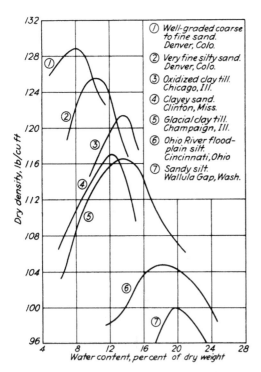

FIGURE 1.7. Moisture-density relations for various types of soil as determined by ASTM Method D-698.

cedure must always be defined when values of w_{opt} or γ_{max} are given.

Nevertheless, the type of soil is the major variable in establishing the moisture-density relations. The influence on the results of tests performed on several different soils in accordance with ASTM Method D-698 is illustrated in Fig. 1.7. It is apparent that not only the shapes but also the positions of the curves change as the texture of the soils varies from coarse to fine.

Since a major purpose of the laboratory moisture-density relations test is to control the compaction of soil in the field, tests of the field-compacted soil must always be carried out to check whether or not the desired density has been achieved. Specifications for placing compacted fill vary depending on the type of loadings to which the soil will be subjected. Commonly, specifications require that dry densities be obtained that are at least equal to 95 per cent of maximum dry density determined on the basis of ASTM Method D-698. This may be referred to as 95 per cent Standard AASHO compaction. Subgrades for heavily loaded airfield pavements usually have to be compacted to 98 per cent Modified AASHO (ASTM D-1557). Seldom can such rigid specifications be met unless the water content of the soil is close to the appropriate optimum. In fact, it should be apparent from a study of Fig. 1.6 that no amount of compaction applied to a soil can produce at a given water content a density any greater than that indicated by the γ_z line. Thus, if the glacial till were at a natural water content of 16 per cent, the upper limit of its dry density would be about 118 lb/cu ft. To produce a density near the standard AASHO maximum, the soil would have to be dried to about 14 per cent or for a density near the modified to about 10 per cent.

The density of a compacted soil in place is determined by a field density check test. Two procedures are most commonly employed. In both procedures the surface of the soil at the test site is first leveled and a hole of 3- to 5-in. diameter is carefully dug nearly through the compacted layer. The sides of the hole should be made as smooth as possible and all of the soil removed must be carefully saved. The moist soil removed should be accurately weighed before any water can evaporate and a relatively large sample taken for water content determination. The volume of the hole is then determined by (1) filling it with a calibrated dry uniform sand through a special sand-cone device (ASTM D-1556, AASHO T-147) or (2) by forcing into it a water-filled rubber balloon from a calibrated container reading volume directly (ASTM D-2167). In the sand-cone method the volume of the hole is determined from the difference in weight of the container and sand-cone before and after filling the hole and from a knowledge of the unit weight that the sand assumes when it falls freely from the container. Thus, a balance and a supply of clean dry sand must be available at the test site. The in-place wet density of the soil is computed by dividing the weight of the soil removed by the volume of the hole. This is converted to dry density by eq. 1.6a. The *per cent compac-*

tion is then defined as a ratio, multiplied by 100, of the field dry density to γ_{max} for the soil as determined in the laboratory. It is emphasized that, unlike the definition of density index, the definition of per cent compaction is arbitrary in that it depends on the details of the test procedure. Moreover, the strength of a given soil, either during or after compaction, is not related in any simple way to the per cent compaction.

The field moisture content and density can also be determined with nuclear moisture and density meters placed on a smooth surface of compacted soil. Such instruments have a considerable advantage over the more conventional methods in the short time required to make a test. As yet such instruments are quite expensive and are often plagued with errors due to faulty calibration or lack of proper adjustment. In spite of these disadvantages the use of nuclear meters is growing rapidly since they make possible many more tests in a given time and thus provide closer control of field compaction.

Relationships Among Soil Aggregate Properties. The various aggregate properties discussed in this article are interrelated and can be computed in terms of each other by algebraic expressions. However, it is usually more expeditious to carry out any necessary computations with the aid of the diagram, Fig. 1.4, and the defining equations for the various quantities. The simplicity of this procedure will become apparent upon study of the illustrative problems that follow.

ILLUSTRATIVE PROBLEMS

1. A sample of soft saturated clay has a natural water content of 43 per cent. The specific gravity of the solid matter is 2.70.

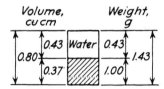

What are the void ratio, the porosity, and the saturated unit weight of the soil?

Solution. Inasmuch as the soil is completely saturated, the accompanying sketch represents the weight-volume relationships. Since the water content is known, the quantities on the right-hand side of the diagram may be set down immediately, on the assumption that $W_s = 1.0$ g.

The volume of solids is then computed and recorded on the left side. It is equal to $1.00/2.70 = 0.37$ cu cm. Since water has a specific gravity of unity, the volume of water may be recorded as 0.43 cu cm. Hence, the total volume may be taken as $0.43 + 0.37 = 0.80$ cu cm. Then, by definition,

$$e = V_v/V_s = 0.43/0.37 = 1.16$$
$$n = V_v/V = 0.43/0.80 = 0.51$$
$$\gamma_{sat} = W/V = 1.43/0.80 = 1.79 \text{ g/cu cm}$$

or
$$\gamma_{sat} = 1.79 \times 62.5 = 112 \text{ lb/cu ft}$$

2. A sample of sand above the water table was found to have a natural moisture content of 15 per cent and a unit weight of 120 lb/cu ft. Laboratory tests on a dried sample indicated values of $e_{min} = 0.50$ and $e_{max} = 0.85$ for the densest and loosest states, respectively. Compute the degree of saturation and the density index. Assume $G = 2.65$.

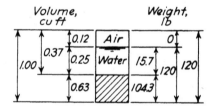

Solution. In this problem the total volume may conveniently be taken as 1 cu ft, whence the total weight is 120 lb. This is also the weight of soil plus water. These values may be written in a sketch, as shown.

Since the water content is 15 per cent, we may write, by definition,

$$w = W_w/W_s = 0.15$$
or $$W_w = 0.15 W_s$$

Moreover, since $W_w + W_s = 120$ lb, we may write

$$0.15 W_s + W_s = 120$$

whence $\quad W_s = 120/1.15 = 104.3$ lb

and $\quad W_w = 120 - 104.3 = 15.7$ lb

These values may now be inscribed on the right side of the diagram.

The volume of solids is

$$\frac{104.3}{2.65 \times 62.5} = 0.63 \text{ cu ft}$$

and the volume of water is

$$15.7/62.5 = 0.25 \text{ cu ft}$$

By difference, the volume of air is $1.0 - (0.63 + 0.25) = 0.12$ cu ft. The volume of voids is, therefore, 0.37 cu ft.

By definition, the degree of saturation is

$$S_r = 100\frac{V_w}{V_v} = 100 \times \frac{0.25}{0.37} = 68 \text{ per cent}$$

By definition, the void ratio is

$$e = V_v/V_s = 0.37/0.63 = 0.59$$

and the density index is

$$I_d = \frac{e_{\max} - e}{e_{\max} - e_{\min}} = \frac{0.85 - 0.59}{0.85 - 0.50} = \frac{0.26}{0.35}$$
$$= 0.74$$

3. A sample of soil compacted according to the Standard Proctor test has a unit weight of 131.1 lb/cu ft at 100 per cent compaction and at the optimum water content of 14.0 per cent. What is the dry density? What is the dry unit weight at zero air voids? If the voids became filled with water, what would be the saturated unit weight? Assume that the solid matter has a specific gravity of 2.67.

Solution. The conditions may be represented by the accompanying sketch drawn on the assumption that the total volume is 1 cu ft.

Since the water content is 14.0 per cent, we may write, by definition,

$$W_w = 0.140 W_s$$

and since

$$W_w + W_s = 131.1 \text{ lb}$$
$$0.140 W_s + W_s = 131.1$$
$$W_s = 131.1/1.140 = 115.0 \text{ lb}$$

and $\quad W_w = 131.1 - 115.0 = 16.1$ lb

by definition

$$\gamma_d = W_s/V = 115/1 = 115 \text{ lb/cu ft}$$

(This value may also be determined by use of eq. 1.6a.) The volume of solids is

$$\frac{115}{2.67 \times 62.5} = 0.69 \text{ cu ft}$$

The volume of voids is then $1.00 - 0.69 = 0.31$ cu ft.

The volume of water is $16.1/62.5 = 0.26$ cu ft.

The volume of air is then $0.31 - 0.26 = 0.05$ cu ft.

If all of the air were squeezed out of the sample the dry unit weight at zero air voids would be, by definition,

$$\gamma_z = \frac{115}{0.69 + 0.26} = 121.1 \text{ lb/cu ft}$$

On the other hand, if the voids were filled with water, the weight of the water would be

$$0.31 \times 62.5 = 19.4 \text{ lb}$$

and the saturated unit weight would be

$$\gamma_{sat} = \frac{115 + 19.4}{1} = 134.4 \text{ lb/cu ft}$$

1.7. Structure and Consistency of Soil Aggregate

Primary and Secondary Structure. The *primary structure* of a soil refers to the arrangement of the grains. This arrangement is usually developed during the processes of sedimentation or rock weathering. In addition, various discontinuities may arise subsequent to the deposition or formation of the soil. These

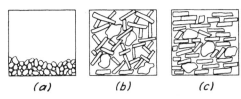

FIGURE 1.8. Diagram illustrating (a) dense single-grained structure, (b) a flocculated structure, and (c) a dispersed structure.

constitute the *secondary structure* of the deposit. They correspond to such phenomena as the development of systems of joints in sedimentary rocks.

The primary structure of a soil may be described as *single-grained*, *flocculated*, or *dispersed*. In a single-grained structure (Fig. 1.8a), each grain touches several of its neighbors in such a way that the aggregate is stable even if there are no forces of adhesion at the points of contact between the grains. The arrangement may be dense or loose, and the properties of the aggregate are greatly influenced by the denseness or looseness.

Figures 1.8b and 1.8c represent concepts of the structure of fine-grained soils. The oval-shaped particles represent silt grains whereas the flat-sided particles represent clay mineral platelets. In the *flocculated* structure (Fig. 1.8b), the edge or corner of one clay platelet tends to be attracted to the flat face of another. Consequently, the particles assume a loose but fairly stable structure that can be maintained as long as the electrical charges on the edges of the platelets remain opposite in sign to those on the faces. The degree of looseness of this arrangement depends at least in part on the nature and amount of electrolytes present during sedimentation. In the *dispersed* structure (Fig. 1.8c), the edges, corners, and faces of the clay platelets have like electrical charges. Thus, the particles repel each other and assume nearly parallel positions. Even though the dispersed structure may be quite loose at the time of sedimentation, pressure can force the adjacent platelets into a dense state more readily than if they possessed the flocculated structure (Fig. 1.8b).

The principal types of secondary structure are *cracks, joints, slickensides,* and *concretions.* Cracks and joints are commonly formed as a result of desiccation sometime after the deposition of the material. Slickensides are polished surfaces in stiff clays that have experienced differential movement or expansion. Concretions are accumulations of carbonates or iron compounds. All these features disrupt the continuity of the soil mass and may impart to it properties significantly different from those of intact samples taken from the deposit.

Consistency and Sensitivity. Undoubtedly the most significant index property of fine-grained soils in the natural state is the *consistency.* The consistency of natural cohesive soil deposits is expressed qualitatively by terms such as *soft, medium, stiff,* and *hard.* The meaning of these terms, however, varies widely in different parts of the country, depending on whether the local soils are generally hard or generally soft. Rather than rely on such vague terms, the engineer should develop his ability to estimate the compressive strength of the soil.

Quantitatively the consistency of an undisturbed cohesive soil may be expressed by the *unconfined compressive strength* q_u. In the unconfined compressive strength test a prism or cylinder of soil having a height of $1\frac{1}{2}$ to 2 times the average diameter is loaded to failure, in simple compression, quickly enough that the water content of the soil does not change. The equipment illustrated in Fig. 1.9 is commonly used for the test. Simultaneous observations are made of the load and of the vertical shortening. The failure load or, if the sample does not fail outright, the load required to produce 20 per cent strain is expressed as the load per unit of cross-sectional area, in tons per square foot or kilograms per square centimeter. Table 1.5 indicates the relationship between the qualitative terms describing consistency and the quantitative values of the unconfined compressive strength. If equipment for making compression tests is not available, a rough estimate can be based

FIGURE 1.9. Type of apparatus commonly used to determine the unconfined compressive strength of samples of soil.

on the simple field tests suggested in the table.

When a sample of undisturbed clay is kneaded in the hands at unaltered water content, it usually becomes softer. This characteristic may be investigated by remolding without loss of water a sample of soil previously tested in unconfined compression, packing the remolded material into a cylindrical mold, extruding the remolded sample, and testing it in unconfined compression. The ratio

$$S_t = \frac{q_u \text{ undisturbed}}{q_u \text{ remolded}} \qquad 1.11$$

is known as the *sensitivity* of the clay.

Some clays with such secondary structural characteristics as cracks, joints, or slickensides may have sensitivities less than 1. The sensitivity of most other clays exceeds 1 but is not greater than about 8. Natural soils having values of sensitivity greater than 4 are known as *sensitive* clays, and, if the sensitivity exceeds 8, as *extrasensitive* clays. If the sensitivity exceeds 15 such clays are described as *quick*. These should be treated with care during construction operations because disturbance tends to transform them, at least temporarily, into viscous liquids. Before remolding, highly sensitive clays may have a loose flocculated structure (Fig. 1.8b), whereas remolding may produce a dispersed structure (Fig. 1.8c).

If a sample of remolded clay is allowed to stand without further disturbance and without change in water content, it may regain at least part of its original strength and stiffness. This increase in strength is due to the gradual reorientation of the adsorbed molecules of water, and is known as *thixotropy*.

Atterberg Limits. If the water content of a thick suspension of clay is gradually reduced, the clay-water mixture passes from a

Table 1.5 Qualitative and Quantitative Expressions for Consistency of Clays

Consistency	Field Identification	Unconfined Compressive Strength q_u (tons/sq ft)
Very soft	Easily penetrated several inches by fist	Less than 0.25
Soft	Easily penetrated several inches by thumb	0.25–0.5
Medium	Can be penetrated several inches by thumb with moderate effort	0.5–1.0
Stiff	Readily indented by thumb but penetrated only with great effort	1.0–2.0
Very stiff	Readily indented by thumbnail	2.0–4.0
Hard	Indented with difficulty by thumbnail	Over 4.0

liquid state through a plastic state and finally into a solid state.

It has been found that the water contents corresponding to the transitions from one state to another usually differ for clays having different physical properties in the remolded state, and are approximately equal for clays having similar physical properties. Therefore, the limiting water contents may serve as index properties useful in the classification of clays.

The significance of the limiting water contents for each physical state was first suggested by A. Atterberg in 1911. Hence, these limits are commonly known as the *Atterberg limits*, and the tests required to determine them are the *Atterberg-limit tests*. Actually, as the soil–water mixture passes from one state to another, there is no abrupt change in the physical properties. The limit tests, therefore, are arbitrary tests that have been adopted to define the limiting values.

Above the *liquid limit* w_L, the soil-water system is a suspension. Below the liquid limit and above the *plastic limit* w_P, the soil–water system is said to be in a plastic state. In this state the soil may be deformed or remolded without the formation of cracks and without change in volume. The range of water content over which the soil–water system acts as a plastic material is frequently referred to as the *plastic range*, and the numerical difference between the liquid limit and the plastic limit is called the *plasticity index* I_P (often designated PI):

$$\text{Plasticity index, } I_P = w_L - w_P \quad 1.12$$

Somewhat below the plastic limit the soil-water system reaches the *shrinkage limit* w_S. Reduction of the water content by drying below the shrinkage limit is not accompanied by decrease in volume; instead, air enters the voids of the system and the material becomes unsaturated.

The Atterberg limits vary with the amount of clay present in a soil, on the type of clay mineral, and on the nature of the ions adsorbed on the clay surface. It has been pointed out (Art. 1.5) that differences in the atomic structure of clay minerals lead to differences in the electric charges on the clay surfaces. The existence of such charges is indicated by the ability of clays to adsorb ions from solution. Cations (positive ions) are more readily adsorbed than anions (negative ions); hence, negative charges must be predominant on the clay surfaces. A cation, such as Na^+, is readily attracted from a salt solution and attached to the clay surface. However, the adsorbed Na^+ ion is not permanently attached; it can be replaced by K^+ ions if the clay is placed in a solution of potassium chloride KCl. The process of replacement by the excess cation is called *cation exchange*.

The cation exchange capacity of the different types of clay minerals may be measured by washing a sample of each with a solution of a salt not commonly found in nature, such as ammonium chloride NH_4Cl, and determining the amount of adsorbed NH_4^+ by measuring the difference between the original and final concentration of the washing solution. It is convenient to express the cation exchange capacity in terms of the number of equivalent weights of an ion adsorbed per 100 g of clay mineral, since this factor is independent of the weight of each adsorbed ion and the number of charges associated with it.

Typical ranges of cation exchange capacities of various clay minerals are given in Table 1.6.

TABLE 1.6 Cation Exchange Capacity of Clay Minerals

Clay Mineral	Cation Exchange Capacity Milliequivalents per 100 g
Kaolinite	3–15
Illite	10–40
Montmorillonite	70–100

Table 1.6 indicates that montmorillonites are about 10 times as active in adsorbing cations as kaolinites. Hence, a much smaller amount of montmorillonite than of kaolinite is required to impart the typical properties of a clay to a mixed-grained soil.

Table 1.7 Atterberg-Limit Values of Clay Minerals with Various Adsorbed Cations

Cation	Na^+		K^+		Ca^{++}		Mg^{++}	
	w_L	I_P	w_L	I_P	w_L	I_P	w_L	I_P
Clay mineral								
Kaolinite	29	1	35	7	34	8	39	11
Illite	61	27	81	38	90	50	83	44
Montmorillonite	344	251	161	104	166	101	158	99

After W. A. White (1958).

Although the ability to adsorb water onto the surface of a clay mineral varies in the same order as the exchange capacity, it is also affected by the type of cations present. Therefore, certain relationships exist among soil–water properties such as the Atterberg limits, the type of clay mineral, and the nature of the adsorbed ions. Table 1.7 presents the liquid limits and plasticity indices of representatives of each group of clay minerals made homoionic to various ions. A comparison of Tables 1.6 and 1.7 reveals that there is indeed a correspondence between the cation exchange capacity of a clay and its soil–water properties as measured by the Atterberg limits. It is also apparent that the greater the cation exchange capacity of a clay the greater the effect of changing the adsorbed cation.

The type of clay mineral and the nature of the adsorbed ions are of significance in the chemical stabilization of soils, in drainage by electrical methods, and in other rather specialized problems. The necessary information can be obtained by such techniques as X-ray analysis. For most problems of foundation engineering, however, the influence of the mineralogical and chemical characteristics is reflected adequately in the values of the Atterberg limits.

The liquid limit and the plasticity index together constitute a measure of the plasticity of a soil. Soils possessing large values of w_L and I_P are said to be highly plastic or *fat*. Those with low values are described as slightly plastic or *lean*. The interpretation of liquid and plastic limit tests is greatly facilitated by the use of the *plasticity chart* developed by A. Casagrande. In this chart (Fig. 1.10) the ordinates represent values of the plasticity index, and the abscissas represent values of the liquid limit. The chart is divided into six regions by the inclined line A having the equation $I_P = 0.73 (w_L - 20)$, and the two vertical lines $w_L = 30$ and $w_L = 50$. All soils represented by points above line A are inorganic clays; the plasticity ranges from low ($w_L < 30$) to high ($w_L > 50$) with increasing values of the liquid limit. Soils represented by points below line A may be inorganic silts, organic silts, or organic clays. If they are inorganic, they are said to be of low, medium, or high compressibility, depending on whether the liquid limit is below 30, between 30 and 50, or above 50. If they are organic silts, they are represented by points in the region corresponding to a liquid limit between 30 and

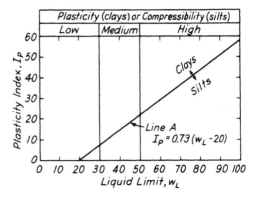

FIGURE 1.10. Plasticity chart (after A. Casagrande, 1948).

50 and, if they are organic clays, to a liquid limit greater than 50.

The distinction between organic and inorganic soils can usually be made by performing two liquid-limit tests on the same material, one starting with moist or air-dried soil, and the other with oven-dried soil. Oven-drying produces irreversible changes in organic constituents that significantly lower the liquid limit. If the liquid limit of the oven-dried sample is less than about 0.75 times that for the undried sample, the soil may usually be classed as organic. A few inorganic clay minerals and other fine-grained soil constituents also experience irreversible changes on oven-drying; hence, the identification cannot always be based on the results of the limit tests.

The natural water content of a clay is itself a useful index property. Of even greater significance, however, is the relation of the water content to the liquid and plastic limits. Those deposits having water contents close to the liquid limit are usually much softer than those with moisture contents close to the plastic limit. One of the most important index properties of natural clay deposits is, therefore, the *liquidity index*, defined by the equation
Liquidity index,

$$I_L = \frac{w - w_P}{w_L - w_P} = \frac{w - w_P}{I_P} \qquad 1.13$$

It may be seen that I_L is negative for soils having water contents less than the plastic limit. As the water content increases from the plastic limit to the liquid limit, the value of I_L increases from 0 to 1.0. If the water content is greater than the liquid limit, the liquidity index is greater than 1.0. The consistency of a clay in the remolded state may be estimated when the natural water content and limit values are known. The relationships are illustrated in Fig. 1.11.

None of the Atterberg-limit tests is difficult to perform, although a certain amount of experience is required to develop the technique necessary to obtain reproducible results. The liquid-limit test is commonly made by means of the mechanical apparatus designed by A. Casagrande (Fig. 1.12). A mixture of soil and water is placed in the cup, and a groove 2 mm wide at its base and 8 mm high is made in the center of the soil pat. The operator then turns the crank which lifts the cup to a height such that the point of contact between cup and base is 1 cm above the base. The cup then falls freely from this position. The soil is at the liquid limit if 25 blows are required to cause the lower edges of the groove to come into contact with each other for a length of about $\frac{1}{2}$ in. The water content at this number of blows is the liquid limit.

The plastic limit test is performed by

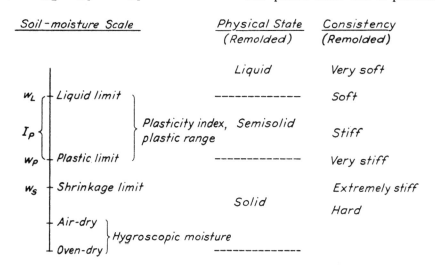

FIGURE 1.11. Diagram of the soil-moisture scale showing Atterberg Limits, corresponding physical state, and approximate consistency of remolded soil.

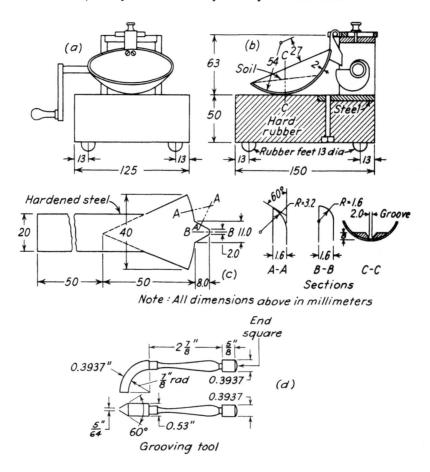

FIGURE 1.12. (a) and (b) Mechanical liquid limit device. (c) Casagrande's grooving tool. (d) Standard ASTM grooving tool.

rolling a sample of plastic soil into a thread with a diameter of ⅛ in. If the soil does not crumble, the thread is picked up, remolded, and rolled out again. This process is repeated until the thread just begins to crumble when it reaches the diameter of ⅛ in. The water content at which crumbling takes place is defined as the plastic limit.

The shrinkage limit of a soil is determined by preparing a sample of known volume at a moisture content above the liquid limit and by drying the sample in an oven. The weight and volume of the oven-dry sample are measured. From these data and the initial water content, a computation is made of the water content at which the dried sample would be just saturated. This water content is considered to be the shrinkage limit.

More detailed instructions for performing liquid limit, plastic limit, and shrinkage limit tests are given in standard ASTM Methods, D-423, D-424, and D-427.

1.8. Soil-Classification Systems

Introduction. Because the soil deposits of the world are infinitely varied, it has not been found possible to create a universal system of soil classification for dividing soils into various groups and subgroups on the basis of their important index properties. However, useful systems based on one or two index properties have been devised. Some of these systems are in such common use by workers in various fields involving soils that the engineer must have at least a general knowledge of them. At the same time it is essential to keep in mind that no system can

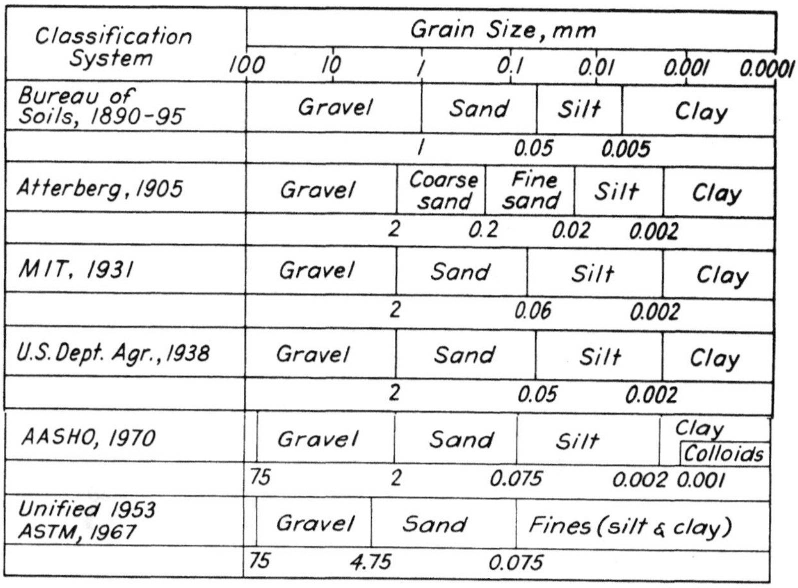

Classification System	Grain Size, mm					
	100	10	1	0.1	0.01 0.001	0.0001
Bureau of Soils, 1890-95	Gravel		Sand	Silt	Clay	
			1	0.05	0.005	
Atterberg, 1905	Gravel		Coarse sand	Fine sand	Silt	Clay
			2	0.2	0.02 0.002	
MIT, 1931	Gravel		Sand	Silt	Clay	
			2	0.06	0.002	
U.S. Dept. Agr., 1938	Gravel		Sand	Silt	Clay	
			2	0.05	0.002	
AASHO, 1970	Gravel		Sand	Silt	Clay	Colloids
	75		2	0.075	0.002	0.001
Unified 1953 ASTM, 1967	Gravel		Sand	Fines (silt & clay)		
	75	4.75		0.075		

FIGURE 1.13. Comparison of several common textural classification systems.

adequately describe any soil for all engineering purposes. Indeed, many systems ignore the properties that are the most important from the standpoint of the foundation engineer.

Textural Systems. Since the particle size is probably the most obvious characteristic of a soil, it is natural that the earliest classification systems should have been based on texture alone. Indeed, many such systems have been suggested. Several of the more common are shown in Fig. 1.13. The MIT and Unified systems are commonly used by foundation engineers, the AASHO system by highway engineers, and the Unified system by engineers charged with the design of dams and airfields.

To classify a soil according to a particular textural system, the particle-size distribution curve is usually plotted and the percentages by weight are calculated of the particles contained within each of the ranges of size specified in the system. Thus, a mixed-grained soil might be described as "3 per cent gravel, 46 per cent sand, 17 per cent silt, and 34 per cent clay, according to the MIT classification."

In the textural method of classification used by soil scientists of the U. S. Department of Agriculture, only three ranges of particle size are specified and material coarser than 2.0 mm is excluded. Hence the percentages of sand-, silt-, and clay-size particles can be represented by a triangular chart (Fig. 1.14). After these percentages have been determined for a given sample, the point representing this mechanical composition is located on the triangular chart and the soil is given the name assigned to the area in which the point is located. If the soil contains a significant quantity of material coarser than 2.0 mm, an appropriate adjective, such as gravelly or stony, is added to the textural name. Although the triangular chart does not reveal any properties of the soil other than particle-size distribution, it is widely used in various modified forms by workers in the fields of agriculture and highway engineering. Unfortunately the textural name derived from the chart does not always correctly express the physical characteristics of the soil. For example, since some clay-size particles are much less active than others, a soil described as clay on the basis of a textural system may have physical properties more typical of silt.

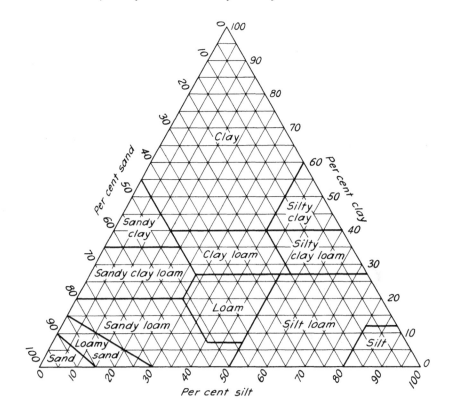

FIGURE 1.14. Triangular textural classification chart used by the U. S. Department of Agriculture.

AASHO System. About 1928 the Bureau of Public Roads introduced a soil-classification system still widely used by highway engineers. All soils were divided into eight groups designated by the symbols A-1 through A-8. Since it was believed that the soil best suited for the subgrade of a highway is a well-graded material composed largely of sand and gravel, but containing a small amount of excellent clay binder, such a material was given the designation A-1. All other soils were grouped roughly in decreasing order of stability. The system has undergone many revisions. In the beginning, neither the percentages of the various size fractions nor the plasticity characteristics of the clay fraction were definitely specified.

In 1945 an extensive revision of the Public Roads system was made by a committee of highway engineers for the Highway Research Board. In 1949 and again in 1966 revisions were adopted by the American Association of State Highway Officials and the method is now known as the AASHO system. The characteristics of the various groups and subgroups and the classification procedure are given in Table 1.8. In the AASHO system the inorganic soils are classified in 7 groups corresponding to A-1 through A-7. These in turn are divided into a total of 12 subgroups. Highly organic soils are classified as A-8. Any soil containing fine-grained material is further identified by its *group index*; the higher the index, the less suitable the soil.

The group index is calculated from the formula

Group index
$$= (F - 35)[0.2 + 0.005(w_L - 40)] \\ + 0.01(F - 15)(I_P - 10) \quad 1.14$$

in which

F = percentage passing No. 200 sieve, expressed as a whole number
w_L = liquid limit
I_P = plasticity index

Table 1.8 American Association of State Highway Officials
Classification of Soils and Soil-Aggregate Mixtures
AASHO Designation M-145

General Classification [a]	Granular Materials (35 per cent or Less Passing No. 200)							Silt-Clay Materials (More than 35 per cent Passing No. 200)			
Group Classification	A-1		A-3	A-2				A-4	A-5	A-6	A-7
	A-1-a	A-1-b		A-2-4	A-2-5	A-2-6	A-2-7				A-7-5, A-7-6
Sieve analysis per cent passing:											
No. 10	50 max										
No. 40	30 max	50 max	51 min								
No. 200	15 max	25 max	10 max	35 max	35 max	35 max	35 max	36 min	36 min	36 min	36 min
Characteristics of fraction passing No. 40:											
Liquid limit				40 max	41 min	40 max	41 min	40 max	41 min	40 max	41 min
Plasticity index	6 max		N.P. [b]	10 max	10 max	11 min	11 min	10 max	10 max	11 min	11 min
Usual types of significant constituent materials	Stone fragments —gravel and sand		Fine sand	Silty or clayey gravel and sand				Silty soils		Clayey soils	
General rating as subgrade	Excellent to good							Fair to poor			

[a] Classification procedure: With required test data in mind, proceed from left to right in chart; correct group will be found by process of elimination. The first group from the left consistent with the test data is the correct classification. The A-7 group is subdivided into A-7-5 or A-7-6 depending on the plastic limit. For $w_P < 30$, the classification is A-7-6; for $w_P \gtreqless 30$, A-7-5.
[b] N.P. denotes nonplastic.

The group index is always reported to the nearest whole number unless its calculated value is negative whereupon it is reported as zero. The group index is appended to the group and subgroup classification. For example, a clay soil having a group index of 25 might be classified as A-7-6(25).

Unified System. The soil-classification system most widely used by foundation engineers in North America today was developed by Arthur Casagrande for the Corps of Engineers, U. S. Army. First designated as the *Airfield Classification* (AC) system, it was originated to assist in the design and construction of military airfields during World War II. After the war it was adopted with minor revisions by the Corps and by the U. S. Bureau of Reclamation as the *Unified* system. In 1969, the Unified system was adopted by the American Society for Testing and Materials as a Standard Method for Classification of Soils for Engineering Purposes, ASTM D-2487.

According to the Unified system the coarse-grained soils are divided into

1. Gravel and gravelly soils; symbol G.
2. Sands and sandy soils; symbol S.

The gravels and sands are each subdivided into four groups:

a. Well-graded, fairly clean material; symbol W.
b. Well-graded material with excellent clay binder; symbol C.
c. Poorly graded, fairly clean material; symbol P.
d. Coarse materials containing fines, not included in preceding groups; symbol M.

Fine-grained soils are divided into three groups:

1. Inorganic silty and very fine sandy soils; symbol M.
2. Inorganic clays; symbol C.
3. Organic silts and clays; symbol O.

Each of these three groups of fine-grained soils is subdivided according to its liquid limit into

a. Fine-grained soils having liquid limits of 50 or less; that is, of low to medium compressibility; symbol L.
b. Fine-grained soils having liquid limits greater than 50; that is, of high compressibility; symbol H.

Highly organic soils, usually fibrous, such as peat and swamp soils of very high compressibility, are not subdivided and are placed in one group, symbol Pt, on the basis

Table 1.9 Unified System Classification of Soils for Engineering Purposes ASTM Designation D-2487

Major Divisions			Group Symbols	Typical Names	Classification Criteria	
Coarse-Grained Soils (More than 50% retained on No. 200 sieve)	Gravels (50% or more of coarse fraction retained on No. 4 sieve)	Clean Gravels	GW	Well-graded gravels and gravel-sand mixtures, little or no fines	$C_u = D_{60}/D_{10}$ Greater than 4 $C_z = \dfrac{(D_{30})^2}{D_{10} \times D_{60}}$ Between 1 and 3	
			GP	Poorly graded gravels and gravel-sand mixtures, little or no fines	Not meeting both criteria for GW	
		Gravels with Fines	GM	Silty gravels, gravel-sand-silt mixtures	Atterberg limits plot below "A" line or plasticity index less than 4	Atterberg limits plotting in hatched area are borderline classifications requiring use of dual symbols
			GC	Clayey gravels, gravel-sand-clay mixtures	Atterberg limits plot above "A" line and plasticity index greater than 7	
	Sands (More than 50% of coarse fraction passes No. 4 sieve)	Clean Sands	SW	Well-graded sands and gravelly sands, little or no fines	$C_u = D_{60}/D_{10}$ Greater than 6 $C_z = \dfrac{(D_{30})^2}{D_{10} \times D_{60}}$ Between 1 and 3	
			SP	Poorly graded sands and gravelly sands, little or no fines	Not meeting both criteria for SW	
		Sands with Fines	SM	Silty sands, sand-silt mixtures	Atterberg limits plot below "A" line or plasticity index less than 4	Atterberg limits plotting in hatched area are borderline classifications requiring use of dual symbols
			SC	Clayey sands, sand-clay mixtures	Atterberg limits plot above "A" line and plasticity index greater than 7	
Fine-Grained Soils (50% or more passes No. 200 sieve)	Silts and Clays (Liquid limit 50% or less)		ML	Inorganic silts, very fine sands, rock flour, silty or clayey fine sands	Plasticity chart for classification of fine-grained soils and fine fraction of coarse-grained soils. Atterberg limits plotting in hatched area are borderline classifications requiring use of dual symbols. Equation of A-line: $PI = 0.73(LL-20)$	
			CL	Inorganic clays of low to medium plasticity, gravelly clays, sandy clays, silty clays, lean clays		
			OL	Organic silts and organic silty clays of low plasticity		
	Silts and Clays (Liquid limit greater than 50%)		MH	Inorganic silts, micaceous or diatomaceous fine sands or silts, elastic silts		
			CH	Inorganic clays of high plasticity, fat clays		
			OH	Organic clays of medium to high plasticity		
Highly organic soils			Pt	Peat, muck and other highly organic soils	Visual-manual identification	

Classification on basis of percentage of fines: Less than 5% Pass No. 200 sieve — GW, GP, SW, SP; More than 12% Pass No. 200 sieve — GM, GC, SM, SC; 5% to 12% Pass No. 200 sieve — Borderline classification requiring use of dual symbols

of visual identification. The pertinent characteristics of the various groups are given in Table 1.9.

Shortcomings of Engineering Classifications. The various textural systems, the AASHO system, and the Unified system are based on the properties either of the grains themselves or of remolded material; they do not take into consideration the properties of the intact material as found in nature. It is primarily the properties of the intact material that determine the behavior of the soil during and after construction. Hence, none of the systems of classification can serve as more than a starting point for adequate description of soils in the conditions under which they are encountered in the field. Nevertheless, even with these limitations, much information concerning the general characteristics of a soil can be inferred as a consequence of its proper classification according to one of the systems described under the preceding subheadings. The engineer who deals with soils and foundations should commit to memory the details of at least the engineering classification system that seems most appropriate to his area of activity. He should constantly train himself to identify and classify soils in the field correctly by comparing his field descriptions of soil samples with the corresponding laboratory test results. Since all systems of soil classification just described are in common use, it is advantageous to be thoroughly familiar with each.

Still further useful information can be obtained from sources outside the field of civil engineering, particularly geology and pedology. The foundation engineer should possess a knowledge of at least the descriptive terminology of these two sciences. Descriptive information about the most common earth deposits and their important engineering characteristics is given in Chap. 6.

ILLUSTRATIVE PROBLEMS

1. A sample of inorganic soil has the following grain-size characteristics:

Size mm	Per Cent Finer
2.0 (No. 10)	100
0.075 (No. 200)	71
0.050	67
0.005	31
0.002	19

The liquid limit is 53 per cent, and the plasticity index is 22 per cent.

Classify this soil according to the following systems: U. S. Department of Agriculture; AASHO; Unified.

Solution. U. S. Department of Agriculture:

Clay size 19 per cent, silt size $67 - 19 = 48$ per cent

Classification (Fig. 1.14): Loam

AASHO:
Computation of group index:

$$F = 71$$
$$w_L = 53$$
$$I_P = 22$$

Index
$= (71 - 35)[0.2 + 0.005(53 - 40)]$
$\quad + 0.01(71 - 15)(22 - 10)$
$= 16.26$, say 16

On the basis of limit values and Table 1.8, the soil is either A–7–5 or A–7–6. Since $w_P = 53 - 22 = 31$, which is greater than 30, classification is A–7–$5(16)$.

Unified:
On the basis of Table 1.9 (including Plasticity Chart) classification is MH.

2. A sample of residual soil developed on shaly sandstone has the following grain-size characteristics:

Size, mm	Per Cent Finer
2.0 (No. 10)	100
0.075 (No. 200)	45
0.050	39
0.005	20
0.002	16

The liquid limit is 27 per cent and the plasticity index is 6 per cent. Classify the soil according to the three systems mentioned in Prob. 1.

Solution.

U. S. Department of Agriculture:

Clay size 16 per cent, silt size $39 - 16 = 23$ per cent

Classification (Fig. 1.14): Sandy loam

AASHO:
Computation of group index:

$$F = 45$$
$$w_L = 27$$
$$I_P = 6$$

Index
$= (45 - 35)[0.2 + 0.005(27 - 40)]$
$\quad + 0.01(45 - 15)(6 - 10)$
$= 0.15$, say 0

On the basis of limit values and Table 1.8, the classification is A-$4(0)$.

Unified:

Since more than 50 per cent is retained on No. 200 sieve, soil is coarse-grained and since more than 50 per cent of the coarse fraction is finer than the No. 4 sieve, it is a sand (Table 1.9). Furthermore, since more than 12 per cent is finer than the No. 200 sieve the soil must be either SM or SC. On the basis of the limit values and plasticity chart, classification is borderline SM-SC.

3. Classify the soil represented by curve 6 (Fig. 1.2), according to the MIT classification. What are its effective grain size and uniformity coefficient?

Solution. Soil is 12 per cent clay size, $82 - 12 = 70$ per cent silt size, and 18 per cent sand size, according to the MIT classification.

Effective grain size, $D_{10} = 0.0014$ mm
Uniformity coefficient,
$$C_u = 0.04/0.0014 = 28.6$$

1.9. Description and Classification of Rocks

The behavior of soils in connection with foundations is determined primarily by the characteristics of the soils themselves and only secondarily by such features of the entire deposit as cracks, joints, or slickensides. Therefore, the classification of soils and the determination of their index properties are essential steps in foundation engineering. On the other hand, most intact rocks have strengths and rigidities considerably in excess of the requirements imposed by ordinary foundations. The adequacy of a foundation on rock is governed almost exclusively by discontinuities such as sinkholes, joints, bedding planes, zones of weathering and hydrothermal alteration, faults, and shear zones. These features are discussed in Chap. 6.

Because of the inherent differences in the characteristics that determine the behavior of soil and rock masses, an engineering system for classification of rock materials assumes relatively less importance in foundation engineering than that for soil materials. Indeed, no such system is in general use. The principal attributes that should be included in an engineering classification or rock materials are, according to Deere (1963), the strength and the fabric of the rock. Commonly accepted terms for strength are related in Table 1.10 to the unconfined compressive strength of cylindrical samples of intact rock having heights equal to at least twice their diameters.

Table 1.10 Classification of Intact Rocks According to Strength

Description Strength	Unconfined Compressive Strength lb/sq in
Very high	Over 32,000
High	16,000–32,000
Medium	8000–16,000
Low	4000–8000
Very low	Less than 4000

On the basis of their fabric, rocks can be divided into four categories: *interlocking, cemented, laminated,* and *foliated.* Rocks with interlocking fabric usually consist of crystals intergrown into a fairly homogeneous mass with similar properties in all directions. In cemented rocks, individual grains or particles have been joined together by chemical precipitates usually carried by circulating groundwater; such rocks are also likely to have fairly similar properties in all directions. Laminated rocks, on the other hand, are strongly directional in their properties as a result of sedimentation in thin layers. (If the layers are thicker than about 1 cm, each layer is usually considered to be a stratum with its own, perhaps isotropic, characteristics.) Foliated rocks also have strongly directional properties because of the parallel orientation of platy minerals, but their anisotropy is a consequence of deformations while in a plastic state. In addition to these four attributes based on their fabric, some rocks may possess other properties that should be included in an engineering description, such as solubility and the tendency to change volume upon exposure.

Table 1.11 Engineering Characteristics of Common Unweathered Rock Materials

Lithologic Type	Usual Range of Strength	Usual Fabric and Other Attributes
Igneous		
Basalt	Very low to very high	Interlocking, may be vesicular
Diabase	High to very high	Interlocking
Granite	Medium to high	Interlocking
Pegmatite	Medium to high	Interlocking
Syenite	High to very high	Interlocking
Diorite	Medium to high	Interlocking
Gabbro	Low to very high	Interlocking
Peridotite	Medium to high	Interlocking
Dolerite	Medium to high	Interlocking
Aplite	High to very high	Interlocking
Felsite	High to very high	Interlocking
Volcanic glass	High to very high	Glassy
Pumice	Very low to low	Porous
Tuff	Very low to medium	Cemented, fragmental
Andesite	Medium to very high	Interlocking
Dacite	Medium to very high	Interlocking
Rhyolite	Medium to very high	Interlocking
Metamorphic		
Gneiss	Medium to high	Foliated
Marble	Medium	Interlocking
Quartzite	High to very high	Interlocking
Schist	Very low to high	Foliated, often highly directional
Granite gneiss	Medium to high	Foliated
Serpentine	Very low to high	Foliated
Soapstone	Very low to low	Massive to foliated
Phyllite	Low to very high	Foliated, strongly directional
Slate	Low to high	Foliated, strongly directional
Hornfels	High to very high	Interlocking, isotropic
Anthracite coal	Medium	May be directional
Sedimentary		
Dolomite	Medium to very high	Isotropic unless thin-bedded
Limestone	Low to very high	Isotropic unless thin-bedded, may contain solution features
Rock Salt	Very low to low	Interlocking, readily soluble
Sandstone	Very low to very high	Cemented
Siltstone	Low to high	Cemented
Conglomerate	Very low to medium	Cemented
Breccia	Very low to medium	Cemented
Arkose	Very low to very high	Cemented
Graywacke	Medium to very high	Cemented
Shale	Very low to high	Often laminated, may be thick-bedded; may swell under reduced pressure
Gypsum	Low	Interlocking, moderately soluble
Bituminous coal	Very low to low	Interlocking

Data largely from D. U. Deere (1968) in Stagg and Zienkiewicz, *Rock Mechanics in Engineering Practice*, Wiley, N. Y., and R. P. Miller (1965), Engineering Classification and Index Properties for Intact Rock, Ph.D. thesis, University of Illinois.

As a practical matter, engineers generally understand and use classification systems developed by geologists. Although not ideally suited to the needs of the foundation engineer, they permit him to make use of the extensive geological literature. Moreover, they suggest considerable information about the pertinent engineering properties.

The principal geologic classification of rock materials is based on *lithology*, a term that reflects the mineralogy, texture, and fabric of the rock. Such names as granite, limestone, schist, quartzite, and gypsum are well-known examples. In many instances, features of the mass of rock of particular interest to the foundation engineer are associated with a specific rock type. For example, limestone, gypsum, or rock salt are likely to be associated with such solution features as sinkholes and caves; basalt is often subdivided into fragments by columnar jointing. Such associations are considered in Chap. 6. They emphasize the significance of lithology.

Rocks are also classified geologically, according to their origin, as *igneous*, *sedimentary*, and *metamorphic*. Such a classification is in itself of little significance to the foundation engineer, but when the origin is considered along with the lithology, useful engineering generalizations are possible. For instance, strongly directional characteristics are exhibited by many metamorphic rocks such as gneiss, schist, and slate, whereas most igneous rocks are roughly isotropic. A summary of the engineering characteristics of the more common rock materials is given in Table 1.11.

PROBLEMS

1. The following data were obtained from simple identification tests on two different soils. What typical name should be assigned to each?

    a. Medium dry strength; no visible dilatancy reaction; medium tough plastic thread; when dispersed, more than 50 per cent of the sample settles in less than 1 min.

    b. Low dry strength; rapid dilatancy reaction; weak plastic thread; when dispersed, most of the sample settles in 30 to 45 min.

    Ans. Sandy clay; silt

2. A 50 cu cm sample of moist clay was obtained by pressing a sharpened hollow cylinder into the wall of a test pit. The extruded sample had an initial weight of 85 g and after oven-drying a weight of 60 g. Compute w, e, S_r, and γ_d, if $G = 2.70$.

    Ans. 41.7; 1.25; 89.9%; 75 lb/cu ft.

3. A sample of saturated clay has a water content of 56 per cent. Assume $G = 2.72$ and compute e, γ_{sat}, and n.

    Ans. 1.52; 105.0 lb/cu ft; 60.3%

4. A soil sample weighing 11.23 lb has a volume of 0.092 cu ft. If $w = 13.4$ per cent and $G = 2.65$, compute γ, γ_d, e, n, and S_r.

    Ans. 122.1 lb/cu ft; 107.6 lb/cu ft; 0.538; 34.8%; 66.1%.

5. A hand-carved sample of soft saturated clay has a volume of 100 cu cm and weighs 175 g. If the oven-dry weight is found to be 120 g, compute w, e, and G.

    Ans. 45.8%; 1.22; 2.67.

6. A field-compacted sample of sandy loam was found to have a wet density of 136 lb/cu ft at a water content of 10 per cent. The maximum dry density of the soil obtained in a Standard Proctor test was 125.0 lb/cu ft. Assume $G = 2.65$ and compute γ_d, γ_z, S_r, n, and per cent compaction of the field sample.

    Ans. 123.6 lb/cu ft; 130.9 lb/cu ft; 78.0%; 25.4%; 98.9%.

7. A sample of moist quartz sand was obtained by carefully pressing a sharpened hollow cylinder with a volume of 150 cu cm into the bottom of an excavation. The sample was trimmed flush with the ends of the cylinder and the total weight was found to be 325 g. In the laboratory the dry weight of the sand alone was found to be 240 g and the weight of the empty cylinder 75 g. Laboratory tests on the dry sand indicated $e_{max} = 0.80$ and $e_{min} = 0.48$. Estimate G and compute w, e, S_r, γ_d, and I_d of the sand in the field.

Ans. 2.66; 4.17%; 0.663; 16.7%; 100.0 lb/cu ft; 0.428.

8. The natural water content of a saturated sample of Mexico City clay is 400 per cent. If $G = 2.40$, compute γ_{sat}, e, and n.

Ans. 70.8 lb/cu ft; 9.60; 90.6%.

9. The dry density of a compacted sand sample is 125 lb/cu ft. Estimate G and determine the water content of the material in a saturated condition. What would be the unit weight if $S_r = 20$ per cent?

Ans. 2.65; 12.2%; 128.1 lb/cu ft.

10. Specifications for a proposed earth fill require that the soil be compacted to 95 per cent of Standard Proctor dry density. Tests on a glacial-till borrow material indicate $\gamma_{max} = 124.0$ lb/cu ft at an optimum water content of 12 per cent. The borrow material in its natural condition has a void ratio of 0.60. If $G = 2.65$, what is the minimum volume of borrow required to make 1 cu ft of acceptable compacted fill?

Ans. 1.14 cu ft.

11. A sample of uniform sand has a porosity of 43 per cent and a water content of 12 per cent. From lab tests $e_{max} = 0.85$ and $e_{min} = 0.55$. Assume that $G = 2.65$ and compute the void ratio, density index, degree of saturation, and dry unit weight in lb/cu ft.

Ans. 0.754; 0.32; 41.9%; 94.4 lb/cu ft.

12. A saturated clay soil has a void ratio of 2.5; estimate its water content and dry density if $G = 2.70$.

Ans. 92.6%; 48.2 lb/cu ft.

13. A shrinkage limit test on a clay soil gave the following data. Compute the shrinkage limit. (*Hint*: Assume that the total volume of the dry soil cake is equal to its total volume at the shrinkage limit.)

Weight of shrinkage dish and saturated soil	38.78 g
Weight of shrinkage dish and oven-dry soil cake	30.46 g
Weight of shrinkage dish	10.65 g
Volume of shrinkage dish	16.29 cu cm
Total volume of oven-dry soil cake	10.00 cu cm

Ans. 10.2%.

14. The following data were obtained from a field-density test on a compacted fill of sandy clay. Laboratory moisture-density tests on the fill material indicated a maximum dry density of 120 lb/cu ft at an optimum water content of 11 per cent. What was the per-cent compaction of the fill? Was the fill water content above or below optimum?

Weight of moist soil removed from test hole	1038 g
Weight of soil after oven-drying	914 g
Volume of test hole from rubber-balloon apparatus	0.0169 cu ft

Ans. 99.1%; above optimum.

15. The following data were obtained from a standard moisture-density test on a glacial till. Plot the moisture—dry density curve and determine the values of maximum dry density and optimum water content. On the assumption that $G = 2.72$, compute the void ratio and degree of saturation for each trial. Plot the zero air voids (100 per cent saturation) curve on the same graph sheet as the moisture-density curve.

Trial	Weight of Moist Soil in Mold (lb)	Moist Weight of Water Content Sample (g)	Dry Weight of Water Content Sample (g)
1	4.25	116.8	105.4
2	4.39	130.3	116.8
3	4.48	114.4	101.7
4	4.52	129.5	114.2
5	4.49	125.3	109.6
6	4.44	140.2	121.5

Ans.

Trial	w (%)	γ_d (lb/cu ft)	e	S_r (%)	γ_z (lb/cu ft)
1	10.82	115.3	0.473	61.7	131.5
2	11.56	118.3	0.435	71.3	129.6
3	12.49	119.5	0.423	80.8	126.8
4	13.40	119.8	0.417	86.7	124.7
5	14.32	118.0	0.442	88.2	122.4
6	15.39	115.6	0.473	87.9	120.3

From graph $\gamma_{max} = 119.9$ lb/cu ft; $w_{opt} = 13.1\%$.

16. The following data were summarized from tests on a light gray silty clay that was assumed to be saturated in the undisturbed condition. On the basis of these data compute the liquidity index, sensitivity, and void ratio of the saturated soil. Classify the soil according to the Unified and AASHO systems.

Index Property	Undisturbed	Remolded
Unconfined compressive strength, tons/sq ft	2.55	1.50
Water content, per cent	22	22
Liquid limit, per cent		45
Plastic limit, per cent		20
Shrinkage limit, per cent		12
Per cent passing No. 200 sieve (washed and air dried)		90

Ans. 0.08; 1.7; 0.594; CL; A–7–6(24).

17. Using the grain-size curves in Fig. 1.2, classify soils 2, 3, 4 and 7 according to the MIT system and the USDA triangular chart.

Ans.

	MIT Classification				USDA
Soil	Gravel	Sand	Silt	Clay	
2	31	64	5	—	Gravelly sand
3	7	30	45	18	Loam
4	—	100	—	—	Sand
7	—	—	72	28	Silty clay loam

18. Given the following additional data, classify soils 2, 3, 4, 5, and 7 according to the Unified and AASHO systems.

Description	w_L	w_P
2. River gravel		N.P.
3. Glacial till	22.8	13.6
4. Sand		N.P.
5. Glacial rock flour	24.0	N.P.
7. Silty clay	41.6	24.7

Ans. 2. SW–SM; A–1–b(0)
3. CL; A–4(3)
4. SP; A–3(0)
5. ML; A–4(0)
7. CL; A–7–6(18)

19. Several samples of soils were obtained from a boring made for a major structure. Using the following data, determine the liquidity index and the classification of each sample according to both the Unified and AASHO systems.

Depth (ft)	w (natural) (%)	w_L (%)	w_P (%)	Per Cent Passing Sieve			
				No. 4	No. 10	No. 40	No. 200
2	60	54	31		100	98	93
8	42	46	24	88	78	46	36
15	16	21	15	97	94	88	67
25	25	32	18			100	59
35	60	66	24			100	99

Ans.

Depth (ft)	I_L	Unified	AASHO
2	1.26	MH	A-7-5(26)
8	0.82	SC	A-7-6(3)
15	0.17	CL-ML	A-4(1)
25	0.50	CL	A-6(6)
35	0.86	CH	A-7-6(48)

20. Using the grain-size curves in Fig. 6.6, classify soils 1, 2, 5, and 7 according to the MIT system and the USDA triangular chart. Compute the uniformity coefficient and coefficient of curvature of soils 1, 2, and 7.

Soil Number	MIT Classification				USDA Classification	C_u	C_z
	Gravel	Sand	Silt	Clay			
1	53	35	10	2	Very gravelly loamy sand	320	0.08
2	8	66	21	5	Sandy loam	20	3.6
5	—	2	35	63	Clay	—	—
7	18	57	23	2	Gravelly loamy sand	23	1.4

SUGGESTED READING

Numerous textbooks, monographs, and journal articles deal with the identification, basic physical properties, mineralogical characteristics, and classification of soil. Most standard texts on soil mechanics and foundation engineering cover essentially the same material as has been treated here; however, more detailed treatments may be found in T. W. Lambe and R. V. Whitman (1969), *Soil Mechanics*, New York, Wiley, and in R. N. Yong and B. P. Warkentin (1966), *Introduction to Soil Behavior*, New York, Macmillan.

A comprehensive review of the knowledge of the physico-chemical properties of soils is contained in the proceedings of a symposium on this subject published in the *Journal of the Soil Mechanics and Foundations Division*, ASCE, *85*, SM2, April 1959. Included are the following papers: R. E. Grim, (1959), "Physico-Chemical Properties of Soils: Clay Minerals," pp. 1–17; A. W. Taylor (1959), "Physico-Chemical Properties of Soils: Ion Exchange Phenomena," pp. 19–30; I. Th. Rosenqvist (1959), "Physico-Chemical Properties of Soils: Soil-Water Systems," pp. 31–53; and T. W. Lambe (1959), "Physico-Chemical Properties of Soils: Role of Soil Technology," pp.

55–70. The proceedings also contain several significant discussions of the main papers.

Two classic papers on engineering properties of soils are: A. Casagrande (1932a), "Research on the Atterberg Limits of Soil," *Public Roads*, *13*, 121–136; and R. R. Proctor (1933), "Design and Construction of Rolled-Earth Dams," *Engineering News-Record*, Vol. 111. The latter paper describes the development of the compaction test.

Standardized procedures for testing soils are found in "Bituminous Materials for Highway Construction, Waterproofing and Roofing; Soils; Skid Resistance," Part 11, *Standards*, ASTM, issued annually in April and *Standard Specifications for Highway Materials and Methods of Sampling and Testing*, 10th Edition (1970), Am. Assn. of State Highway Officials. Other procedures suggested for soil testing are to be found in the *Earth Manual* (1963), U. S. Bureau of Reclamation, U. S. Govt. Printing Office; and "Special Procedures for Testing Soil and Rock for Engineering Purposes," (1970) *Special Tech. Publ. No. 479*, ASTM.

A thorough review of soil classification systems with an excellent bibliography and numerous discussions is given in A. Casagrande (1948), "Classification and Identification of Soils," *Trans.*, ASCE, *113*, 901–991.

Other important works on soil identification and classification are: "Unified Soil Classification System" (1953), *Tech. Mem. 3-357*, U. S. Corps of Engineers, Waterways Experiment Station.

Soil Survey Manual (1951), Handbook No. 18, U. S. Dept. of Agriculture. Describes the pedologic system.

Manual of Photographic Interpretation (1960), Am. Society of Photogrammetry. Describes procedures of photointerpretation in geology, soils, and engineering.

D. M. Burmister (1951), "Identification and Classification of Soils," Symposium on Identification and Classification of Soils, *Special Tech. Publ. 113*, ASTM, 3–24.

Numerous textbooks on geology contain information on the classification and identification of rocks and rock minerals. A reference especially useful to engineers is the discussion of pertinent rock properties given in D. U. Deere (1968), "Geological Considerations," Chap. 1, in *Rock Mechanics in Engineering Practice*, edited by K. G. Stagg and O. C. Zienkiewicz, New York, Wiley.

Procedures for identifying and describing rocks for engineering purposes are given in:

D. O. Woolf (1950), *The Identification of Rock Types*, Bureau of Public Roads, U. S. Govt. Printing Office.

D. U. Deere (1963), "Technical Description of Rock Cores for Engineering Purposes," *Felsmechanik und Ingenieurgeologie*, *1*, (1), 16–22.

Allen Hazen (1868–1930)

From 1888 to 1893, while head of the Experiment Station of the Massachusetts State Board of Health at Lawrence, he conducted extensive studies of methods of measuring the physical properties of sands and gravels and of evaluating their suitability for practical use in water-supply and sewage filters. In his long career he became recognized as the preeminent sanitary engineer of the United States. He contributed notably to the technology of hydraulic-fill and other types of dams as well as to hydraulic engineering and hydrology.

PLATE 2.

CHAPTER 2

Hydraulic Properties of Soil and Rock

2.1. Introduction

In Chap. 1 only the properties necessary for the identification and classification of soils and rocks were discussed. However, the foundation engineer must also have a quantitative knowledge of the physical properties of the materials with which he deals. If, for example, the construction of a foundation requires that the water table be lowered, the engineer must be informed concerning the hydraulic properties and the drainage characteristics of the subsurface materials. These matters will be discussed in this chapter. Since all earth materials are compressible to some degree, the engineer may need to estimate the amount of compression that will take place under a given load. Such an estimate requires consideration of the stress-deformation characteristics of the material (Chap. 3). Finally, to avoid the complete failure of his foundations, the engineer must have information in regard to the ultimate strength of the supporting material. This subject is considered in Chap. 4.

2.2. Permeability of Soil

Definitions and Darcy's Law. A material is said to be permeable if it contains continuous voids. Every soil and every rock satisfy this condition. Nevertheless, there are large differences in the degree of permeability of the various earth materials. The quantity of flow through a dense rock may be so small as to pass unnoticed because evaporation prevents the accumulation of water on the exposed face; yet the flow of water through such a material may produce seepage pressures between the mineral grains as large as those exerted in more permeable materials under the same conditions of hydraulic head.

In order to understand the forces that govern the flow of water through earth materials, it is necessary to know the meaning of certain terms used in the field of hydraulics. In Fig. 2.1 points a and b represent the extremities of the path along which part of the water flows through a sample of earth material. At each extremity a standpipe known as a *piezometric tube* has been installed to permit observation of the level to which the water rises at these points. The water level in the tube at b is known as the *piezometric level* at b and the vertical distance from this level to point b is the *piezometric head* at b. The vertical distance between a and b represents the *position head*, ΔH, at b with respect to a. If the piezometric level at a is equal to the piezometric level at b, the system is in a state of rest, regardless of the magnitude of ΔH, and there is no flow from

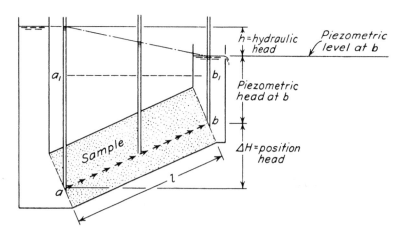

FIGURE 2.1. Diagram illustrating the meaning of hydraulic head, piezometric head, and position head.

a to b. Flow occurs only if there is a difference in the piezometric levels at a and b. This difference is known as the *hydraulic head h*, or as the *difference in piezometric level* between a and b. In the figure the two points a_1 and b_1 are at the same elevation. Under the conditions represented, the pressure at a_1 exceeds that at b_1 by a quantity equal to the unit weight of water times the difference in piezometric level. This quantity, $\gamma_w h$, is referred to as the *excess hydrostatic pressure* at a with respect to b and is designated by the symbol u. This pressure causes the water to move through the soil from a to b. The ratio

$$i_p = \gamma_w(h/l) = u/l \qquad 2.1$$

is known as the *pressure gradient* from a to b, and the ratio

$$i = i_p/\gamma_w = h/l \qquad 2.2$$

is defined as the *hydraulic gradient* between a and b. It is a pure number.

The flow of water through permeable substances is governed by an empirical relationship

$$v = ki \qquad 2.3$$

first stated by H. Darcy in 1856. In this expression, v is the *discharge velocity*, defined as the quantity of water that percolates in a unit time across a unit area of a section oriented at right angles to the direction of flow; i is the hydraulic gradient; and k is a coefficient known as the *coefficient of permeability*. The value of k, which has the units of a velocity, depends primarily on the characteristics of the permeable substance, but it is also a function of the unit weight and the viscosity of the fluid. Since water is the only fluid with which the foundation engineer is concerned, the influence of differences in the unit weight of the fluid is negligible. Moreover, the changes in viscosity within the ordinary range of temperature of groundwater are relatively unimportant and can usually be ignored in the solution of practical problems. Hence, it is customary and justifiable for civil engineers to regard the coefficient of permeability as a property of the soil or rock.

In a general way, the coefficient of permeability increases with increasing size of voids, which in turn increases with increasing grain size. However, the shape of the void spaces also has a marked influence on the permeability. As a consequence, no simple relationships have been found between permeability and grain size except for fairly coarse soils with rounded grains. For example, in his studies of the permeability of loose filter sands, Allen Hazen found that

$$k = CD_{10}^2 \qquad 2.4$$

where C is equal approximately to 100/cm sec and D_{10} is expressed in centimeters.

Similarly, it has been found that the coefficient of permeability of coarse-grained soils varies approximately as the square of the void ratio. No such simple relation has been established for soils containing flake-shaped particles.

Permeability Tests. The lack of simple and general relationships between the coefficient of permeability and the results of classification tests often leads to the necessity for performing permeability tests, either in the field or in the laboratory. Permeability tests on soil samples are usually made with a falling-head permeameter or a constant-head permeameter. The constant-head permeameter (Fig. 2.2a) gives reliable results for highly permeable materials such as clean sands and gravels. The value of k is computed by means of the equation

$$k = QL/hAt \qquad 2.5$$

in which Q is the quantity of discharge, L the length of the sample in the direction of flow, A the cross-sectional area of the sample, h the hydraulic head, and t the time. All these quantities are measured in the test.

The falling-head permeameter (Fig. 2.2c) is more suitable for tests on materials of low permeability because the dimensions of the apparatus can be adjusted so that the measurements of head and time may be carried out with high accuracy over a wide range in the values of the coefficient of permeability. The value of k may be calculated from the quantities measured during the test by means of the equation

$$k = 2.3 \frac{aL}{At} \log_{10} \frac{h_0}{h_1} \qquad 2.6$$

In this equation, a is the cross-sectional area of the standpipe, L the length of the sample, A the cross-sectional area of the sample, t the time, and h_0 and h_1 are the original and final hydraulic heads, respectively.

The results of permeability tests on cohesionless materials are often misleading because it is impracticable to obtain representative samples of permeable material and to place them in the permeameter without disturbance. Not only is it difficult to obtain undisturbed samples, but since most granular deposits are quite erratic, it is difficult even to obtain representative ones.

Permeability tests are also subject to various types of experimental errors. One of the most important of these arises from the formation of a filter skin of fine material on or just below the surface of the sample. The skin may be caused by segregation of the silty constituents of the sand during place-

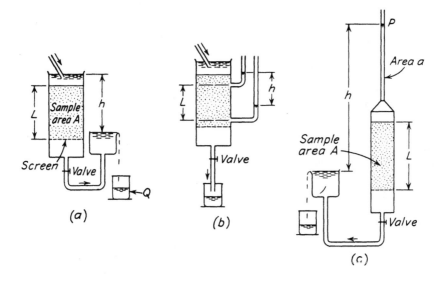

FIGURE 2.2. (a) and (b) Constant-head permeameters; (c) Falling-head permeameter.

ment of the sample in the permeameter or by migration of loose fine particles at the surface into the voids of the sand just beneath. The segregated layer of fine material greatly reduces the measured permeability. Figure 2.2b shows a type of constant-head apparatus that may be used to eliminate the effect of the surface skin. In this apparatus the loss in head is measured through a distance in the interior of the sample, and the drop in head across the filter skin has no effect on the results.

In conducting permeability tests the technician must be careful to saturate the sample completely and to make sure that no air bubbles are released by the water during the test. The air bubbles block the voids and thus lower the permeability. At best, permeability tests on laboratory specimens can serve only as a basis for a rough estimate of the permeability of an actual soil deposit. For this reason an estimate of the permeability of a coarse-grained deposit based on grain size may often be as useful as the results of laboratory tests.

The permeability of clay samples can best be determined indirectly from the data obtained by performing consolidation tests. The basis for the computations is discussed more fully in Art. 3.12.

Although standard procedures (ASTM D-2434) have been developed for the conduct of permeability tests and are useful guides, the multiplicity of factors that may influence the results demands experience and judgment on the part of the technician who should depart from the standards when he considers them inapplicable.

Permeability of Stratified Deposits. Many deposits consist of layers or lenses of materials that differ in grain size and permeability. The average coefficient of permeability of such deposits differs significantly in the horizontal and vertical directions. In the horizontal direction the average permeability may be almost as great as the permeability of the most permeable layer or lens, whereas in the vertical direction the average may be almost as small as the permeability of the least permeable layer or lens. The ratio of the average coefficients of permeability in the horizontal and vertical directions ranges for most natural deposits between 1 or 2 and about 10.

Field Pumping Tests. The most reliable information concerning the permeability of a deposit of coarse-grained material below the water table can usually be obtained by conducting pumping tests in the field. Although such tests have their most extensive application in connection with dam foundations, they may also prove advisable on large bridge or building foundation jobs where the water table must be lowered. A test is made by drilling one well from which water is pumped and several others in which the position of the water table is observed. The observation wells are drilled at various distances from the pumping well along two straight lines, one oriented approximately in the direction of the natural groundwater flow and the other at right angles to it. Once a steady state of flow has been established and the water levels in all the wells are nearly stationary, the coefficient of permeability may be calculated from the results of the observations. The appropriate equations depend on the boundary conditions and, for some practical problems, may be quite complex. For the simple conditions illustrated in Fig. 2.3, representing a horizontal bed of homogeneous sand extending from the ground surface to an impervious stratum and penetrated for its full depth by the pumping well and the observation wells,

$$k = \frac{2.3q}{\pi(h_2{}^2 - h_1{}^2)} \log_{10} \frac{r_2}{r_1} \qquad 2.7$$

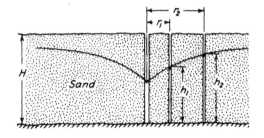

FIGURE 2.3. Illustration of flow of water toward a well during a pumping test.

Table 2.1 Coefficient of Permeability of Various Soils

k (cm/sec)	Drainage	Soil Type	Determination of k				
10^2	Good	Clean gravels	Pumping tests. Reliable if properly conducted	Constant head permeameter-reliable	Falling-head permeameter	Reliable	Computation from grain size
10^1	Good	Clean gravels					
1.0	Good	Clean sands					
10^{-1}	Good						
10^{-2}	Good	Clean sand and gravel mixtures					
10^{-3}	Good						
10^{-4}	Good						
10^{-5}	Poor	Very fine sands				Unstable. Much experience required	
10^{-6}	Poor	Organic and inorganic silts, mixtures of sand silt and clay, glacial till, stratified clay deposits.					
10^{-7}	Practically impervious	Impervious soils, for example, homogeneous clays below zone of weathering.				Fairly reliable	Computation from consolidation data (reliable)
10^{-8}							
10^{-9}							

After Casagrande and Fadum (1940).

in which q equals the rate of flow across the boundary of any cylindrical section having a radius r and, therefore, the quantity of water pumped from the well per unit of time.

Approximate values of the coefficient of permeability for various types of soils and the recommended method of determining these values are assembled in Table 2.1. As the table indicates, it is difficult to obtain reliable values for the coefficient of permeability of soils that contain appreciable quantities of silt or very fine sand. There is no indirect method of computing the permeability of such materials, and laboratory tests are likely to be extremely unreliable unless made by experienced technicians.

ILLUSTRATIVE PROBLEMS

1. A loose uniform sand with rounded grains has an effective grain size D_{10} equal to 0.3 mm. Estimate the coefficient of permeability.

Solution. The estimate can be made by means of eq. 2.4. Thus,

$$k = CD_{10}^2 = 100 \times (0.03)^2$$
$$= 9 \times 10^{-2} \text{ cm/sec}$$

2. In a constant-head permeability test a sample 8 cm long was tested in a permeameter with an inside diameter of 5 cm. After a state of steady flow was established under a head of 50 cm, a discharge of 120 cu cm was collected in 30 sec. Compute the value of k.

Solution. According to eq. 2.5,

$$k = \frac{QL}{hAt} = \frac{120 \times 8}{50 \times \pi \times \frac{25}{4} \times 30}$$
$$= 3.3 \times 10^{-2} \text{ cm/sec}$$

3. A falling-head permeability test was performed in a permeameter with an inside diameter of 5 cm. The inside diameter of the standpipe was 2 mm. The sample had a length of 8 cm. During a period of 6 min

the head on the sample decreased from 100 to 50 cm. Compute the value of k.

Solution. According to eq. 2.6,

$$k = 2.3 \frac{aL}{At} \log_{10} \frac{h_0}{h_1}$$
$$= 2.3 \frac{0.2^2 \times 8}{5^2 \times 6 \times 60} \log \frac{100}{50}$$
$$= 8.2 \times 10^{-5} \log 2.0$$
$$= 2.5 \times 10^{-5} \text{ cm/sec}$$

2.3. Permeability of Rock

Even sound rocks such as granite, marble, or slate cannot be regarded as impervious; their coefficients of permeability range from 10^{-8} to 10^{-10} cm/sec. However, the permeability of sound rock is of little more than academic interest to the engineer engaged in the design and construction of building or bridge foundations, because almost every deposit of rock contains extremely permeable channels such as joints, bedding planes, or solution cavities. The influence of these features on the permeability of rock masses is considered in Chap. 6.

2.4. Effective and Porewater Pressures

Definitions. The total stress p that acts at any point on a section through a mass of saturated soil or rock may be divided into two parts. One part, known as the *porewater pressure* or *neutral stress* u_w, acts in the water and in the solid in every direction with equal intensity. The other part, known as the *effective stress* $\bar{p}$, represents an excess over the neutral stress and acts exclusively between the points of contact of the solid constituents. That is,

$$p = \bar{p} + u_w \quad 2.8$$

Experience has demonstrated that only effective stresses can induce changes in volume of a soil mass. Likewise, only effective stresses can produce frictional resistance in soils or rocks. Neutral stresses, on the other hand, cannot in themselves cause volume change or produce frictional resistance.

This statement is known as the *principle of effective stress* (Terzaghi, 1925, 1936; Skempton, 1960). Its implications are among the most important in soil mechanics and foundation engineering. They will be emphasized in Chaps. 3 and 4 and throughout Part C of this book.

Indirectly, changes in neutral stresses may lead to changes in volume or frictional resistance, but only under circumstances such that the changes in neutral stress cause corresponding changes in effective stress. The distinction between effective and neutral stresses can be illustrated by means of Fig. 2.4, which shows a container partly filled with granular material and completely filled with water. To the bottom of the container is attached a flexible tube connected to a reservoir. In Fig. 2.4a the water level in the reservoir is at the same elevation as that in the container, so that no flow takes place. On plane ab at depth $(H_1 + z)$ below the top of the container, the vertical pressure p is

$$p = H_1 \gamma_w + z \gamma_{sat} \quad 2.9$$

where γ_w is the unit weight of water, and γ_{sat} the unit weight of the saturated soil. Since p depends on the weight of all the overlying soil and water, it is referred to as the *total pressure*. It consists of the effective pressure $\bar{p}$ and the porewater pressure u_w.

The water above ab extends in continuous voids to the height z and in a continuous mass above the soil for a distance H_1. Hence, according to the laws of hydraulics, the porewater pressure u_w at ab must be

$$u_w = (H_1 + z)\gamma_w \quad 2.10$$

The effective pressure $\bar{p}$ is then
$$\bar{p} = p - u_w = H_1 \gamma_w + z\gamma_{sat} - (H_1 + z)\gamma_w$$
or

$$\bar{p} = z(\gamma_{sat} - \gamma_w) \quad 2.11$$

The quantity $\gamma_{sat} - \gamma_w$ is known as the *submerged unit weight*, γ'. Therefore, for the conditions illustrated in Fig. 2.4a,

$$\bar{p} = z\gamma' \quad 2.12$$

Thus, the effective pressure is independent of the depth H_1 of water above the submerged soil or rock.

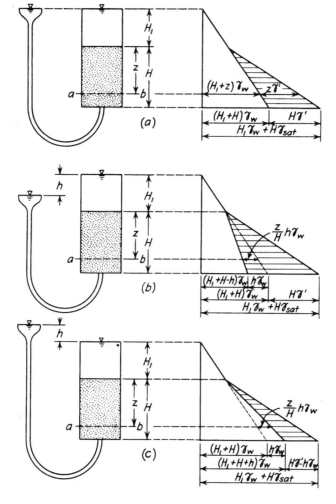

FIGURE 2.4. Diagram illustrating meaning of effective and porewater pressure.

Under the conditions of Fig. 2.4a there is no flow of water through the voids. In the pressure diagram to the right of the figure the change in width of the unshaded area represents the distribution of the porewater pressure with depth, and the change in width of the shaded area represents the distribution of the effective pressure. If the level of the water in the reservoir is different from that in the container, then flow occurs and the conditions are similar to those illustrated in Fig. 2.4b or 2.4c. Thus, eq. 2.10 is no longer valid, and eqs. 2.11 and 2.12 are no longer applicable.

Seepage Pressures and Critical Hydraulic Gradient. If the free water surface is maintained at the top of the container and the reservoir is lowered, a state of steady flow is soon established in accordance with the conditions illustrated in Fig. 2.4b. Since the reservoir and the flexible tube represent a standpipe freely communicating with the bottom of the container, the porewater pressure at the bottom of the container may be expressed as

$$u_w = (H_1 + H - h)\gamma_w \qquad 2.13$$

This represents a decrease of $h\gamma_w$ in the porewater pressure as compared to the condition shown in Fig. 2.4a. It may be noted that this decrease cannot be attributed to the velocity of the flowing water, because the velocity head $v^2/2g$ is negligible for velocities comparable to the highest ones encountered in natural soil deposits. Hence, the total pressure on the bottom of the container is determined solely by the weight of the soil and water above it. As a result the effective pressure must be increased by $h\gamma_w$ over that indicated in Fig. 2.4a. At plane ab the proportionate increase in the effective pressure is therefore equal to $h\gamma_w(z/H)$. The increase in effective pressure due to the flow of water through the voids is known as *seepage pressure*. It is the result of the frictional drag of the flowing water on the soil grains.

It may be noted that the loss in head between the top of the granular material and the depth z is hz/H. The corresponding hydraulic gradient i is h/H. Thus, the seepage pressure may be expressed as $iz\gamma_w$, and the effective pressure at depth z is

$$\bar{p} = z\gamma' + iz\gamma_w \qquad 2.14$$

If the reservoir is raised above the container, so that water flows from the reservoir upward through the granular material (Fig. 2.4c), the porewater pressure at the bottom of the container is increased by the amount $h\gamma_w$. Thus the effective pressure on plane ab is reduced to

$$\bar{p} = z\gamma' - iz\gamma_w \qquad 2.15$$

By increasing the upward hydraulic gradient the seepage pressure can be increased to $z\gamma'$, whereupon the effective pressure $\bar{p}$ becomes zero. This occurs when

$$z\gamma' - i_c z\gamma_w = 0 \qquad 2.16$$

and

$$i_c = \gamma'/\gamma_w \qquad 2.17$$

The hydraulic gradient at which the effective pressure becomes zero is known as the *critical hydraulic gradient* i_c. Under these conditions a cohesionless soil cannot support any weight on its surface. Moreover, as the hydraulic gradient approaches the critical value, the soil becomes much looser and the coefficient of permeability k increases. Hence, if an excavation is made into a cohesionless soil below groundwater level to such a depth that the effective pressure is reduced to zero, a visible agitation of the grains of soil is observed. This phenomenon is known as *boiling* or the *quick condition*.

Most quicksands are the result of these particular hydraulic conditions.

ILLUSTRATIVE PROBLEMS

1. Compute the vertical effective pressure at a depth of 30 ft in the deposit represented by the accompanying figure.

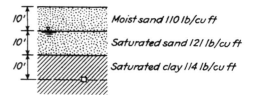

Solution. On the basis of submerged weights (eq. 2.12), we may write

$$
\begin{aligned}
10 \times 110 &= 1100 \text{ lb/sq ft} \\
10 \times (121 - 62.5) &= 585 \\
10 \times (114 - 62.5) &= \underline{515} \\
\bar{p} &= 2200 \text{ lb/sq ft}
\end{aligned}
$$

On the basis of eq. 2.9, we may write
$$
\begin{aligned}
p &= 10 \times 110 + 10 \times 121 + 10 \times 114 \\
&= 3450 \text{ lb/sq ft} \\
u_w &= 20 \times 62.5 = \underline{1250} \\
\bar{p} &= 2200 \text{ lb/sq ft}
\end{aligned}
$$

2. In the deposit shown in the accompanying figure the groundwater level was originally at the ground surface. The water table was lowered by drainage to a depth of 20 ft, whereupon the degree of saturation of the sand above the lowered water table decreased to 20 per cent. Compute the vertical effective pressure at the middle of the clay layer before and after lowering the water table.

Solution. Before the water table is lowered, the effective pressure is

$$
\begin{aligned}
50 \times (135 - 62.5) &= 3625 \text{ lb/sq ft} \\
12.5 \times (120 - 62.5) &= \underline{719} \\
\bar{p} &= 4344 \text{ lb/sq ft}
\end{aligned}
$$

After the water table is lowered, the moist weight of the sand in the upper 20 ft becomes

$$116 + 0.2(135 - 116) = 116 + 3.8$$
$$= 119.8 \text{ lb/cu ft}$$

Therefore, the effective pressure at 20 ft becomes

$$\begin{aligned} 20 \times 119.8 &= 2396 \\ 30 \times (135 - 62.5) &= 2175 \\ 12.5 \times (120 - 62.5) &= \underline{719} \\ \bar{p} &= 5290 \text{ lb/sq ft} \end{aligned}$$

Note that lowering the water table *increases* the effective pressure in the layer of clay.

2.5. Soil Moisture, Drainage, and Frost Action

Water Table. The level assumed by the water in an observation well extending into a deposit of soil is known as the *groundwater level*, *free water surface*, or *water table*. This definition is valid irrespective of the coefficient of permeability of the deposit, although the determination of the groundwater level in very fine-grained soils by means of observation wells may be very inaccurate unless specialized techniques are used.

In general, the water table beneath the ground surface of an area even as small as a building site is not horizontal but consists of a gently undulating surface. Water supplied by rainfall, melting snow, or artificial sources such as irrigation flows continually through the soil from higher to lower elevations under the influence of gravity. The intersection of the water table and a vertical surface oriented in the direction of the steepest slope of the water table is often referred to as the *line of seepage* (Fig. 2.5). The piezometric head at any point P on the line of seepage is zero; that is, in a piezometric tube with its bottom at P the water level would coincide with the level of the free water surface. Similarly the piezometric head at Q is zero. The hydraulic head h at P with respect to Q, which is the difference in piezometric levels between P and Q, is therefore identical with the difference in actual elevation between points P and Q. This relation is an inherent property of the free water surface and of the line of seepage. The hydraulic gradient that drives a particle of water along the line of seepage from P to Q is then h/l.

Capillary Phenomena. Below the water table, most soils are completely or almost completely saturated. Above the water table, the degree of saturation depends on the climatic conditions, the grain size of the soil, and the distance to the water table. Coarse-grained soils are only partially saturated even at elevations close to the water level, whereas fine-grained soils may be saturated for a considerable distance above it. In the latter instance, the free water surface may be defined as the level at which the pore-water pressure is equal to atmospheric pressure; that is, $u_w = 0$.

If gravity were the only force acting upon the water in the soil, the soil above groundwater level would always be completely dry except during percolation resulting from recent precipitation. Water, however, exhibits the force of surface tension due to the attraction between its molecules at an air-water interface. The surface tension combined with the attraction between water and most solid substances, as demonstrated by the ability of water to wet these substances, constitutes a force that tends to draw or retain moisture above the water table. This force is a manifestation of *capillarity*.

Capillarity may be demonstrated by immersing the lower end of a glass tube of small diameter into a vessel containing water. The water rises in the tube to a height governed primarily by the diameter of the tube and the cleanliness of its inner surface. This height is known as the *height of capillary rise*. It is given in centimeters by the expression

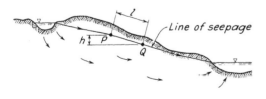

FIGURE 2.5. Diagram illustrating line of seepage.

Height of capillary rise, $h_c = \dfrac{2T_s}{r\gamma_w} \cos \alpha$

$$\text{2.18}$$

in which T_s is the surface tension of the water in grams per centimeter, r the radius of the tube in centimeters, and α the contact angle between the surface of the water and the wall of the tube. Above the free water surface the pressure in the water is negative with respect to atmospheric pressure. Thus, at elevation z above the water table,

$$u_w = -z\gamma_w \qquad 2.19$$

The conditions for capillary rise in a mass of soil are not strictly analogous to those in a glass tube because the voids in the soil have variable diameters. Nevertheless, the average diameter of the voids is related to the average grain size and thus to the height of capillary rise. Hence, as a rough approximation for the maximum height h_c (cm) to which water can be lifted by capillarity in a given soil, we may write

$$h_c = C/eD_{10} \qquad 2.20$$

in which e is the void ratio, D_{10} is Hazen's effective size in centimeters, and C is an empirical coefficient that depends on the shape of the grains and the surface impurities. In general, C has a value between 0.1 and 0.5 sq cm.

The height of capillary rise is greatest for very fine-grained soil materials, but the rate of rise in such materials is slow because of their low permeability. Thus, the capillary rise that occurs in a given length of time, such as 24 hr, is a maximum for materials of intermediate grain size such as silts and very fine sands. Figure 2.6 shows the relationship between the grain size of a uniform quartz powder and the height of capillary rise for a 24-hr period.

The water that rises by capillarity above the free water surface attains the maximum height h_c only in the smaller voids. A few large voids may effectively stop capillary rise in certain parts of the soil mass. Hence the height h_{cc} (Fig. 2.7), to which the soil is completely saturated by capillary rise, is likely to be considerably smaller than h_c.

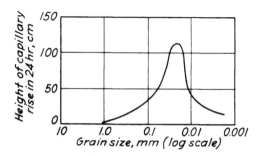

FIGURE 2.6. Relation between grain size of uniform quartz powder and height of capillary rise for 24-hr period (after Atterberg, 1908).

Since the threads of water are continuous, however, the stress in the water is everywhere governed by eq. 2.19.

Some of the water that percolates into the ground from the surface does not reach the free water table but is held by surface tension either in the smaller voids or as annular rings surrounding the points of contact between soil grains. Additional moisture may rise above the free water table by the processes of evaporation and condensation. Soil moisture from these sources is known as *contact moisture*. The surface tension in the films of water tends to pull the soil grains together and gives rise to the *apparent cohesion* of sands and silts. The cohesion is called apparent because immersion destroys the films of water and the tendency of the grains to stick together. The apparent cohesion between the grains of a loose moist sand inhibits the ability of the grains to slide with respect to each other and to assume a denser configuration. Hence, an uncompacted fill of such material may exist in a

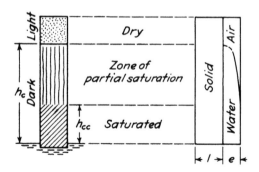

FIGURE 2.7. Capillary rise of water into dry sand.

very loose, metastable state known as a *bulked* condition. Upon submergence the very loose structure collapses and the grains move into a loose but stable state.

Drainage by Gravity. The difficulties of construction below water level in soil deposits may be avoided, and the stability of the materials increased, by lowering the water table. In coarse-grained materials this can usually be done by draining the water into ditches or galleries or by pumping from sumps or wells. Since gravity is the force that causes the flow of the water toward the drains, the process is known as *drainage by gravity*.

As the water flows toward the drains, it produces seepage pressures that tend to move the grains toward the outlet. If the soil is a mixed-grained material containing fine sand or silt, the finer constituents are likely to be washed out. This may lead to settlement or to the development of erosion tunnels in the drained soil and, in addition, may clog the drains. Hence, it is usually necessary to prevent the migration. This can be accomplished most effectively by covering the drained material where the water emerges with a coarser granular material, known as a *filter*, which does not impede the flow of water but which contains voids too small to be invaded by the fine particles of the materials being drained.

For a satisfactory filter the particle-size curve of the filter material must bear a suitable relation to that of the material to be protected. The requirements have been investigated experimentally (Terzaghi, 1922; USBR, 1947). They can be expressed most conveniently by means of two *filter ratios* R

$$R_{50} = \frac{D_{50} \text{ of filter}}{D_{50} \text{ of protected soil}} \quad 2.21a$$

$$R_{15} = \frac{D_{15} \text{ of filter}}{D_{15} \text{ of protected soil}} \quad 2.21b$$

Appropriate values of R are listed in Table 2.2. In addition to satisfying the requirements with respect to the 50 per-cent and 15 per-cent sizes, the particle-size curve representing the filter material should have a smooth shape without pronounced breaks and should be roughly parallel to that of the soil being protected.

The requirements expressed in Table 2.2 were developed for earth dams and other permanent installations. For temporary protection or less exacting uses, the limits on gradation may be exceeded somewhat.

It may not prove possible to satisfy the requirements of Table 2.2 with a single filter material that can prevent the fines in the protected soil from being washed out while being able to drain freely all the water transmitted to it. It is then necessary to construct a series of filters that become increasingly coarser as the water approaches the outlet. Each layer of the system must meet the grain-size requirements for a filter with respect to the layer of material adjacent to it. Such a system is called a *graded filter*.

The grain-size requirements of a filter may not be satisfied by available local materials without washing, screening, or blending. The cost of these operations may be considerable. Consequently, the layers of graded filters are often made as thin as is

Table 2.2 Particle-Size Requirements for Filters

Grading of Filter Material	R_{50}	R_{15}
Uniform (C_u up to 4)	5 to 10	No requirement
Nonuniform, subrounded particles	12 to 58	12 to 40
Nonuniform, angular particles	9 to 30	6 to 18

The particle-size curve representing the filter material should have a smooth shape without pronounced breaks, and should be roughly parallel to that of the soil being protected.

compatible with the operation of construction equipment. Such thin filters are vulnerable to damage, especially during construction. In most instances it is preferable to use appreciably thicker layers of natural materials even if they only approximately satisfy the requirements. Segregation of filter materials is difficult to prevent but must be kept to the practicable minimum.

Not all the water can be removed from a soil by gravitational processes. The amount that remains in the voids is of little practical importance compared with the rate at which the drainage can be accomplished. For clean sands and gravels only a few hours or days are required to lower the water table several feet, whereas for silty soils several weeks may be needed.

As the water flows out of a coarse-grained soil, part of the void space is invaded by air. In fine-grained soils, however, the void space continually decreases, and there is practically no air invasion until the shrinkage limit is reached. This process is known as *drainage by consolidation*.

Drainage by Desiccation. It is not possible to drain fine-grained soils in a reasonable length of time by gravity alone. However, in nature, drainage may be effected by desiccation. During this process, moisture is gradually lost by evaporation. As the water evaporates, the solid soil particles are drawn into closer and closer contact on account of the surface tension of the water. At the shrinkage limit, the resistance of the soil is great enough to withstand the forces of surface tension without further consolidation and, if further desiccation takes place, air invasion begins. The moisture content does not decrease to zero but attains a value in equilibrium with the relative humidity of the soil atmosphere. Clay soils may retain as much as 7 per cent moisture even in the air-dry state.

The process of desiccation increases the strength of fine-grained soils because the increasing tension in the water pulls the colloidal particles into close contact. Clay soils that have become desiccated in the field and are subsequently saturated usually retain a large portion of the strength imparted by desiccation. Layers of such clay are not uncommon and may provide adequate foundations for light structures.

When an air-dried or oven-dried sample of fine-grained soil is placed in water, the soil *slakes*. The force of surface tension draws the water into the voids and compresses the air trapped in the interior until the pressure exceeds the tensile strength of the soil skeleton and slaking occurs.

Frost Action. If the temperature of the surface of the soil is below freezing and if the water table is not too deep with respect to the height of capillary rise, moisture moving upward continually by capillary action and by evaporation and condensation collects and freezes to form lenses of ice in the upper part of the ground. This phenomenon, known as *frost action*, is most prevalent in very fine sands and silts because these soils are capable of lifting the greatest amount of water in the shortest time by capillary action (Fig. 2.6). On the other hand, it does not occur in clean sands or gravels or mixed-grained soils in which less than 3 per cent of the particles are smaller than 0.02 mm. Such materials are designated as *frost free*.

Clay soils are also affected by freezing, but their low permeability often acts to prevent the formation of ice lenses. However, if the clay has a plasticity index of less than about 10 to 12, it is highly susceptible to frost action and may act much like a silt.

In the northern United States and in Canada it is not uncommon for ice lenses to reach an aggregate thickness of 6 to 18 in. during a single winter season. They are responsible for much damage to highway pavements because when they melt the soil becomes a thick suspension with negligible bearing capacity. The damage may be prevented by removing all highly frost-susceptible materials within the depth of frost penetration and by replacing them with frost-free materials. Where the expense of such treatment would be prohibitive, frost-free materials are placed only in the base directly under the pavement and slightly or moderately frost-susceptible materials are placed in the rest of the subgrade.

2.6. Seepage and Flow Nets

The pattern of flow of water through real soils and of the corresponding pore pressures is extremely complex because of the erratic manner in which the permeability is likely to vary from point to point and in different directions. Hence, exact analyses of such common problems as the effect of a dewatering system or the flow beneath a cofferdam into the excavation for a bridge pier are rarely possible. In spite of the complexities of real problems, however, the engineer can greatly improve his judgment with respect to seepage and its effects by a study of flow under simple, idealized conditions.

For example, let us consider the flow of water through the pervious materials (Fig. 2.8a) into which a sheet pile cutoff has been driven. It is assumed that the permeability of the soil is the same at every point and is equal in all directions; moreover, the cutoff wall and the bedrock underlying the soil are assumed to be completely impervious. Darcy's law is considered to be valid, and both soil and water are considered to be incompressible.

Water entering the soil upstream of the sheet pile wall flows to the downstream ground surface along smooth curved paths, such as AB (Fig. 2.8a), known as *flow lines*. The flow is caused by the hydraulic head h, which drives the water from A to B. As a particle of water proceeds from A toward B it exerts a frictional drag on the soil particles; this drag, in turn, produces a seepage pressure in the soil structure; the seepage pressure at any point acts in the direction of the flow line at that point. Because of the frictional drag, the hydraulic head decreases steadily from the upstream to the downstream end of every flow line. Consequently, the piezometric level at some point such as C is intermediate between those corresponding to A and B. Between the extremes of any other flow line, such as $A'B'$, the hydraulic head is also h and there is a point C' at which the piezometric level is the same as at C. A line, such as LM, connecting points of equal piezometric level, is known as an *equipotential line*.

If the permeability is constant and the same in all directions, theory demonstrates that equipotential lines must be perpendicular to flow lines. This conclusion makes possible the solution of problems involving flow through such materials by a graphical procedure in which flow lines and equipotential lines are sketched and their positions revised until the necessary geometrical relations are satisfied. The resulting diagram, exemplified by Fig. 2.8a, is known as a *flow net*.

The first step in constructing a flow net is to note whatever boundary conditions must be satisfied; that is, to determine if any flow lines or equipotential lines are known in advance. In Fig. 2.8a, for example, the sheet pile wall itself constitutes a flow line. Water entering the soil just to the left of the piling moves vertically downward to the tip of the pile, passes to the right beneath the pile, and rises vertically along the downstream face of the pile. The top of the impermeable stratum also constitutes a flow line. Water entering the formation infinitely far to the left flows along this surface until it has passed infinitely far to the right. It is apparent that these two flow lines mark the boundaries of the flow pattern. All other flow lines must be located between them. Furthermore, it is noted that the upstream ground surface is an equipotential line, because the water level in any standpipe with its lower end at the upstream ground surface would coincide with the upstream water surface. Similarly, the downstream ground surface is also an equipotential line; the piezometric level coincides with the downstream water surface. All other equipotential lines must be located between these two.

The boundary conditions for the problem are summarized in Fig. 2.8b. They are represented by the equipotential lines ab and cd, and by the flow lines bec and fg. Construction of the remainder of the flow net is started by sketching a small number of flow

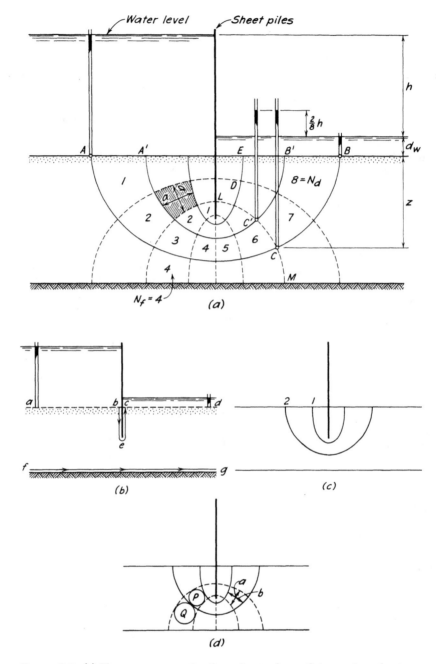

FIGURE 2.8. (a) Flow net representing flow of water beneath impervious sheet-pile wall. (b) Boundary conditions to be satisfied by flow net. (c) First trial flow lines in construction of flow net. (d) First trial equipotential lines in construction of flow net.

lines, perhaps only two; each flow line starts on *ab* and ends on *cd*. Since *ab* and *cd* are equipotential lines, the flow lines must intersect these lines at right angles. The sketched flow lines should be smooth curves with shapes making gradual transitions from one boundary flow line (*bec*) to the other (*fg*). The initial trial may resemble Fig. 2.8*c*.

Next, an attempt is made to sketch equipotential lines consistent with the requirements of the problem. These lines are also smooth curves, and they must cross flow lines at right angles. Moreover, to simplify interpretation of the completed flow net, the spacing between equipotential lines should be such that the drop in piezometric level is the same between each successive pair of equipotential lines. It is also convenient to space the flow lines in such a manner that the quantity of flow through each flow channel bounded by two successive flow lines is equal to that through every other flow channel. Both of these requirements can be satisfied by making each area bounded by two adjacent flow lines and two adjacent equipotential lines approximately equidimensional. That is, the distances *a* and *b* (Fig. 2.8*d*) should be equal. As an aid in judging whether a curved-sided area satisfies this criterion, a circle may be inscribed in the area. Thus, in Fig. 2.8*d* it is apparent that area *P* is reasonably equidimensional, but that area *Q* does not satisfy the requirements.

The first trial set of equipotential lines should be sketched in an effort to provide right-angle intersections with the flow lines insofar as possible, and to subdivide the space into areas that may differ from each other in size but that should each be equidimensional. The first attempt will ordinarily not be satisfactory, but a study of the sketch will suggest appropriate revisions in both flow and equipotential lines. Since the shape and position of each set of lines depend on those of the other, a series of adjustments is commonly necessary. Considerable facility in drawing flow nets can be attained with practice and by the study of well-drawn nets for various boundary conditions. Several flow nets for problems pertinent to foundation engineering are shown in Fig. 2.9.

When the flow net has been refined so that it satisfies the boundary conditions and the geometric criteria, it provides precisely the same information as would a rigorous analytical solution of the same problem. Indeed, flow nets can often be drawn readily for problems too complex or difficult for analytical treatment.

From the completed flow net the pore pressure can be determined at any point in the pervious material. The conditions at point *C*, Fig. 2.8*a*, will serve as an example. According to the flow net, a particle of water following the flow line *AB* (or any other flow line) from the upstream to the downstream boundary would cross eight spaces bounded by successive equipotential lines. Each space represents an *equipotential drop* Δh. If N_d represents the number of equipotential drops along any flow line,

$$\Delta h = \frac{h}{N_d}$$

In Fig. 2.8*a*, $\Delta h = \frac{1}{8}h$. By the time the water has reached point *C*, the head loss is $6\Delta h$, or $\frac{6}{8}h$. The piezometric level at *C* is then $\frac{6}{8}h$ below headwater level, or $\frac{2}{8}h$ above tailwater level. The piezometric head at *C*, is, therefore,

$$z + d_w + \tfrac{2}{8}h$$

and the porewater pressure at *C* is

$$u_c = \gamma_w(z + d_w + \tfrac{2}{8}h)$$

The *excess porewater pressure* at *C* with respect to tailwater level is the pressure available at *C* to drive the water the remaining distance to *B*, and is equal to $\frac{2}{8}h\gamma_w$.

The quantity of flow beneath the sheet piles for a unit length of the wall may be calculated readily. Let us consider the flow Δq through the shaded area in Fig. 2.8*a*. According to Darcy's law, the quantity of flow per unit of time is

$$\Delta q = kiA$$

where *A* is the cross sectional area of the flow channel. The channel has a width *a* and a thickness of unity in the direction of

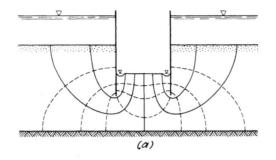

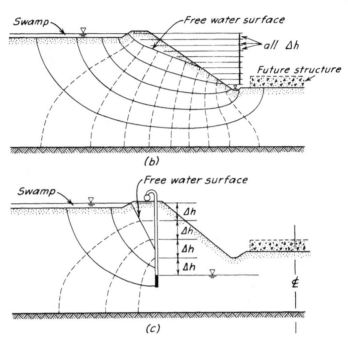

FIGURE 2.9. Flow nets for various conditions. (a) Cofferdam for construction of bridge pier. (b) Excavation for foundation below water table in sand. (c) Excavation predrained by well points from which water is pumped.

the wall. Therefore,

$$\Delta q = kia$$

The hydraulic gradient across the shaded area is $\Delta h/a$. However,

$$\Delta h = \frac{h}{N_d}$$

whence

$$i = \frac{h}{aN_d}$$

and

$$\Delta q = k\frac{h}{N_d}$$

If the number of flow channels is N_f, the total flow q per unit length of wall is

$$q = N_f \Delta q = kh\frac{N_f}{N_d} \qquad 2.22$$

An examination of the flow net (Fig. 2.8a) shows that the seepage emerges at points such as E or B' in a vertically upward direction. The upward hydraulic gradient at E, for example, can be estimated as Δh divided by the distance DE. If this gradient exceeds the critical value (eq. 2.17), the soil just downstream of the sheet pile wall will become quick and a failure may occur. Study of the flow nets (Fig. 2.9) will disclose several conditions under which quick conditions may occur in practice unless suitable precautions are taken, such as increasing the length of sheet piles or adding berms, filters, or drains.

In many practical problems the flow of water is not confined by an artificial upper boundary, but takes place below a free water surface (Figs. 2.9b and 2.9c). The uppermost flow line is then a line of seepage (Art. 2.5). Since the line of seepage is a flow line, equipotential lines meeting it must do so at right angles, but each such line terminates at the line of seepage. To satisfy the special hydraulic requirements of the line of seepage, the vertical component of the distance between the terminations of two adjacent equipotential lines must equal the equipotential drop Δh, as indicated in Figs. 2.9b and 2.9c. To draw the flow net one must assume the position of the free water surface, construct a trial flow net, check all the criteria previously discussed as well as the special conditions along the free surface, and revise the diagram until all conditions are satisfied. Construction of such flow nets is considerably more difficult than if the upper boundaries are fixed, but the principles are unaltered.

The procedures just described may be modified to take account of stratification or of differing values of permeability in horizontal and vertical directions (A. Casagrande, 1935). These conditions are, of course, encountered far more frequently in practice than the simple ones considered in the preceding paragraphs. As a matter of fact, the pattern of permeability is often so variable that no flow net can satisfactorily represent the actual conditions. Nevertheless, extremely useful practical conclusions can be drawn from studying the flow patterns for various simplified conditions.

PROBLEMS

1. A sand with rounded grains has an effective size of 0.080 mm and a uniformity coefficient of 3.5. Estimate its coefficient of permeability.
   Ans. 6.4×10^{-3} cm/sec.

2. During a constant-head permeability test on a sample of sand, 150 cu cm of water were collected in 2 min. The sample had a length of 10 cm and a diameter of 2 in. The head was maintained at 20 cm. Compute the coefficient of permeability.
   Ans. 3.08×10^{-2} cm/sec.

3. A falling-head permeability test was performed on a sample of uniform clean sand. The standpipe consisted of a graduated burette, and it was observed that 1 min was required for the water level to fall from the 0 cu cm to the 50 cu cm graduation. The initial hydraulic head was 90 cm and the final head 40 cm. The sample was 20 cm long and had a diameter of 4.0 cm. Calculate the coefficient of permeability.
   Ans. 2.15×10^{-2} cm/sec.

4. A pumping test was made in pervious gravels and sands extending to a depth of 50 ft where a bed of clay was encountered. The normal groundwater level was at the ground surface. Observation wells were located at distances of 10 and 25 ft from the pumping well. At a discharge of 48 gpm from the pumping well a steady state was attained in about 24 hr. The drawdown at 10 ft was 5.5 ft, and at 25 ft was 1.2 ft. Compute the coefficient of permeability.
   Ans. 2.37×10^{-3} cm/sec.

5. A pumping test was made in a fine sand extending to a depth of 50 ft where a bed of clay was encountered. The normal groundwater level was at the ground surface. Observation wells were established at distances of 10 and 20 ft from the pumping well. A steady state was reached at a rate of pumping equal to 1.2 gpm. The drawdown at the outer well was 0.5 ft, and at the inner well 1.5 ft. What was the coefficient of permeability of the sand?
   Ans. 1.83×10^{-4} cm/sec.

6. A clay layer 12 ft thick rests beneath a deposit of submerged sand 26 ft thick. The top of the sand is located 10 ft below the surface of a lake. The saturated unit weight of the sand is 125 lb/cu ft

and of the clay is 117 lb/cu ft. Compute the total vertical pressure, the pore-water pressure, and the effective vertical pressure at midheight of the clay layer.

Ans. 4577 lb/sq ft; 2625 lb/sq ft; 1952 lb/sq ft.

7. The surface of a saturated clay deposit is located permanently below a body of water. Laboratory tests have indicated that the average natural water content of the clay is 47 per cent and that the specific gravity of the solid matter is 2.74. What is the vertical effective pressure at a depth of 37 ft below top of clay?

Ans. 1754 lb/sq ft.

8. If the water level (Prob. 7) remains unchanged and an excavation is made by dredging, how many feet of clay must be removed to reduce the effective pressure at 37 ft by 1000 lb/sq ft?

Ans. 21 ft.

9. The water table is lowered from a depth of 10 ft to a depth of 20 ft in a deposit of silt. All the silt is saturated, even after the water table is lowered. Its water content is 26 per cent. Estimate the increase in effective pressure at a depth of 34 ft on account of lowering the water table.

Ans. 625 lb/sq ft.

10. Compute the critical hydraulic gradient for the following materials:
    a. Coarse gravel, $k = 10$ cm/sec; $G = 2.67$; $e = 0.65$.
    b. Sandy silt, $k = 10^{-6}$ cm/sec; $G = 2.67$; $e = 0.80$.

Ans. (a) 1.01; (b) 0.93.

11. The depth of water outside the cofferdam in Fig. 2.9a is 10 ft and the penetration of the sheet piles below the original surface of the sand is 18 ft. The water level in the ditch is 20 ft below the outside water level. What is the pressure in the water at the tips of the sheet piles? If $k = 5 \times 10^{-3}$ cm/sec, what will be the seepage into the ditches, per foot of length of the entire cofferdam?

Ans. 1036 lb/sq ft; 1.26 gal/min/ft of cofferdam.

12. The elevation of the water in the swamp (Fig. 2.9b) is 30 ft above that in the ditch inside the excavation. The soil is a sand having a coefficient of permeability k equal to 2×10^{-2} ft/min. What will be the seepage into the ditch on one side of the excavation?

Ans. 0.3 cu ft/min/lin ft of ditch.

SUGGESTED READING

The classic presentation of the flow-net method and its applications is A. Casagrande (1935), "Seepage Through Dams," *J. New England Water Works Assoc.*, 51, 2, 131-172. Reprinted in "Contributions to Soil Mechanics 1925-1940," Boston Soc. Civ. Eng., 1940, and as Harvard Univ. Soil Mech. Series No. 5.

The fundamentals of seepage, with applications, are well presented in H. R. Cedergren (1967), *Seepage, Drainage, and Flow Nets*, New York, Wiley, 489 pp.

The history and significance of the concept of effective stress are discussed by A. W. Skempton (1960) in "Terzaghi's Discovery of Effective Stress," included in *From Theory to Practice in Soil Mechanics—Selections from the Writings of Karl Terzaghi*, New York, Wiley, pp. 42-53.

Techniques of permeability testing may be found in the following references:

A. Casagrande and R. E. Fadum (1940), "Notes on Soil Testing for Engineering Purposes," Harvard Univ. Grad. School of Engineering Publ. No. 8, 74 pp.

T. W. Lambe (1951), *Soil Testing for Engineers*, Chap. 6, "Permeability Tests," pp. 52-62, New York, Wiley, 165 pp.

A modern discussion of the mechanism of frost action is contained in Chap. 12, "Soil Freezing and Permafrost," in R. N. Yong and B. P. Warkentin (1966), *Introduction to Soil Behavior*," New York, Macmillan, pp. 391-428. Somewhat less theoretical is C. W. Kaplar (1970), "Phenomenon and Mechanism of Frost Heaving," *Hwy. Res. Rec.*, 304, 1-13. Criteria for frost susceptibility are summarized in K. A. Linell, F. B. Hennion, and E. F. Lobacz (1963), "Corps of Engineers' Pavement Design in Areas of Seasonal Frost," *Hwy. Res. Rec.*, 33, 76-136.

Settlement Due to Consolidation

Shrine of our Lady of Guadalupe, Mexico City. The part on the left was built about 1709, that on the right about 1622. A ridge of rock beneath the junction of the two parts has permitted little settlement, whereas adjacent portions, located above deposits of highly compressible clay as much as 60 ft thick, have settled more than 7 ft. The structure on the left was underpinned in 1969.

PLATE 3.

CHAPTER 3

Consolidation Characteristics of Soils

3.1. Significance of Stress–Strain Characteristics of Earth Materials

The stress–strain characteristics of a soil or rock determine the settlement that a given structure will experience. In some instances, they may also serve as an indication of construction difficulties that may arise during excavation into soil masses.

The settlements of structures above beds of soft clay, sometimes buried deeply beneath stronger and less compressible materials, may take place slowly and may reach large magnitudes. Because of the time lag between the end of construction and the appearance of cracking, such settlements were once considered to be of mysterious origin. The first successful efforts to explain the phenomena on a scientific basis were made by Terzaghi in 1919.

Terzaghi's studies dealt with the amount and rate of settlement originating in a layer of clay prevented from experiencing lateral displacements and capable of expelling water upward or downward as the particles tended to squeeze together. In many instances these conditions are approached in practice. For this reason, and because an understanding of the phenomena is basic to the solution of more complex practical problems, a consideration of the behavior of various soils when subjected to stress is preceded by a discussion of the one-dimensional consolidation of saturated clays.

3.2. Consolidation Tests on Remolded Clays

The relations among vertical pressure, settlement, and time are investigated in the laboratory by the *confined compression, oedometer,* or *consolidation test.* During such a test the specimen is completely confined by a metal ring (Fig. 3.1). The load is transmitted to the upper and lower faces of the specimen through two porous disks that permit water to flow into or out of the clay. The deformation is measured by means of a dial gage.

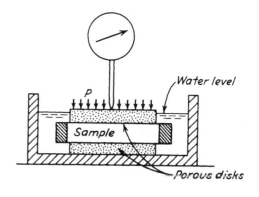

FIGURE 3.1. Apparatus for consolidation test.

The pressure p is increased in steps. After each increment the load is allowed to remain constant until the deformation practically ceases. This usually requires several hours, even for a specimen having a thickness as little as $\frac{3}{4}$ in., because the deformation takes place only as fast as water can be squeezed out of the clay. When the rate of deformation under a given pressure has become very small, a new increment is applied and the procedure is repeated.

The results are plotted in the form of a curve representing the final void ratio corresponding to each increment of pressure as a function of the accumulated pressure. It is convenient to plot the pressure to a logarithmic scale. The diagram is then known as an *e*-log *p* curve.

As an introduction to the characteristics of *e*-log *p* curves for naturally sedimented clays, we shall first consider the results of a consolidation test on a sample completely remolded in the laboratory at a water content near the liquid limit. At high void ratios the curve is concave downward but, as shown by segment km (Fig. 3.2), it soon becomes almost a straight line. The nearly straight portion of the curve is known as the *virgin consolidation line* or *virgin branch*. If loading is discontinued at a pressure p_0' corresponding to point m and the load is then decreased in successive increments, the sample *swells* as indicated by the *rebound curve* mm'. If loading is resumed, the initial part of the *reloading curve* lies slightly above the rebound curve. The reloading curve then bends downward rather abruptly at a pressure close to p_0', passes below point m, and approaches the extension of the virgin branch. If at a pressure p_0'' the sample is again unloaded and reloaded, another rebound curve (nn') and another reloading curve are obtained; the slopes of these curves are roughly the same as those determined previously. If the pressure is increased beyond p_0'', the *e*-log *p* curve again bends sharply downward and approaches the virgin branch. As the pressure is further increased, the virgin curve tends to become slightly concave in an upward direction.

At the time when the sample is in the

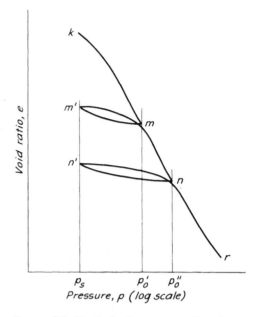

FIGURE 3.2. Typical *e*-log *p* curve for clay remolded near liquid limit.

state represented by m' it is said to be *preloaded* or *overconsolidated* because it has previously been consolidated under a pressure p_0' greater than the pressure p_s under which it is now in equilibrium. The value p_0' is known as the *preconsolidation pressure*. The degree of overconsolidation is measured by the ratio p_0'/p_s, known as the *overconsolidation ratio*. It is evident that the break in the curve $m'n$ is closely related to the magnitude of the preconsolidation pressure. A useful graphical procedure has been devised (A. Casagrande, 1936) to estimate the value of p_0' if only the curve $m'n$ is available. It is illustrated in Fig. 3.3.

According to this procedure, the point c of maximum curvature of the *e*-log *p* curve is selected by eye. From c, a tangent is drawn to the curve, and a horizontal line is also constructed. The angle between these two lines is then bisected. The point of intersection of this bisector with the upward extension of a tangent to the straight part of the curve is denoted by d. The abscissa of d corresponds to the pressure p_0'.

The pressures p_0' and p_s are effective stresses, inasmuch as the corresponding void ratios are determined after the sample has

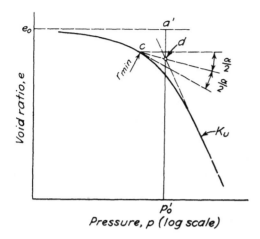

FIGURE 3.3 Graphical construction for determining former maximum pressure p_0' from the e-log p curve (after A. Casagrande, 1936).

come to equilibrium and water no longer tends to be expelled from or drawn into it. For simplicity, however, the bars indicating effective stress, such as $\bar{p}_0'$ and $\bar{p}_s$, are omitted unless the omission would lead to confusion.

3.3. Consolidation Characteristics of Normally Loaded Deposits

The compressibility of a clay in the field may be investigated by making consolidation tests on samples recovered in a state as nearly undisturbed as possible. Two different conditions of practical importance must be recognized, namely, whether the stratum from which the sample was taken is normally loaded or preloaded. A stratum is said to be *normally loaded* if it has never been acted on by vertical pressures greater than those existing at present. On the other hand a *preloaded* or *overconsolidated* stratum was at one time in its history subjected to greater vertical pressures than are now active.

A typical e-log p curve for an undisturbed sample of a normally loaded clay of low sensitivity is shown in Fig. 3.4. It is designated by K_u. Like the portion $m'n$ of the e-log p curve in Fig. 3.2, it characteristically consists of two branches, a fairly flat initial portion and a nearly straight inclined portion. For comparison, K_r represents the results of a test on the same material remolded at a water content near the liquid limit. The lower, nearly straight portion of K_r has a slope flatter than that of K_u.

The coordinates of point a (Fig. 3.4) represent the void ratio and effective pressure corresponding to the state of the clay in the field. When a sample is extracted by means of the best techniques, the water content of the clay does not increase significantly. Hence, the void ratio e_0 at the start of the test is practically identical with that of the clay in the ground. When the pressure on the sample reaches p_0, the e-log p curve should pass through point a unless the test conditions differ in some manner from those in the field. In reality, K_u always passes below point a, because even the best sample is at least slightly disturbed.

The engineer is interested in the e-log p relationship for the clay in the field, not in the laboratory. Therefore, he requires some procedure for extrapolating from the results of the laboratory tests to those representative of conditions in the field. One procedure for accomplishing this end can be developed on the basis of a comparison between curves K_u and K_r. It has been observed (Schmertmann, 1955) for many clays that the extensions of the straight-line lower portions of curves K_u and K_r inter-

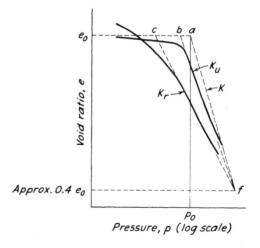

FIGURE 3.4. Typical e-log p curves for undisturbed (K_u) and remolded (K_r) samples of normally loaded clay of low sensitivity.

sect at a point f corresponding to a void ratio roughly equal to $0.4e_0$. The extensions of e-log p curves for samples with intermediate degrees of disturbance also pass through or close to point f. Hence, it is reasonable to assume that the field e-log p relation may be approximated by a straight line K extending from a to f.

If a clay deposit is known to be normally loaded, the value of p_0 can be computed from a knowledge of the water content, the unit weight of the solid constituents, and the degree of saturation of the overlying strata, provided the position of the water table is known. The initial void ratio of the clay can also be computed readily. Hence, the coordinates of point a can be determined. The procedures for computing p_0 and e_0 have been discussed in Arts. 1.6 and 2.4.

Point f can be determined by performing a consolidation test on an undisturbed sample or even on a partly disturbed or remolded sample. Points a and f can then be connected by a straight line to establish the field relationship K upon which a computation of settlement can be based.

The slope of K on the semilogarithmic diagram is designated as the *compression index* C_c, defined by the equation

$$C_c = \frac{e_0 - e_1}{\log_{10} p_1 - \log_{10} p_0} = \frac{e_0 - e_1}{\log_{10}(p_1/p_0)} \qquad 3.1$$

The numerical value of C_c can be determined readily from the diagram. If e_1 corresponds to any arbitrary pressure p_1 and e_2 to the pressure $p_2 = 10p_1$, then $C_c = e_1 - e_2$.

It has been found (Skempton, 1944) that C_c is closely related to the liquid limit of normally loaded sedimented clays. The relationship between the two quantities is approximately

$$C_c = 0.009(w_L - 10) \qquad 3.2$$

where w_L is expressed as a percentage. Equation 3.2 is of great practical importance because it permits one to compute the approximate settlement of a structure above a deposit of fairly insensitive normally loaded sedimented clay if only the liquid limit is known, and even if no consolidation tests have been performed.

3.4. Computation of Settlement

A cross section through a layer of clay with a thickness H and with its midheight located at a depth D below the original ground surface is shown in Fig. 3.5a. The original effective pressure at a point A is equal to p_0, and the increase in pressure is Δp. The void ratio of the clay before construction is equal to e_0.

Figure 3.5b shows a prismatic element containing point A. The element can be assumed to consist of solid matter with a height equal to unity and a void space with an additional height equivalent to e_0. The total height of the element is, therefore, $1 + e_0$. If the void ratio decreases an amount Δe on account of consolidation, the unit strain of the element is $\Delta e/(1 + e_0)$. On the assumption that this strain is constant from top to bottom of the clay layer, the decrease in thickness of the layer, or the settlement S above point A, is given by the equation

$$S = H \frac{\Delta e}{1 + e_0} \qquad 3.3$$

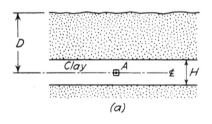
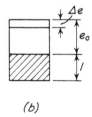

FIGURE 3.5. (a) Section through compressible clay layer. (b) Compression of element of layer.

This equation is general and can be used to calculate the settlement whenever the original void ratio and change in void ratio are known. If the clay is normally loaded, according to eq. 3.1,

$$\Delta e = e_0 - e_1 = C_c \log_{10} \frac{p_1}{p_0}$$

$$= C_c \log_{10} \frac{p_0 + \Delta p}{p_0}$$

By substituting this quantity into eq. 3.3, we find the following expression for the settlement of the ground surface above point A due to consolidation of a normally loaded layer with thickness H:

$$S = \frac{C_c}{1 + e_0} H \log_{10} \frac{p_0 + \Delta p}{p_0} \qquad 3.4$$

3.5. Consolidation Characteristics of Preloaded Deposits

The e-log p diagram for a typical preloaded clay is shown in Fig. 3.6. It is assumed that the sample has been taken with a minimum of disturbance; moreover, at an advanced stage of the test the sample has been permitted to rebound and has been reloaded.

Point a represents the void ratio of the sample and the effective pressure that acted on it before its removal from the ground. When the sample is placed in the consolidation device in the presence of water, it is likely, under low applied pressures, to tend to swell. It is customary to prevent such swelling by the prompt addition of load. Thus, the first part of the e-log p curve is usually quite flat; it is likely to pass slightly below point a, whereas the field curve must pass through it.

Point a' corresponds to the void ratio e and pressure p_0' of the clay in the field when the stratum was under its maximum load. By some process such as erosion the pressure was reduced and the void ratio increased slightly by swelling until the present state, represented by a, was reached. (The slight swelling is neglected in the figure.) The field curve should pass close to a', because before the erosion the field relationship should have been represented by $a'f$. We cannot construct this curve precisely, because the pressure p_0' corresponding to point a' is no longer known; the present overburden pressure is only p_0. However, we note that the curve K_u at pressures greater than p_0 often consists of two quite distinct segments resembling the segments of curve $m'n$ in Fig. 3.2. By analogy to the behavior of the rebounded and reloaded sample of remolded material, we may conclude that the first segment of K_u (Fig. 3.6), at a pressure greater than p_0, is a reloading curve whereas the second has the slope of the virgin curve. The break between the two segments should, therefore, correspond to the preconsolidation pressure p_0' and could be estimated by the procedure of A. Casagrande. Moreover, the slope of the reloading portion of the curve should approximate that of the reloading segment $n'g$ (Fig. 3.6).

These considerations suggest the following procedure for approximating the e-log p curve that corresponds to field conditions for a precompressed material. The consolidation test should be conducted on the best obtainable sample. After the pressure has been increased sufficiently to define the virgin branch of the e-log p curve, a rebound-reloading loop should be determined. The preconsolidation pressure should be approximated by A. Casagrande's proce-

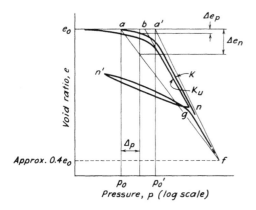

FIGURE 3.6. Typical e-log p curve (K_u) for undisturbed sample of preloaded clay and reconstructed field relation (K).

dure. The field curve from which the settlements can be estimated should then be sketched by starting at point a and continuing parallel to the reloading curve $n'g$ to a point about halfway to p_0'. The curve should then be directed toward point f in such a manner as to follow the shape of K_u. This procedure does not eliminate the judgment of the interpreter, but it leads to reasonable results. The settlement corresponding to a pressure between the present overburden pressure and the preconsolidation pressure can be computed by means of eq. 3.3, but the value of Δe must be determined from the sketched field e-log p curve K (Fig. 3.6).

If we fail to recognize that the clay is preloaded and base a settlement computation on af, the computed settlement will be far too great. The change Δe_n in void ratio computed for an increase Δp in pressure is likely to be 4 to 10 times as great as the real change Δe_p, provided Δp is not greater than about half of $p_0' - p_0$. As Δp approaches $p_0' - p_0$, the error becomes smaller.

In connection with practical problems, the most important consideration is to be able to recognize whether or not a clay is preloaded. It almost certainly is if the water content is closer to the plastic limit than to the liquid limit. Moreover, by performing a consolidation test on a very carefully taken sample, one can often obtain the data required to make the decision. In Fig. 3.4 it may be noticed that the upward projection of the straight part of K_u intersects the line $e = e_0$ at point b, which is located on the left of point a. This is always true of a normally loaded clay of ordinary sensitivity. On the other hand, in Fig. 3.6, which refers to a preloaded clay of ordinary sensitivity, b is located on the right side of a. Unfortunately, the influence of disturbance tends, by displacing b to the left, to destroy the evidence of preloading. Hence, if it is probable that a clay deposit may be preloaded and samples are to be obtained for consolidation tests, the best possible techniques for sampling should be used. An additional procedure for investigating the state of preloading of a soft sedimented clay will be discussed in Art. 4.8.

Preloaded clays are widespread. Many deposits were at one time subjected to the weight of overlying soils subsequently removed by erosion. In some instances glacial ice overrode and consolidated the underlying soils; when the ice melted, the deposits were then overconsolidated with respect to the remaining overburden pressure. The surface of many deposits, especially in flood plains, was exposed from time to time during deposition, whereupon evaporation and desiccation occurred (Art 2.5). The shrinkage stresses may have been large compared to the present effective overburden pressures and such deposits are likely to contain layers or lenses of overconsolidated clays. Long-term lowering of groundwater levels by pumping or construction activities may also have induced consolidation and preloaded the clay with respect to the conditions existing after restoration of the groundwater levels. In assessing the possibility or the degree of preloading at a site, geological and historical considerations may provide valuable insight.

3.6. Consolidation Characteristics of Sensitive Clays

The results of a consolidation test on a carefully taken sample of a clay of high sensitivity (S_t greater than about 8) resemble curve K_u (Fig. 3.7). The flat initial branch of the e-log p curve merges rather abruptly into a steep segment that appears to represent a structural breakdown of the clay such that a slight increase of pressure leads to a large decrease in void ratio. The curve then passes through a point of inflection and its slope decreases. If the statistical relationship, eq. 3.2, is used to determine the compressibility of such a material, the computed settlement is too small. As the error is on the unsafe side, settlement calculations must be based on tests of undisturbed samples.

If a tangent is drawn to the steep part of the curve at its point of inflection c, it intersects the line e_0 at b. The pressure corre-

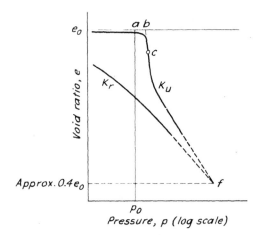

FIGURE 3.7. Typical e-log p curves for undisturbed (K_u) and remolded (K_r) samples of clay of high sensitivity.

sponding to b is approximately equal to that at which the structural breakdown takes place. In most extrasensitive clays, b lies at least slightly to the right of point a, which represents the effective overburden pressure p_0. The increment of pressure $b - a$ may represent the extent to which the soil has been preloaded, or it may represent a *bond strength* originating in physicochemical processes subsequent to deposition. In either event, the increment represents the maximum pressure that may be added to the soil without triggering the large and often spectacular settlements associated with the steep branch of the curve. In areas underlain by soft highly sensitive clays, such as Mexico City or parts of Scandinavia, experience has indicated the necessity for limiting the pressure added by a new building to only a fraction of the bond strength.

Disturbance during sampling and handling may completely mask the significant characteristics of the soil; the e-log p curves for completely remolded specimens of highly sensitive clays are in no respect different from those of clays of ordinary sensitivity. Even the completely remolded clays are not so compressible as the undisturbed soil at pressures somewhat above the breakdown pressure. Hence, unlike estimates of settlement for clays of ordinary sensitivity, those for highly sensitive clays are, if based on the results of remolded tests, on the unsafe side. Therefore, good undisturbed samples are mandatory.

3.7. Consolidation Characteristics of Residual Soils

The foregoing comments refer to the behavior of samples from deposits of sedimented clay. Cohesive residual soils, although products of markedly different geological processes, are often characterized by e-log p curves similar to those of transported clays of moderate to high sensitivity. They may appear to be preloaded, but the apparent preconsolidation pressure is a consequence of residual cohesive bonds between the particles rather than of effective pressures produced by a previous overburden or by desiccation.

A zone of soft, highly compressible material with a natural water content near the liquid limit is sometimes encountered in the lower parts of residual deposits, especially if the underlying less weathered bedrock is impermeable enough to cause the groundwater to flow laterally through the weathered material above. Within this zone the destruction of the cohesive bonds by chemical alteration may advance to the stage that the bonds are exactly in equilibrium with the weight of the overlying materials. Consequently, the soil may have the compressibility characteristics of a normally loaded clay, sometimes of high sensitivity.

3.8. Consolidation Characteristics of Collapsible Soils

Partially saturated soils may possess bonds that endow them with cohesion and considerable rigidity. If the bonds consist of soluble compounds, such as the chemical precipitates that accumulate under semiarid conditions, they may be destroyed upon prolonged submergence. If the bonds consist of clay minerals or various amorphous claylike substances, they may adsorb water and become weakened when submerged; true loess belongs in this category, as do

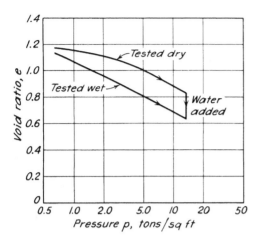

FIGURE 3.8. Decrease in void ratio upon addition of water to sample of loess while under vertical pressure at natural moisture content in consolidation test apparatus. Results may be compared with those of sample saturated before test (after Holtz and Gibbs, 1951).

certain loose slightly clayey sands. Even fills compacted at water contents below the optimum moisture content are likely to display an apparent cohesion due to capillary moisture and to possess a rigidity that can readily be destroyed by flooding.

The e-log p curves of such materials tested without permitting water to gain access to the samples through the porous stones display no unusual features, but if water is admitted at any stage of a test a spectacular decrease in void ratio at constant external vertical pressure may occur, as illustrated in Fig. 3.8. Soils exhibiting such behavior are said to be *collapsible*. The magnitude of the settlement of a deposit of collapsible soil may be estimated by subjecting samples at their field moisture content to the vertical stresses anticipated in the ground and by observing the change in void ratio due to submergence while the vertical stresses are maintained.

Collapsible natural soils are widespread in many parts of the world where there is a long dry season and where the groundwater table is at great depth. Even the irrigation of lawns around newly constructed houses in such localities may cause serious settlements. The engineering implications of collapsible soils with respect to foundations are discussed in Art. 20.2.

3.9. Consolidation Characteristics of Sands

The e-log p curves for a laterally confined sand (Fig. 3.9a) resemble those for preloaded clays. The corresponding e-p curves (Fig. 3.9b) are concave upward; rebound and reloading curves are very flat. Figure 3.9b indicates that the compressibility of a given sand depends strongly on the density index. Loose sands are far more compressible than dense sands; moreover, even very large pressures cannot decrease the void ratio of a loose sand to that of the same sand in a dense condition. Under large pressures the grains are likely to crush. On the other hand, the void ratio of the same loose sand can readily be decreased to that of a dense sand by vibration alone.

Well-graded sands are less compressible than uniform or gap-graded sands of the same grain shape and density index. Sands consisting of rounded particles are usually less compressible than those of otherwise comparable angular particles. The addition of even small percentages of plate-shaped particles, such as mica flakes, may greatly increase the compressibility.

3.10. Evaluation of Compressibility in Practice

If a deposit of clay is known to be normally loaded and not extrasensitive, the compressibility can be evaluated with reasonable accuracy by any one of three procedures. Undisturbed samples may be subjected to consolidation tests and the field curve K determined by the construction illustrated in Fig. 3.4. Disturbed or remolded samples may be subjected to consolidation tests and the field curve K similarly determined as shown in Fig. 3.4. Or, without the performance of consolidation tests, the compression index may be evaluated on the basis of the liquid limit and the statistical relationship, eq. 3.2.

If the clay is extrasensitive, the real compressibility is likely to exceed greatly that

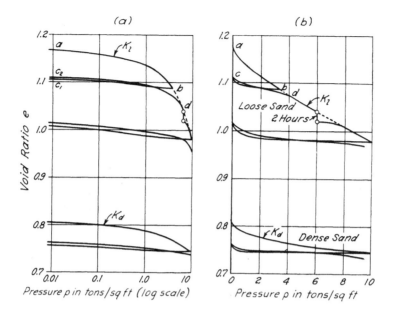

FIGURE 3.9. Typical e-log p and e-p curves for initially loose (K_l) and initially dense (K_d) laterally confined sand.

determined by any of the foregoing procedures unless consolidation tests are performed on the best of undisturbed samples.

If the clay is preloaded, the compressibility at pressures smaller than the preconsolidation pressure p_0 is likely to be much less than that of normally loaded clays of otherwise similar characteristics. Even with the best undisturbed samples, the settlements estimated on the basis of consolidation tests are likely to be too great.

A sample of clay subjected to a consolidation test in the laboratory is for practical reasons subjected to changes in applied stress far more rapidly than is the deposit during, for example, the excavation for and the construction of a building. Moreover, in laboratory tests the time during which the clay sustains a single load increment is at most a few days, whereas the lifetime of a structure located above the clay deposit may be several years. Because of these and other differences, predicted and actual settlements may differ appreciably, but experience has demonstrated that forecasts of ultimate settlement, particularly above soft deposits, are essential steps in design and, when made with judgment, can be relied on.

The ultimate settlement of a clay deposit subjected to a change in load does not, however, occur immediately; often it develops only gradually, over many months or years. Hence, in practice an estimate of the rate of settlement may also be needed. The lag of settlement with respect to increase of load is largely a consequence of the resistance to the flow of water from the pores of the clay.

3.11. Swelling Clays and Clay-Shales

In Art. 3.2 it was pointed out that, if loading is discontinued in a consolidation test on a clay and the load is then decreased, the sample swells as indicated by the rebound curve mm' (Fig. 3.2). The swelling, of course, requires the presence of enough water to permit the increase in void ratio. In some clays and shales the swelling caused by reduction of stress is so great that it disrupts roadways or structures. Moreover, if the moisture content of these materials is normally kept low because, for instance, of the aridity of the locale, a decrease in the rate of evaporation, possibly caused by the presence of the floor slab of a building, may lead to an accumulation of moisture and to energetic swelling. If the swelling is restricted by the construction, extremely large forces may develop. Soils that behave in this manner are said to be *swelling*. They lead to serious foundation difficulties in many parts of the world. The characteristics of these materials to be evaluated for foundation design are discussed in Chap. 20.

3.12. Rate of Consolidation

Theory of Consolidation. Clays are so impervious that the water is almost trapped in the pores. When an increment of load is applied, the pore water cannot escape immediately. Since the clay particles tend to squeeze together, pressure develops in the pore water as it does in the oil filling a hydraulic jack when a weight is placed on the ram. This pressure tends to make the fluid flow out. The flow is rapid at first but, as it continues, the pressure drops and the rate of flow decreases. As the water is forced out of a clay sample, the particles can move closer together. Therefore, the surface of the specimen settles. The rate of settlement, rapid at first, decreases to a small value.

The progress of consolidation can be observed by measuring the decrease of porewater pressure in different parts of the sample. If the consolidation apparatus (Fig. 3.1) were equipped with a number of very small standpipes (Fig. 3.10a), it would be noticed that at the instant t_0 of application of an increment of pressure Δp the water would rise in every standpipe to a height $h = \Delta p/\gamma_w$ above the level of the free surface of the water surrounding the sample. The horizontal line representing the locus of all the standpipe levels at time t_0 is known as the *initial isochrone*. The pressure in the water in any pore of the sample at this time is equal to Δp in excess of the hydrostatic pressure of the water in the container at the elevation of the pore. Hence, the water pressure Δp is known as an *excess pressure*. It is also known as the *initial consolidation pressure* because it initiates the process of consolidation.

The pressure in the pore water cannot remain equal to Δp at the upper and lower boundaries of the sample, $z = 0$ and $z = 2H$, because the excess pressure in the water in the porous disks is zero. On account of the difference in pressure there is a large hydraulic gradient toward each porous disk; the gradient causes water to flow from the sample into each disk, whereupon the pressure decreases in the pores of the clay near the disks. After the lapse of a time t_1, the height of water in the standpipes near the middle ($z = H$) of the sample may have dropped only slightly, whereas the heights corresponding to points closer to the boundaries may have dropped substantially. The isochrone corresponding to this time is designated as curve t_1 in Fig. 3.10a.

Point A_1 on this isochrone corresponds to the excess porewater pressure u at depth z_1 in the sample, at time t_1. The excess pressure is represented by the vertical distance A_1B. The original excess pressure at time t_0 is represented by A_0B. Hence, of the original excess pressure Δp, only the fraction A_1B/A_0B remains in the pore water. Since the total stress to be carried by the sample at depth z_1 remains Δp, the principle of effective stress leads to the conclusion that the fraction A_0A_1/A_0B must now be an effective stress carried by the soil skeleton. The ratio A_0A_1/A_0B is known as the *degree of consolidation* U_z at depth z_1 and at time t_1.

At time t_2 the standpipe level corresponding to depth z_1 has dropped still further to A_2 and the degree of consolidation at z_1 has increased to $U_z = A_0A_2/A_0B$. The corresponding isochrone is designated as t_2 (Fig. 3.10a). Finally, after a very long time, no excess pressures remain in the sample, the corresponding isochrone is indicated by t_∞, and U_z at all depths is 100 per cent.

The slope of the tangent to an isochrone at a point such a A_1 represents the hydraulic gradient in the pore water at the corresponding values of z and t. When the slope is downward and to the left, the flow of water is upward; when it is downward and to the right, the flow is downward. At midheight of the sample where $z = H$, the slope

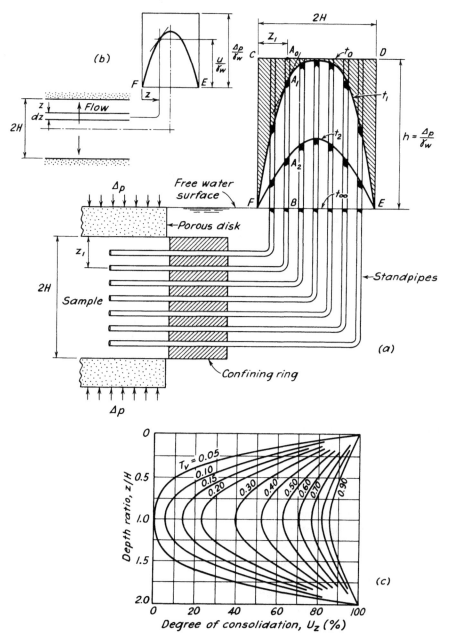

FIGURE 3.10. Process of consolidation in sample of clay drained top and bottom. (*a*) Section through sample and confining ring showing imaginary standpipes for observation of porewater pressures and isochrones at time t_1 and t_2. (*b*) Simplified representation of (*a*) indicating basis for differential equation of consolidation. (*c*) Solution of differential equation in terms of dimensionless coefficients.

of all the isochrones is zero. As a consequence there is never any flow of water across a horizontal plane at $z = H$. The midplane could as well be regarded as an impervious boundary. Hence, the isochrones for a sample with thickness H but drained only at the top are identical to those for a sample with thickness $2H$ but drained top and bottom.

The process of consolidation may be described by the successive positions of the isochrones that define the relative amounts of the initial consolidation pressure that are carried by effective and neutral stresses at a given time. On the assumptions that the laws of hydraulics govern the decrease of porewater pressure, and that the decrease in volume of the soil is proportional to the increase in effective stress and is equal to the quantity of water squeezed out, a theory has been developed (Terzaghi, 1923, 1943) for computing the degree of consolidation at any depth and time. It is known as the *theory of consolidation*.

The differential equation governing the process of consolidation for flow in one direction can be derived by a consideration of Fig. 3.10b, which is a simplified representation of Fig. 3.10a. It is assumed that:

1. Flow takes place only along vertical lines.
2. The coefficient of permeability k of the material is constant throughout the mass and does not change with time during any one load increment.
3. A constant ratio exists during any one load increment between the change in volume per unit of volume and the change in effective pressure. That is,

$$\frac{dV}{V} = m_v \, d\bar{p}$$

where m_v (cm^2/g) is known as the *coefficient of volume compressibility*. Since the volume change takes place only in the pore space, $dV/V = dn$, where n is the porosity, and

$$m_v = \frac{dn}{d\bar{p}} \qquad 3.5$$

At any time t, at depth z, the upward hydraulic gradient i across an element with thickness dz (Fig. 3.10b) is the slope of the isochrone for the appropriate values of t and z. Thus

$$i = -\frac{\partial h}{\partial z} = -\frac{1}{\gamma_w} \frac{\partial u}{\partial z}$$

where u is the excess porewater pressure. The negative sign indicates that i causes an upward flow whereas z increases in the downward direction. The velocity of flow is

$$v = ki = -\frac{k}{\gamma_w} \frac{\partial u}{\partial z}$$

The quantity of water traversing a unit area of a horizontal section through the soil per unit of time is also numerically equal to v. Hence, the difference in the quantity of water flowing into and out of an element with thickness dz in a given time dt is

$$\frac{\partial v}{\partial z} dz \, dt = -\frac{k}{\gamma_w} \frac{\partial^2 u}{\partial z^2} dz \, dt$$

During the same time dt, the change in volume of the element with unit area and thickness dz is

$$\frac{\partial n}{\partial t} dt \, dz = m_v \frac{\partial \bar{p}}{\partial t} dt \, dz$$

However, the change in volume is precisely equal to the difference between the inflow and outflow of water. Hence,

$$-\frac{k}{\gamma_w} \frac{\partial^2 u}{\partial z^2} = m_v \frac{\partial \bar{p}}{\partial t} = -m_v \frac{\partial u}{\partial t}$$

or

$$\frac{\partial u}{\partial t} = \frac{k}{m_v \gamma_w} \frac{\partial^2 u}{\partial z^2} = c_v \frac{\partial^2 u}{\partial z^2} \qquad 3.6$$

Equation 3.6 is known as the differential equation of consolidation for one-dimensional flow, and

$$c_v = \frac{k}{m_v \gamma_w} \qquad 3.7$$

is the *coefficient of consolidation*.

Equation 3.6 can be solved for u as a function of z and t; that is, the isochrones

(Figs. 3.10a and 3.10b) can be determined. The solution must satisfy the hydraulic boundary conditions of the problem. For the particular case represented by Fig. 3.10a, these conditions are:

1. The excess porewater pressure u is equal to Δp at any depth z for $t = 0$.
2. The excess porewater pressure u is zero at any time t at the drainage surfaces $z = 0$ and $z = 2H$.
3. At any time t the hydraulic gradient i is zero at depth $z = H$.
4. After a very long time, $u = 0$ at all depths.

The solution for the preceding boundary conditions may be expressed by the following relationships:

$$U_z(\%) = f\left(T_v, \frac{z}{H}\right) \quad 3.8$$

where

$$T_v = \frac{c_v}{H^2} t \quad 3.9$$

In this expression, T_v is a dimensionless number called the time factor; c_v is a property of the soil known as the coefficient of consolidation; H is the half-thickness of the sample; and t is the time corresponding to the degree of consolidation U_z. From this expression it may be observed that the time required to reach a given degree of consolidation varies as the square of H.

The theoretical relationship stated symbolically in eq. 3.8 is shown graphically in Fig. 3.10c (Taylor, 1948). This figure conveys exactly the same information as the family of isochrones shown in Fig. 3.10a, except that the data are plotted in terms of the dimensionless coefficients z/H, U_z, and T_v. Therefore, the theory permits us to predict the isochrones, provided the coefficient of consolidation c_v is known.

It has been pointed out that the degree of consolidation corresponding to z_1 and t_1 (Fig. 3.10a) is represented by the ratio A_0A_1/A_0D. The average degree of consolidation of the entire sample at time t_1 is, therefore, represented by the shaded area in this figure divided by the area of the rectangle $CDEF$. The average degree of consolidation is designated simply as the *degree of consolidation* U at time t. The shaded area is a measure of the fraction of the initial consolidation pressure throughout the sample that has, at time t_1, become an effective stress. If it is assumed that the settlement of the upper surface of the sample is roughly proportional to the increase in effective stress, then U may be regarded as approximately the amount of settlement that has occurred at a given time compared with the total amount that will take place as a consequence of the increment of pressure Δp.

The theory of consolidation may be extended to provide the means for computing the value of U. It is found that

$$U\% = f(T_v) \quad 3.10$$

This relationship is shown graphically in Fig. 3.11, plotted to a semilogarithmic scale. If the values of c_v and H are known for a given layer of clay, the theoretical curve can be replotted on the basis of eq. 3.9 to show the per cent consolidation as a function of time. Moreover, if the ultimate settlement due to the consolidation of the clay layer has been computed, the curve can be further modified into a settlement-time curve.

The factor H in eq. 3.9 is defined as the half-thickness of a layer drained on both sides as in a consolidation test. Thus, the time required to reach a given degree of consolidation is four times as great for a layer of given thickness drained on one side only as it is for one drained on both sides.

The coefficient of consolidation can be evaluated on the basis of information obtained in the consolidation test. Under each increment of pressure Δp applied to the sample, the height of the sample decreases with increasing time. The experimental curves, such as that shown in Fig. 3.12, if time is plotted to a logarithmic scale, resemble the theoretical relationship between U and T_v (Fig. 3.11), except that they approach an inclined rather than a horizontal asymptote. Up to a degree of consolidation of about 60 per cent the divergence is in-

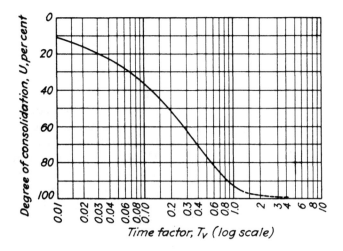

FIGURE 3.11. Theoretical relationship between degree of consolidation U and time factor T_v.

significant and the shape of the experimental curve is governed by eq. 3.10. The value of c_v is determined by fitting the experimental curve to the theoretical curve for values of U less than 60 per cent. The first step in the procedure is to determine the deformations corresponding to $U = 0$ per cent and $U = 100$ per cent. This can most readily be done by a method proposed by A. Casagrande. First, the deformation (dial reading) is plotted against the logarithm of time for the appropriate load increment in the consolidation test. The curve has the shape shown in Fig. 3.12.

To determine the line $U = 0$ per cent,

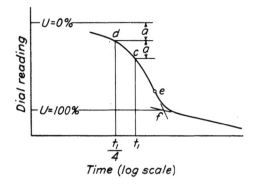

FIGURE 3.12. Graphical construction for determining deformations corresponding to $U = 0$ per cent and $U = 100$ per cent in consolidation test (after Casagrande and Fadum, 1940).

we select any time t_1 at which less than half the consolidation for the entire load increment has taken place. We locate the corresponding point c on the curve and then establish point d corresponding to a time $t_1/4$. The vertical distance between c and d is a. We then lay off a horizontal line at a distance a above d. The ordinate of this line is $U = 0$ per cent.

To obtain the line $U = 100$ per cent, we select the point of inflection e of the experimental curve and draw a tangent downward from this point. We then draw to the left a tangent to the lower, straight-line part of the experimental curve. The two tangents intersect at f, at a dial reading corresponding to $U = 100$ per cent.

To fit the curves we commonly select the values of t for $U = 50$ per cent. The corresponding value of $T_v = 0.197$. Hence, by eq. 3.9, we can evaluate c_v as

$$c_v = \frac{T_v}{t} H^2 \qquad 3.11$$

where H is the half-thickness of the test specimen.

Once c_v for the clay is known, eq. 3.9 and Fig. 3.11 can be used to plot the theoretical time-settlement relation or the primary consolidation for any structure located above a stratum of the clay.

On the basis of the consolidation test data the coefficient of permeability k of the sample during any given load increment may be computed by means of the equation

$$k = c_v \gamma_w m_v \qquad 3.12$$

In eq. 3.12 the coefficient of volume compressibility m_v may be evaluated as

$$m_v = \frac{\Delta n}{\Delta p} = \frac{e_0 - e_1}{1 + e_0} \cdot \frac{1}{p_1 - p_0}$$

$$= \frac{1}{1 + e_0} \cdot \frac{e_0 - e_1}{p_1 - p_0} = \frac{a_v}{1 + e_0} \qquad 3.13$$

in which

$$a_v = \frac{e_0 - e_1}{p_1 - p_0} \qquad 3.14$$

the *coefficient of compressibility*, is defined as the ratio between the change in void ratio and the change in effective stress for the given increment.

According to eq. 3.11, the rate of consolidation depends on the square of the half-distance between drainage layers. In a clay deposit containing thin layers or lenses of sand or silt, the actual rate of consolidation is therefore governed by the spacing and degree of continuity of the more pervious inclusions. These factors often cannot be assessed reliably by sampling and testing (Chap. 6). Hence they introduce practical limitations to the accuracy of predictions of rate of settlement.

Secondary Consolidation. The time-consolidation curve obtained for a given load increment in a consolidation test differs from the theoretical relation (Fig. 3.11) in advance stages of consolidation. Whereas the theoretical curve approaches a horizontal asymptote corresponding to 100 per cent consolidation, an actual curve approaches an inclined tangent with a nearly constant slope if time is plotted to a logarithmic scale (Fig. 3.12). The portion of the total settlement that takes place in accordance with eq. 3.6 and the assumptions on which it is based is called the *primary consolidation*. The additional settlement is designated as *secondary consolidation*.

Secondary consolidation also occurs in

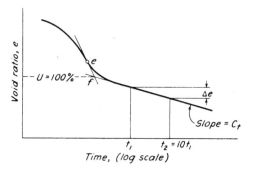

FIGURE 3.13. Typical log time-void ratio curve for single load increment in consolidation test or in field. Slope of straight-line portion represents secondary compression index.

the field. In highly compressible deposits it may account for a substantial portion of the long-term settlement. The rate of secondary consolidation may be expressed by the *coefficient of secondary consolidation*:

$$C_\alpha = \frac{\Delta e}{1 + e_0} \frac{1}{\log_{10}(t_2/t_1)} = \frac{C_t}{1 + e_0} \qquad 3.15$$

where C_t, the slope of the straight-line portion of the e-log t curve, is known as the *secondary compression index*. Numerically, C_t is equal to the value of Δe for a single cycle of time on the curve (Fig. 3.13).

The value of C_α for normally loaded compressible soils increases in a general way with the compressibility and, hence, with the natural water content, in the manner shown in Fig. 3.14 (Mesri, 1972). Although the range in values for a given water content is extremely large, the relation gives a conception of the upper limit of the rate of secondary settlement that may be anticipated if the deposit is normally loaded or if the stress added by the proposed construction will appreciably exceed the preconsolidation load. The rate is likely to be much less if the clay is strongly preloaded or if the stress after addition of the load is small compared to the existing overburden pressure. The rate is also influenced by the length of time the preload may have acted, by the existence of shearing stresses, and by the degree of disturbance of the samples. The effects of these various factors have not yet been adequately evaluated.

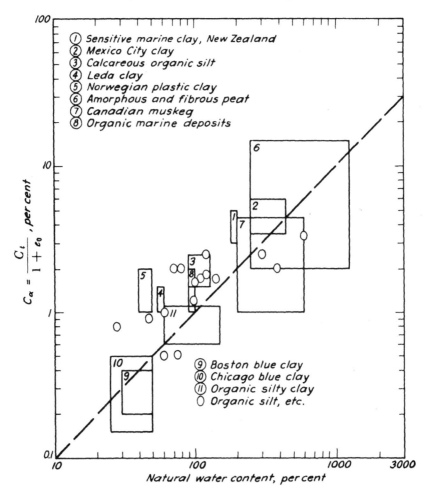

FIGURE 3.14. Relation between coefficient of secondary consolidation and natural water content of normally loaded deposits of clays and various compressible organic soils (after Mesri, 1973).

ILLUSTRATIVE PROBLEMS

1. A normally loaded soft clay of low sensitivity has a liquid limit of 57 per cent. Estimate the compression index.

Solution. Since the clay is soft and normally loaded, its water content is likely to be close to the liquid limit, and eq. 3.2 is applicable. Therefore,

$$C_c = 0.009(w_L - 10)$$
$$= 0.009 \times (57 - 10) = 0.42$$

2. The accompanying time-dial reading curve was obtained during a consolidation test on a soft glacial clay ($w_L = 43\%$, $w_P = 21\%$, $w = 39\%$) when the pressure was increased from 1.66 kg/sq cm to 3.33 kg/sq cm. The void ratio after 100 per cent consolidation under 1.66 kg/sq cm was 0.945 and that under 3.33 kg/sq cm was 0.812. The dial was set at zero at the beginning of the test, and the initial height of the sample was 0.75 in. Drainage was permitted at both faces of the sample. Compute the coefficient of permeability corresponding to the stated increment of pressure.

Solution. To determine dial reading R_d at 0 per cent consolidation, choose $t_1 = 1.0$ min

For $t_1 = 1.0$, $\quad R_d = 645$
For $t_1/4 = 0.25$, $\quad R_d = 602$
$\quad\quad\quad\quad a = 43; \quad 2a = 86$
(see Fig. 3.12)

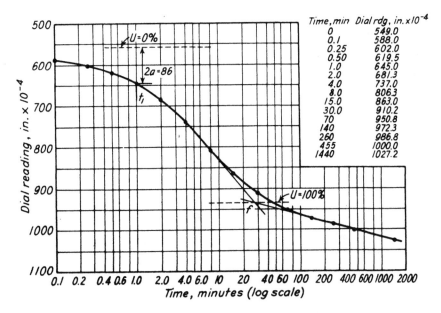

Time, min	Dial rdg, in. ×10⁻⁴
0	549.0
0.1	588.0
0.25	602.0
0.50	619.5
1.0	645.0
2.0	681.3
4.0	737.0
8.0	806.3
15.0	863.0
30.0	910.2
70	950.8
140	972.3
260	986.8
455	1000.0
1440	1027.2

For $U = 0\%$, $R_d = 645 - 86$
$= 559$

For $U = 100\%$, $R_d = 936$ (from graphical construction, Fig. 3.12)

For $U = 50\%$, $R_d = 559 + (936 - 559)/2 = 748$

For $U = 50\%$, $t = 4.5$ min
$= 270$ sec (from accompanying plot)

For $U = 50\%$, $T_v = 0.197$ (from Fig. 3.11)

For $U = 50\%$,
$2H = 0.7500 - 0.0748 = 0.6752$ in.;
$H = 0.3376$ in.

Hence,

$$c_v = \frac{T_v H^2}{t} = \frac{0.197}{270} \times (0.3376 \times 2.54)^2$$
$$= 5.37 \times 10^{-4} \text{ sq cm/sec}$$

$$a_v = \frac{e_0 - e_1}{p_1 - p_0} = \frac{0.945 - 0.812}{3330 - 1660}$$
$$= \frac{0.133}{1670} = 7.96 \times 10^{-5} \text{ sq cm/g}$$

$$m_v = \frac{a_v}{1 + e_0} = \frac{7.96 \times 10^{-5}}{1.945}$$
$$= 4.09 \times 10^{-5} \text{ sq cm/g}$$

$$k = c_v \gamma_w m_v = 5.37 \times 10^{-4} \times 1.0 \times 4.09 \times 10^{-5}$$
$$= 2.20 \times 10^{-8} \text{ cm/sec}$$

3. In a laboratory consolidation test a sample of clay with a thickness of 1 in. reached 50 per cent consolidation in 8 min. The sample was drained top and bottom. The clay layer from which the sample was taken is 26 ft thick. It is covered by a layer of sand through which water can escape and is underlain by a practically impervious bed of intact shale. How long will the clay layer require to reach 50 per cent consolidation?

Solution. According to eq. 3.9

$$\frac{t_t}{t_l} = \frac{H_t^2}{H_l^2}; \qquad t_l = t_t \frac{H_l^2}{H_t^2}$$

where t refers to the test and l to the layer. For the test, $H_t = 0.5$ in.; for the layer, $H_l = 26$ ft. Then

$$t_l = 8 \times \frac{(26 \times 12)^2}{0.5^2} \times \frac{1}{60 \times 24 \times 365}$$
$$= 5.93 \text{ years}$$

4. For the load increment in Prob. 2 compute the secondary compression index C_t. Assume $\gamma_s = 2.80$ g/cu cm.

Solution. Compute the void ratio at the start of the test, corresponding to the original height of the sample of 0.75 in.

$$e_0 = wG = 0.39 \times 2.80 = 1.092$$

By means of the sketch, compute the reduced height of solids H_s to facilitate conversion from dial readings to void ratio

$$\frac{H_s}{1} = \frac{0.75}{2.092}$$

$$H_s = 0.359 \text{ in.}$$

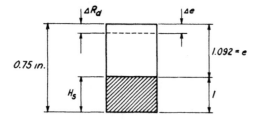

If R_d is expressed in inches, $\Delta e = \Delta R_d/H_s$. At $t = 100$ min, the dial reading R_d is 0.0963 in., and at $t = 1000$ min, the dial reading is 0.1020 in. Hence,

$$C_t = \frac{0.1020 - 0.0963}{0.359} = \frac{0.0057}{0.359}$$

$$= 0.0159 = 1.59\%$$

5. The coefficient of consolidation of a normally loaded clay has been judged, on the basis of consolidation tests on undisturbed samples, to be 6×10^{-5} sq ft/min. The average natural water content of the deposit is 40 per cent, the unit weight of the solid matter is 2.8 g/cu cm, and the compression index C_c is 0.36. If the clay deposit is 20 ft thick and is drained top and bottom ($2H = 20$ ft), sketch an approximate settlement-log time curve. The existing effective overburden pressure at the center of the clay layer is 2.000 tons/sq ft, and the increment of pressure causing the expected settlement is 0.289 ton/sq ft.

Solution. The settlement at the end of primary consolidation will be

$$S = \frac{2H}{1 + e_0} C_c \log_{10} \frac{p_0 + \Delta p}{p_0}$$

$$= \frac{20 \times 0.36}{1 + 0.40 \times 2.8} \log_{10} \frac{2.000 + 0.289}{2.000}$$

$$= 0.199 \text{ ft} = 2.39 \text{ in.}$$

The time for various degrees of consolidation can be determined from

$$t = \frac{H^2}{c_v} T_v = \frac{10^2}{6 \times 10^{-5}} T_v \text{ (min)}$$

$$= 3.17 \, T_v \text{ (years)}$$

Points on the primary portion of the curve can then be determined from the following table constructed with the aid of Fig. 3.11.

U (%)	T_v	t (year)	$S = U \times 2.39$ in. (in.)
0.1	0.008	0.025	0.24
0.2	0.031	0.098	0.48
0.3	0.071	0.23	0.72
0.4	0.126	0.40	0.96
0.5	0.197	0.62	1.20
0.6	0.287	0.91	1.43
0.7	0.403	1.28	1.67
0.8	0.567	1.80	1.91
0.9	0.848	2.69	2.15
0.95	1.20	3.80	2.27

The coefficient of secondary consolidation C_α is estimated from Fig. 3.14 to be 0.004.

To get the slope of the secondary part of the settlement-log time curve, we note that primary consolidation is nearly complete at about three years. At 30 years (one cycle), the void ratio will decrease by C_t and the corresponding increase in settlement will be

$$\Delta S = 2H \frac{C_t}{1 + e_0}$$

$$= 2H C_\alpha$$

$$= 20 \times 12 \times 0.004$$

$$= 0.96 \text{ in.}$$

The dashed line in the figure is drawn with a value of ΔS equal to 0.96 in. for one cycle on the time scale. The continuation of the settlement-log time curve is drawn by constructing the full line parallel to the dash line and tangent to the primary portion of the curve. The assumption that a

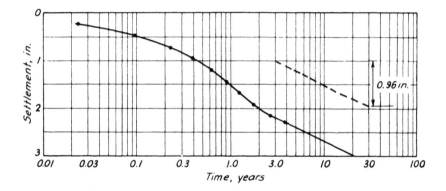

straight line in the e-log t curve is also straight in the settlement-log t curve is an approximation.

PROBLEMS

1. The water content of a soft clay is 54.2 per cent and the liquid limit 57.3 per cent. Estimate the compression index.
Ans. 0.43.

2. A consolidation test was made on a sample of saturated glacial clay in a circular ring with a height of 1.5 in. and an area of 90.18 sq cm. The sample weighed 645.0 g at the start of the test and 477.8 g in the oven-dry state after the test. An independent test indicated a specific gravity of 2.74 for the solid matter. Primary consolidation under each addition of load was complete at 1000 min. The dial readings corresponding to this time are recorded in the following table:

Pressure (tons/sq ft)	Dial Reading (in. $\times 10^{-4}$)
0.0000	0
0.0665	70
0.133	110
0.266	216
0.5325	386
1.064	734
2.13	1333
4.26	1995
8.52	2630

Compute the void ratio at the start of the test and after each increment of load. Plot the e-log p curve.

Ans. 0.959; 0.950; 0.945; 0.931; 0.909; 0.863; 0.785; 0.699; 0.616.

3. Using the graphical construction proposed by A. Casagrande, estimate the maximum overburden pressure that has ever acted upon the soil represented by the sample in Prob. 2. Determine the compression index C_c.
Ans. 0.9 to 1.1 tons/sq ft; 0.285.

4. From the data in Prob. 2, compute a_v and m_v for the increment of pressure from 0.5325 to 1.064 tons/sq ft.
Ans. 8.8×10^{-5} sq cm/g; 4.6×10^{-5} sq cm/g.

5. The dial readings observed during the consolidation of the sample (Prob. 2) under the pressure of 1.064 tons/sq ft are tabulated below.

Time (min)	Dial Reading (in. $\times 10^{-4}$)	Time (min)	Dial Reading (in. $\times 10^{-4}$)
0.00	396.0	15	549.0
0.10	427.0	30	594.7
0.25	436.0	60	636.8
0.50	443.8	135	670.7
1.0	453.5	240	689.9
2.0	466.0	1180	729.0
4.0	485.0	1600	737.0
8.0	511.8	2625	749.2

Plot the dial reading-time curve, and determine the readings corresponding to 0 and 100 per cent consolidation. Compute the value of the coefficient of consolidation c_v and the coefficient of permeability k.
Ans. 418.5×10^{-4} in.; 671.0×10^{-4} in.; 7.9×10^{-4} sq cm/sec; 3.7×10^{-8} cm/sec.

6. The sample (Prob. 2) was taken from a layer of clay 12 ft thick, drained top and bottom. After how many days would this layer reach 30 per cent consolidation under a suddenly applied load of about 1 ton/sq ft?
Ans. 35 days.
7. If the layer of clay (Prob. 6) contained a thin freely draining parting of sand at a depth of 3 ft below its upper surface, how many days would be required for it to reach an average degree of consolidation equal to 30 per cent?
Ans. 9 days.
8. The time to reach 60 per cent consolidation is 32.5 sec for a sample 0.5 in. thick tested in the laboratory under conditions of double drainage. How long will the corresponding layer in nature require to reach the same degree of consolidation if it is 15 ft thick and drained on one side only?
Ans. 195 days.
9. A layer of clay beneath a building has consolidated and caused a settlement of 9.2 in. in 200 days since the weight of the building was added. According to the results of laboratory consolidation tests this corresponds to 30 per cent consolidation of the layer. Plot the time-settlement curve for the structure for the first five years of its existence.
Ans. Settlement at the end of each year in inches: (1) 12.3; (2) 17.0; (3) 20.9; (4) 23.3; (5) 25.4.
10. For the dial reading-time curve plotted in Prob. 5, determine the coefficient of secondary compression C_α.
Ans. 0.38%.

SUGGESTED READING

The theoretical aspects of the consolidation of saturated soils are considered in an enormous number of technical papers and many books. The basic treatise, however, remains K. Terzaghi (1943), *Theoretical Soil Mechanics*, New York, Wiley, 510 pp. The theory of consolidation is considered in pp. 265-296.

The original work on determination of the preconsolidation pressure by means of consolidation tests is A. Casagrande (1936), "The Determination of the Pre-Consolidation Load and its Practical Significance," *Proc. 1 Int. Conf. Soil Mech.*, Cambridge, Mass., 3, 60–64. The reconstruction of the field e-log p curve from laboratory tests has been considered most thoroughly by J. H. Schmertmann (1955), "The Undisturbed Consolidation Behavior of Clay," *Trans. ASCE*, 120, 1201-1227.

Familiarity with the consolidation characteristics of various deposits is an asset of great value to the foundation engineer. Examples may be found in the following:

A. Casagrande (1932b). "The Structure of Clay and Its Importance in Foundation Engineering," *J. Boston Soc. Civ. Eng.*, 19, 4, 168–221. Reprinted in *Contributions to Soil Mechanics, 1941–1953*, Boston Soc. Civ. Eng. (1940), pp. 72-125. First paper in which the difference between e-log p curves for undisturbed and remolded samples was appreciated; the data pertain to sensitive clays from Boston and extrasensitive clays from the valley of the St. Lawrence River. The concepts of the structure of clay and the conclusions concerning the practical effects of remolding pointed the way to further investigation and have subsequently been modified.

L. Bjerrum (1967), "Engineering Geology of Norwegian Normally-Consolidated Marine Clays as Related to Settlements of Buildings," *Géotechnique*, 17, 2, 81–118. The Seventh Rankine Lecture of the British Geotechnical Society, and a model for papers regarding the in-situ behavior of clays.

J. I. Adams (1965), "The Engineering Behavior of a Canadian Muskeg." *Proc. 6th Int. Conf. Soil Mech.*, Montreal, 1, 3–7. Highly organic soils.

E. T. Hanrahan (1954), "An Investigation of Some Physical Properties of Peat," *Géotechnique*, 4, 3, 108–123.

G. G. Meyerhof and G. Y. Sebastyan (1970), "Settlement Studies on Air Terminal Building and Apron, Vancouver International Airport, British Columbia." *Canadian Geot. Jour.* 7, 4, 433–456. Normally consolidated clays of low sensitivity in delta of Fraser River.

C. B. Crawford (1964), "Some Characteristics of Winnipeg Clay," *Canadian Geot. Jour.*, *1*, 4, 227–235. Precompressed highly plastic lacustrine clay.

P. C. Rutledge (1944), "Relation of Undisturbed Sampling to Laboratory Testing," *Trans. ASCE*, *109*, 1155–1183. Effect of remolding on e-log p curves for various organic soils, glacial clays from Chicago, and Mexico City clays. Discussion by R. F. Dawson concerning expansive clays from Texas.

A. W. Skempton (1914–   )

One of a small group, at the Building Research Station in Britain, who were largely responsible for the rapid development of soil mechanics in that country after World War II. Moving to Imperial College in 1946, he attacked the problem of understanding the shear strength of clays. To him we owe the concept of the pore-pressure coefficients, the first sound explanation of the undrained shear strength of saturated clay and of the $\phi = 0$ method of analysis, and the useful relation between undrained shear strength and effective overburden pressure in normally loaded deposits. His deep interest in history has enriched our knowledge of many of the creators of the art of foundation engineering.

PLATE 4.

CHAPTER 4

Stress-Deformation-Strength Characteristics of Soil and Rock

4.1. Behavior of Soils Under Complex States of Stress

The one-dimensional consolidation test and Terzaghi's theory of consolidation, discussed in Chap. 3, illustrate significant principles. The specific results, however, are applicable only to those conditions in which lateral deformations of the consolidating soil are negligible and, if the rate of settlement is to be predicted, to circumstances in which the flow of the escaping water is primarily vertical. Only a few of the problems encountered by the foundation engineer fall into these narrow categories. In most instances lateral as well as vertical deformations occur, and lateral as well as vertical flows take place. Moreover, the engineer needs to investigate the degree to which his structure will tax the strength of the foundation material. Hence, he must evaluate not only the relations between stress and deformation, but also the conditions under which the soil may fail.

The stress–strain relationships for soils and rocks are usually too complex to be represented by constants such as the modulus of elasticity and Poisson's ratio, which suffice to define the behavior of steel within the range of working stress. The stress–strain curves are commonly not linear or even independent of the stress history. The configurations of the curves and the ultimate strengths of the materials are dependent on the lateral pressure. Inasmuch as real soils are acted on by lateral pressures due to the overburden and to other loads supported by the soil, it is often necessary to take this dependence into account. The stress-deformation characteristics of soils are also time-dependent, in some instances on account of the phenomenon of consolidation, and in others because of a tendency to creep or deform under constant stress.

Although the many complexities of even the most routine practical problems cannot be taken into account with precision, an understanding of the behavior of soil and rock under several well-defined conditions permits the development of practicable procedures for designing and predicting the performance of real foundations.

4.2. Behavior in Shear of Idealized Granular Mass

Like other engineering materials, soils decrease in volume when subjected to an equal all-around pressure. When subjected to shearing stresses they distort; if the distortion becomes great enough the particles slip over one another and the soil is said to fail in shear. Inasmuch as most soils can withstand only small tensile stresses or even none at all, significant tension rarely develops in masses of soil and, consequently, most failures of soil take place in shear. Hence, a knowledge of the shear-strength characteristics of soils is a prerequisite to the solution of many problems in foundation engineering.

Unfortunately even the knowledge required for solving the more common problems is somewhat complex. As most of the complexities arise because of the particulate nature of soils, considerable insight into the behavior of real soils can be gained by investigating the behavior of an assemblage of roughly equidimensional particles, such as those shown in Fig. 4.1a. The particles are assumed to be confined between two large horizontal plates with roughened inner surfaces. The plates permit a vertical pressure p per unit of gross area to be applied to the assemblage of particles. The actual pressure between grains at their points of contact is, of course, many times greater than p. The plates also permit application of a shearing stress t per unit of gross area to the assemblage. The behavior upon the application of t depends greatly on the looseness or denseness to which the grains are originally packed.

The grains, although touching each other at several points per particle, may be arranged in a very loose array, such as that shown in Fig. 4.1a. Upon application of the pressure p the distance between plates decreases slightly. If then the shearing stress t is gradually increased, the distortion, measured by the angle δ, also increases (Fig. 4.1b). The distortion is associated with slipping between the grains and a gradual rearrangement of the particles into a denser packing; consequently, the distance h between the plates decreases. The decrease Δh in distance is likely to be much greater than that resulting from the application of the pressure p.

If the grains are initially arranged in a very dense packing (Fig. 4.1c), the same pressure p also reduces the distance between plates, although by a smaller amount than for the loose aggregation. On the other hand, as the distortion δ increases, the particles cannot move with respect to each other without breaking unless the distance h between the plates increases. If the in-

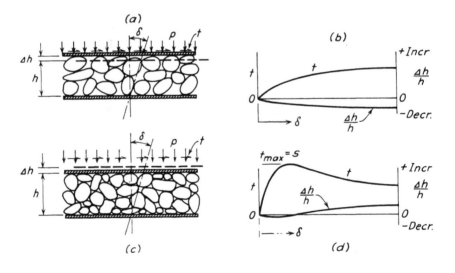

FIGURE 4.1. (a) Shear test on idealized mass of equidimensional particles in initially loose state. (b) Shearing stress t and volume change $\Delta h/h$ as functions of shearing distortion δ. (c) and (d) Corresponding relations for same material in initially dense state.

dividual particles are assumed to be strong, Δh can be expected to increase with increasing δ, as shown in Fig. 4.1d. The shearing stress t at a given value of δ is, in the early stages, much greater than that of the loose particles, but when δ becomes very large the particles have spread apart to about the same degree of density as that approached by the loose array at comparably large distortions. Hence, the relation between t and δ for the initially dense array shows a peak.

We see, therefore, that an initially loose array of strong particles becomes denser during shear, whereas an initially dense array becomes looser. This behavior is a fundamental characteristic of all more or less equidimensional grains strong enough not to be crushed. Sands and gravels closely approximate to this behavior. For quite different reasons soft clays tend to decrease in volume during shear, whereas stiff clays tend to expand. Hence, the behavior of clays is, to this extent, analogous to that of sands.

The stress-deformation characteristics of real soils may be investigated by means of direct shear tests in apparatus closely resembling the arrangement shown in Fig. 4.1. Better control over several of the test conditions can usually be achieved, however, by performing so-called triaxial compression tests. The arrangement and interpretation of such tests are discussed in the next section.

4.3. Triaxial Tests and Mohr's Circle of Stress

In a triaxial test, a cylindrical specimen is subjected to an equal all-around pressure and, in addition, to an axial pressure that may be varied independently of the all-around pressure.

The essential features of the triaxial apparatus are shown diagrammatically in Fig. 4.2. The cylindrical surface of the sample is covered by a rubber membrane sealed to a pedestal at the bottom and to a cap at the top. The assemblage is contained in a chamber into which water may be admitted under any desired pressure; this pressure acts laterally on the cylindrical surface of the sample through the rubber membrane and vertically through the top cap. The additional axial load is applied by means of a piston passing through the top of the chamber.

Porous disks are placed against the top and bottom of the sample and are connected to the outside of the chamber by

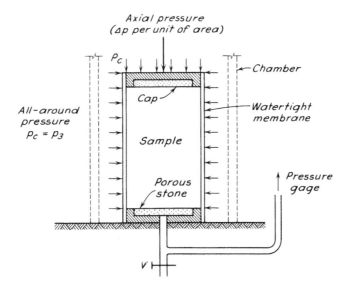

FIGURE 4.2. Principal features of triaxial apparatus.

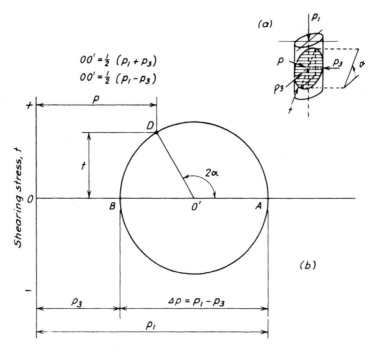

FIGURE 4.3. (*a*) Stresses on triaxial specimen and inclination α of failure plane to horizontal. (*b*) Mohr's circle of stress.

tubing. By means of the connections the pressure in the water contained in the pores of the sample can be measured if drainage is not allowed. Alternatively, if flow is permitted through the connections, the quantity of water passing into or out of the sample during the test can be measured. As the loads are altered, the vertical deformation of the specimen is measured by a dial gage.

A test is usually conducted by holding the all-around pressure constant and increasing the vertical pressure. Since no shearing stresses are exerted by the fluid pressure against the cylindrical boundary of the sample, the pressures acting on vertical planes in the specimen are principal stresses. Because of symmetry, all these pressures are equal and are designated as p_3 (Fig. 4.3*a*). The vertical stress on horizontal planes is also a principal stress, designated as p_1. The stress p_1 may also be considered as $p_3 + \Delta p$, where Δp is called the *stress difference*. As Δp is usually positive in routine triaxial tests, p_1 is the major and p_3 the minor principal stress. The stress–strain curve consists of a plot of the stress difference Δp against the axial strain ϵ.

At any stage of the test, since p_1 and p_3 are known, the normal pressure p and shearing stress t on any plane making an angle α with the plane on which the major principal stress acts can be calculated, according to the principles of equilibrium, from the equations

$$p = \tfrac{1}{2}(p_1 + p_3) + \tfrac{1}{2}(p_1 - p_3) \cos 2\alpha \quad 4.1a$$

$$t = \tfrac{1}{2}(p_1 - p_3) \sin 2\alpha \quad 4.1b$$

These equations represent points on a circle in a rectangular system of coordinates (Fig. 4.3*b*) in which the horizontal axis is that of principal stresses and the vertical axis is that of shearing stresses. The circle is designated as the *circle of stress*.

Every point, such as D, on the circle of stress represents the normal stress and shearing stress on a particular plane inclined at an angle α to the direction of the plane on which the major principal stress acts. From the geometry of the figure it can be shown that the central angle $AO'D$ is equal to 2α.

4.4. Stress–Strain Relations for Dry Sands and Gravels

The results of two triaxial tests on a coarse dry sand are shown in Fig. 4.4. At the start of each test, the specimen is subjected to the equal all-around cell pressure p_3 (Fig. 4.4a). The strains associated with the stress p_3 represent a volume change and are equal in all directions. They are not plotted in Fig. 4.4. The axial stress is then increased by small increments. The relation between the axial strain and the vertical stress difference Δp for the sand in an initially loose state is shown in Fig. 4.4b. With increasing strain, Δp increases continuously without reaching a peak, but approaches a limiting value designated the *compressive strength*. The compressive strength is usually taken for convenience as the value of Δp at 20 per cent strain. The volume changes ΔV during the increase of Δp, as a function of the original volume V, are shown in Fig. 4.4c. The volume becomes smaller with increasing values of Δp; in the later stages of the test, however, the rate of volume decrease approaches zero.

The corresponding relationships for an

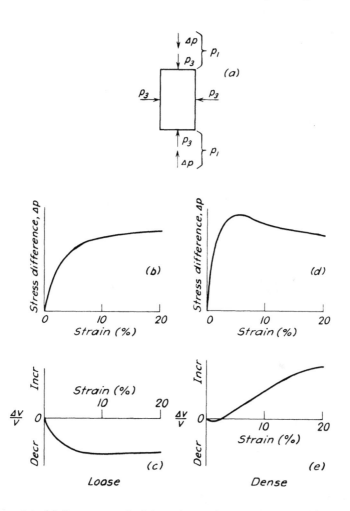

FIGURE 4.4. (a) Stresses on triaxial specimen of coarse dry sand. (b) Relation between stress difference Δp and vertical strain for initially loose sample. (c) Relation between volume change and vertical strain for same initially loose sample. (d) and (e) Corresponding relations for initially dense sample.

initially dense specimen of the same sand subjected to the same cell pressure p_3 are shown in Figs. 4.4d and 4.4e. With increasing axial strain the stress difference Δp increases to a maximum or peak value and thereafter reduces gradually to a lower limiting value approximately equal to the value approached by the loose specimen at an advanced state of strain. The peak value of Δp is commonly considered the compressive strength of the dense material. At very small values of Δp the sample may experience a slight decrease in volume, but as Δp increases, the volume increases and the sample is said to *dilate*.

Since a loose specimen experiences a decrease in volume during the test whereas a dense one dilates, it is possible to prepare a specimen at an intermediate initial relative density or void ratio such that it experiences practically no volume change. Such a sample is said to be at the *critical void ratio*. The critical void ratio decreases somewhat with increasing values of the confining pressure p_3.

An increase in confining pressure does not alter the general shape of the curves shown in Fig. 4.4, but the values of stress difference corresponding to a given strain increase approximately in proportion to the increase in p_3. Thus, increasing confinement increases the strength of the sand. It also increases the rigidity that, for very low stresses, may be represented in a crude way by the tangent to the stress–strain curve at the origin. The slope of this tangent (stress per unit of strain) is designated as the *initial tangent modulus* E_i. For loose sands, E_i increases directly with p_3, as shown in Fig. 4.5. For dense sands, E_i increases rapidly with p_3 when p_3 is small, but for large lateral pressures the rate of increase drops to a value comparable with that for loose sands.

4.5. Mohr's Rupture Diagram

The shearing strength of the soil is not measured directly by means of triaxial tests, but must be determined by calculation from the observed principal stresses p_1 and p_3. The calculation can be performed most readily with the aid of Mohr's circle of stress. If the principal stresses p_1 and p_3 correspond to a state of failure in the specimen, at least one point on the circle of stress (Fig. 4.3b) must represent a combination of normal and shearing stresses that caused failure on some plane through the specimen. Moreover, if the coordinates of that point are known, the inclination of the plane on which failure took place can be determined from a knowledge of the angle α.

If a series of tests at different values of p_3 is performed and the circle of stress corresponding to failure is plotted for each of the tests, at least one point on each circle must represent the normal and shearing

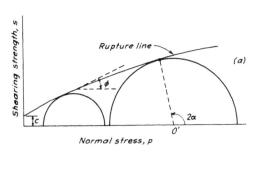

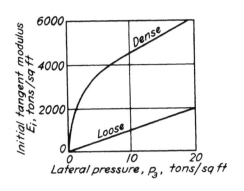

FIGURE 4.5. Relation between initial tangent modulus and all-around pressure for sand (after Scheidig, 1931).

FIGURE 4.6. (a) Mohr's rupture diagram. (b) Relation between angles ϕ and α.

stresses associated with failure. As the number of tests increases indefinitely, it is apparent that the envelope of the failure circles (Fig. 4.6a) represents the locus of points associated with failure of the specimens. The envelope is known as the *rupture line* for the given material under the specific conditions of the series of tests. For materials in general, the rupture line may be curved, and it may have an intercept c on the axis of shearing stress. Since the values of shearing strength t corresponding to the rupture line all represent failure, they are designated as values of shearing strength s, and the vertical axis in Fig. 4.6a is called the axis of shearing strength. If the rupture line is considered to be straight, it may be represented by

$$s = c + p \tan \phi \quad\quad 4.2$$

known as *Coulomb's equation*.

From the geometry of Fig. 4.6b, it may be seen that for any failure circle

$$2\alpha = 90° + \phi$$

Therefore, the angle between the planes on which failure occurs and the plane on which the major principal stress acts is

$$\alpha = 45° + \frac{\phi}{2} \quad\quad 4.3$$

4.6. Shearing Strength of Dry Sands and Gravels

The rupture lines for dry sands and gravels pass through the origin of the rupture diagram; hence, the intercept c is equal to zero. If the material is in a loose state, the rupture line is linear and may be represented accurately by the equation

$$s = p \tan \phi_d \quad\quad 4.4$$

where ϕ_d is the angle between the rupture line and the p-axis. For the same materials in a dense state, the rupture line has a slight downward curvature, but for practical purposes in foundation engineering it may also be represented by eq. 4.4.

For gravels, sands, silty sands, and inorganic cohesionless silts the value of ϕ_d depends primarily on the relative density,
the grain-size distribution, and the shape of the grains. It may be estimated with the aid of Table 4.1.

Table 4.1 Representative Values of ϕ_d for Sands and Silts

Material	Degrees	
	Loose	Dense
Sand, round grains, uniform	27.5	34
Sand, angular grains, well graded	33	45
Sandy gravel	35	50
Silty sand	27–33	30–35
Inorganic silt	27–30	30–34

ILLUSTRATIVE PROBLEM

A drained triaxial test is to be performed on a uniform dense sand with rounded grains. The all-around pressure p_3 is to be 2 tons/sq ft. At about what vertical pressure should the sample fail?

Solution. If $s = p \tan \phi$, it can be seen from the sketch that

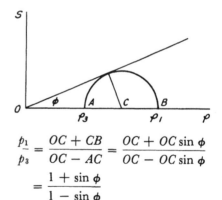

$$\frac{p_1}{p_3} = \frac{OC + CB}{OC - AC} = \frac{OC + OC \sin \phi}{OC - OC \sin \phi}$$

$$= \frac{1 + \sin \phi}{1 - \sin \phi}$$

Whence by trigonometric transformation

$$\frac{p_1}{p_3} = \tan^2 \left(45° + \frac{\phi}{2}\right) = \frac{1}{\tan^2 [45° - (\phi/2)]}$$

According to Table 4.1, the value of ϕ_d is likely to be about 34°. Therefore,

$$\tan^2 \left(45° + \frac{\phi_d}{2}\right) = \tan^2 (45° + 17°)$$

$$= 1.881^2 = 3.54$$

Hence

$$p_1 = 2.0 \times 3.54 = 7.08 \text{ tons/sq ft}$$

Therefore, the sample should fail at a vertical pressure of about $7.08 - 2.00 = 5.08$ tons/sq ft in excess of the all-around pressure.

4.7. Influence of Water in Voids

The mere presence of water in the voids of a sand or gravel does not ordinarily produce significant changes in the values of ϕ_d given in Table 4.1. On the other hand, if stresses develop in the porewater, they may cause changes in the effective stresses between the particles, whereupon the shear strength and the stress–strain relations may be radically altered. Whether or not the pore pressures develop depends on the drainage characteristics of the mass of soil and on the tendency of the soil to dilate or contract. The loose dry sand with volume-change characteristics represented by Fig. 4.4c, for example, tends to decrease in volume as Δp increases. If the voids are filled with water and if each increment of Δp is small and applied long after its predecessor, the water hardly impedes the tendency of the volume to decrease. The pore pressure then remains essentially zero during the entire test and has no influence on the behavior of the sample. A test conducted under these conditions is known as an *S-test* or a *drained test*. Indeed, *S*-tests provide the most reliable means for obtaining the volume-change curve (Fig. 4.4c) by measurement of the volume of water expelled from a saturated specimen as the test proceeds. If an *S*-test is conducted on a dense specimen, water is drawn into the material as it dilates but, again, because ample time is afforded for the inflow after each small increment of Δp, the pore pressure remains essentially zero. The stress-deformation characteristics are therefore identical to those shown in Figs. 4.4d and 4.4e for the same material, under the same cell pressure p_3, in a dry state.

On the other hand, a triaxial test may be conducted in which no dissipation of pore pressure is permitted after the sample is initially brought into hydraulic equilibrium under the all-around pressure p_3. Such a test is called an *R-test* or a *consolidated-undrained test*. When an *R*-test is carried out on an initially saturated sample, the cell pressure p_3 is first applied and the valve *V* (Fig. 4.2) left open until no pore pressure remains in the sample. The valve is then closed, whereupon no further movement of water into or out of the sample is possible.

For a loose sand comparable to that discussed in connection with Figs. 4.4b and 4.4c, the relation between the stress difference Δp and axial strain is represented by Fig. 4.7b. The ordinates are always smaller than those corresponding to the *S*-test (Fig. 4.4b). If the valve *V* were not closed, the sample would experience a decrease in volume. Since the water cannot escape from the voids, however, no volume decrease can occur and, in exchange, pressure builds up in the porewater. The relation between axial strain and the pore pressure u_d associated with the stress difference Δp is shown in Fig. 4.7c. If the sample is loose enough, u_d at large strains may approach Δp; that is, almost the entire stress difference may be carried by the porewater and very little increase in effective stress may occur.

A similarly conducted *R*-test on dense sand leads to the results shown in Figs. 4.7d and 4.7e. The ordinates of the curve relating stress difference and strain (Fig. 4.7d) are greater than those corresponding to the *S*-test (Fig. 4.4d). If the valve *V* were not closed, the sample would, after a small initial compression, experience an increase in volume. Since the closed valve, however, prevents the water from flowing into the voids, the volume increase cannot occur and a negative pore pressure develops. The relation between axial strain and the pore pressure u_d is shown in Fig. 4.7e. The pore pressure is measured with respect to that in the water before the valve is closed and any stress difference applied. Thus a negative value does not necessarily imply tension in the water but, instead, a stress lower than the original value.

If the values of p_1 and p_3 at failure are used to plot Mohr's circle of stress for ob-

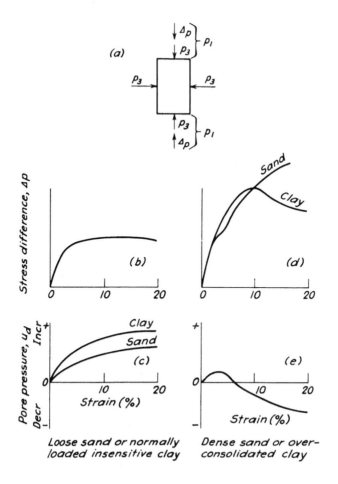

FIGURE 4.7. Results of consolidated-undrained or R-tests in triaxial apparatus. (a) Stresses acting on specimen. (b) Relation between stress difference Δp and axial strain for loose sand or normally loaded insensitive clay. (c) Pore pressures in same specimens as function of strain. (d) and (e) Corresponding relationships for dense sand or overconsolidated clay.

taining the rupture line corresponding to a series of R-tests on sand or gravel, the results are likely to resemble those shown by the dashed lines in Fig. 4.8. For loose specimens (Fig. 4.8a) the rupture line is curved slightly, whereas for dense specimens (Fig. 4.8b) the curvature is pronounced. With increasing relative density a linear relation between shearing strength and normal stress becomes an increasingly poor approximation; for dense sands the discrepancy is too great to be tolerated. On the other hand, if the pore pressures u_{df} have been measured at failure during the tests, the Mohr circles may be plotted in terms of the effective stresses at failure. The major principal stress $\bar{p}_1$ equals $p_1 - u_{df}$, and the minor principal stress $\bar{p}_3$ equals $p_3 - u_{df}$. Such circles and the corresponding rupture lines are indicated by the solid lines in Fig. 4.8. It has been found that the rupture lines based on effective stresses are for all practical purposes identical with those obtained on the same materials in S-tests. Hence, if eq. 4.3 is written

$$s = \bar{p} \tan \phi_d = (p - u_{df}) \tan \phi_d \quad 4.5$$

in terms of effective stresses, it is equally valid for expressing the results of S- or R-tests. Moreover, the inclination ϕ_d of the rupture line may be considered a property

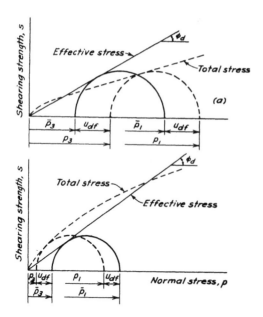

FIGURE 4.8. (a) Failure circles and rupture lines for triaxial tests on loose saturated sand under consolidated-undrained or R conditions. Dashed lines indicate results plotted in terms of total stresses, solid lines in terms of effective stresses. (b) Corresponding results for dense saturated sand.

of the soil, independent of the test conditions. It is designated the *angle of internal friction* or, more specifically, the *drained angle of internal friction*.

Figure 4.8 indicates that, over a considerable range of pressure, the shear strength of a loose sand in R-tests is less than that in S-tests, whereas the converse applies to dense sands. The carefully controlled conditions of drainage in the laboratory represent two extreme limits: complete pore-pressure adjustment in S-tests and no adjustment whatsoever in R-tests. In the field, intermediate degrees of adjustment, with intermediate values of shear strength, are likely to prevail. The engineer must select values appropriate to the conditions of his problem. For example, if a small footing on coarse sand below the water table is to be loaded gradually as the building is erected, there will surely be time enough for pore pressures due to the loading to dissipate; hence, the results of S-tests are applicable. On the other hand, if a mass of very fine saturated sand is to be crossed quickly by a heavy self-propelled machine, the results of R-tests may be more appropriate. If the sand is loose, failure may occur before there is time for appreciable drainage. If the sand is dense, it may be possible for the machine to cross quickly with safety, whereas it may sink excessively if it moves so slowly that the sand can dilate. The practical aspects of selecting suitable values of shear strength for design are summarized at the end of the chapter.

4.8. Behavior of Fine-Grained Soils

Pore Pressures During Shear. Most fine-grained soils in nature contain appreciable amounts of water; many are nearly or fully saturated. Hence, the strength of saturated materials is a matter of considerable practical importance.

Although the physical causes of the phenomena are different, the stress-deformation relationships for silts and normally loaded clays of low to moderate sensitivity, in both S-tests and R-tests, are similar to those for loose sand (Figs. 4.4b, 4.4c, 4.7b, and 4.7c). The relations for overconsolidated clays closely resemble those for dense sands, except that the stress difference for an overconsolidated clay in an R-test reaches a peak value and then decreases as shown in Fig. 4.7d. Positive pore pressures develop in R-tests on normally loaded clays, whereas overconsolidated clays tend to dilate and develop negative pore pressures. The rupture line for normally loaded silts and clays may still be expressed, in terms of effective stresses, by eq. 4.5. Overconsolidated clays, on the other hand, may possess significant shearing strength even in S-tests in which the cell pressure p_3 is zero. As a rough approximation, the rupture line corresponding to cell pressures less than the preconsolidation load may be expressed as

$$s = c_1 + \bar{p} \tan \phi_d = c_1 + (p - u_{df}) \tan \phi_d$$
4.6

in which, for a given clay, ϕ_d is found to be nearly constant but c_1 is found to depend on the preconsolidation load.

For most foundation problems involving the failure of saturated fine-grained soils, eqs. 4.5 and 4.6 have little direct applicability because the pore pressure u_{df} cannot be evaluated readily. It is usually more convenient to conduct tests in such a way as to incorporate the influence of the pore pressures into the test results. In many instances the engineer's judgment is aided greatly by a knowledge of the behavior of the soil under conditions of no pore-pressure dissipation. Hence, this subject deserves detailed consideration.

$\phi = 0$ *Condition.* In Fig. 4.9a the circle of stress designated A represents the results of an R-test on a saturated soft clay. The abcissa of the left-hand extremity of the diameter is the total stress p_3 at the time of failure; that of the right-hand extremity is the total axial stress at failure, p_1. The diameter of the circle is Δp_f, where the subscript denotes that the stress conditions correspond to failure.

The pore pressure in the sample at failure may be evaluated in two different ways. It may be determined by direct measurement, in which event the test is designated as an $\bar{R}$-test. As the pore pressure u_{df} acts with equal intensity in all directions, the effective minor principal stress is $\bar{p}_3 = p_3 - u_{df}$ and the effective major principal stress is $\bar{p}_1 = p_1 - u_{df}$. The circle of stress plotted in terms of effective stresses at failure is, therefore, displaced to the left of the total-stress circle A by the distance u_{df}. The effective-stress circle corresponding to the failure conditions is represented by the solid circle E. Since $\Delta p_f = p_1 - p_3 = \bar{p}_1 - \bar{p}_3$, the failure circle for a given test has the same diameter whether it is plotted in terms of effective stresses or total stresses.

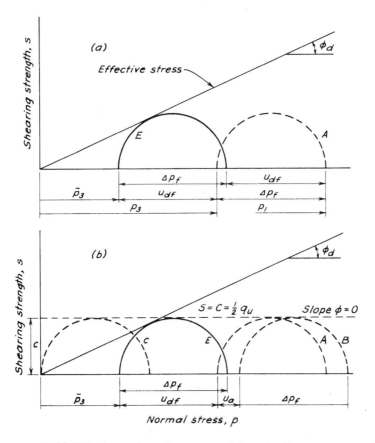

FIGURE 4.9. (a) Effective and total stresses at failure in R-test on saturated fine-grained soil. (b) Diagram illustrating $\phi = 0$ concept.

The pore pressure can also be determined by eq. 4.5 if ϕ_d is independently known, for instance, by means of a series of S-tests. The rupture line in terms of effective stress is represented by the solid straight line rising from the origin at the angle ϕ_d. The rupture circle for the R-test, plotted in terms of effective stress, must be tangent to this rupture line. Hence, the effective-stress circle corresponding to the R-test is circle E. The horizontal distance by which circle E is displaced from circle A represents the pore pressure u_{df}.

The total stress circle A corresponds to an R-test in which the pore pressure before application of the axial load was zero and the pore pressure at the end of the test was u_{df}. If, however, after the initial consolidation under the cell pressure $\bar{p}_3$ the cell pressure had been increased by an amount u_a without allowing drainage, the pore pressure in the sample before application of the axial load would have been u_a and the pore pressure at failure would have been $u_a + u_{df}$. The corresponding failure circle would have been B (Fig. 4.9b). The effective stress circle would, nevertheless, still be E. Since any change u_a in cell pressure could have been chosen, it follows that if several samples are consolidated under the same cell pressure $\bar{p}_3$ and then are tested under undrained conditions at different cell pressures, the rupture line *with respect to total stresses* is horizontal. It may be expressed simply by

$$s = c = \tfrac{1}{2}(p_1 - p_3) \quad \quad 4.7a$$

Since the slope of the line is horizontal, the foregoing circumstances are often known as the $\phi = 0$ *condition* (Skempton 1948). Inasmuch as an unconfined compression test is merely a triaxial test in which the total minor principal stress p_3 is zero (circle C in Fig. 4.9b), the shearing strength under $\phi = 0$ conditions may be evaluated on the basis of unconfined compression tests as

$$s = c = \tfrac{1}{2} q_u \quad \quad 4.7b$$

The $\phi = 0$ conditions are satisfied only if the change in cell pressure u_a is not associated with any flow of porewater into or out of the sample. Since the soil is saturated, this restriction is equivalent to the requirement that, after equilibrium is achieved under the cell pressure $\bar{p}_3$, no change is permitted in the water content of the sample. It may also be concluded that the shearing strength of a given saturated clay at a given degree of disturbance has the same value regardless of the method of testing, provided the water content of the clay remains constant.

In connection with soils having permeabilities as low as those possessed by most clays and some silts, there are many practical problems in which the water content of the soil does not change significantly for an appreciable time after the application of a stress. That is, undrained or $\phi = 0$ conditions prevail. Thus, if an undisturbed sample is extracted without alteration of its water content, and is then tested at the same water content, either in unconfined compression or without consolidation under a cell pressure $p_3 + u_a$, the shearing strength of the soil in situ may be taken as half the compressive strength. Tests satisfying the foregoing conditions are designated as Q-tests. Hence, as a consequence of the $\phi = 0$ concept, Q-tests and particularly the unconfined compression test assume unusual practical import. Field and laboratory vane tests and torvane tests, if carried out rapidly enough, are also types of Q-tests from which the ultimate strength $s = c$ may be determined.

The shear strength determined by means of Q-tests on intact clays is always conservative if the conditions in the field would lead ultimately to a decrease in water content or consolidation of the clay. Beneath a footing, for example, consolidation is likely to occur under the applied load and, consequently, the strength is likely to increase with time. If positive pore pressures are also induced by the shear stresses, a further tendency for consolidation is created. Hence, a determination of the bearing capacity of the footing on the basis of Q-tests is on the side of safety. On the other hand, the clay beneath a deep excavation may tend to swell because of the decrease in stress due to the removal of the overburden. If the swelling tendency exceeds the tendency to consolidate as a result of positive pore pressures induced by the shearing stresses, the water content of the clay may eventually increase and the shear strength correspondingly decrease. These conditions are not likely to occur beneath excavations one or two stories deep in normally loaded or slightly overconsolidated clays. If they do, however, the shear strengths derived from Q-tests no longer err on the side of safety.

The $\phi = 0$ concept and the use of Q-tests would also be valid for overconsolidated clays if in the field no opportunity existed for change in water content. However, the strong negative pore pressures associated with high overconsolidation ratios create a tendency for the soil to swell, whereupon the strength is reduced. Thus, in most practical problems the $\phi = 0$ concept for an overconsolidated clay leads to results on the unsafe side. Hence, except for overconsolidation ratios as low as possibly 2 to 4, the $\phi = 0$ concept should not be used for preloaded clays.

Many stiff saturated clays contain networks of hair cracks or slickensides. The shearing strength of deposits of this kind depends on the influence of such defects. Triaxial Q-tests on large specimens that include a representative number of defects have in some instances been found useful in determining the shearing strength of the mass. The cell pressure is usually taken as the overburden pressure on the sample as it existed in the ground. More reliable data can be obtained by means of large-scale loading tests or test excavations in the field.

c/p Ratio. The $\phi = 0$ concept leads to a useful corollary. According to eq. 4.5 the strengths of normally consolidated samples are defined by the rupture line

$$s = \bar{p} \tan \phi_d \qquad 4.5$$

An effective-stress rupture circle for one of a series of undrained tests is shown in Fig. 4.10a. The cell pressure under which all the samples in the series were consolidated is $\bar{p}_3$. The value of s corresponding to the $\phi = 0$ concept is the radius c of the circle. It is ap-

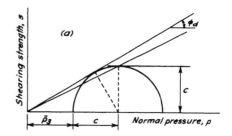

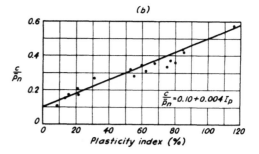

FIGURE 4.10. (a) Rupture diagram illustrating constancy of ratio $c/\bar{p}_3$ for normally loaded samples of clay consolidated under different cell pressures. (b) Relation between ratio $c/\bar{p}_n$ and plasticity index.

parent that, for samples of a given material consolidated under different confining pressures, the ratio $c/\bar{p}_3$ is a constant.

In a natural deposit of normally loaded sedimented soil the consolidation stresses differ in horizontal and vertical directions. This condition introduces a complication into the interpretation, but it has nevertheless been found that a constant ratio exists between values of c determined by Q-tests and the effective vertical overburden pressure on horizontal planes. This ratio is designated as $c/\bar{p}_n$ or, for short, as the c/p ratio. Furthermore, a broadly valid statistical relation has been found between $c/\bar{p}_n$ and the plasticity index for normally loaded sedimentary clays (Skempton, 1948; Bjerrum and Simons, 1960). The relation is shown in Fig. 4.10b. It may be approximated by the equation

$$\frac{c}{\bar{p}_n} = 0.10 + 0.004 I_P \qquad 4.8$$

where I_P is expressed in per cent.

This relation is useful in at least two ways. If a deposit is known to be normally loaded, values of c for the various layers in the deposit can be estimated roughly on the basis of the Atterberg-limit tests on disturbed samples. On the other hand, if values of c and I_P have been determined by test, the relation can be used to judge whether the deposit is preloaded and, in a qualitative way, what the degree of overconsolidation may be.

4.9. Shearing Resistance of Unsaturated Soils

The relations between effective normal stress and shear strength for unsaturated materials are not significantly different from those for saturated soils. However, evaluation of the shear strength on the basis of these relations requires a knowledge of the pore pressure not only in the water contained in the voids but also in the air that occupies the remainder of the voids. The pore-air pressure and the porewater pressure may have quite different values on account of the surface tension at the air-water interfaces. Because of the difficulties in evaluating these pressures, it is current practice to investigate the strength of partly saturated soils by means of triaxial tests in which only total stresses are measured and in which the laboratory test conditions are made to duplicate, as closely as possible, those anticipated in the field. In many instances Q-tests are appropriate. The water content of each sample is kept constant. Volume changes occur, nevertheless, because of the compression of the air in the voids.

Typical results of several series of Q-tests on samples of an inorganic clay (CL) are shown in Fig. 4.11 (Casagrande and Hirshfeld, 1960). All samples were compacted to the same dry density. The rupture line for samples having a relatively low initial degree of saturation is markedly curved. For increasingly greater initial degrees of saturation, the strengths decrease. Moreover, for a given initial degree of saturation, increases in pressure cause compression of the air in the voids and, in addition, increase the solubility of air in water. Consequently, the

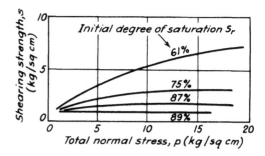

FIGURE 4.11. Results of Q-tests on partially saturated samples of an inorganic clay compacted to equal dry densities.

degree of saturation increases. For those samples with high initial degrees of saturation, S_r may reach 100 per cent at a comparatively low pressure, whereupon the $\phi = 0$ conditions are satisfied and the rupture line with respect to total stresses becomes horizontal.

A compacted fill is ordinarily placed at a moisture content close to the optimum value; this value corresponds to a partially saturated condition. The strength at the

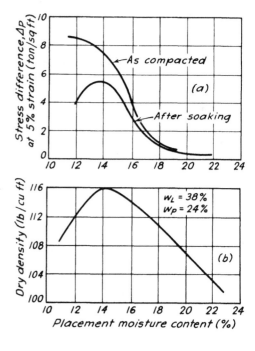

FIGURE 4.12. (a) Results of Q-tests on samples of a compacted silty clay tested as compacted and after soaking. (b) Standard Proctor moisture-density curve for material.

time of compaction depends, for a given compaction procedure, on the placement moisture content. This is illustrated by the results of Q-tests on a silty clay (Fig. 4.12a), for which the moisture-density curve is shown in Fig. 4.12b. Ultimately, however, the fill may become nearly or completely saturated. The strength after saturation may differ significantly from that at placement, as shown in Fig. 4.12a, and requires investigation by tests appropriate for saturated soils. The relationships exemplified by Fig. 4.12 differ greatly for different soils, and for the same soils subjected to different compaction procedures (Seed et al., 1960). They also depend on whether the moisture change occurs with or without volume change.

4.10. Effects of Repetitive Loads and Time

Repetitive Loading. The repeated application and removal of vertical stress on a sample of laterally confined sand under drained conditions leads to a stress–strain diagram such as that shown in Fig. 4.13. The total deformation increases with each application of the load, but the magnitudes of the increase become successively smaller.

Under undrained conditions, however, each application of load is accompanied by

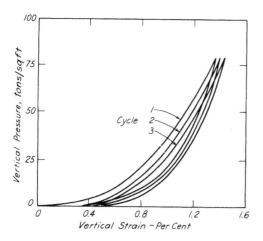

FIGURE 4.13. Relation between stress and vertical strain for confined moderately dense coarse uniform sand subject to repeatedly applied vertical load (after Hendron, 1963).

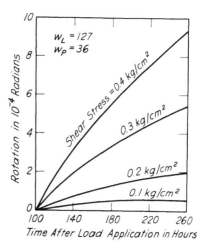

FIGURE 4.14. Relation between angular deformation and time for drained specimen of remolded unconsolidated plastic clay in shear test (after Hvorslev, 1937).

an increase in pore pressure. As the pore pressure accumulates, the effective stresses caused by the applied load correspondingly decrease and the strength of the sample also decreases. If the sample is initially loose the strength may reduce to zero, at least locally in the vicinity of stress concentrations. The sand may then flow and is said to exhibit *cyclic mobility* (Casagrande, 1971).

Creep. If the shearing stress acting on a sample of undisturbed clay is less than a value known as the *creep strength*, the clay deforms almost instantaneously upon application of the shearing stress and thereafter experiences no progressive deformation. On the other hand, if the creep strength is exceeded, the clay deforms continuously under constant shearing stress. The rate of creep increases with increasing values of the shearing stress, as shown for a remolded clay in Fig. 4.14 (Hvorslev, 1937).

Rate of Loading. In the tests from which values of shear strength are determined, the stress difference producing failure is usually reached within a few minutes to a few hours; in some drained tests several days may be required. During many construction operations in the field, the rate at which shearing stresses increase is very much slower. Hence,

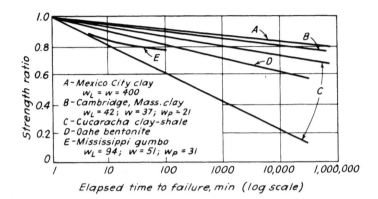

FIGURE 4.15. Undrained strength of clay soils reaching failure at various times compared to strength when tested to failure in 1 min (after Casagrande and Wilson, 1951).

the influence of the rate of load application on the shear strength is of practical concern.

Investigations of the shear strength of undrained samples at very slow rates of loading have been carried out in the laboratory. In general, no influence of decrease in rate could be detected for sands. Several clays and clayey shales, however, exhibited appreciable reductions in strength as the rate of loading decreased (Fig. 4.15). These findings (Casagrande and Wilson, 1951) demonstrate that the results of conventional laboratory tests may be unconservative and should be used with caution; wherever possible, local experience under full-scale field conditions should be utilized to investigate the applicability of the results obtained in the laboratory. On the basis of studies of embankments on soft-ground foundations that failed under undrained conditions, it has been concluded (Bjerrum et al., 1972) that a reduction factor C_r should be applied to the results of undrained laboratory tests or field vane tests on clays of high plasticity. The factor may be approximated by the equation

$$C_r = 1.0 - 0.5 \log (I_P/20) \quad (I_P \gtreqless 20) \quad 4.9$$

In many practical problems, consolidation occurs simultaneously with increase in load and the corresponding increase in strength may more than compensate for the decrease associated with slow rate of loading.

At very rapid rates of loading, amounting to a few thousandths of a second to failure, the undrained strength of some soils appears to increase, possibly by as much as 100 per cent. However, it is likely that a large part of the observed increase is associated with localized negative pore pressures arising from certain peculiarities of the triaxial apparatus itself, and it would appear unwise to depend on such increases in strength until a better understanding of the phenomenon has been obtained.

4.11. Selection of Test Procedures for Determining Shear Strength of Soils in Practice

Introduction. According to the preceding articles, it is evident that the shear strength of any soil depends primarily on the pore pressures that exist at the time shear failure may occur. Excess pore pressures may be caused by the direct stresses to which the soil is subjected and by the tendency of the volume of the soil to change during shear. On the other hand, the excess pore pressures tend to dissipate by drainage. The rate at which they can dissipate, and hence the shear strength that can be developed in the field, depend to a considerable extent on the permeability and on the dimensions of the mass of soil influenced by the shear stresses. They also depend on the rate at which the stresses are applied; a very slow change in the stress applied to a soil mass of low permeability may not produce any greater pore pressures than a rapidly applied stress in a highly permeable soil. These considerations provide a basis for estimating the shear strength in practical problems or for selecting test procedures appropriate to the problems.

Sands and Gravels. Sands and gravels, with coefficients of permeability greater than about 10^{-4} cm/sec, will under most circumstances experience enough drainage to eliminate excess pore pressures due to the application of footing loads with the possible exception of such transient loads as those due to gusts of wind or earthquakes. Hence, the shear strength can be determined by the equation

$$s = (p - u) \tan \phi_d \qquad 4.10$$

where p represents the full normal pressure associated with the applied loads, u is merely the pore pressure due to the groundwater conditions, and ϕ_d is the angle of internal friction. The value of ϕ_d can be determined by means of drained or S-tests but it is usually sufficiently accurate to estimate the value on the basis of Table 4.1 or on the basis of correlations with the results of simple field tests such as the standard penetration test (Art. 5.4). For rapidly applied transient loads on loose materials it may be advisable to increase the required factor of safety.

However, if the stresses are applied very rapidly, if the permeability is in the range 10^{-3} to 10^{-4} cm/sec, and especially if the mass of sand has large dimensions, the stresses may induce pore pressures that cannot be dissipated quickly enough to maintain the shearing strength corresponding to eq. 4.10. If the sand is dense, its strength may be temporarily increased; it is then conservative to use the S-test value. However, if it is loose its strength may be temporarily reduced to the R-test value. A series of R-tests on samples of the sand, reconstituted at the density index corresponding to field conditions, can be carried out at successively increasing cell pressures. The envelope to the failure circles, plotted in terms of total stress, permits assessment of the reduced shearing strength. If extremely loose deposits of such materials are subjected to sudden shocks or to earthquakes, they may lose their shearing strength temporarily. They are then said to *liquefy*.

General Considerations Regarding Silts and Clays. For less permeable soils, such as silts and clays, no simple statements such as those in the preceding two paragraphs are universally applicable. In principle, it should be possible to obtain an undisturbed sample from the ground for triaxial testing, to restore by means of the cell pressure the original state of effective stress that existed in the ground, and then to increase the stress difference under conditions of loading and drainage representing as closely as possible those likely to prevail in the field. In reality, however, these procedures are often somewhat impractical. In the ground, the lateral and vertical effective stresses are commonly not equal. The vertical stress can easily be calculated, but the lateral stress cannot readily be determined (Skempton, 1961). For normally loaded or slightly overconsolidated deposits the lateral effective stress is usually less than the vertical; consequently, if in the triaxial cell the sample is initially brought to equilibrium under an all-around cell pressure p_3 equal to the effective vertical pressure in the ground, the sample consolidates and its strength becomes greater than that of the deposit it represents. The inevitable disturbance associated with sampling also leads to consolidation under the cell pressure and to an alteration of the strength of the soil. Hence, although triaxial tests may give valuable information about the shear strength of such soils, considerable judgment and interpretation are required for their best use.

Saturated Soils of Low Permeability. Fortunately, however, the shearing strength can be determined for many practical problems without recourse to triaxial tests. For saturated or nearly saturated soils having coefficients of permeability less than about 10^{-6} cm/sec, the time required for pore-pressure adjustment is usually very long compared to that within which the loads are applied. Even the period of construction for a building, during which the footing loads are built up, may be relatively short in comparison with that required for consolidation of the soil. Hence, virtually undrained or Q conditions prevail and the $\phi = 0$ concept is applicable, at least during and for a short time after the application of the load. The shear strength is then readily determined as half the unconfined compressive strength of undisturbed samples. Alternatively, shear tests may be carried out in situ by means of the vane (Art. 5.4). In either event, the analysis must be done in terms of total stresses (Art. 4.8). Some slightly fissured clays tend to break apart along the fissures when tested without confinement; under these circumstances it may be desirable to perform triaxial Q-tests with a cell pressure p_3 roughly equal to the in-situ effective overburden pressure, but consolidation should not be permitted under the cell pressure.

If the change in stress due to construction or loading will ultimately produce consolidation, the use of the unconfined compression test or its equivalents leads to conservative results; ultimately the shear strength may considerably exceed the immediate value. On the other hand, if the change in stress will ultimately lead to swelling, the shear strength may decrease with time; the use of the unconfined compression or triaxial Q-tests may give satisfactory results for the construction period and immediately thereafter, but as time goes by the results may become increasingly less conservative. Hence, for evaluating the long-term stability of cut slopes in stiff clays, Q-tests are not suitable.

If a normally loaded deposit is likely to consolidate, possibly under the weight of a fill, and then be subjected to shearing stresses from footings or other loads, it may be necessary to estimate the value of the shearing strength after consolidation has been completed. For this purpose it is convenient to determine the $c/\bar{p}_n$ ratio for the deposit, usually by means of unconfined compression or vane tests corresponding to different depths and, hence, to different values of effective vertical overburden pressure $\bar{p}_n$. Alternatively, the value of $c/\bar{p}_n$ may be estimated by means of the approximation, eq. 4.8. The required undrained shear strength c may then be calculated for the effective stress that will be reached after consolidation of the deposit.

Overconsolidated Clays. Strongly overconsolidated clays (overconsolidation ratio greater than about 6) with plasticity indices greater than about 40 require special consideration. Such materials almost always contain joints and slickensides; the presence of these defects may control the strength of the entire deposit. Excavation often causes strains sufficient to induce swelling and degradation of the materials; even the small shearing strains caused by the addition of stress may open the joints and slickensides and lead to softening. Hence, undrained or Q conditions do not apply. Although samples can be allowed to swell in the triaxial chamber under the stress conditions anticipated in the field, and can then be subjected to stress differences under conditions of drainage chosen to represent the field conditions, the results can do no more than provide guidance for judgment, because the laboratory and field behaviors of such materials differ for reasons not yet well understood. Less plastic stiff materials (plasticity index below about 40) behave more predictably and the results of triaxial tests can be used with considerable confidence.

Saturated Soils of Intermediate Permeability. The shear strength of saturated soils of intermediate permeability (k between about 10^{-4} and 10^{-6} cm/sec) is difficult to evaluate because conditions are likely to approximate neither undrained nor drained states. In many instances it is uneconomical or impracticable to carry out the studies necessary to take advantage of the decrease of pore pressure resulting from consolidation and drainage; under these circumstances, triaxial Q-tests give conservative values of shear strength. Unconfined compression tests are not likely to be suitable because capillary stresses may have a considerable influence on the results. In some instances it is possible to measure the pore pressures in the field while the stresses are being applied by the loading and to restrict the rate of loading so as to keep the shear strength at or above the desired value.

Partially Saturated Soils. The practical determination of the shear strength of partially saturated soils depends largely on whether the soil is coarse-grained or fine-grained. For gravels and sands (grain size greater than about 0.06 mm) the apparent cohesion due to capillary moisture is usually neglected for permanent construction, and values of ϕ_d are determined from drained triaxial tests or from information such as that shown in Table 4.1. For soils with smaller grain size, a series of triaxial Q-tests is appropriate in which samples are subjected without drainage to different cell pressures p_3 and then, also without drainage, to the stress differences Δp. The resulting relationship (represented by Fig. 4.11) is used with the estimated values of total pressure and degree of saturation anticipated in the field.

If it is likely that partially saturated soils may become flooded or submerged, the strength should also be evaluated as for saturated soils.

4.12. Strength and Deformability of Rock

The unconfined compressive strength of the concrete used for the construction of most footings and piers ranges from 2500 to 5000 lb/sq in. (180 to 360 tons/sq ft). Intact specimens of most commonly encountered sound rocks except weak shales exhibit strengths greatly exceeding these values, as shown in Table 4.2. The initial tangent modulus of the concrete used in foundations is likely to vary from about 2,500,000 to 5,000,000 lb/sq in. (180,000 to 360,000 tons/sq ft). According to Table 4.2, the moduli of intact rocks are likely to exceed these values by a substantial margin. Therefore, intact and sound bedrock is usually more than adequate for support of ordinary foundations.

Unfortunately, most rock masses are not intact and, moreover, their upper portions are usually weathered. Consequently, the values listed in Table 4.2 or similar values determined by testing intact samples from a specific site are seldom significant or useful in connection with the design of a particular foundation on rock. Joints, bedding surfaces, shear zones, and even faults may be encountered beneath a site. Weathering usually penetrates into the rock mass from such defects; consequently, the extent of weathering may vary radically from one portion of the bedrock to another. The weathered rock may often be investigated by the techniques appropriate to soils.

The presence of such features as subsurface cavities, wide joints, altered shear zones, and zones of intense local weathering may determine the behavior of an entire foundation, particularly if loads of high intensity are applied to the rock by a small number of piers or piles. The properties of the defective zones, in contrast to those of of the intact rock, may require detailed investigation.

ILLUSTRATIVE PROBLEM

Compute the shearing resistance against sliding along a horizontal plane at a depth of 20 ft in the deposit of sand shown in the accompanying figure.

Assume that the sand can drain freely and that ϕ_d for the wet sand is 32°.

Solution. The total stress p at depth 20 ft is

$$7 \times 118 = 826 \text{ lb/sq ft}$$
$$13 \times 128 = 1664$$
$$p = 2490 \text{ lb/sq ft}$$

The neutral stress is

$$u = 13 \times 62.5 = 812.5 \text{ lb/sq ft}$$

Therefore, from eq. 4.9

$$s = (p - u) \tan \phi_d = (2490 - 812) \tan 32°$$
$$= 1050 \text{ lb/sq ft}$$

Table 4.2 Physical Properties of Typical Sound, Intact Rocks[a]

Rock Type	Unconfined Compressive Strength		Tangent Modulus[b]	
	lb/sq in. × 10^3	ton/sq ft	lb/sq in. × 10^6	ton/sq ft × 10^3
Basalt, Pullman, Wash.	33	2400	11	790
Basalt, Arlington, Ore.	51	3700	16	1100
Diabase, West Nyack, N. Y.	35	2500	14	970
Diabase, Culpeper, Va.	47	3400	12	840
Dolomite (Oneota)	13	910	6.9	500
Dolomite (Lockport)	13	940	2.8	200
Dolomite (Bonne Terre)	22	1600	9.8	710
Gneiss, Orofino, Idaho	24	1700	4.0	290
Granite, Pikes Peak	13–33	940–2400	2.7–8.4	195–610
Granite, Barre, Vt.	28	2000	6.3	450
Limestone (Bedford)	7.4	540	3.9	280
Limestone (Solenhofen)	36	2600	9.3	670
Marble, W. Rutland, Vt.	9.4	680	6.8	490
Quartzite, Baraboo, Wis.	46	3400	10	740
Sandstone (Berea)	11	770	1.0	70
Sandstone (Crab Orchard)	31	2200	3.4	250
Sandstone (Navajo)	6.3	450	1.4	99
Siltstone, Hackensack, N. J.	18	1300	4.4	310
Tuff, Mercury, Nev.	3.5	250	0.8	59

Note: Values for rock masses in nature are always smaller.
[a] After R. P. Miller, 1965.
[b] From unconfined tests at axial stress of about 100 lb/sq in.

PROBLEMS

1. It is believed that the shearing strength of a soil under certain conditions in the field will be governed by eq. 4.2, wherein $c = 0.2$ ton/sq ft and $\phi = 22°$. What minimum lateral pressure would be required to prevent failure of the soil at a given point if the vertical pressure were 4.5 tons/sq ft?
   Ans. 1.78 tons/sq ft.

2. A sample of dry sand was tested in a direct shear device under a vertical pressure of 20 lb/sq in. on the horizontal plane of shear. Failure occurred at a shearing stress of 14 lb/sq in. Compute the angle of internal friction of the sand.
   Ans. 35°

3. The sand in a deep natural deposit has an angle of internal friction of 40° in the dry state, a dry density of 110 lb/cu ft, and a saturated unit weight of 131 lb/cu ft. If the water table is at a depth of 20 ft, what is the shearing resistance of the material to sliding along a horizontal plane at a depth of 10 ft?
   Ans. 0.462 ton/sq ft.

4. Compute the shearing resistance under the conditions specified in Prob. 3 if the water table is at the ground surface.
   Ans. 0.287 ton/sq ft.

5. In a drained triaxial test on dense sand the all-around pressure was 1.5 tons/sq ft and the added vertical pressure to cause failure was 5.43 tons/sq ft. Compute the angle of internal friction ϕ and the angle of inclination α of the failure planes, on the assumption that eq. 4.2 is valid.
   Ans. 40°, 65°.

6. In a triaxial test on a saturated sample of dense sand, the sample was consolidated under an all-around pressure of

1.5 tons/sq ft. Further drainage was then prevented. During the addition of vertical load the porewater pressure in the sample was measured. At the instant of failure it amounted to 1.20 tons/sq ft. The added vertical pressure at this time was 1.45 tons/sq ft. What was the value of ϕ for the sand?

Ans. 45°.

7. In an undrained triaxial test on a sample of saturated clay the all-around pressure is maintained at 1.0 ton/sq ft. The unconfined compressive strength is 3.62 tons/sq ft. At what vertical load in addition to the all-around pressure should the sample fail?

Ans. 3.62 tons/sq ft.

8. A normally loaded clay is under an effective vertical pressure of 1.20 tons/sq ft. Its plasticity index is 45. Approximately what is its undrained shearing resistance c?

Ans. 0.336 ton/sq ft.

9. The effective overburden pressure on a clay having a plasticity index of 30 is 1860 lb/sq ft. The unconfined compressive strength of an undisturbed sample of the clay was 1.1 tons/sq ft. Is the clay represented by the sample normally loaded?

Ans. No. The unconfined compressive strength would have been only about 0.41 ton/sq ft had the clay been normally loaded.

SUGGESTED READING

The standard work on triaxial testing is A. W. Bishop and D. J. Henkel (1962), *The Measurement of Soil Properties in the Triaxial Test*, 2nd ed., London, Edward Arnold, 228 pp.

The shear strength of soils, in spite of extensive research, remains a complex and controversial subject; the vast literature is confusing even to the specialist. A significant step in clarifying the issues was taken by the ASCE when, in 1960, it convened a Research Conference on Shear Strength of Cohesive Soils at Boulder, Colorado. The Proceedings of that conference remain a valuable source of information. Possibly the paper of greatest influence from a practical point of view is A. W. Bishop and L. Bjerrum (1960), "The Relevance of the Triaxial Test to the Solution of Stability Problems," pp. 437–501. It discusses the circumstances under which drained and undrained tests are appropriate, the corresponding use of effective- and total-stress analyses, and the limitations of both. An authoritative guide to the fundamental problems and research thereon is contained in M. J. Hvorslev (1960), "Physical Components of the Shear Strength of Saturated Clays," pp. 169–273. Both papers include extensive bibliographies.

References to the stress-deformation-strength characteristics of various types of natural deposits are listed at the end of Chap. 6.

M. Juul Hvorslev (1895–   )

Foremost authority on methods of boring and sounding and of obtaining samples of soil for engineering purposes. His report on this subject, prepared after more than a decade of intensive and meticulous research, is an invaluable contribution to the literature of foundation engineering. His earlier studies of the shearing resistance of clays significantly advanced knowledge in this aspect of soil mechanics.

PLATE 5.

CHAPTER 5

Techniques of Subsurface Investigation

5.1. Methods of Exploration

Before the engineer can design a foundation intelligently, he must have a reasonably accurate conception of the physical properties and the arrangement of the underlying materials. The field and laboratory investigations required to obtain this essential information are called the *soil exploration* or the *exploratory program*. Because of the complexity of natural deposits, no one method of exploration is best for all situations.

The method most suitable under a wide variety of conditions consists of drilling holes into the ground and extracting samples for identification and, in some instances, for testing. Several procedures are in common use for drilling the holes. Likewise, a variety of methods is available for obtaining the samples. The choice depends on the nature of the material and on the purpose of the exploratory program. Since the method used for drilling the holes does not necessarily dictate the method of sampling, the descriptions of procedures for drilling and for sampling are discussed in separate articles.

After preliminary borings have disclosed the general characteristics of the underlying material, a more extensive program of boring and sampling may be indicated. Or it may prove more expedient to investigate the consistency or the relative density of the weaker parts of the deposit by means of penetration tests or other direct methods that do not require sampling. The specific procedure must be chosen in consideration of the character of the soil deposit and the type of information needed for design or construction.

Other procedures less common than those just mentioned may be used under suitable conditions. For example, the character of the subsoil is occasionally investigated by direct inspection of the materials disclosed on the walls of test pits, shafts, or tunnels. In some instances, load tests are performed at the bottom of the test pits. When general information is needed concerning the location of boundaries between hard material such as bedrock and overlying softer deposits, geophysical methods may sometimes be used to advantage.

In the following articles, the various procedures for soil exploration are discussed in greater detail.

5.2. Exploratory Borings

Auger Borings. The simplest device for making a hole in the ground is an auger. Two varieties of hand-operated augers are illustrated in Fig. 5.1. Although hand-drilled auger holes can be made to depths of more

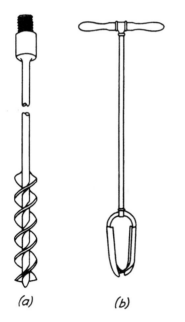

FIGURE 5.1. Hand-operated augers. (*a*) Helical auger. (*b*) Iwan or post-hole auger.

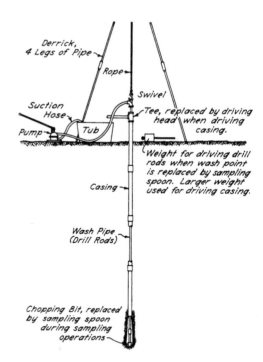

FIGURE 5.2. Apparatus for making wash borings (after Mohr, 1943).

than 100 ft by adding successive sections to the stem of the auger, they are most commonly made in connection with soil investigations for railroad, highway, or airport construction where it is usually unnecessary to drill deeper than about 12 ft. Moreover, portable power-driven helical augers are available in diameters ranging from 3 to 12 in. or more. These are often used for making deeper holes in soils having sufficient cohesion to prevent caving as the material is removed. Auger-type and bucket-type truck-mounted drilling machines are also in wide use for inspection shafts up to several feet in diameter. These are described in Art. 5.5.

If the walls of the hole will not stand unsupported, the soil may be prevented from caving by means of a pipe known as *casing*. The casing is driven for a short distance into the ground and cleaned out with the auger. Lengths of casing are then added, the casing is driven again, and again cleaned out. Casing is rarely used with hand-operated augers, and its use is inconvenient in connection with power-operated augers because the auger must usually be removed while the casing is being driven. Therefore,

augers are not commonly used in materials that require lateral support. An exception is the hollow-stem auger, which acts as its own casing, as described subsequently.

Wash Borings. A simple procedure for making relatively deep holes in soil deposits is wash boring. The most rudimentary form of the necessary equipment is shown in Fig. 5.2.

The hole is started by driving a piece of casing, with a diameter of 2 to 4 in., to a depth of 5 or 10 ft. The casing is then cleaned out by means of a chopping bit fastened to the lower end of a wash pipe that is inserted inside the casing. Water is forced down through the wash pipe and emerges at high velocity through small openings in the bit. It then rises, carrying fragments of soil, through the annular space between the casing and the wash pipe. It overflows at the top of the casing through a T connection into a tub from which it is again pumped through a hose into the wash pipe. The connection between the water hose and the wash pipe is provided with a swivel joint so that the wash pipe and the chopping bit can be twisted as they are raised and dropped on the soil at the bottom of the hole. This facilitates the cutting. The wash pipe is extended and additional casing is driven as the hole progresses. If the material will stand without caving, however, the casing need not extend more than 10 or 15 ft below the surface of the ground.

Although this procedure has been largely replaced by the use of mechanized truck-mounted drilling machines, it permits penetrating all but the most resistant soil strata. Because the equipment is simple and lightweight, it can be used at relatively inaccessible sites that cannot be reached with large and heavy trucks.

Hollow-Stem Auger Borings. Wash-boring equipment is often replaced by truck- or tractor-mounted drilling rigs that can turn an auger into the soil. Holes may be drilled rapidly to depths of over 200 ft using continuous flights of augers with hollow stems through which sampling tools may be operated. Thus, the auger is not intended to act as a sampling device, but is used to advance and case the hole simultaneously.

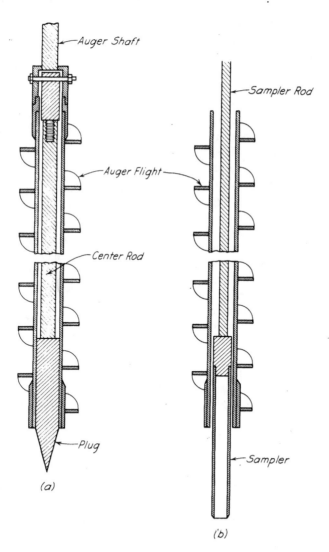

FIGURE 5.3. Hollow-stem auger. (*a*) Plugged while advancing auger. (*b*) Plug removed and sampler inserted to sample soil below auger.

A removable plug attached to a center rod is used to block the entry of soil into the stem until the desired sampling depth is reached. The plug and center rod are then retracted and the sampler and sampler rod are inserted (Fig. 5.3). In cohesive soils the hollow-stem auger is often used without a plug, since these materials will usually enter the mouth of the lowest auger section only 2 to 4 in. before forming their own plug. The sampler can then be driven or washed through the earth plug into undisturbed material. Augers with inside diameters of $2\frac{1}{2}$ or $3\frac{3}{8}$ in. are most commonly used.

When the hollow-stem auger is used in cohesionless soils below the water table, excess hydrostatic pressures may force water-saturated sand several feet into the stem as the plug is withdrawn. This action loosens the material below the stem and results in too low an indication of its relative density. Special cleanout procedures are also required to wash the material from the stem before sampling can proceed. Therefore, under such conditions the plug should not be used and water should be maintained inside the stem at a level above the ground water table.

If the hollow-stem auger is used in deposits of loose silt or granular material, it may decrease the natural void ratio and increase the confining pressure near the mouth of the lowest auger section. Both processes lead to misconceptions on the unsafe side concerning the compressibility and strength of the material.

Rotary Drilling. Rotary drilling may be used in rock, in clay, or even in sand. It is the most rapid method for penetrating highly resistant materials unless the deposit is very loose or badly fissured. In this method, a rapidly rotating drilling bit cuts or grinds the material at the bottom of the hole into small particles. The particles are removed by circulating water or drilling fluid in a manner similar to that in wash borings. To obtain a sample, the drilling bit is removed and replaced by a sampler (Art. 5.3). A sketch of a rotary drilling rig is shown in Fig. 5.4.

In rotary drilling for site exploration, casing is usually unnecessary except near the ground surface. Collapse of the hole is normally prevented by drilling fluid, which consists of a slurry of clay and water to which bentonite is often added. This slurry, known as *drilling mud*, coats and supports the sides of the drill hole and seals off permeable strata. The diameters of rotary borings for foundation exploration usually range from about 2 to 8 in. Holes of larger diameter suitable for direct inspection of the substrata are discussed in Art. 5.5.

Percussion Drilling. If the drill hole must be advanced through exceptionally hard strata of soil or through rock, auger borings or wash borings cannot be used. One method for drilling holes through such deposits is known as *percussion* or *cable-tool drilling*. In this method a heavy drilling bit is alternately raised and dropped in such a manner that it grinds the underlying material to the consistency of a sand or silt. If possible, the bore hole is kept dry except for a small amount of water that forms a slurry with the material ground up by the bit. When the accumulation of slurry interferes with drilling, the drilling tools are removed from the hole and the slurry cleaned out with a bailer. The hole may be cased if the formation will not stand without collapse. Although percussion drilling is often used for drilling water wells, it is not generally favored for site exploration where intact samples must be obtained for identification and testing.

5.3. Sampling

Types of Samples and Soil Sampling Tools. The kind of samples that should be obtained from an exploratory drill hole depends on the purpose for which the exploration is made. Auger samples may be used to identify soil strata and for some classification tests even though the physical state of the material is completely altered by the sampling process. The cuttings or choppings from wash borings are of small value except for indicating changes in stratification to the boring foreman. The material brought

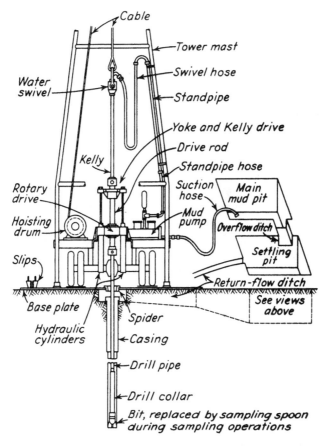

FIGURE 5.4. Rotary drilling rig (after Hvorslev, 1948).

up in drilling mud is contaminated and usually unsuitable even for identification. The rock chips obtained by bailing the slurry from percussion drill holes reveal little regarding the natural condition of the rock. However, they may indicate gross mineralogical characteristics, such as the presence of mica or calcium carbonate. In general, the fragments of soil or rock obtained as a by-product of the drilling or boring process are rarely satisfactory for determining or even indicating the physical characteristics so important to the foundation engineer.

For proper identification and classification of a soil or rock, *representative* samples are required. They should contain all the constituents in their proper proportions. Such samples are adequate for visual classification, for the performance of mechanical analyses, and for determining the Atterberg limits, the unit weight of the solid constituents, the carbonate content, and the organic content. The mechanical properties of the soil, however, may be appreciably altered by sampling. If so, such samples are not suitable for the determination of the stress–strain characteristics or density index of the materials. If information of this type is required, the samples must be taken in such a manner that they experience negligible deformation during sampling. Such samples are called *undisturbed*, although a certain amount of disturbance must be regarded as inevitable.

Representative samples can usually be obtained by driving or pushing into the ground an open-ended cylindrical tube known as a *sampling spoon*. Spoons having inside diameters from $1\frac{3}{8}$ to $2\frac{1}{2}$ in. usually consist of four parts, a cutting shoe at the bottom, a barrel consisting of a length of

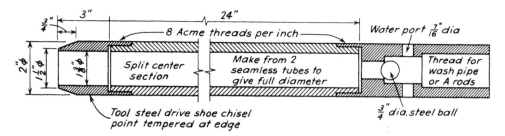

FIGURE 5.5. Split barrel sampler for standard penetration test.

pipe split longitudinally into two halves, and a coupling at the top for connection to the drill rods. Such a device is known as a *split spoon* or a *split barrel sampler* (Fig. 5.5). After a sample has been taken, the cutting shoe and the coupling are unscrewed and the two halves of the barrel separated to expose the material.

If spoon samples are to be transported to the laboratory without examination in the field, the barrel is often cored out to hold a cylindrical thin-walled tube known as a *liner*. After a sample has been obtained, the liner and the sample it contains are removed from the spoon, and the ends are sealed with caps or with metal disks and wax. Liner spoons are available for samples up to about 8 in. in diameter, but the largest samples commonly taken in this manner have a diameter of 5 in.

The degree of disturbance of spoon samples depends on the way in which the force is applied to the spoon, whether by pushing or driving, on the rate of penetration, and on the dimensions of the sampler. If other conditions are equal, the degree of disturbance is roughly indicated by the *area ratio*

$$A_r(\%) = 100 \frac{D_e^2 - D_i^2}{D_i^2} \quad 5.1$$

where D_e is the external diameter and D_i the internal diameter of the cutting shoe through which the sample must pass. If the area ratio is not greater than about 10 per cent, the distortion of the sample is small in almost any type of soil. The degree of disturbance also is less if the sampler is advanced with a rapid steady motion instead of by intermittent pushing or driving.

In order to obtain samples of soil several feet long, it is necessary to reduce the friction between the core and the inside of the tube. This is accomplished by crimping the cutting edge so that its inside diameter D_i is slightly smaller than the inside diameter of the tube D_{it}. The degree of sampling disturbance is also affected by the *inside clearance ratio*,

$$C_r(\%) = 100 \frac{D_{it} - D_i}{D_i} \quad 5.2$$

If this ratio becomes too large, the sample may expand excessively as it passes into the sampling tube and its strength may be considerably decreased. For undisturbed samples of high quality the inside clearance ratio should not exceed 1 per cent.

Samples of cohesionless materials, such as sand below the water table, cannot be retained in conventional sampling spoons without the addition of a spring core catcher (Fig. 5.6a). In deposits of very fine sand, or in sand containing small pebbles that may prevent the springs from closing, no recovery may be possible. Disturbed samples can sometimes be obtained by inserting a one-way (flapper) valve between the cutting shoe and barrel of the split sampling spoon; however, such samples may not always contain the various particle-size fractions in their proper proportions. Representative samples, but in a completely altered state, can usually be obtained by means of a *scraper bucket* (Fig. 5.6b). The bucket is driven below the casing and rotated. As it turns, scrapings from the sides of the hole fall into the lower compartment.

The cost of undisturbed samples increases rapidly with increasing diameter. It has

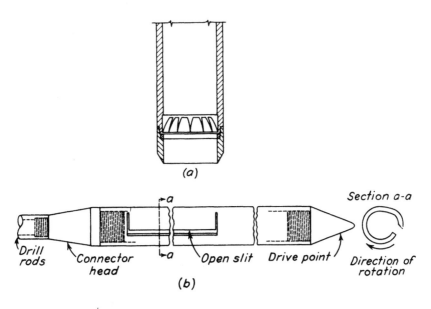

FIGURE 5.6. (a) Spring core catcher. (b) Scraper bucket.

been found by experience that most soft or moderately stiff cohesive soils can be sampled without excessive disturbance in *thin-walled tubes* of seamless steel having a diameter not less than 2 in. and an area ratio of about 10 per cent. Tubing with a diameter of 2 or 3 in. is commonly used, in lengths varying from 2 to 3 ft. The lower end of the tube is sharpened and slightly crimped to form the cutting edge. The upper end is machined for attachment to the drill rods.

In tube sampling, the entire tube is pushed into the ground at the bottom of the hole and is removed with the sample inside. The two ends of the tube are sealed, and the sample is shipped to the laboratory. In the laboratory the sample is extruded with the least possible disturbance. In some instances, disturbance is reduced by cutting the tube into short lengths of about 6 in. before extrusion.

To improve the quality of samples and to increase the recovery of soft or slightly cohesive soils, a *piston sampler* may be required. Such a sampler consists of a thin-walled tube fitted with a piston that closes the end of the sampling tube until the apparatus is lowered to the desired depth (Fig. 5.7a). The sampling tube is then pushed past the piston, which remains on the surface of the soil at the bottom of the hole, as shown in Fig. 5.7b. The presence of the piston prevents soft soils from squeezing rapidly into the tube and thus eliminates most of the distortion of the sample. The piston also helps to increase the length of sample that can be recovered by creating a slight vacuum that tends to retain the sample if the top of the column of soil begins to separate from the piston. During the withdrawal of the sampler, the piston also prevents water pressure from acting on the top of the sample and thus increases the chances for recovery. The design of piston samplers has been refined to the extent that it is sometimes possible to take undisturbed samples of sand from below the water table (Bishop, 1948).

Cores. Short intact samples of rock can be obtained during percussion drilling by means of a special core barrel. However, the necessity for taking a core greatly slows drilling progress. Therefore, percussion drilling is rarely used in connection with coring operations for exploratory purposes.

In rotary drilling, the drilling bit may be replaced by a core drill attached to the lower end of a core barrel, as shown in Fig. 5.8. The drill cuts an annular ring in the

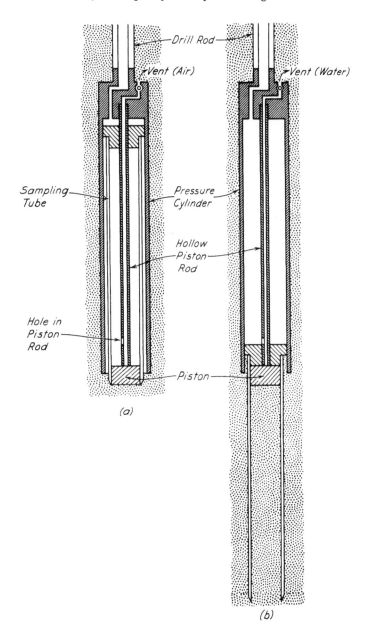

FIGURE 5.7. Piston sampler of hydraulically operated type. (*a*) Lowered to bottom of drill hole, drill rod clamped in fixed position at ground surface. (*b*) Sampling tube after being forced into soil by water supplied through drill rod.

rock and leaves a central core that enters and is retained in the core barrel. The cutting element may consist of diamonds, chilled shot, tungsten carbide inserts, or steel cutters. Core drilling does not greatly increase the time for advancing the hole and is widely used for sampling resistant soils and rocks.

The cuttings from the annular ring are commonly washed away with the drilling fluid, but in some materials the action of the fluid alters the character of the material.

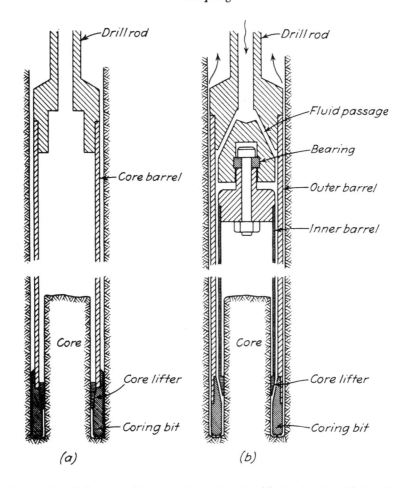

FIGURE 5.8. Schematic diagram of core barrels. (*a*) Single tube. (*b*) Double tube.

Under these circumstances, it may be possible to remove the cuttings with an air jet or a ring-shaped auger located directly above the cutting tool.

The core barrel may consist of a *single tube* (Fig. 5.8*a*) or a *double tube* (Fig. 5.8*b*). Samples taken in a single-tube barrel are likely to experience considerable disturbance due to torsion, to swelling, and to contamination with the drilling fluid. The double-tube barrel is designed to protect the core against the action of the circulating fluid. In some assemblies the inner barrel, including the core lifter, can be removed through the drill string. The operation is then known as *wire-line drilling*. In addition, the wire-line equipment may be modified to include a longitudinally split third tube nested within the inner barrel. After the inner barrel is removed from the drill hole, the split tube is pushed out by a hand-operated hydraulic pump. The core is retained in an essentially undisturbed condition by the split tube. Use of this assembly, known as a *triple-tube* core barrel, is especially advantageous for sampling fractured rock, brittle rock of low shear strength, or hard clay.

Standard rock cores range from about $1\frac{1}{4}$ to nearly 6 in. in diameter. The more common sizes are listed in Table 5.1. Most core barrels are capable of retaining cores at least 5 ft long. The *recovery ratio*, defined as the percentage ratio between the length of core recovered and the length of core drilled on a given run, is related to the quality of

Table 5.1 Standard Sizes of Core Barrels, Drill Rods, and Compatible Casing[a]

	Core Barrel		Drill Rod		Compatible Casing	
Symbol	Hole Diameter (approx., in.)	Core Diameter (approx., in.)	Symbol	Outside Diameter (in.)	Symbol	Outside Diameter (in.)	Inside Diameter (in.)
EWX,[b] EWM[c]	1½	13/16	E	15/16	—	—	—
AWX, AWM	1 15/16	1 3/16	A	1 5/8	EX[d]	1 13/16	1½
BWX, BWM	2 3/8	1 5/8	B	1 7/8	AX	2¼	1 29/32
NWX, NWM	3	2 1/8	N	2 3/8	BX	2 7/8	2 3/8
2¾ × 3 7/8	3 7/8	2 11/16	—	—	NX	3½	3

[a] As standardized by the Diamond Core Drill Manufacturer's Association, Bulletin No. 2 (ASTM D 2113).
[b] Symbol X indicates single-tube barrel.
[c] Symbol M indicates double-tube barrel.
[d] EX casing will fit into a hole drilled by AWX or AWM barrel and EWX or EWM barrels will fit inside EX casing.

rock encountered in a boring, but it is also influenced by the drilling technique and the type and size of core barrel used. Generally, the use of a double-tube barrel results in higher recovery ratios than can be obtained with single-tube barrels.

A better estimate of in-situ rock quality is provided by a modified core recovery ratio known as the *rock quality designation* (RQD). This ratio is determined by considering only pieces of core that are at least 4 in. long and are hard and sound. Breaks obviously caused by drilling are ignored. The diameter of the core should preferably be not less than $2\frac{1}{8}$ in. (NWX, NWM). The percentage ratio between the total length of such core recovered and the length of core drilled on a given run is the RQD. Table 5.2 gives the rock quality description as related to the RQD.

Hand-carved Samples. Practically undisturbed samples may be carved from soils possessing at least a trace of cohesion, provided the material is exposed in a test pit, shaft, or tunnel. Two methods of obtaining such samples are illustrated in Fig. 5.9.

5.4. Direct Measurements of Consistency and Relative Density

Penetrometers. Several methods have been developed for investigating the consistency of cohesive deposits or the relative density of cohesionless deposits without the necessity for making a drill hole and extracting samples. Most of these procedures are based on measuring the resistance offered by the soil to the advancement of a device known as a *penetrometer*. If the penetrometer is pushed steadily into the soil, the procedure is known as a *static penetration test*. If driven into the soil, it is known as a *dynamic penetration test*. As a rule, static tests are preferable in connection with soft cohesive deposits and dynamic tests with very hard deposits. Both static and dynamic tests have been found

Table 5.2 Relation of RQD and in situ Rock Quality

RQD (%)	Rock Quality
90–100	Excellent
75–90	Good
50–75	Fair
25–50	Poor
0–25	Very poor

After Deere, 1963.

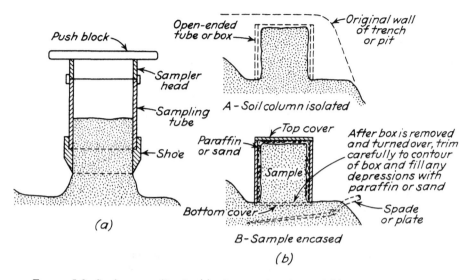

FIGURE 5.9. Surface sampling by (a) advance trimming and (b) block sampling.

useful in connection with cohesionless deposits.

Many varieties of penetrometers have been designed, each especially adapted to certain kinds of material. Three of these are shown in Fig. 5.10.

Standard Penetration Test and N-Values. In the United States, the most commonly used penetrometer is the ordinary split sampling spoon, and the most widespread test of this kind is known as the *standard penetration test* (ASTM D-1586). It is made by dropping a

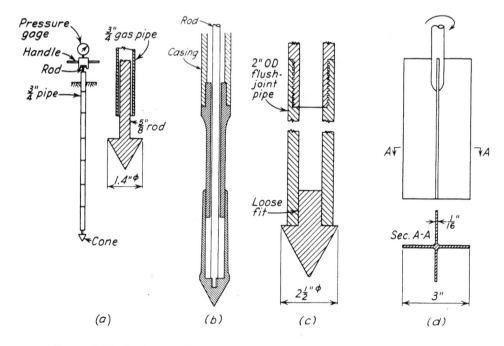

FIGURE 5.10. Devices for direct measurement of consistency of soil under field conditions. (a) Dutch cone penetrometer. (b) Refined Dutch cone penetrometer. (c) Conical drive point for sand and gravel. (d) Vane-shear apparatus.

hammer weighing 140 lb onto the drill rods from a height of 30 in. The number of blows N necessary to produce a penetration of 1 ft is regarded as the penetration resistance. The sampler has the dimensions shown in Fig. 5.5. To avoid seating errors, the blows for the first 6 in. of penetration are not taken into account; those required to increase the penetration from 6 to 18 in. constitute the N-value.

The results of the standard penetration test can usually be correlated in a general way with the pertinent physical properties of the soil. Table 5.3 shows such a correlation. However, the scattering of individual results from the conservative values given in the table may be quite large, and it is preferable to make direct comparisons with the results of other appropriate tests in any given locality. The correlation for clays can be regarded as no more than a crude approximation, but that for sands is often reliable enough to permit the use of N-values in foundation design.

In saturated, fine or silty, dense or very dense sands, the N-values may be abnormally great because of the tendency of such materials to dilate during shear under undrained conditions (Art. 4.2). Hence, in such soils, the results of standard penetration tests should be interpreted conservatively.

In addition, the value of N in cohesionless soils is influenced to some extent by the depth at which the test is made. Because of the greater confinement caused by increasing overburden pressure, N-values at increasing depths may indicate larger relative densities than actually exist. If the N-value at a depth corresponding to an effective overburden pressure of 1 ton/sq ft is considered to be a standard, the correction factor C_N to be applied to field N-values for other pressures is given approximately by

$$C_N = 0.77 \log_{10} \frac{20}{\bar{p}} \qquad 5.3$$

where $\bar{p}$ is the effective vertical overburden pressure in tons/sq ft at the elevation of the penetration test. The equation is valid for $\bar{p} \gtrless 0.25$ ton/sq ft.

By far the most common error in connection with the standard penetration test in sand or silt occurs, however, when drilling is being done below the water table. If the water level in the drill hole is allowed to drop below groundwater level, as may easily occur, for instance, when the drill rods are removed rapidly, an upward hydraulic gradient is created in the sand beneath the drill hole. Consequently, the sand may become quick and its relative density may be greatly reduced. The N-value will accordingly be much lower than that corresponding to the relative density of the undisturbed sand. Care is required to see that the water level in the drill hole is always maintained at or slightly above that corresponding to

Table 5.3 Penetration Resistance and Soil Properties on Basis of the Standard Penetration Test

Sands (Fairly Reliable)		Clays (Rather Unreliable)	
Number of Blows per ft, N	Relative Density	Number of Blows per ft, N	Consistency
		Below 2	Very soft
0–4	Very loose	2–4	Soft
4–10	Loose	4–8	Medium
10–30	Medium	8–15	Stiff
30–50	Dense	15–30	Very stiff
Over 50	Very dense	Over 30	Hard

the piezometric level at the bottom of the hole. For the reasons noted in Art. 5.2, use of the plugged hollow-stem auger as a means for drilling in cohesionless soils below water table almost inevitably leads to alteration of the relative density; hence, N-values determined under these circumstances should not be relied on.

In deposits containing many boulders the results of standard penetration tests may be unreliable because of the small size of the sampling spoon compared to that of the boulders.

In highly sensitive clays the standard penetration test may lead to a gross misconception of the consistency. Moreover, it is far too crude a test to justify its use for even approximating numerical values representing the strength of soft or very soft saturated clays. The ease of penetration of the sampler depends not only on the strength of the soil but also on its compressibility. Thus a strong cohesive soil with a high air content may have a substantially lower N-value than an equally strong saturated soil in which the voids cannot collapse as the sampler advances.

Although the standard penetration test cannot be regarded as a refined and completely reliable method of investigation, the N-values give a useful preliminary indication of the consistency or relative density of most soil deposits. The information is in some instances even sufficient for final design. In any event, it provides data for more intelligent planning of whatever additional exploration may be desirable at the site.

Dutch Cone Penetrometer. The most widely used static penetrometer is the *Dutch Cone*, developed by the Soil Mechanics Laboratory at Delft, the Netherlands. In its simplest form, the apparatus consists of a 60° cone, with a base area of 10 sq cm, attached to the bottom of a rod protected by a casing (Fig. 5.10a). The cone is pushed by the rod at a rate of 2 cm/sec; the cone resistance q_p is the force required to advance the cone divided by the base area. A refined form of the apparatus, considered as a standard in many localities, is shown in Fig. 5.10b. With this equipment, the point resistance and the friction on the outside of the casing can each be measured. In a further modification the friction is measured on a separate sleeve of limited length, located just above the point.

In localities where there has been considerable experience with the Dutch Cone, values of the penetration resistance have been related to such properties as the angle of shearing resistance ϕ of sand or the consistency of clays. In the United States such detailed correlations have not yet been developed, but rough statistical relations have been found useful between the cone-penetration resistances and the N-values obtained from the standard penetration tests in granular materials. This information is summarized in Table 5.4 (Sanglerat, 1972).

Table 5.4 Approximate Relation Between Dutch-Cone and Standard Penetration Resistances

Soil Type	$\dfrac{q_p{}^a}{N}$
Silts, sandy silts, slightly cohesive silt-sand mixtures	2.0
Clean fine to medium sands and slightly silty sands	3 to 4
Coarse sands and sands with little gravel	5 to 6
Sandy gravels and gravels	8 to 10

[a] q_p = Dutch cone resistance, kg/sq cm; N = standard penetration resistance.

Improvised or Special-Purpose Penetrometers. A simple penetrometer that can be used for investigating the relative density of sand and gravel deposits without the use of casing is shown in Fig. 5.10c. It consists of a cone that fits loosely into the bottom of a string of pipes driven into the ground by means of a drop hammer with constant height of fall. A continuous record is made of the number of blows required to advance the point through each foot of depth. When the point has reached its final elevation, the pipe is withdrawn and the cone left at the bottom of the hole. Friction on the pipe is minimized by making the diameter of the cone somewhat greater than the outside diameter of the pipe. Tests of this type can be made rapidly and economically. At one site it was found that two soundings each to a depth of 50 ft could be made in one day. On the other hand, nearly three days were required for one wash boring with normal split spoon sampling to the same depth. The results of cone soundings can be made more meaningful by performing several series of standard penetration tests in borings adjacent to the locations of the cone penetration tests. The cone driving resistance can be correlated with the N-values, and Table 5.3 can then be used as a guide to foundation design. Figure 6.12 shows a comparison obtained during the exploration of a gravel deposit in Denver, Colorado.

Direct Measurement of Shearing Strength. Various devices have been developed for direct measurement of the shearing resistance of cohesive soils, particularly in the undrained or Q condition. One of the more widely used is known as the *vane-shear apparatus* (Fig. 5.10d). In its simplest form it consists of a four-bladed vane fastened to the bottom of a vertical rod. The vane and the rod can be pushed into the soil without appreciably disturbing the material. The assembly is then rotated and the torque required to turn the vane is measured. Since the soil fails along a cylindrical surface passing through the outer edges of the vane, the shearing resistance can be computed if the dimensions of the vane and the torque are known. If the vane is rotated rapidly through several revolutions, the soil becomes remolded and the shearing strength can again be determined. Hence, it is possible to measure not only the shearing resistance but also the sensitivity of the clay. In more refined versions of the vane test, a casing is used to eliminate skin friction on the rod; in some equipment a special housing protects the vane while the apparatus is being pushed into the ground. In spite of these refinements the vane cannot be used satisfactorily in clays that have unconfined

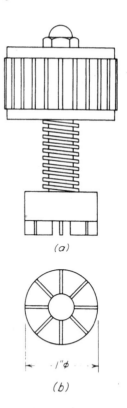

FIGURE 5.11. Torvane for determining shear strength of materials for which $s = c$. (a) Side view. (b) Bottom view of vanes.

compressive strengths much greater than 1.0 ton/sq ft, that contain sand layers or many pebbles, or that possess a secondary structure.

A modification known as the *torvane* (Fig. 5.11), permits the rapid performance of many vane tests on a freshly cut surface such as that of a sample of clay sliced longitudinally, or the freshly exposed wall of a test pit. By means of adapters, reliable determinations of shear strength from 0.1 to 5 tons/sq ft can be made.

5.5. Miscellaneous Methods of Soil Exploration

Inspection Pits and Shafts. Under some circumstances it is advantageous to inspect the subsurface formations in their natural state. This may be done by excavating open pits and large-diameter shafts or by driving tunnels through the materials. The section exposed in such openings should be examined not only by engineers, but also by competent engineering geologists. Ordinarily, it is not economical to conduct a full program of exploration by such means, but direct inspection of an extremely variable deposit may furnish a more valid impression of its nature than can be obtained from many borings. Such openings provide a means for obtaining hand-carved undisturbed samples and are essential for performing load tests on the soil or rock.

Relatively inexpensive inspection holes 3 to 4 ft in diameter can be drilled through rock by means of shot core barrels. Holes of this type have been used chiefly for exploration of dam foundations, but under some circumstances they have proven useful in connection with other types of structures. Large power augers and bucket augers, with diameters up to 5 ft or more, are commonly used for exploring soil deposits to depths of well over 50 ft, provided the holes will stand open at least briefly (Art. 13.3). A large-diameter casing containing ports that can be closed is sometimes used for protection.

Direct inspection of rock formations in bore holes with diameters as small as $3\frac{1}{2}$ in. has been made possible by the development of television cameras and film cameras that may be lowered into the holes. The rock surface can be observed on a television tube or recorded for later viewing on a cylindrical screen where it may be examined as if the viewer were in the hole.

Load Tests. The stress–strain characteristics of soils and soft rocks are sometimes investigated by means of field load tests. A square or circular bearing plate is established at a suitable elevation on the bottom of a test pit or shaft. Vertical load is applied to the plate in increments, and the settlement is observed after each application of load. The results are plotted in the form of a load-settlement diagram (Fig. 5.12a).

The results of load tests require careful interpretation and may in many instances be misleading. Unless the subsoil is uniform for a considerable depth below the base of a proposed foundation, the results of load tests are likely to give a false conception of the supporting capacity of the underlying material. The limitations of such tests will be discussed in connection with the various types of foundations (Part C).

However, load tests may represent a convenient method for investigating the relative density of sand deposits, especially for the calibration of the results of penetration tests. Load tests made for this purpose and carried out in the following manner are

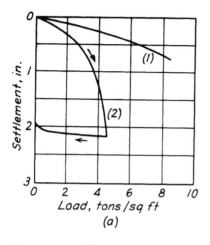
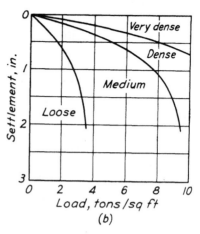

FIGURE 5.12. (*a*) Load-settlement diagrams for (1) a dense clean sand in a caisson 26 ft below the bottom of a river and (2) a sand of medium density at the bottom of a 30-ft shaft. (*b*) Chart for estimating relative density of dry sand on basis of results of standard load test on bearing plate 1 ft square.

known as *standard load tests*. The test pit is at least 5 ft square. The loading plate is 1 ft square, and no surcharge is placed on the ground within a distance of 2 ft from the plate. Inasmuch as capillary moisture produces an apparent cohesion that has an influence on the test results, standard load tests should not be performed within the capillary zone. In coarse sands the effect of capillarity is negligible, but in medium or finer sands it may lead to seriously overestimating the relative density. Under these circumstances the tests should preferably be carried out at the water level.

The results of a standard load test may be interpreted by means of Fig. 5.12*b*, which shows the boundaries of the areas occupied by load-settlement curves for sands of different relative densities. If the test has been performed at the water table, the values of settlement should be reduced 50 per cent before the load-settlement curve is plotted for comparison with those in Fig. 5.12*b*.

In some instances, it is convenient to drive piles as a means for determining the resistance of the soil. The piles may then be loaded to determine their supporting capacity. The appropriate use of such tests is described in Part C.

Groundwater Observations. In deposits of permeable granular materials, some indication of the position of the groundwater level can be obtained from measurements in bore holes within about 24 hr after the borings are completed. However, if only fine-grained materials are present, the permeability of the surrounding ground may be so low that the true position of the water table must be determined by the installation of specially constructed piezometers that require the flow of only minute quantities of water for proper operation. The best-known instruments of this type consist of a porous stone connected to a standpipe preferably made of plastic tubing. The porous stone, or piezometer tip, is carefully placed on a cushion of sand in the bottom of a cased bore hole. It is then surrounded by sand, whereupon the casing is withdrawn a few feet, and a seal is formed between the tubing and the casing by means of a nearly impermeable layer of bentonitic clay. During and after installation, the piezometer and surrounding sand are saturated. Since the introduction of the piezometer itself produces changes in the pore pressures within the strata, there is a time lag before the pore pressures at the piezometer tip reach equilibrium with those in the soil. The lag may vary from a few minutes if the surrounding soil is a fine sand to several weeks if it is a silty clay. Figure 5.13 is a sketch of a typical piezometer assembly for measurement of pore pressure in fine-grained soils.

Geophysical Methods. The boundaries between different elements of the subsoil may sometimes be located by geophysical methods. These procedures are based on differences in the gravitational, magnetic, electrical, radioactive, or elastic properties of the different elements of the subsoil. Geophysical methods were developed primarily for the mining and petroleum industry, and some of the procedures are not well suited for the purposes of civil engineering. Differences in unit weight, magnetic characteristics, and radioactivity of deposits near the surface of the earth are rarely great enough to permit use of the corresponding geophysical methods of exploration.

Seismic Surveys. Seismic exploration requires (1) equipment to produce an elastic wave, such as a small explosive charge and detonator or even a sledge hammer used to strike a plate on the surface; (2) a series of detectors, or geophones, spaced at intervals along a line from the point of origin of the wave; and (3) a time-recording mechanism, such as an oscillograph, to record the time of origin of the wave and the time of arrival at each detector (Fig. 5.14). If the stratigraphy is simple, the depth of each of the layers can be determined from the data. In addition, some knowledge can be obtained of the nature of the strata from the velocities with which the seismic waves advance through them. In areas that have not been previously explored, however, one or more borings must be made to correlate the seismic data with the soil and rock profile.

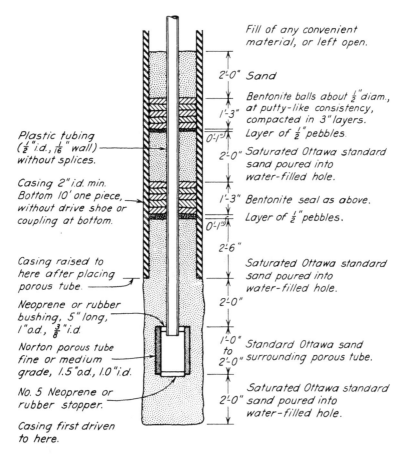

FIGURE 5.13. Casagrande open-standpipe porous tube piezometer (after A. Casagrande, 1949).

Seismic surveys are particularly useful in determining the depth to sound rock overlain by soft or loose strata. Where boulders or broken rock overlie sound rock, the seismic data may give a somewhat better indication of the sound rock surface than can be obtained from borings. On the other hand, the presence of a soft material beneath a stiff one cannot be detected. Table 5.5 gives typical seismic velocities of various materials.

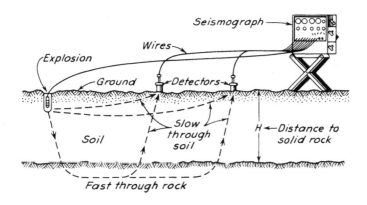

FIGURE 5.14. Simplified diagram of seismic refraction test (after Moore, 1961).

Table 5.5 Typical Seismic Velocities of Earth Materials

Material	Velocity (ft/sec)
Dry silt, sand, loose gravel, loam, loose rock, talus, and moist fine-grained topsoil	600–2500
Compact till; indurated clays; gravel below water table,[a] compact clayey gravel, cemented sand, and sand-clay	2500–7500
Rock, weathered, fractured, or partly decomposed	2000–10,000
Shale, sound	2500–11,000
Sandstone, sound	5000–14,000
Limestone, chalk, sound	6000–20,000
Igneous rock, sound	12,000–20,000
Metamorphic rock, sound	10,000–16,000

[a] Velocity of sound in water is about 4700 ft/sec and all fully saturated materials should have velocities equal to or exceeding this value.

Resistivity Surveys. The results of an electrical resistivity survey can delineate a well-defined boundary between a low-resistance material such as fine-grained soil or residuum, and a high-resistance material such as sound rock, no matter which one is on top. Resistivity methods are particularly useful in locating pockets of clean gravel (high resistance) within deposits of mixed-grained glacial drift or fine-grained soil (low resistance). A common procedure makes use of four electrodes that are driven into the ground at equal spacing along a straight line. An electric current I, usually direct current, is sent into the ground through the two outside electrodes and the induced potential E is measured across the two interior electrodes. The earth resistivity ρ is computed for a given spacing A from the formula

$$\rho = 2\pi A \frac{E}{I} \qquad 5.4$$

It is customary to express A in cm, E in volts, I in amperes, and ρ in ohm-cm and to assume that ρ represents the average resistivity to depth A. As in the case of seismic surveys, occasional borings are required to confirm interpretations or investigate anomalies; however, the method allows rapid reconnaissance of an area. Table 5.6 gives some representative resistance values for earth materials.

Table 5.6 Representative Values of Resistivity

Material	Resistivity (ohm–cm)
Clay and saturated silt	0–10,000
Sandy clay and wet silty sand	10,000–25,000
Clayey sand and saturated sand	25,000–50,000
Sand	50,000–150,000
Gravel	150,000–500,000
Weathered rock	100,000–200,000
Sound rock	150,000–4,000,000

Geological Information. Geologic and pedologic maps and reports constitute a summary of data concerning previous investigations of the earth materials in a given area. Many of the more recent publications of both types contain specific data on the engineering properties and correlations of direct value in engineering. Even when they do not, however, the general descriptions of the geology and characteristics of the various strata often provide a basis for planning a detailed subsurface investigation and even for preliminary design. Chapters 6 and 7 will emphasize the importance of such preliminary information and its value as a basis for engineering correlations.

5.6. Record of Field Exploration

The records of subsurface explorations and sampling operations should be clear and accurate. Field notes should contain the date the work was done, the location with respect to a permanent system of coordinates, the elevation of the ground surface with respect to a permanent bench mark, the elevation of the water table, the elevation of the upper boundary of each successive stratum of soil or rock, a field classification of the strata encountered, and whatever values of penetration resistance or other measures of consistency were obtained. The records of special investigations during which large-diameter undisturbed samples or hand-carved samples were taken should be as complete as possible.

The engineer or boring foreman in charge of any exploration should always keep in mind that seemingly insignificant details of the procedure required to advance the hole and keep it open may yield information as valuable as that obtained from the samples. For this reason the records should include the types of equipment and tools used in the investigation, together with any changes made. The depth at which a change was made and the reason for the change should also be noted. Furthermore, the methods used to stabilize the walls of the bore hole or to sheet the test pit should be recorded. All this information should be condensed and presented for every exploratory hole whether successfully completed or not.

After samples have been examined and tested in the laboratory, the materials should be classified according to the most appropriate method, and the field notes together with the laboratory results should be assembled in such a form that the boundaries between different materials are plotted at their correct elevations on a suitable vertical scale.

SUGGESTED READING

The outstanding reference on the techniques of subsurface investigation is the report of M. J. Hvorslev (1948), "Subsurface Exploration and Sampling of Soils for Civil Engineering Purposes," Waterways Experiment Station, Vicksburg, Miss., 465 pp. The author describes practically every method of exploration and gives details of the sampling equipment and techniques used. The bibliography contains nearly 1000 references.

Several excellent papers deal with the uses and abuses of the standard penetration test:

H. A. Mohr (1943), *Exploration of Soil Conditions and Sampling Operations*, Soil Mechanics Series No. 21, 3rd ed., Graduate School of Engineering, Harvard University, 63 pp.

H. J. Gibbs and W. G. Holtz (1957), "Research on Determining the Density of Sands by Spoon Penetration Testing," *Proc. 4 Int. Conf. Soil Mech., London*, 1, 35–39.

G. F. A. Fletcher (1965), "Standard Penetration Test: Its Uses and Abuses," *ASCE J. Soil Mech.*, 91, SM4, 67–75.

H. O. Ireland, O. Moretto, and M. Vargas (1970), "The Dynamic Penetration Test: a Standard that is not Standardized," *Géotechnique*, 20, 2, 185–192.

An example of the use of static cone tests is given in:

J. H. Schmertmann (1970), "Static Cone to Compute Static Settlement Over Sand," *ASCE J. Soil Mech.*, 96, SM3, 1011–1043.

An excellent review and analysis of all types of penetration testing is contained in G. Sanglerat (1972), *The Penetrometer and Soil Exploration*, Elsevier Publ. Co., Amsterdam, 464 pp.

Several useful papers deal with piston-sampling procedures for fine-grained soils and with adaptations designed to sample sands below the water table:

A. W. Bishop (1948), "A New Sampling Tool for Use in Cohesionless Sands Below Ground Water Level," *Géotechnique*, 1, 2, 125–131.

"Undisturbed Sand Sampling Below the Water Table" (1950), *Bull. 35*, Waterways Experiment Station, Vicksburg, Miss., 19 pp.

J. O. Osterberg (1952), "New Piston Type Soil Sampler," *Eng. News-Rec.*, 148, pp. 77–78.

T. Kallstenius (1963), "Studies on Clay Samples Taken with Standard Piston Sampler," *Proc. Swedish Geotechnical Inst.*, No. 21, Stockholm, 210 pp.

Published information on the techniques of rock coring is rather scarce; however, much pertinent information can be obtained in the catalogs of such suppliers as Sprague and Henwood, Inc., Scranton Pa.; Acker Drill Co., Inc., Scranton, Pa.; and

Longyear Co., Minneapolis, Minn. The following reference, directed primarily to mining engineering, is useful:

L. W. Le Roy and H. M. Crain (eds.) (1949). "Subsurface Geologic Methods, a Symposium," *Colorado School of Mines Quarterly*, 44, 3, Golden, 826 pp.

Numerous papers discuss the uses of the vane-shear apparatus for the exploration of soft soil deposits. A few of the more significant are the following:

L. Cadling and S. Odenstad (1950), "The Vane Borer," *Proc. Swedish Geotechnical Inst.*, No.2, Stockholm, 88 pp.

"Vane Shear Testing of Soils" (1957). *ASTM Spec. Tech. Publ. 193*, 70 pp.

G. Aas, (1965). "A Study of the Effect of Vane Shape and Rate of Strain on the Measured Values of *in-situ* Shear Strength of Clays," *Proc. 6 Int. Conf. Soil Mech.*, Montreal, *1*, 141–145.

An excellent discussion and a good bibliography on installations to measure various types of soil behavior in the field are to be found in:

W. L. Shannon, S. D. Wilson, and R. H. Meese (1962), "Field Problems: Field Measurements," *Foundation Engineering*, G. A. Leonards (ed.), McGraw-Hill, New York, pp. 1025–1080.

Applications of the geophysical methods of exploration are found in a number of publications of the Highway Research Board and American Society for Testing and Materials. A basic source of information is the text:

C. A. Heiland (1940), *Geophysical Exploration*, Prentice-Hall, Englewood Cliffs, N. J., 1013 pp.

Specific examples of the use of the procedures in civil engineering are given in:

"Symposium on Surface and Subsurface Reconnaissance" (1952): *ASTM Spec. Tech. Publ. 122*, 228 pp.

"Geophysical Methods and Statistical Soil Surveys in Highway Engineering, 6 Reports" (1965). *Hwy. Res. Rec.*, No. 81, 60 pp.

A comprehensive review and bibliography on applications of geologic and pedologic data are included in:

T. H. Thornburn (1969). "Geology and Pedology in Highway Soil Engineering," *Reviews in Engineering Geology*, II, D. J. Varnes and G. Kiersch (eds.), Geological Society of America, Boulder, Colo., pp. 17–57.

Several publications of the American Society for Testing and Materials contain information on sampling as well as field and laboratory testing of soil and rock. Especially useful are the following:

Annual Book of ASTM Standards, Part 11.

"Testing Techniques for Rock Mechanics" (1966). *ASTM Spec. Tech. Publ. 402*, 297 pp.

"Special Procedures for Testing Soil and Rock for Engineering Purposes" (1970). *ASTM Spec. Tech. Publ. 479*, 630 pp.

"Sampling of Soil and Rock" (1971). *ASTM Spec. Tech. Publ. 483*, 193 pp.

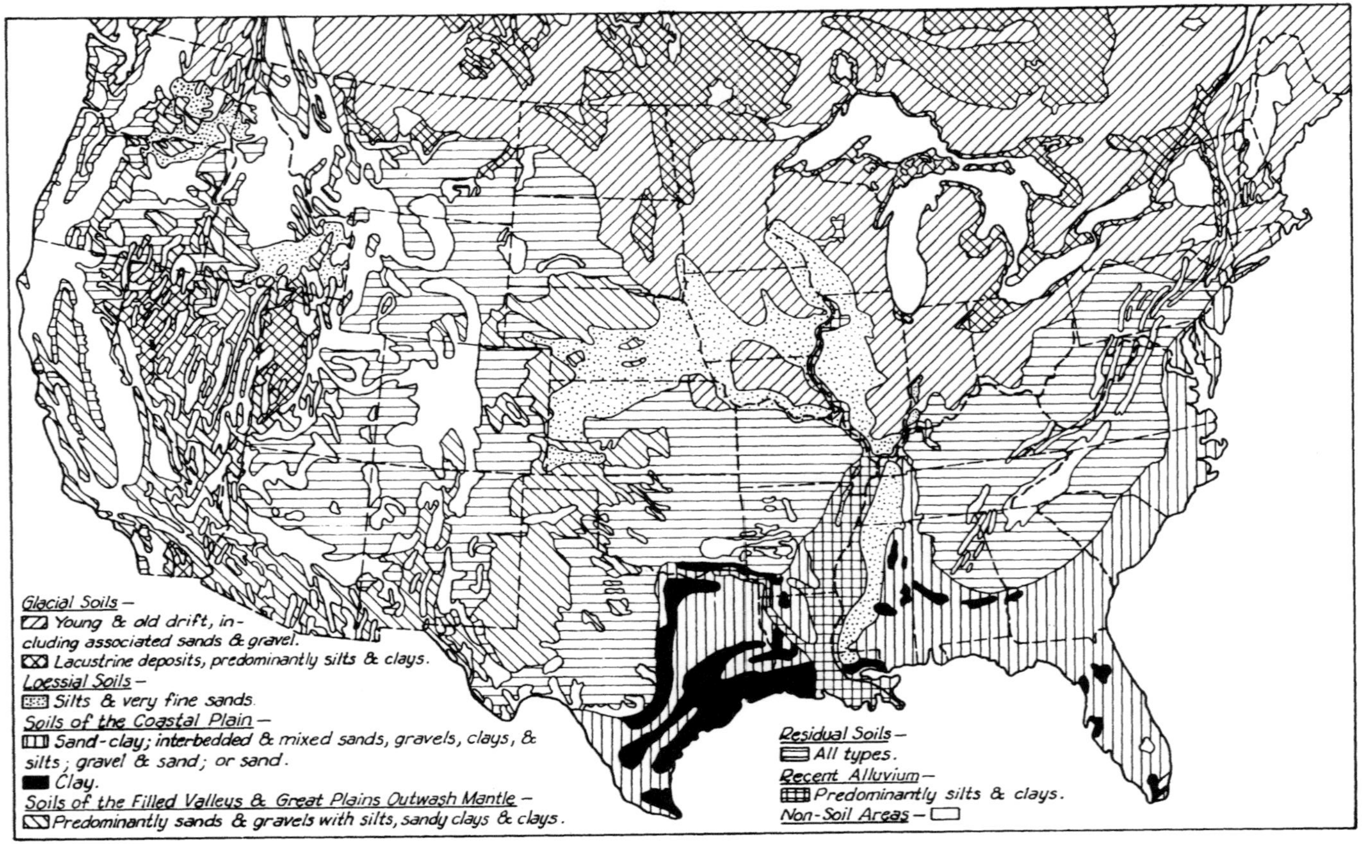

Soil deposits of the United States and Canada (after Belcher, et al., 1946; Flint, 1945).

PLATE 6.

CHAPTER 6

Character of Natural Deposits

6.1. Origin of Natural Deposits

Engineering Significance of Geologic Processes. The program of subsurface exploration for any foundation project must be adequate to disclose the essential character of the deposit and particularly its possible variations from point to point. Yet economy and the limitations of time dictate that no greater expenditure than necessary should be made to produce the desired results. This end cannot be achieved if the engineer does not have at least a rudimentary knowledge of the anatomy of various kinds of deposits. Such knowledge will assist him in interpreting information as it is obtained from the field and laboratory and in recognizing the stage at which further information would not be worth the added cost.

The difficulties of the problems to be solved by the foundation engineer increase with decreasing strength and increasing compressibility of the supporting materials and, for material of a given strength and compressibility, with increasing variability. Hence, sound rock, even if its structure may have been altered by folding, faulting, or metamorphism, generally constitutes a satisfactory foundation material. The undesirable features of rock foundations are usually associated with such defects as joints, solution channels, and zones altered by chemical or physical weathering.

Probably the most variable deposits are those associated with glaciation. In many parts of the world, the topography is the direct result of the action of glaciers that plowed up old soils, ground up rocks, and deposited the materials in random fashion partly on land and partly under water. Indirectly, the events of the glacial age influenced foundation conditions far beyond the maximum limits of glaciation itself. During the glacial epoch the climate was colder and the rainfall greater than now; consequently, there were large inland lakes, and the flow of rivers was abnormally great. At the peaks of glacial activity much of the earth's water was in the form of ice; hence, sea level was lowered as much as 300 ft. The fluctuations of sea level induced erosion of the beds of rivers near the sea shores, frequently modified the shore lines, and led to the formation of shore deposits, especially in bays and estuaries.

Because of the extraordinary significance of the events of the glacial epoch with respect to the work of the foundation engineer, deposits associated with glaciation are the first to be considered in the remaining articles of this chapter. They illustrate that the engineer dares not assume uniformity of the subsurface conditions. He must learn the character of the deposit at each site in order to forecast the most unfavorable conditions that may be encountered.

Geological Terminology. Knowledge of the geological origin of a soil deposit often furnishes insight into its physical characteristics. Therefore, the engineer should consult whatever geological data are available. In the following articles, some of the more common types of natural deposits are described, and representative examples are presented of the results of exploratory programs to determine their characteristics.

From the geological standpoint, soils can be divided into two major groups, transported and residual. *Transported* soils no longer cover the rock material from which they were derived. They may be further classified according to the mode of transportation and deposition as follows: *alluvial* soils, transported by running water; *lacustrine* soils, deposited in quiet lakes; *marine* soils, deposited in sea water; *aeolian* soils, transported by wind; *colluvial* soils, deposited primarily through the action of landslides and slopewash; and *glacial* soils. *Residual* soils have been developed from the parent rock over which they now lie. Deep deposits of residual soil are common in the southeastern United States, Hawaii, Puerto Rico, and generally in the humid tropics. They are rare in the northern half of the United States and in Canada, because the continental glaciers removed most of the products of weathering that had formed on the bedrock.

In a very general way, soils tend to be arranged in *profiles* or systems of layers. The most significant of these are profiles of weathering and profiles of deposition. In many instances, one of the former is superimposed on one of the latter and a rather complex system of soil layers may be found near the surface.

Processes of Weathering. The oldest rocks exposed on the face of the earth appear to be metamorphosed sediments. They have been extensively deformed by heat and pressure, and in places have recrystallized into granite and other typical igneous rocks. Yet, their sedimentary origin bespeaks the effectiveness of the agents of weathering throughout geological time, inasmuch as these ancient rocks are composed of the products of weathering of still older rocks.

Some of the agents of weathering are purely physical. One of the most important is the *differential expansion* and *contraction* that arises on account of heating and cooling masses of rock containing minerals of different thermal properties. Others are the *abrasive action* of wind and water, especially when these agents transport sediments that aid the grinding action; the *expansive force* of freezing water; and in many localities, especially in the past, the *plucking and grinding* action of glaciers. Important as these physical agents are, however, they are not capable of reducing the size of individual fragments below about 0.01 mm. More complete disintegration can be accomplished only by chemical processes.

The principal agents of chemical weathering are *oxidation, hydration, carbonation,* and *solution*. Almost all materials combine with oxygen, especially in humid climates, but iron compounds are particularly susceptible. The combination with oxygen is usually associated with an increase in volume, whereupon the oxidized portion is likely to split away from the unaltered material. Similarly, an increase in volume accompanies the formation of new minerals by hydration and aids in the disintegration of the original minerals. Falling rain dissolves small quantities of carbon dioxide from the atmosphere and thus becomes a weak solution of carbonic acid. The chemical combination of certain minerals with carbonic acid leads to the formation of softer and weaker minerals; more important, however, is that weak carbonic acid is a better solvent than pure water. It attacks not only materials containing calcium carbonate but also those containing silicates.

As a rule, physical weathering predominates in frigid or dry climates, whereas chemical weathering is more important in warm humid climates. Plant and animal life contribute to both processes. However, rocks may also be subjected to chemical decomposition when far below the surface, because of the circulation of magmatic waters charged with chemicals. Granites and gneisses thus may locally have been transformed into soft rocks of claylike character, and hard quartzites into cohesionless sand.

After the processes of weathering have eliminated the cohesive bonds between most of the constituents of the rock, chemical weathering leads to the formation of clay minerals. Nevertheless, weathering still continues. Ultimately, under humid tropical conditions the disintegration may be so complete that even the clay minerals are broken down.

Pedological Terminology. Pedology is the branch of the science of geology that deals with the outer 4 or 5 ft of the earth's crust. Within this zone the rock or disintegrated rock material has become altered by weathering to an extent and in a manner that depends on the climate, the presence of organisms, the topography, the character of the parent material, and the length of time the processes of weathering have been active. To the pedologist, only this altered zone is called *soil*. Hence, the pedologist's definition is considerably more restricted than the engineer's.

Wherever mature soils have developed, a succession of several distinct layers is found within a few feet of the surface. These layers are called *soil horizons* and the complete system of horizons is referred to as the *soil profile*. The uppermost horizon has usually lost by leaching a significant amount of fine-grained mineral matter although it may contain a rather large quantity of finely divided organic matter derived from decaying vegetation. It is often called the zone of depletion. The second horizon is the zone in which the fine-grained mineral matter is accumulated; in most profiles it is more plastic than the horizons either above or below. In many instances it can best be recognized by the aggregation of particles into a well-developed structural arrangement referred to as blocky or prismatic. Together the two upper horizons constitute the solum, the zone in which the soil-forming processes are most active. Below the solum is a third horizon composed of relatively unaltered parent material. The parent material may be partially decomposed and disintegrated bedrock, or it may be soil material or rock material that has been transported and deposited at a new location. A hypothetical profile of soil weathering is shown in Fig. 6.1. The major horizons are designated as A, B, and C. These horizons are usually subdivided on the basis of minor variations that occur in the transitional zones between horizons or on the basis of special physical or chemical characteristics. No one profile would exhibit all the horizons shown; from the engineering standpoint it is usually sufficient to deal in terms of only the major horizons. Where fresh materials are still being deposited, as on the flood plains of rivers, there is no opportunity for a typical pedological profile to form.

The pedological classification of a soil is determined principally on the basis of the geology of the parent material and certain inherent characteristics of the profile. The latter include the number, color, texture, structure, thickness, chemical composition, and relative arrangement of the horizons.

All soils that have similar profiles except for the texture of the surface horizon are grouped together as a *soil series* and given a series name. These names are usually taken from the locality where the soil was first mapped. Ordinarily the series name is further modified by terms indicating the texture of the surface soil as determined by a triangular classification chart (Fig. 1.14). The engineering characteristics of a soil profile are determined primarily by the texture, plasticity, thickness, and drainage characteristics of the horizons. Several different pedological series may have similar engineering properties.

Pedology has been utilized extensively by agronomists and to a considerable degree by highway and airport engineers. Because of correlations relating pedology to the behavior of shallow foundations, to the position of the water table, and to performance of septic tank disposal systems, it also enters

Horizons[a]		Descriptions	
O_1	Undecomposed organic debris.	L.[b] Loose leaves.	
O_2	Partly decomposed organic debris.	F.[b] Organic structure evident. H.[b] Amorphous humus.	
A_1	A dark-colored horizon with a relatively high content of organic matter mixed with mineral matter.	A. Zone of eluviation. Darker in color and/or lower in clay content than the underlying horizon.	Solum
A_2	A light-colored horizon where maximum eluviation and leaching have occurred.		
A_3	Transition to underlying horizon but more like A.		
B_1	Transition from overlying horizon but more like B.	B. Zone of illuviation. A zone of structural development underlying the A horizon.	
B_2	Horizon with greatest clay content and/or greatest structural development.		
B_3	Transition to underlying layer but more like B.		
C_1	Slightly altered material. Soluble salts may be precipitated in this horizon in arid climates.	C. Does not show soil structure but, in residual soils, will reflect the structural characteristics of the parent rock.	
C_2	Relatively unweathered material.		
R	Underlying consolidated bedrock; not necessarily material from which C horizons are derived.		

[a] Roman numerals (II, etc.) are prefixed to the appropriate horizon designations to indicate successive layers of contrasting material from the surface downward.
[b] Terminology used by foresters.

FIGURE 6.1. Hypothetical soil profile showing the major characteristics and relationships among horizons.

increasingly into land-use planning. Moreover, since the pedological description of a soil includes a description of the parent material, the foundation engineer can obtain from such a description useful general information about the nature of the soils in a locality with which he is not familiar.

In the United States, most pedological information is contained in county soil survey reports that have been published by the U. S. Department of Agriculture since the early part of the century. Only those reports published since about 1940, however, can be depended on for information of high accuracy. Such reports include detailed descriptions of the soil profile and geology of each soil series in the county, and a map to the scale of 1 in./mile or larger showing the boundaries of each series. Most reports published since 1960 contain an engineering section and a soil map to a scale of about 3 in./mile. They also usually include a tabulation of test data and engineering classifications, according to the AASHO and Unified systems, of samples taken from typical profiles of the more prominent soil series.

In the preparation of modern pedologic maps the soil surveyor uses aerial photographs; all recent maps are published on an airphoto base. The elements revealed on the photographs often permit more accurate delineation of the soil boundaries than can be achieved by field surveys alone. Moreover, under given climatic conditions, the character of the parent material is revealed by such elements of the photographic pattern as *land form, soil color, erosion, surface drainage, vegetative cover, slope, land use,* and *microrelief.* The principal elements of the pattern are often greatly influenced by soil texture and moisture conditions. Hence, each pattern must be correlated with the corresponding soil profile as determined by ground surveys, but after this has been done rapid studies of very large areas can be carried out by airphoto interpretation.

Vertical aerial photographs are available from such government agencies as the U. S. Department of Agriculture, U. S. Geological Survey, and U. S. Forest Service. The ability to interpret photographs for engineering purposes depends on one's background in geology, pedology, and plant ecology as well as his knowledge of the characteristics of natural soil deposits. In areas where neither geologic nor pedologic reports are available, the interpretation of aerial photos provides a valuable tool for the location of soil boundaries, for predicting the engineering characteristics of the soil areas, and for planning an intelligent program of soil exploration to determine the details of the stratigraphy.

6.2. Deposits Associated with Glaciation

Introduction. The great continental glaciers covered much of the surface of the land north of the 40th parallel. Elsewhere most mountains were buried or partly covered by Alpine glaciers, even in the tropics. The ice excavated, transported, and redeposited loose rocks and soils. All materials laid down by glaciers are known as *drift*. Those deposited directly out of the ice are called *till*. The meltwater flowing away from the ice also carried debris and deposited it in broad sheets known as *outwash*. The concentration of meltwater into torrential streams of variable rates of flow depending on the temperatures gave rise to *glaciofluvial deposits*. In some instances, the meltwater was dammed between high ground and the glacier itself. Thus lakes were formed in which sediments known as *glacial lake deposits* were laid down.

Moraines. Ice, carrying drift, continually flows toward the outer edges of every active glacier. Near the edges, melting takes place and the drift is concentrated at the bottom of the ice where part of it becomes fixed to the frozen ground. This part constitutes *ground moraine*. It consists primarily of till of erratic composition. A small amount of ground moraine is deposited when an ice sheet is growing, a larger amount when it is shrinking. Where there has been alternate growth and shrinkage of a glacier, there may be several distinct sheets of till. Figure 6.2 represents a cross section through a part of downtown Chicago, where there are at least three successive ground moraines.

If the edge of the ice remains stationary for at least a few years, the drift accumulates in a ridge at the face of the glacier. Such accumulations are known as *terminal* or *end moraines*. They consist largely of till, but they may be stratified in places once occupied by pools of meltwater, and they may contain deposits of outwash having irregular shapes. Commonly, there is considerable outwash of sand, gravel, or silt on the side of the moraine that slopes away from the ice. Figure 6.3 shows a cross section through such a moraine.

Figure 6.4 is a vertical airphoto of the front or southern edge of a portion of the Iroquois Moraine of Wisconsinan Age in eastern Illinois. The mottled pattern of light and dark tones evident in area 3 is typical

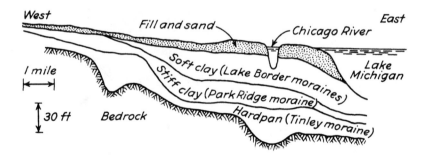

FIGURE 6.2. Simplified cross section through glacial deposits beneath Lake Street in Chicago showing successive ground moraines.

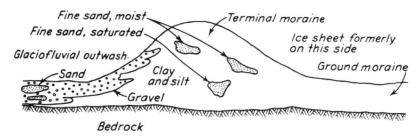

FIGURE 6.3. Idealized cross section through typical terminal moraine.

of much of the young morainal region in central United States. The variation of this pattern having a hazy or cloudlike appearance, shown in area 5, is a fine-textured till in contrast to the medium-textured till of area 3. Close inspection of the figure shows parallel drainage channels leading to the outwash, area 8; their parallelism indicates that they lie on the sloping face of the terminal moraine. The more uniform tones of area 8, broken mostly by the presence of old channel markings and alluvium (area 9) along the present stream, contrast sharply with the glacial till of the moraine.

In the central part of the United States are vast areas of ground moraine with fairly level surfaces. These are known as *till plains*. The different till sheets beneath such plains are sometimes separated by old topsoils and by buried channels. The photo pattern of young (Wisconsinan) till plains is similar to that of area 3 in Fig. 6.4 except that the mottling is usually coarsely rather than finely stippled. On the other hand, much of the glacial material exposed in the extreme southern part of the glaciated region is estimated to be more than 200,000 years older than the Wisconsinan till. It has,

FIGURE 6.4. Airphoto mosaic of a young till moraine area showing patterns of medium-textured till (3), fine-textured till (5), glacial outwash (8), and recent alluvium (9) (U. S. Dept. Agr. photos).

FIGURE 6.5. Airphoto mosaic of an old drift plain (U. S. Dept. Agr. photos).

Table 6.1 Index Properties and Engineering Classification of Seven Typical Young Glacial Tills

Soil Type and Location	w_L	I_P	γ_{max} (lb/cu ft)	w_{opt} (%)	AASHO Classification	Unified Classification
1. Gloucester stony, sandy loam, Rockingham Co., N. H.	—	NP[a]	121	9	A-1-b(0)	GM
2. Gogebic sandy loam, Bayfield Co., Wis.	—	NP	126	8	A-2-4(0)	SM
3. Miami silt loam, Fairfield Co., Minn.	30	13	121	12	A-6(5)	CL
4. Le Sueur silt loam, Scott Co., Minn.	54	29	96	24	A-7-6(24)	CH
5. Clarence silt loam, Livingston Co., Ill.	67	43	99	21	A-7-6(49)	CH
6. Ash River till, Vancouver Island, B. C., Canada	25	10	135	10	A-2-4(0)	SM
7. Lake St. Anne till, Quebec, Canada	—	NP	128	8	A-2-4(0)	SM

[a] NP = Nonplastic.

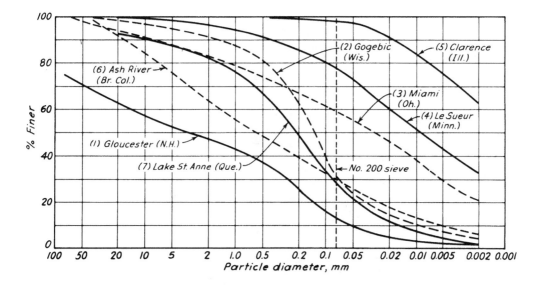

FIGURE 6.6. Particle-size distribution curves for typical glacial tills. (1) Stony till, New Hampshire. (2) Sandy till, Wisconsin. (3) Silty till, Ohio. (4) Silty clay till, Minnesota. (5) Clay till, Illinois. (6) Gravelly till, British Columbia. (7) Sandy till, Quebec.

therefore, been much more deeply weathered, to depths of 6 ft or more, whereas the soil profile in the young drift region seldom exceeds 3 to 4 ft. The photo pattern shown in Fig. 6.5 is typical of the older (Illinoian) drift regions in southern Illinois. Here the mottled drift pattern has been almost completely obliterated by the surface weathering and, in addition, a well-integrated drainage pattern has developed, even on nearly level areas.

Glacial tills vary widely in their texture, plasticity, and engineering properties. Data for seven typical young tills, selected to exemplify a range in texture from coarse to fine, are given in Table 6.1; their particle-size curves are shown in Fig. 6.6. The names of the first five are the pedologic soil types

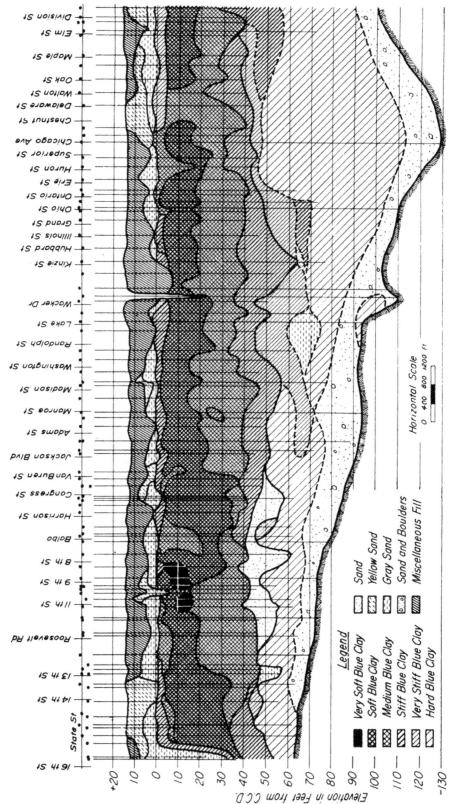

FIGURE 6.7. Cross section through glacial deposits along State Street in Chicago showing variations in consistency of till (After Peck and Reed, 1954).

developed on the parent materials to which the tabulated properties correspond. Tills 6 and 7 are representative of soils that have demonstrated excellent performance in earth dams in Canada (MacDonald et al., 1961). All seven particle-size curves are quite similar and indicate well-graded materials (Fig. 6.6), even though the clay-size fraction varies from 2 to more than 60 per cent. The strength of tills may vary both vertically and horizontally as shown in the generalized cross section (Fig. 6.2) and in the more detailed profile (Fig. 6.7), which pertains to the subsoils of downtown Chicago.

All types of moraines are likely to contain some waterlaid clays and silts deposited in temporary ponds. They may also contain uniform sands and gravels laid down in channels and tubes in the ice. These appear as irregularly shaped inclusions such as those shown in Fig. 6.8. Some moraines are composed of clay till having exceptionally uniform water content whereas others, formed under nearly similar conditions, may exhibit extremely erratic variations. The contrast between two such tills, one overlying the other, is shown in Fig. 6.9.

Morainic areas are likely to be poorly drained, especially if the till was deposited during one of the more recent glacial advances. In the poorly drained pockets, deep beds of peat are often encountered.

Glaciofluvial Deposits. Tremendous quantities of water, flowing from the faces of the continental glaciers during the warm seasons, carried coarse material for short distances and transported sands, silts, and clays for greater distances. Temporary channels quickly became choked with debris, and new ones were created elsewhere. The resulting deposits, especially if they were formed close to the glaciers, consist of lenses of coarse and fine materials, some loose and some dense. They are among the most variable natural coarse-grained sediments. Figure 6.10 is a photograph of a cut in a glaciofluvial deposit in Central Illinois. A record of the penetration of steel rails driven by a pile driver into glaciofluvial gravel at Port Alberni, B. C., is shown in Fig. 6.11. Wide variations in resistance to penetration and in grain size are apparent, even though the explorations were made within relatively small distances of each other. Similar variations in resistance are apparent in Fig. 6.12, which shows the results of standard penetration tests and dynamic cone penetration tests in a glacio-

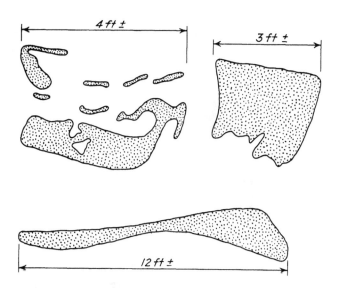

FIGURE 6.8. Sketches of coarse sand inclusions encountered during tunneling operations in ground moraine consisting of soft clay.

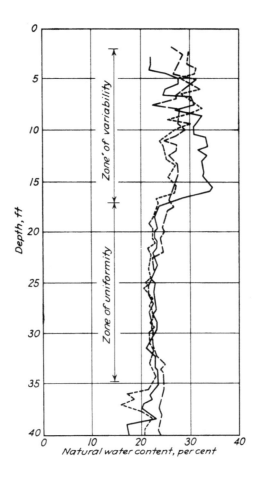

FIGURE 6.9. Variations in water content along vertical lines extending from variable moraine above to uniform moraine below surface of contact. The two moraines are indistinguishable to the eye.

FIGURE 6.10. Cut bank of a pit in a glaciofluvial deposit, central Illinois.

fluvial deposit in Denver. On the basis of these exploratory data, the site for the structure was shifted to avoid the loose materials indicated by the low blow counts on the left.

Peat and marsh deposits are even more common on outwash plains than on till plains. Usually such organic deposits can be readily located on geologic or pedologic maps or by the inspection of airphotos, as shown in Fig. 6.13, where they appear as dark blotches in the light-colored outwash. They are ordinarily avoided as sites for building construction, or even highways, because of their high compressibility. If they

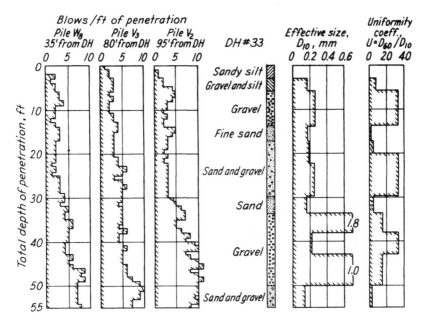

FIGURE 6.11. Penetration resistance of glaciofluvial gravel deposit at Port Alberni, B. C. (Courtesy of K. Terzaghi).

are utilized, the peat must be excavated or consolidated (Chap. 11).

Some glaciofluvial materials were deposited by streams flowing within the body of the ice or in crevasses near the ice front. Because of their confinement until the melting of the ice, the deposits may have the form of conical hills, called *kames*, or sinuous ridges, known as *eskers*. Such deposits, although not usually extensive, can be readily recognized on airphotos; they provide ideal sources of coarse granular material. In some parts of Canada they constitute the only elevated well-drained sites for construction. Their physical properties are variable and are similar to those of outwash deposits.

Glacial Lake and Marine Deposits. In contrast to the glaciofluvial deposits, those laid down in the relatively quiet waters of glacial lakes show a high degree of uniformity. Many, however, are laminated or *varved*. As meltwater flowed into the basins the coarser fraction was dropped near shore whereas the finer sediments were carried into the open water. During warm periods both silt and clay settled to the bottom. When melting and inflow ceased during cold periods, the finer clay fraction still in suspension continued to settle. As a consequence, a banded deposit was formed (Fig. 6.14a). The silty and clayey layers of a varved sample from Marathon, Canada were tested separately and provided the following substantially different results:

Silty layers:
$w = 24\%; \quad w_L = 28\%; \quad I_P = 24\%$

Clayey layers:
$w = 35\%; \quad w_L = 42\%; \quad I_P = 35\%$

Where the bottom was shallow enough to be influenced by currents, the details of laminations sometimes became very intricate as illustrated by a sample from glacial Lake Agassiz (Fig. 6.14b).

Glacial lake deposits are common around the Great Lakes and smaller inland lakes in northeastern United States and southern Canada. They are also widespread in the Pacific Northwest and the Canadian Rockies. If they were never exposed to desiccation they are likely to be soft, compressible, and sometimes quite sensitive. Since the lacustrine deposits are often associated with morainal or outwash deposits

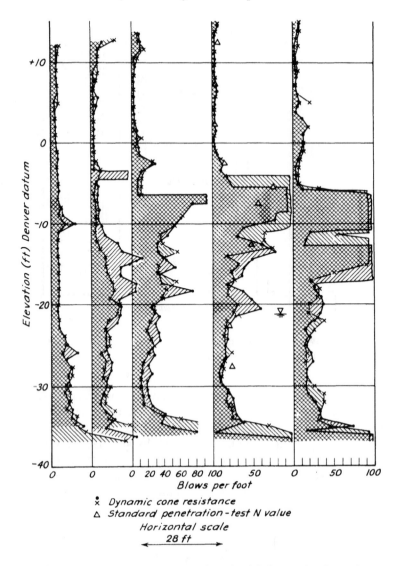

FIGURE 6.12. Penetration resistance of glaciofluvial deposit by dynamic cone test and standard penetration test, Denver, Colo.

of fairly high bearing capacity, their low strength has sometimes been overlooked with disastrous results.

The particle size and plasticity characteristics of glaciolacustrine samples depend on the proximity of the shore line at the time of deposition as well as on the source of the sediments. Data on three representative near-surface parent materials of glaciolacustrine origin, identified by pedologic names, are given in Table 6.2 and Fig. 6.15. The Paulding clay is a deep water deposit derived from drift consisting predominantly of shale. The Salmon and Trout River soils represent, respectively, shore line and beach ridge deposits formed where the drift was derived from more resistant bedrock.

Where the glacial meltwaters flowed into marine embayments, the saline waters tended to flocculate the silts and clays so that they settled simultaneously and varves were not formed. Many of these deposits were uplifted with respect to sea level because of the isostatic rise associated with the removal of the weight of glacial ice. Consequently the salt water originally in the pores

FIGURE 6.13. Airphoto of pitted glacial outwash and peat, northern Indiana (U. S. Dept. of Agr. photo).

FIGURE 6.14. (a) Typical varved clay (courtesy of M. J. Hvorslev). (b) Laminated clay from bed of glacial Lake Agassiz.

of the soil was gradually replaced by fresh water originating in rainfall. The physicochemical changes associated with the leaching resulted in the development of unusually high sensitivities. Such quick clays (Art. 1.7) occasionally liquefy and flow on very gentle slopes. A slide of this type took place near St. Thuribe, Quebec in 1898. The plasticity characteristics given in Table 6.2 and Fig. 6.15 do not suggest unusual characteristics; however, the natural water content exceeds the liquid limit and the sensitivity exceeds

Table 6.2 Index Properties and Engineering Classification of Glacial Lake and Marine Deposits

Soil Type and Location	w_L	I_P	γ_{max} (lb/cu ft)	w_{opt} (%)	AASHO Classification	Unified Classification
1. Paulding Clay, Paulding Co., Ohio	63	35	101	23	A–7–6(41)	CH
2. Salmon very fine sandy loam, Franklin Co., N. Y.	—	NP[a]	114	12	A–4(7)	ML
3. Trout River gravelly sandy loam, Franklin Co., N. Y.	—	NP	121	13	A–1–b(0)	SP
4. Mexico City clay, depth 45 ft, natural $w = 300\%$	360	235	—	—	A–7–5(308)	MH
5. St. Thuribe clay, depth 15 ft, natural $w = 46\%$	33	12	—	—	A–6(7)	CL

[a] NP = Nonplastic.

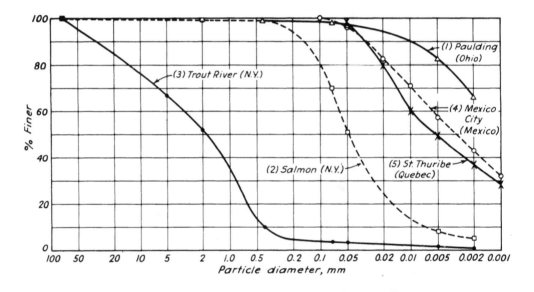

FIGURE 6.15. Particle-size distribution curves for lakebed and shoreline sediments. (1) Lacustrine clay, Ohio. (2) Lacustrine silt and sand, New York. (3) Beach gravel, New York. (4) Lacustrine clay, Mexico City. (5) Quick clay, Quebec.

150. Figure 6.16 is a stereogram of a flow slide area in Leda clay along the Maskinongé River, Quebec. Such flow slide scars are common along the St. Lawrence River and its tributaries and also along some of the major rivers and fjords in Scandinavia.

Because of the climatic conditions that prevailed during the glacial epoch, large lakes also formed in regions not directly affected by glacial ice. The inland basin in which Salt Lake, Utah, is now located was once occupied by Lake Bonneville, a much larger body of water. The valley of the City of Mexico is a similar basin in which

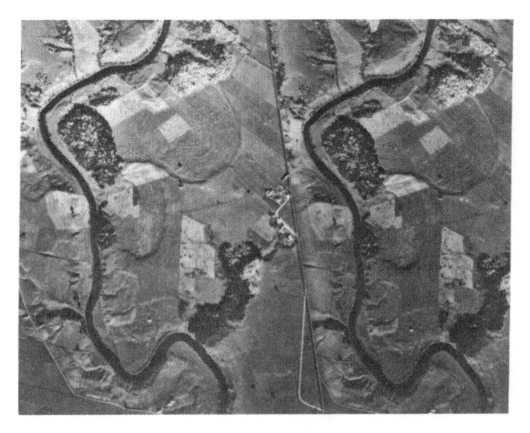

FIGURE 6.16. Airphoto stereogram of flow slide area, Maskinongé River, Quebec (prepared by University of Illinois Committee on Aerial Photography from photos by Canadian Dept. of Energy, Mines and Resources).

lacustrine sediments were deposited, but the principal constituents were of volcanic origin. Data on a sample of Mexico City clay are included in Table 6.2 and Fig. 6.15. A representative boring log from near the center of the city is shown in Fig. 6.17. The water content is remarkably high; moreover, it is close to and sometimes above the liquid limit. Although this clay exhibits appreciable strength in the undisturbed state, it is extremely compressible (Plate 3) and its sensitivity may reach values of 20 or 30.

Buried Valleys. For various reasons, the levels of the glacial lakes fluctuated widely from time to time. While the water level was low, tributary streams eroded valleys in the surrounding lands. As the lakes slowly rose, the mouths of the streams flooded and became filled with sediments, often mixed with the remains of plant life. The entire area may then have been covered by lake sediments or even by till in later glacial advances, and the channels with their soft filling may have been completely buried. The present streams do not necessarily flow in the old valleys, and the subsurface relationships are often complex. A cross section through a buried valley in Jackson County, Ill. is shown in Fig. 6.18a. The old fill is now covered by a high bluff and the present river valley is at a considerable distance from its predecessor. The lower half of the old valley is filled with materials having the characteristics of a lakebed deposit; this condition indicates that the lower end of the valley was at least partially blocked by the rapid filling of the main valley (the Mississippi) into which the old stream

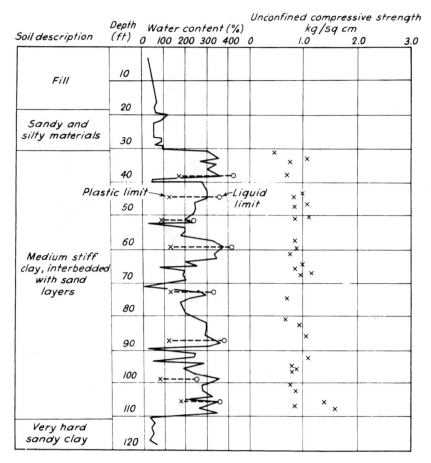

FIGURE 6.17. Variations in water content, Atterberg limits, and unconfined compressive strength of samples from a boring in central Mexico City (after Zeevaert, 1949).

drained. The blockage was probably caused by the rapid deposition of glacial outwash carried by the Mississippi as the ice retreated.

The valley shown in Fig. 6.18b was originally eroded in the bedrock, was enlarged by glacial ice and covered by till, and was then reexcavated in the till, filled with sediments, and covered by loess. The existing stream is very small and is eroding the loess.

Such buried channels are very common near the Great Lakes. They are also found on the seaboard because the level of the ocean was probably as much as 300 ft lower than at present during several periods in the glacial epoch. A rational interpretation of test borings in the vicinity of buried valleys requires at least a rudimentary knowledge of the geological history of the region.

6.3. Windblown Deposits

Introduction. Closely associated with glacial deposits, especially in the vicinity of major glacial drainageways and outwash areas, are deposits of sand and silt sorted by wind. The sweep of the wind across large sand-covered areas, whether outwash plains, beaches, flood plains of broad rivers, or even desert plains, moves the sand and silt-sized particles but leaves the gravel behind. The sand grains are rolled over each other or bounced short distances into the air and piled up to form dunes, whereas the silt-sized grains are blown away.

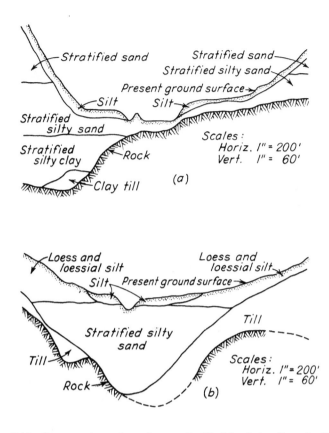

FIGURE 6.18. Cross sections through two silt-filled buried valleys in Illinois. In (a) the entire valley was covered by the sand in which the present stream channel is located; in (b) the valley was covered by till that was subsequently eroded and replaced by alluvium, and then covered by loess.

Dune Sand. Dune sands are among the most homogeneous of natural formations. The process of selection by wind sorts the sand into assemblages of very uniform grain size. Generally, the sand becomes finer with increasing distance from the source. If the source was a glaciofluvial valley deposit, most of the sand has remained in the valley and the adjacent upland deposits are composed predominantly of silt-sized particles (loess). Particle-size curves for several windblown soils are shown in Fig. 6.20. In the transition zone of very fine sand, between dune sand and loess, extremely loose deposits have been encountered. The results of standard penetration tests at one such site in Denver are shown in Fig. 6.19. The corresponding grain-size distribution is shown as curve 8 in Fig. 6.20.

Loess. The particle size of the windblown silts that form a loess deposit decreases in a general way with increasing distance from the source. Nevertheless, loess deposits are usually quite uniform. Their cohesion is due in part to the precipitation of chemicals or colloids leached by rainwater from the upper weathered zone and in part to the presence of clay coatings on the silt particles. Hence, the cohesion may vary considerably from place to place. The deposits tend to be thicker and less weathered and contain less clay closer to the source than at a greater distance. Furthermore, the deposits in the western United States where the rainfall is low usually contain more sand and less clay than those in the more humid regions of the east-central states. Table 6.3 gives the index properties and engineering classifications of

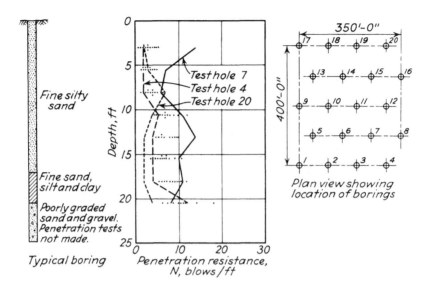

FIGURE 6.19. Results of standard penetration tests in 20 borings on one building site above extremely loose, fine, partly windblown sand in Denver, Colorado.

Table 6.3 Index Properties and Engineering Classification of Representative Windblown Deposits

Soil Type and Location	w_L	I_P	γ_{max} (lb/cu ft)	w_{opt} (%)	AASHO Classification	Unified Classification
1. Plainfield fine sand, Pulaski Co., Ind.	—	NP[a]	107	10	A–3(0)	SP
2. Valentine fine sand, Hall Co., Neb.	—	NP	—	—	A–3(0)	SP–SM
3. Colby loam, Dundy Co., Neb.	24	3	—	—	A–4(1)	ML
4. Natchez silt loam, Claiborne Co., Miss.	27	4	108	16	A–4(3)	ML
5. Palouse silt loam, Walla Walla Co., Wash.	30	6	105	18	A–4(6)	ML
6. Muscatine silt loam, Livingston Co., Ill.	30	10	112	16	A–4(9)	CL
7. Marshall silty clay loam, Washington Co., Neb.	42	18	107	19	A–7–6(20)	CL

[a] NP = Nonplastic.

seven representative near-surface loessial and sandy windblown deposits identified by the pedologic type developed on them. The corresponding grain-size curves are shown in Fig. 6.20.

Unless the deposit has been deeply weathered, loess has a high porosity (50 to 60%) and a low natural unit weight. It also tends to have natural planes of vertical cleavage and a characteristic root-hole structure. The natural structure of loess may be altered appreciably by sampling and it is difficult by means of borings to obtain samples in which no disturbance has oc-

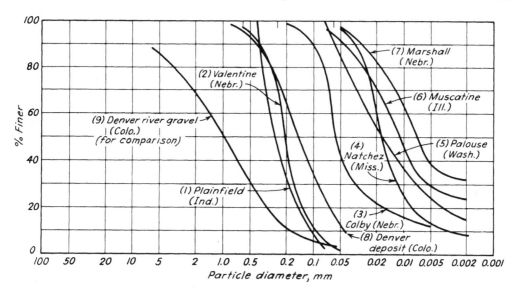

FIGURE 6.20. Particle-size distribution curves for wind-deposited sediments. Dune sand in northern Indiana (1), central Nebraska (2). Sandy loess in western Nebraska (3). Silty loess in western Mississippi (4), southeastern Washington (5). Clayey loess in central Illinois (6), eastern Nebraska (7). Dune sand to loess transition in eastern Colorado (8). River gravel in eastern Colorado (9), for comparison.

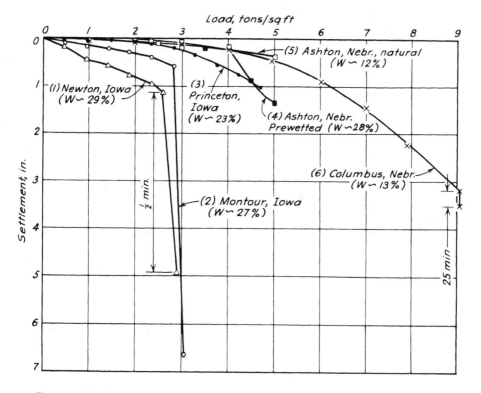

FIGURE 6.21. Results of standard load tests on loess deposits in Nebraska and Iowa.

curred. The structure may partially or completely collapse upon soaking. Hence, drill holes for sampling should be made without water. The load-settlement curves in Fig. 6.21 show that the bond between the particles may break down abruptly when a critical load is applied. They also show the effect of wetting.

Aerial photographs of areas of deep loess tend to have light overall color tones, altered only by vegetation, and to be dominated by a drainage system in which the gullies tend to have U-shaped cross sections, to be nearly parallel to each other, and to enter the tributaries at nearly right angles. These characteristics are illustrated in Fig. 6.22. The ability of loess to be stable on nearly vertical slopes is a well known consequence of its natural cleavage.

6.4. River and Continental Deposits

Stream-Channel Deposits. In its lower reaches, a stream overloaded with suspended matter deposits the material and raises its bed. The water then spills over into an adjacent depression and establishes a new channel. Such a stream is said to be *braided* because it consists of a series of rivulets continually joining each other and then separating. The position of the rivulets constantly shifts. The fill deposited in each of the channels has a relative density and a grain size dependent on the velocity of flow and on the character of the sediments being furnished to the river. Hence, the cross section of such deposits consists of an aggregate of lenticular elements in each of which conditions are fairly uniform but each of which differs from its neighbor in unpredictable fashion.

FIGURE 6.22. Airphoto of Missouri valley loess, central Nebraska (U. S. Dept. of Agr. photo).

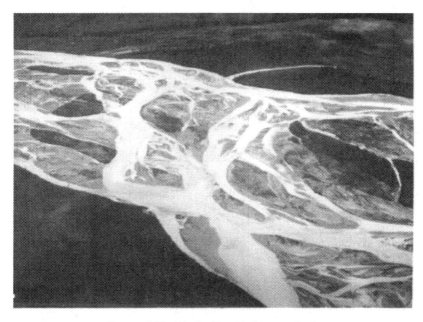

FIGURE 6.23. Typical braided stream bed.

Figure 6.23 is a photograph of a typical braided stream.

The upper reaches of streams that rise in mountainous regions often rest directly on the bedrock. The rock surface is likely to be covered by large boulders and somewhat smaller rock fragments.

Changes in elevation of different parts of the earth's surface, together with changes in climatic conditions, alter the character of the material deposited by a stream at any given point. Consequently, in addition to the minor variations described above, the deposits below the beds of many streams exhibit major variations as well. Variations in the penetration resistance of silts and sands in the valley of the Cedar River in Iowa, once an outlet for glacial meltwater, are shown in Fig. 6.24; they indicate several distinct strata.

Valley Deposits. As a stream matures and subsequently enters old age it meanders back and forth across its valley and creates an extremely complex assemblage of oxbow lakes, abandoned channels, filled meanders, natural levees, and backswamp deposits. Whenever a flood occurs, the sediment-laden water spreads over the adjacent lowlands. Sand and coarse silt are deposited adjacent to the channel and build up ridges known as natural levees. The finer silts and clays are carried farther away onto the flood plain where they settle at different rates and become segregated. Hence, the deposits resulting from successive floods have a laminated character. Usually, they consist predominantly of very fine sand and silt with thin partings of fine sand or clay. Although flood-plain deposits may be quite uniform in horizontal directions, they change vertically in an erratic fashion, as shown in Fig. 6.25. Their permeability in horizontal directions is large compared to that in the vertical direction.

The finest materials are carried into the abandoned channels, meanders, and backswamp areas adjacent to the valley walls; there they are deposited in a shallow lacustrine environment and are mixed with organic matter from the swamp vegetation. Such deposits generally have high natural water contents and are much more plastic and compressible than the deposits on the flood plains and natural levees. Thus, differential settlements are apt to be excessive.

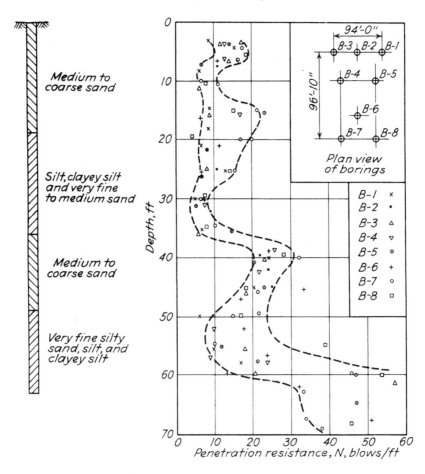

FIGURE 6.24. Variations in penetration resistance of sands and silts in the valley of the Cedar River, Cedar Rapids, Iowa.

The complexity of such deposits in the lower Mississippi Valley is illustrated in the sketches of the Lake Chicot area (Fig. 6.26). Vertical aerial photos are extremely useful in delineating the various types of valley deposits.

Delta Deposits. When the water of a sediment-laden stream enters a lake or sea, it experiences a sudden check in velocity and drops its sediments. The structure of a delta formed under these conditions is simple but there are likely to be many local variations because the burden of the stream depends on the climate and because shore currents continually shift. The variations in permeability of a delta deposit formed during the glacial epoch near Chicopee, Mass. are shown in Fig. 6.27. In contrast to the granular nature of this deposit, many of the world's coastal deltas are composed primarily of compressible fine-grained sediments mixed with organic matter.

Continental Deposits. Large areas of the western United States are covered by deposits up to several hundred feet thick, accumulated largely through the action of streams and rivers. The Great Plains were built by rivers flowing eastward from the Rocky Mountains. The structure of the deposits, often erratic, reflects the braided character of the streams.

Similar deposits derived from the Rockies and the Sierra-Cascades fill the Intermontane Basins, and materials from the posits but frequently intermixed with the debris from mud flows actuated by the heavy precipitation. Extremely coarse materials are likely to be found near the mountain slopes, but farther away the fineness increases. In the temporary lakes near the centers of the basins, fine-grained silts and clays were deposited.

In arid regions the granular materials are often partially cemented with calcium carbonate. Salts of sodium and potassium are also prevalent; they may lead to deterioration of steel and concrete unless suitable precautions are taken.

6.5. Organic and Shore Deposits

Accumulations of highly organic material may be found in association with almost any type of geologic deposit when the environmental conditions are appropriate. They are most commonly formed in depressional areas where the water table is permanently at or above the original ground surface and where climatic conditions are favorable to the growth of aquatic vegetation. Thus, they are frequently found in glaciated regions, coastal areas, and river valleys in the temperate to polar regions. In tropical regions they occur principally in coastal areas.

Organic accumulations, such as *peat*, *muck*, *muskeg*, and *marsh deposits* may vary in depth from a few inches to several tens of feet. Their natural water contents are usually well over 100 per cent. They are all highly compressible. Hence, where possible, they are avoided as foundation materials.

The action of waves and shore currents in lakes and oceans builds up beach and shore deposits primarily of sand and gravel. These deposits may be of relatively uniform grain size and of moderate to high relative density. Results of standard penetration tests in a sandy shore deposit at the southern end of Lake Michigan are shown in Fig. 6.28.

On the other hand, if the shoreline has fluctuated on account of changes in the water level, the deposits of sand may alternate in an erratic manner with organic silts and peats. Such formations are known as

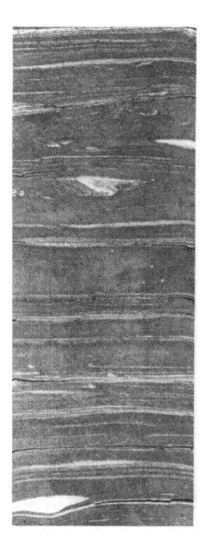

FIGURE 6.25. Two-inch tube sample from floodplain deposit of Mississippi River near Baton Rouge, La.

Sierra-Cascades and the Pacific Coast Ranges occupy the Central Valley of California. Often the streams entering these semiarid valleys carried water only during the occasional periods of heavy precipitation. As they debouched into the basins their waters spread and dropped their burden rapidly in the form of alluvial fans similar in structure to braided-stream de-

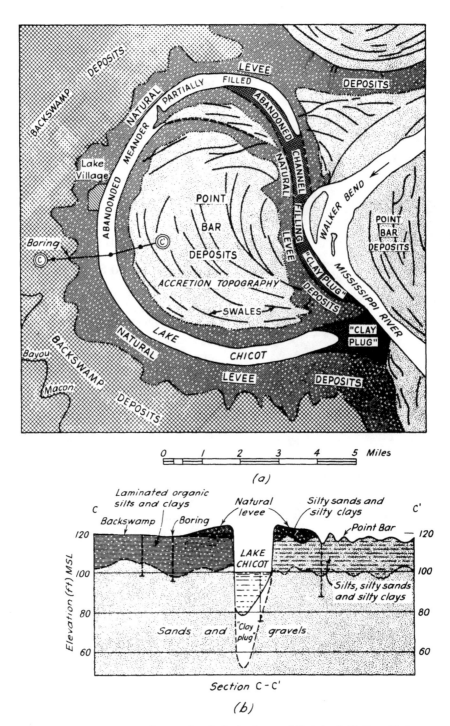

FIGURE 6.26. Portion of meander belt in lower Mississippi Valley (Lake Chicot) showing (a) distribution of deposits at the surface; (b) section through portion of the deposit (after Fisk, 1947).

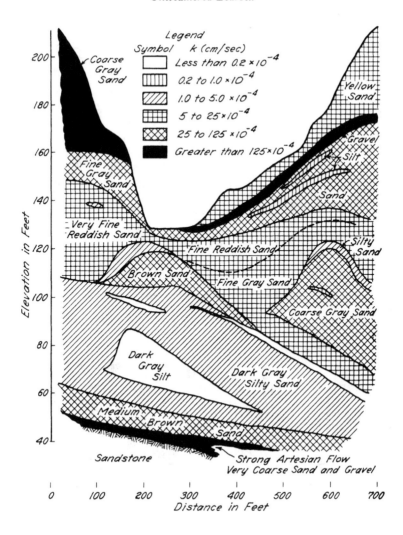

FIGURE 6.27. Permeability profile of glacial delta deposit near Chicopee, Mass. (after Terzaghi, 1929).

composite shore deposits. The one illustrated by Fig. 6.29 is located near the mouth of the Milwaukee River in Wisconsin.

6.6. Unweathered Bedrock

Rock Defects. Unweathered bedrock rarely presents difficulties in the design or construction of foundations except when it must be removed. In some instances, however, discontinuities or defects require consideration. For example, some limestones are composed primarily of a mixture of fossil skeletons, such as corals, crinoids, algae, and shells. Frequently, the fossil fragments are deposited in a loose arrangement, are soft, and are poorly cemented. Consequently, even the unweathered deposits are highly permeable and easily crushed when subjected to heavy loads.

In general, rock defects of concern to the foundation engineer fall into three main categories: bedding planes and joints, faults and shear zones, and solution features.

Bedding Planes and Joints. Bedding planes occur in sedimentary rocks; joints are found in rocks of all types. In their original attitude bedding planes were nearly horizontal but folding and other tectonic disturbances may have changed the attitude radically (Fig. 6.30). Joints occur in several sets,

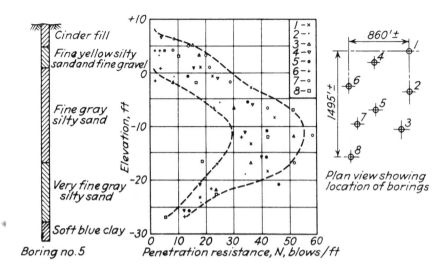

FIGURE 6.28. Results of standard penetration tests in sandy shore deposits at southern end of Lake Michigan.

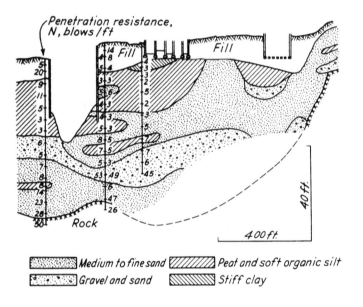

FIGURE 6.29. Cross section through composite shore deposit near mouth of Milwaukee River, Wis.

within each of which they are more or less parallel. They may be open, tight, or filled with secondary mineral deposits. In sedimentary rocks, one set of joints is likely to coincide with the bedding planes.

The compressibility and strength of a mass of rock, in contrast to intact specimens of sound rock, depend primarily on the spacing and number of joints. Hence, appropriate index properties for rock masses must take the jointing into account. The most widely used indicator of the general behavior of rock masses is the rock quality designation or RQD (Art. 5.3). A judgment concerning the quality of the rock may be based on Table 5.2. With experience or by means of suitable correlations, the RQD can be estimated for cores of other than

FIGURE 6.30. Folded sedimentary rock, New York Thruway.

NX size, or for exposed surfaces of fresh rock.

A less informative but widely used index property is the *recovery ratio* (Art. 5.3). Recoveries of 90 per cent or more are considered high and usually indicative of good rock. Nevertheless, recovery may be high in closely jointed rock that, under heavy loadings, might be objectionably compressible.

Since rocks with low RQD's are usually much stronger and less compressible than most soils, only extraordinary structures such as the towers of the World Trade Center (Plate 22) require consideration of the compressibility of the rock on which they are founded. On the other hand, since joints represent surfaces of weakness along which the cohesion of the rock has been destroyed, the presence of such features dipping steeply toward a slope or a proposed excavation requires attention. Weathering along joint surfaces may create clay-filled joints along which the shearing strength may be very small.

Faults and Shear Zones. Most rocks have experienced tectonic disturbances leading to folding and locally to more severe dislocations at which one element of the rock mass has been displaced appreciably with respect to another. Within the zone of dislocation the rock may have been pulverized into gouge, may consist of a matrix of gouge containing pieces of intact rock, or may be finely shattered or sheared. Such zones are likely to be or to have been more permeable than the surrounding rock; circulating groundwaters may have led to further breakdown of the broken rock by facilitating chemical weathering, or may have recemented the fragments into intact strong rock. Few faults are active in a geological sense, but the presence of faults and shear zones deserves study because of the high probability that they may contain weak and compressible soil-like material contrasting with the neighboring rock. Fault gouge and heavily sheared rock frequently contain such alteration products as montmorillonite, chlorite, and serpentine.

Even gentle folding leads to slip or shear along bedding planes, comparable in origin to the horizontal shear along the neutral axis of a beam. The slip may create thin seams of material comparable to fault gouge, with numerous slickensided surfaces. The materials are often altered to highly plastic clays. Such seams, known as *mylonites*, may extend more or less continuously over considerable areas. Since they represent surfaces along which a slip has already occurred, their shearing resistance may be exceptionally low.

Solution Features. To the foundation engineer, by far the most prevalent serious defects are those associated with former solution, particularly of limestones and dolomites. Groundwater circulating in the joint system may have removed the soluble constituents of the adjacent rock, enlarged the joints, and left behind an insoluble residue of soft silty clay. At intersections, particularly of bedding plane joints with those of other systems, solution may have extended laterally to form extensive cavities or caves. Some of these features may have remained to the present time, whereas others may have collapsed to form sinkholes filled with a mixture of blocks of rock, clay residue, and surficial materials. Such topography is said to be *karstic*. Figure 6.31 illustrates the solution pattern at the site for a steam power plant on dolomite. The plan (Fig. 6.31a) shows the degree of support provided by the rock to a heavily reinforced raft foundation designed to bridge the weak zones. Details of the rock-boring information are shown in a cross section (Fig. 6.31b).

The extent and degree of continuity of the solution features are commonly investigated by means of core borings supplemented by more expedient procedures such as the use of pneumatic rock drills. Al-

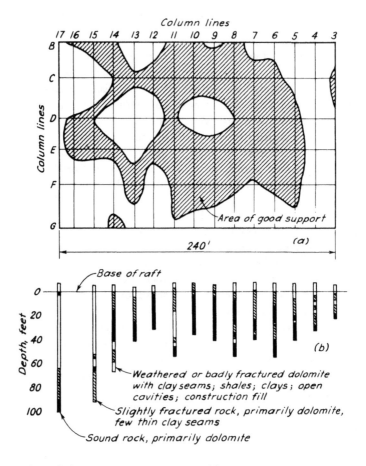

FIGURE 6.31. Solution patterns in dolomite. (*a*) Plan view. (*b*) Profile shown in rock borings.

though no samples are recovered by pneumatic drilling, the rate of descent of the drill rods reflects whether they are penetrating rock or a cavity; usually it is not possible to distinguish among clay-filled, water-filled, or empty cavities. Caliper logs in drill holes and the results of other geophysical investigations may be useful and justified on complex projects.

Since limestones and dolomites are only feebly soluble, active solution at the present time rarely causes appreciable enlargement of cavities. From time to time, however, the roofs of existing cavities become overstressed and collapse. Hence, caution is always justified where this possibility exists. On the other hand, solution phenomena in rock salt and, to a lesser degree, in gypsum, may develop rapidly enough to endanger the foundations of structures during their lifetime. Commercial extraction of salt by solution has led to severe subsidence and even to the formation of sinkholes.

6.7. Weathered Rock and Residual Soil

Significant Features. Mechanical weathering associated with freezing and thawing produces end-products consisting of angular blocks, cobbles, gravel, sand, and even clay-sized rock flour. The mineral constituents of all these products are exactly like those of the original rock. The principal decision required of the engineer is the depth at which to establish the foundation. Depth to intact bedrock is difficult to establish because with increasing depth the size of the blocks of residual material approaches that of the blocks of bedrock established by the spacing of the joints. For support of most ordinary structures the lower part of the residual materials may be entirely adequate.

Chemical weathering, on the other hand, results in the decomposition of the rock and the formation of new minerals. Of primary importance are the depth and nature of the transition from the weathered material to sound bedrock and the engineering properties of the residual materials themselves. These considerations are discussed in detail in the remainder of this article.

Transition from Residual Materials to Sound Bedrock. The nature of residual deposits and the character of the transition between them and sound bedrock depend to a large degree on the solubility of the parent rock.

Many limestones, for example, consist of almost pure $CaCO_3$, which is dissolved and removed by the groundwater. The residual soil consists of insoluble impurities that rest directly on undissolved limestone. Thus the contact is sharp, from residual soil to unweathered rock, without transition. The contact is likely, nevertheless, to be extremely irregular because solution of the limestone occurs preferentially along joints. In some instances the bedrock surface is so irregular that it is characterized as *pinnacled*. The face of an excavation in Georgia is shown in Fig. 6.32. It shows not only the irregular surface but the presence of occasional unweathered boulders in the residuum; such boulders are the remnants of larger blocks once surrounded by joints from which solution progressed. The irregular surface of the bedrock in limestone and dolomite terrains is often further complicated by the collapse features described in Art. 6.6.

In contrast, the constituents of rocks of feeble solubility are likely to remain in place during weathering. In the tropics, where chemical weathering is developed most extensively, the profile of weathering (Art. 6.1) often consists of an upper clayey zone a few feet thick underlain by a sandy or silty zone that in turn passes through a very irregular transition into weathered rock and finally into sound rock. The weathered rock often retains the appearance of the bedrock with such distinctive features as relict joints, dikes, or schistosity. A soil having these relict features is known as a *saprolite*. The thickness of each member of the profile varies greatly from site to site because of the complexity of the interrelationships among the controlling soil-forming factors. Moreover, at a given site there may be great differences in the depth and thickness of the weathered rock within lateral distances of only a few feet. The differences arise because of differences in

FIGURE 6.32. Pinnacled rock surface in excavation in Georgia.

lithology, such as the presence of a dike or a contact, or because of differences in the degree of jointing or extent of shearing. The latter features increase the permeability and the depth to which weathering proceeds.

Engineering Characteristics of Residual Materials. The residue from solution and weathering of limestones and dolomites is characteristically red and is commonly known as *terra rossa*. Just above the sharp bedrock contact, and particularly between rock pinnacles, it may have the properties of a normally loaded plastic clay with its water content close to the liquid limit. At higher levels, because of desiccation, the clay may be stiffer. These conditions are illustrated in Fig. 6.33. Such weak compressible clays beneath a heavy structure may cause large irregular settlements and should always be suspected where terra rossa is encountered.

Residual soils developed on other than soluble rocks usually display a saprolitic structure. Figure 6.34 shows a road cut through a swarm of dikes in volcanic rocks on the island of Oahu. The structure of the bedrock is apparent, but the cut was made with earth-moving equipment. In such soils, joints and other weaknesses in the parent rock remain as weaknesses and, if unfavorably located with respect to the boundaries of an excavation, may lead to slides (Fig. 6.35). Under these conditions, the values of c and ϕ determined by shear tests on representative undisturbed samples of intact material are of no significance. The governing factor is the presence or absence of a relict joint at a critical position.

Many residual deposits have the appearance of bouldery alluvium (Fig. 6.36). The boulders, however, are the central remnants of blocks of the bedrock originally surrounded by joints from which weathering attacked the rock. The process is known as *spheroidal weathering*. Unlike the hard, unweathered residual boulders in terra rossa, spheroidally weathered boulders may be almost as decomposed as the surrounding residuum and can often be cut with a spade. Some may be more resistant, however, and may constitute obstacles to pile driving.

The index properties of residual soils de-

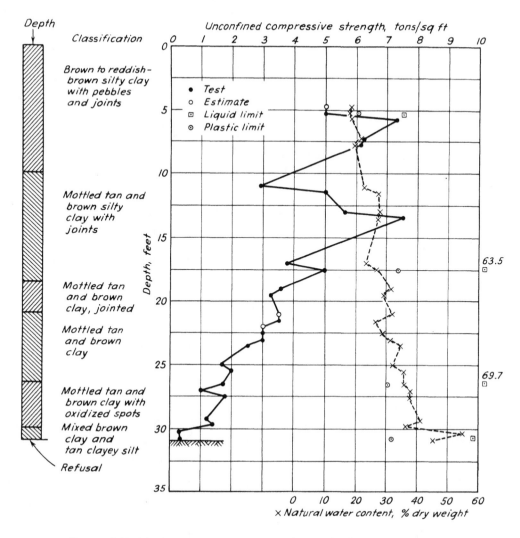

FIGURE 6.33. Variations in water content and unconfined compressive strength of residual clays over limestone.

rived from the nonsoluble rocks vary with depth throughout the zone of transition. Figure 6.37 is an example of the weathering of a granite in North Carolina. Beneath the clayey top layer the soils tend to be somewhat more pervious than transported soils of corresponding grain size and plasticity.

Laterization. Advanced chemical weathering, in tropical and subtropical regions of low relief with abundant rainfall and high temperatures, leads in well-drained areas to deep, strongly leached red, brown, and yellow profiles. If the weathering has been intense enough and has continued long enough, the clay minerals break down, whereupon their silica is released and may be removed by leaching. The remaining soils then consist largely of aluminum oxide or of hydrous iron oxides. This process is known as *laterization*.

All the products of tropical weathering with red or reddish color have much in common with respect to both origin and properties. Those in which the ratio of silica remaining in the soil is small compared to the amount of Fe_2O_3 and Al_2O_3 are often referred to as *laterites*, but the term has been used with such a variety of connotations that it has little significance to the engineer.

FIGURE 6.34. Road cut through dikes in volcanic rock on island of Oahu, Hawaii.

FIGURE 6.35. Slides in tropically weathered soil caused by discontinuities in parent rock.

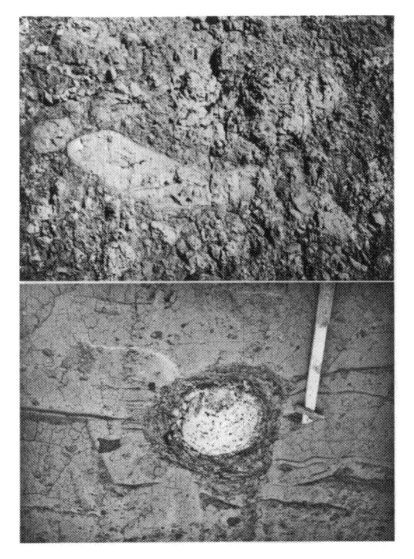

FIGURE 6.36. Exposures of residual soil showing effects of spheroidal weathering.

Distinction among tropically weathered reddish soils should be based on their index and engineering properties.

In continuously wet regions the endproducts of laterization are characterized by high natural water contents, high liquid limits, and irreversible changes upon drying. Classification tests carried out on air-dried or oven-dried samples may lead to radically misleading results, if such soils in the field will not be dried as a consequence of construction, all classification tests in the laboratory should be performed on samples not previously subjected to drying. Moisture-density relations should be determined on samples dried from the natural water content only to the desired moisture content for placement in the compaction mold. The remarkable difference in moisture-density relations for a soil from Hawaii tested in this manner and also after air drying is illustrated in Fig. 6.38. If the climate is so humid that the moisture content of a fill material cannot be reduced by such conventional procedures as disking and harrowing, the fill may have to be placed at

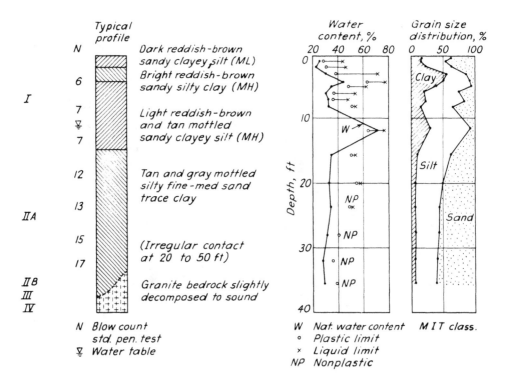

FIGURE 6.37. Profile of weathering on granite in North Carolina (after Deere, 1957).

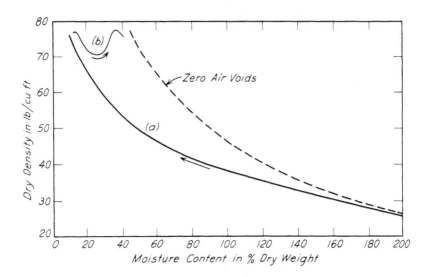

FIGURE 6.38. Moisture density relations for Hawaiian soil (a) tested after partial drying in increments from natural water content and (b) tested after complete air drying (after Willis, 1946).

its natural water content and it may not be feasible to attain the higher densities and strengths associated with drying and placement at or near the optimum moisture content (Art. 1.6).

On the other hand, in regions subject to distinct wet and dry seasons, alternate upward and downward movement of water concentrates the iron or the alumina. The products of laterization under these conditions are likely to exhibit low natural water contents, low plasticity, concretions, and cemented horizons. Frequently associated with the reddish soils of the tropics are very dark-colored soils often called *black cotton* soils. The dark color of these soils is seldom due to organic accumulation, but rather to the presence of iron, manganese, and titanium compounds in the reduced state. These soils are apparently formed under conditions of poor drainage from basic rocks, or sometimes limestone, under alternating wet and dry climatic conditions. They usually show high shrink-swell characteristics with surface cracks, opening during the dry season, which are 2 or more in. wide and several inches deep. The cracks close during the wet season and an uneven soil surface is produced by irregular swelling and heaving. Such soils are especially troublesome as pavement subgrades or under shallow foundations.

SUGGESTED READING

A wide variety of books, papers, and monographs in the fields of geology, soil science, and engineering deals with the characteristics of natural earth deposits. The list that follows includes samples of two types of publications. Some provide general information on various aspects of geology and pedology. Others contain data on index properties and engineering behavior of specific deposits or types of deposits that may pertain directly to foundation conditions at a given site. The latter may appear to be more useful, but they may prove misleading unless the writer and the reader have an adequate understanding of the fundamentals and implications of the earth sciences.

A landmark paper relating earth science and engineering is K. Terzaghi (1955a), "Influence of Geological Factors on the Engineering Properties of Sediments," *Economic Geology*, 50th Anniversary Volume, pp. 557–618; the paper contains more than 100 carefully selected references.

The listing below is arranged by categories that generally follow the order in the preceding chapter.

General geology:

L. D. Leet and S. Judson (1971), *Physical Geology*, 4th ed., Englewood Cliffs, N. J., Prentice-Hall, 687 pp.

L. E. Spock (1962), *Guide to the Study of Rocks*, 2nd ed., New York, Harper, 298 pp.

Geomorphology:

W. D. Thornbury (1954), *Principles of Geomorphology*, New York, Wiley, 618 pp. A basic text on the development of land forms.

W. D. Thornbury (1965), *Regional Geomorphology of the United States*, New York, Wiley, 609 pp. An excellent text on land form characteristics arranged by physiographic divisions. The pertinent references at the end of each chapter provide a good starting point for locating more specific geologic information about a particular area.

N. M. Fenneman (1931), *Physiography of Western United States*, New York, McGraw-Hill, 534 pp.

N. M. Fenneman (1938), *Physiography of Eastern United States*, New York, McGraw-Hill, 714 pp. This reference and the one preceding are classic works describing the various physiographic subdivisions and their characteristics.

K. B. Woods, R. D. Miles, and C. W. Lovell, Jr. (1962), "Origin, Formation, and Distribution of Soils in North America," Chap. 1 in *Foundation Engineering*, G. A. Leonards, ed., New York, McGraw-Hill, pp. 1–65. Follows the regional approach and contains nearly 200 references.

Engineering geology:

S. Paige, ch. (1950), *Application of Geology to Engineering Practice*, Berkey Volume, Geol. Soc. Amer., 327 pp. Contains several significant papers by outstanding engineers and geologists with much practical experience.

P. D. Trask, ed. (1950), *Applied Sedimentation*, New York, Wiley, 707 pp. A collection concerned with a wide variety of topics related to engineering. Useful reference lists.

R. F. Legget (1962). *Geology and Engineering*, 2nd ed., New York, McGraw-Hill, 884 pp. A readable book covering many aspects of engineering geology.

Weathering:

P. Reiche (1950). *A Survey of Weathering Processes and Products*, Univ. of New Mexico Press, 95 pp. A comprehensive and lucid discussion of rock weathering and soil formation. The treatment of the latter topic is somewhat outdated by the new system of pedologic classification.

Pedology:

Soils and Men (1938), U. S. Dept. of Agriculture Yearbook, Wash., D. C., G.P.O., 1232 pp. The first comprehensive discussion of the principles of soil classification and mapping utilized in most agricultural soil survey reports published before 1970.

Soil Survey Manual (1951), U. S. Dept. of Agriculture Handbook No. 18, Wash., D. C., G.P.O., 503 pp. Covers detailed procedure of pedologic classification and mapping in the United States.

Soil Classification, a Comprehensive System (1960), U. S. Dept. of Agriculture, Wash., D. C., G.P.O., 265 pp. This monograph, commonly called the "7th Approximation," outlines the revised system of pedologic classification presently being adopted in the United States and likely to be adopted in many other countries. Numerous modifications to the system have already been made and will be incorporated in a revised version to be published.

T. H. Thornburn, D. J. Hagerty, and T. K. Liu (1970), "Engineering Soil Report, Will County, Illinois," *Univ. of Ill. Eng. Exp. Sta. Bull. 501*, 195 pp. An example of the direct correlation between pedologic map units and engineering properties of soil as determined from a detailed soil sampling and testing program in the glaciated region of northeastern Illinois.

Glacial geology:

R. F. Flint (1971), *Glacial and Quaternary Geology*, New York, Wiley, 893 pp. An excellent text on glaciers, glaciation, and glacial deposits.

R. B. Peck and W. C. Reed (1954), "Engineering Properties of Chicago Subsoils," *Univ. of Ill. Eng. Exp. Sta. Bull. 423*, 62 pp. A compilation of soil boring data assembled in the form of maps and diagrams in a geologic framework. The source of Fig. 6.7. An example of the type of information available in a few major cities.

R. F. Legget, ed., (1961), *Soils in Canada*, Royal Soc. Canada, *Spec. Publ. 3*, Univ. of Toronto Press, 229 pp. An assembly of papers reporting on studies of the geology, pedology, and engineering characteristics of the glacial soils of Canada.

Windblown soils:

G. D. Smith (1942), "Illinois Loess," *Univ. of Ill. Agr. Exp. Sta. Bull. 490*, 45 pp. A classical discussion of the origin, distribution, and characteristics of loess from the pedologic viewpoint.

R. A. Bagnold (1941), *The Physics of Blown Sand and Desert Dunes*, New York, Wm. Morrow, 265 pp. A comprehensive study of the movement of sand and the formation of windblown sand deposits.

"Conference on Loess: Design and Construction" (1968), *Hwy. Res. Rec., 212*, 38 pp. Contains useful information on the behavior and engineering properties of loess.

E. L. Krinitzsky and W. J. Turnbull (1967), "Loess Deposits of Mississippi," *Geol. Soc. Amer. Spec. Paper 94*, 64 pp. Contains geotechnical information.

Waterlaid sediments:

R. B. Peck (1953), "Foundation Exploration—Denver Coliseum," *Proc. ASCE, 79*, Separate No. 326, 14 pp. Discusses the variability of a fluvial deposit as illustrated in Fig. 6.12.

L. Zeevaert (1957a), "Consolidation of Mexico City Volcanic Clay," *ASTM Spec. Tech. Publ. 232*, pp. 28–32.

C. R. Kolb and W. G. Shockley (1959), "Engineering Geology of the Mississippi Valley," *Trans. ASCE, 124*, pp. 633–656. A classic description of the various environments of alluvial deposition and the relationship between the soil formations and engineering problems.

L. Bjerrum (1967), "Engineering Geology of Norwegian Normally-Consolidated Marine Clays as Related to Settlements of Buildings," *Géotechnique*, *17*, 2, 81–118.

Residual soils and rock:

F. J. Pettijohn (1957), *Sedimentary Rocks*, 2nd ed., New York, Harper, 718 pp. The authoritative publication on the characteristics and classification of sedimentary rocks.

K. Terzaghi (1960), "Landforms and Subsurface Drainage in the Gačka Region in Yugoslavia," in "From Theory to Practice in Soil Mechanics," L. Bjerrum, A. Casagrande, R. B. Peck, and A. W. Skempton, eds., New York, Wiley, pp. 81–105. Describes the details of typical karstic topography. First published in Hungary in 1913 and reprinted in 1958 in *Annals of Geomorphology*, *2*, No. 1/2.

E. C. J. Mohr and F. A. van Baren, (1954), *Tropical Soils*, New York, Interscience Publ., 498 pp. An excellent treatise on the pedology and formation of soils under tropical conditions.

G. F. Sowers (1954), "Soil Problems in the Southern Piedmont Region," *Proc. ASCE*, *80*, Separate 416, 18 pp.

P. Lumb, (1962), "The Properties of Decomposed Granite," *Géotechnique*, *12*, No. 3, 226–243.

Engineering Properties of Lateritic Soils (1969). Proc. Specialty Session 1, 7 Int. Conf. Soil Mech., Asian Inst. Tech., Bangkok, 207 pp. Vol. 2 (1970), 203 pp. A compilation of 25 papers and discussions covering a wide variety of lateritic soils; summary of properties as related to engineering.

Wolmar Fellenius (1876–1957)

Chairman and active leader of the Swedish Geotechnical Commission, appointed in 1913 to study the causes of catastrophic landslides on the Swedish State Railways. The report of this Commission, published in 1924, is justly regarded as a milestone in the application of soil mechanics to practical problems. The investigators were among the first to perform laboratory and field tests for determining the shearing strength of soils, and to recognize the essential differences in the properties of undisturbed and remolded clays. (Photo courtesy of B. H. Fellenius.)

PLATE 7.

CHAPTER 7

Program of Subsurface Exploration

7.1. Development of Subsurface Exploratory Program

Introduction. The preceding chapter demonstrates that very few natural soil deposits are even approximately uniform, and many are extremely erratic. It is obvious that in an erratic deposit no program of subsurface exploration can lead to more than a rough idea of the average values for the physical properties of the subsurface material and of the probable variations from these values.

The nature of the deposit is an important factor in determining the method of soil exploration that will yield the greatest amount of useful information. If, for example, the foundation of an important structure is to be established above a fairly homogeneous layer of clay, a considerable amount of testing of undisturbed samples may be justified because the test results permit a relatively accurate forecast of both the amount and the rate of settlement. On the other hand, if the same structure is to be located above a deposit composed of pockets and lenses of sand, silt, and clay, a comprehensive testing program would not be justified because it would provide little more information than could be obtained merely by determining the index properties of representative samples. Much more useful information could be obtained at less cost by making an adequate number of penetrometer measurements that would disclose the pattern of the various soft and stiff elements in the subsoil.

The magnitude and character of the exploratory program should also be chosen in consideration of the importance of the project under construction. If the job involves only a small expenditure, extensive programs of soil exploration cannot be justified economically. It is cheaper to take advantage of whatever information may already be available and to use a liberal factor of safety in the design.

Finally, the program of soil exploration should develop step by step as information accumulates. By this procedure the maximum amount of information can be obtained for a given expenditure and the program can be terminated as soon as adequate data have been collected. Hence, no definite rules can be established for an exploratory program, and even engineers with considerable experience should not attempt to determine the final program before exploration begins.

Preliminary Exploration. The program of subsurface exploration should be preceded by a fact-finding survey. In such a survey the engineer responsible for the exploration should prepare a digest of all available information on soil conditions near the site and on the behavior of other structures in the vicinity. In highly developed regions with unfavorable subsurface conditions, useful information is likely to be available in technical journals and published reports, but in most rural areas or in areas being newly developed for industry, information concerning structural behavior may be lacking. However, the engineer should not overlook the maps and publications of state and federal geological surveys, or reports of soil surveys prepared in connection with agriculture or highway construction. The type of information to be obtained from these sources has been discussed in Art. 6.1.

The preliminary exploration procedure is selected on the basis of the information obtained from the fact-finding survey. Most soil deposits, however, can be appropriately explored by means of a split-barrel sampler and standard penetration tests (Arts. 5.3 and 5.4) carried out in holes made by augers, rotary drills, or wash-boring tools (Art. 5.2). Other methods of exploration are not usually considered in the preliminary phase unless it is known that the underlying material consists of bedrock on the one hand, or of very soft clays, silts, or highly organic materials on the other. Moreover, for many projects, no further subsurface exploration is necessary. This is likely to be the case if the loads on the subsoil will be small and a large factor of safety can be used without excessive cost, if the structure can be founded on rock or strata of high bearing capacity, or if an ordinary structure is to be erected in an area where much practical experience has been summarized in the form of reliable empirical rules or building codes.

Detailed Exploration. When the preliminary exploratory program does not provide sufficient information for design or construction, further investigations are required. The methods should be selected to obtain the most pertinent information at the least cost. Frequently, the properties of fairly uniform deposits of soft clay and plastic silt can be investigated most economically by field vane tests or by obtaining continuous samples in 2- or 3-in. thin-walled tubes and performing appropriate laboratory tests (see Chap. 18). Erratic deposits of soft silt and clay can be examined by means of penetration tests combined with enough tube borings to permit interpretation of the penetrometer data. Standard penetration tests or dynamic cone penetrometer tests are appropriate for sands. Rotary or percussion core barrels are normally used to sample rock, and special peat samplers are available for highly organic deposits. Standard load tests are appropriate for loess and other collapsible soils.

On some jobs involving structures of great importance or foundation conditions of exceptional difficulty, additional information may still be required. It may be advisable to obtain large-diameter undisturbed samples from critical strata, to conduct load tests, to make field pumping tests, or to conduct other special tests. Since such studies are always expensive, they should be undertaken only to investigate specific questions that the cheaper procedures are inadequate to answer.

Number and Depth of Borings. For buildings or structures of ordinary size, it is suitable to plan on making four borings, one at each corner of the structure. Unless bedrock is encountered, the first boring should ordinarily extend to the maximum depth within which the stress caused by the structure could conceivably produce excessive settlement. This depth may have to be established on the basis of approximate stress and settlement computations, as indicated in Part C. For a heavily loaded structure the first boring should ordinarily extend to a depth equal to twice the least width of the structure. Beneath a lightly loaded structure with widely spaced columns, the depth of the first boring should not be less than twice the probable width of the largest footing.

The second boring will serve to indicate whether soil conditions are likely to be relatively uniform or erratic. If the results are similar to those for the first boring, subsequent borings may ordinarily be discontinued when they have penetrated all the soft or compressible strata.

If the borings encounter rock and the conditions are such that the structure may be founded on rock, cores should be obtained for a depth of 5 to 10 ft to make sure that sound bedrock, rather than a boulder or a piece of detached rock, has been reached. If there is evidence of solution channels or deep weathering, the cores should usually be continued into sound rock.

As the exploratory program develops, the engineer should study the implications of all new information. If additional borings, penetration tests, or special investigations are needed, they should be so located and planned that each addition to the program will provide the maximum increase of knowledge at the current stage of the investigation.

Although the program should be developed to provide the information necessary for the project at hand, the engineer should allow for the possibility that there may be changes in preliminary structural layouts, including column spacings and loadings. He should therefore obtain sufficient data to permit consideration of the various practical foundation types for the revised layout if the changes are not too radical. Moreover, the exploration should not be limited to obtaining the information needed for design of foundations of a type the engineer may initially think most suitable; otherwise, he may lack information to select or design foundations of another type that may prove more practical or economical.

Presentation of Results. The arrangement of responsibilities between the owner and the various professional disciplines involved in the design of a project often requires the preparation of a *foundation engineering report*. In such a report, the foundation engineer is responsible for the presentation of the results of all field and laboratory tests in a format most useful and readily comprehensible to the owner and to the other design professionals. Therefore, the data obtained from the overall exploratory program should be presented in a manner that most clearly explains and substantiates the recommendations regarding types of foundations and their predicted ranges of performance and costs. Summary plots of test data should be included and presented in such a way that the behavior of the subsoil and foundation structure as a whole unit can be readily evaluated.

The foundation engineer must be experienced with the design procedures discussed in Parts C and D before he is qualified to develop an exploratory program, prepare a foundation engineering report, or use the results of a program of subsurface exploration for purposes of design.

Conclusion. No matter how complete the program of soil exploration and testing may be, there always remains a large margin of uncertainty concerning the exact nature of subsurface conditions at a given site. This fact is of outstanding practical importance. It makes foundation design fundamentally different in its basic concepts from all other branches of structural design. The engineer cannot proceed as he would with materials having well-defined properties, such as steel, concrete, or timber. Although the latter materials are not perfectly uniform, they can almost always be considered so in design. The foundation engineer must often wait to obtain his final data concerning soil conditions until he can observe what happens in the field, and he should always make use of every fragment of evidence. Soil tests performed on a few samples taken from an erratic deposit do not provide a satisfactory basis for design because the engineer is interested in the behavior of the deposit as a whole rather than of a few specimens taken from it.

In the discussions of foundation design procedures that follow in Part C, natural deposits are grouped according to their general characteristics as clay and plastic silt (Chap. 18), sand and nonplastic silt (Chap. 19), collapsing and swelling soils (Chap. 20), nonuniform soils (Chap. 21), and rock (Chap. 22). In the consideration of each group, particular attention is given to the uncertainties involved in evaluating the probable behavior of the deposit with respect to the bearing capacity and settlement of foundations. The importance of these uncertainties in dictating the appropriate type of foundation for a given type of deposit is emphasized. Furthermore, methods are presented that permit the selection and design of foundations on a rational and economical basis, in spite of the handicap arising from the inevitable uncertainties about subsurface conditions.

SUGGESTED READING

Many papers contain excellent summaries of the results of subsurface explorations, such as those listed at the end of Chaps. 18 through 22, but surprisingly few describe the steps in the development of the exploratory procedure on specific projects. Hence, some of the following references deal with projects carried out several decades ago. The methods are, accordingly, less important than the principles illustrated.

T. L. Condron and E. R. Math (1932), "Investigating a Foundation in Soft Soil," *Civ. Eng. ASCE*, 2, 4, 237–241. The techniques and equipment are somewhat outdated, but the approach remains valid. It was undoubtedly derived from the practice of the senior author who, at the time the article was written, was nearing the end of a long and distinguished career as a foundation and structural engineer.

E. W. Scott, Jr. (1948), "Philadelphia Conducts Extensive Subsurface Exploration Prior to Airport Expansion," *Civ. Eng. ASCE*, 18, 2, 44–46. Brief account of an investigation modified to suit the findings.

J. W. Hunter (1948), "Site Exploration for Foundations at Portsmouth," *Proc. 2nd Int. Conf. Soil Mech.*, Rotterdam, 2, pp. 159–162. Program included observation of settlements of existing buildings.

K. M. Gammon and G. F. Pedgrift (1962), "The Selection and Investigation of Potential Nuclear Power Station Sites in Suffolk," *Proc. Inst. Civil Engrs.*, 21, pp. 139–160. Description of a well-planned site evaluation for a nuclear reactor that could tolerate only limited differential settlements under gross pressures up to 4 tons/sq ft. Preliminary surveys of seven sites included geology, airphoto interpretation, and a minimum number of borings. Detailed foundation exploration required a number of techniques.

R. B. Peck (1969a). "Advantages and Limitations of the Observational Method in Applied Soil Mechanics," *Géotechnique*, 19, 2, 171–187. Although this paper is not directed toward the usual techniques or programs of soil exploration, it describes the use of field observations during construction as a means for improving knowledge of subsurface conditions at a time when alterations in design are still possible, and it points out the rewards and pitfalls associated with the use of observational data.

A manual on subsurface investigations for design of buildings is in preparation by the American Society of Civil Engineers. It has been published for discussion as "Subsurface Investigation for Design and Construction of Foundations of Buildings, Task Committee for Foundation Design Manual, Part I, *ASCE J. Soil Mech.*, 98, SM5, 481–490; Part II, SM6, pp. 557–578; Parts III and IV, SM7, pp. 749–764.

PART **B**

Types of Foundations and Methods of Construction

One of the requisites for the successful practice of foundation engineering is an adequate knowledge of the engineering properties of soils and rocks, the natural materials on which the engineer's structures depend for support. Part A dealt with this aspect of the engineer's background.

Part B is concerned with a second equally important aspect, the commonly used types of foundations and methods of construction. Here, tradition, experience, and industrial competition have played a part more important than that of science, and partly for this reason foundation engineering is properly considered an art.

Under the influence of economic factors such as the relative costs of labor and materials, certain types of foundations and certain construction procedures have evolved and come into general use in the United States and Canada. These are briefly described in the following pages. Although innovation and progress in the state of the art are not to be discouraged, departures from commonly accepted practice should be undertaken with caution because equipment not readily available may be costly and procedures unfamiliar to the workmen and their supervisors may be difficult to enforce.

The broad subject of pile driving is considered in some detail in Part B because many of the procedures and principles are applicable to pile foundations in all the types of soils for which pile foundations are suitable. The selection and evaluation of pile foundations for particular subsurface conditions are discussed in Part C.

Excavating and Bracing Subway Cut

Excavation for station on Washington, D. C., subway system (1971). The flanges of the vertical soldier piles and the timber lagging are shown at the left. The struts, consisting of 24-in. steel pipes, have high inherent resistance against buckling. The horizontal wales, which transfer the load from the soldier piles to the struts, are set out from the soldier piles to allow for the unavoidably imperfect alignment of the piles, and their webs are stiffened to prevent crippling. The size of the cut, 74 ft wide by 63 ft deep, can be inferred from the scale of the bulldozer.

PLATE 8.

CHAPTER 8

Excavating and Bracing

8.1. Introduction

The foundations for most structures are established below the surface of the ground. Therefore, they cannot be constructed until the soil or rock above the base level of the foundations has been excavated.

Ordinarily it is not the function of the foundation engineer to select the equipment for excavating at a given site or to design the bracing if any is required. These are considered to be within the province of the contractor. It is generally the engineer's duty, however, to approve or disapprove the construction procedure proposed by the builder and to check the design of the bracing. In exercising these functions, the engineer is properly concerned only with the ability of the proposed procedures to permit satisfactory construction of the structure as designed; he should not ordinarily presume to dictate the manner in which the construction should be carried out.

On large or complicated jobs, it may be impossible to prepare a design without also deciding upon the method of construction. When the design and the construction procedures are related so closely that they must be considered as a unit, the engineer is obliged to specify the construction method and possibly to design the bracing.

In pervious soils, excavation below the water table usually requires drainage of the site either before or during construction. In this chapter we shall assume that the water table is normally below the bottom of the excavation or that it has been temporarily lowered. The techniques for drainage will be discussed in the next chapter.

Moreover, in this chapter we shall discuss only the general aspects of excavating and of providing support for the sides of the pits or cuts. Details or modifications appropriate in connection with specific foundation conditions or types of structures will be discussed in Part C and the structural design of certain systems of sheeting and bracing will be considered in Part D.

8.2. Open Excavations with Unsupported Slopes

Shallow excavations can be made without supporting the surrounding material if there is adequate space to establish slopes at which the material can stand. The steepness of the slopes is a function of the type and character of the soil or rock, the climatic and weather conditions, the depth of the excavation, and the length of time the excavation must remain open. As a rule, construction slopes are made as steep as the material will permit because the occurrence of a few small slides is generally not serious.

The cost of removing the material affected by the slides may be considerably less than that of the additional excavation required to provide flatter slopes.

The steepest slopes that can be used in a given locality are best determined by experience. Most sands contain a small amount of cementing material or display a small amount of apparent cohesion because of the moisture they contain. This cementation or cohesion cannot safely be depended on for the stability of permanently exposed slopes, but it is ordinarily utilized for the periods during which excavations stand open. Although permanent slopes for sandy soils are seldom steeper than 1 vertical to $1\frac{1}{2}$ horizontal, construction slopes as steep as 1 vertical to $\frac{1}{2}$ horizontal are not uncommon.

The maximum slope at which a clay soil can stand is a function of the depth of the cut and of the shearing resistance of the clay. If the clay is soft below the base level of the excavation, flat slopes may be required to avoid a rise of the bottom. Furthermore, stiff or hard clays commonly possess or develop cracks near the ground surface. If these cracks become filled with water, the hydrostatic pressure greatly reduces the factor of safety and may cause slope failures. The water in the cracks also softens the clay progressively so that the safety of the slope is likely to decrease with time. For these reasons, bracing is often used to support the sides of excavations in clay, even though the clay would stand briefly to the necessary height without lateral support.

8.3. Sheeting and Bracing for Shallow Excavations

Many building sites extend to the edges of the property lines or are adjacent to other sites on which structures already exist. Under these circumstances, the sides of the excavation must be made vertical and must usually be supported. Several methods of bracing are in common use.

If the hole is not to extend to a depth greater than about 12 ft, it is common practice to drive vertical planks known as *sheeting* around the boundary of the proposed excavation. The depth to which the

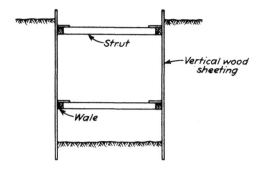

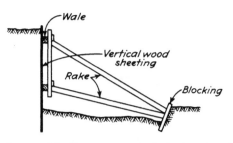

FIGURE 8.1. Common methods for bracing sides of shallow excavations.

sheeting is driven is usually kept near that of the bottom of the hole as excavation progresses. The sheeting is held in place by means of horizontal beams called *wales* that in turn are commonly supported by horizontal *struts* extending from side to side of the excavation. The struts are usually of timber, but, if the excavation is not more than about 5 ft wide, extensible metal pipes called *trench braces* are commonly used. If the excavation is too wide for the use of struts extending across the entire width, the wales may be supported by inclined struts known as *rakes* or *rakers*. Their use requires the soil in the base of the excavation to be firm enough to provide adequate support for the inclined members.

Two typical arrangements for shallow bracing are shown in Fig. 8.1.

8.4. Sheeting and Bracing for Deep Excavations

When the depth of excavation exceeds 15 or 20 ft, the use of vertical timber sheeting generally becomes uneconomical, and other methods of sheeting and bracing are com-

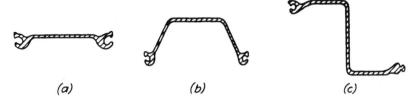

FIGURE 8.2. Types of steel sheet piles commonly used for bracing sides of deep excavations. (*a*) Flat web. (*b*) Arch web. (*c*) Z piling.

monly employed. According to one procedure, steel sheet piles are driven around the boundary of the excavation. As the soil is removed from the enclosure, wales and struts are inserted.

The types of sheet piles commonly used for this purpose are shown in Fig. 8.2. The strength and stiffness of the *arch web* type *b* exceeds that of the *flat web a*, whereas that of the *Z piling c* is the greatest for all rolled sheet-pile sections. Consequently, types *a* and *b* are used for shallow excavations and type *c* for the deepest excavations or those where the heaviest pressures are expected.

As soon as excavation has proceeded for a few feet, wales and struts are inserted as shown in Fig. 8.3. The wales are commonly of steel, and the struts may be of steel or wood. Excavation then proceeds to a lower level, and another set of wales and struts is installed. This process continues until the excavation is complete. In most soils it is advisable to drive the sheet piles several feet below the bottom of the excavation to prevent local heaves. In some instances the embedment may eliminate the necessity for installing a strut at the bottom of the cut. It is important to provide vertical support

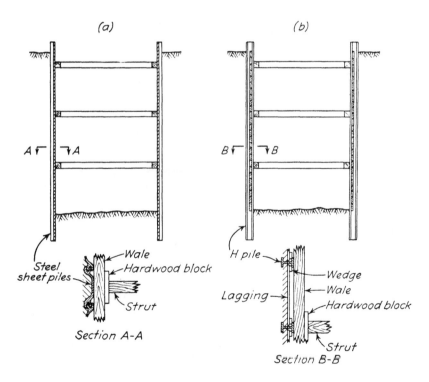

FIGURE 8.3. Cross sections through typical bracing in deep excavation. (*a*) Sides retained by steel sheet piles. (*b*) Sides retained by H piles and lagging.

for the bracing. This may be done by keeping posts beneath the system of bracing to transfer its weight to the underlying soil or by suspending the bracing from beams extending across the top of the cut.

In most types of soil, a vertical face of several square feet can be exposed without danger of collapse of the ground. It may then be possible to eliminate the sheet piles and to replace them with a series of H piles spaced at 4 to 8 ft. The H piles, known as *soldier piles*, are driven with their flanges parallel to the sides of the excavation, as shown in Fig. 8.3b. As the soil next to the piles is removed, horizontal boards known as *lagging* are introduced as shown in the figure and are wedged against the soil outside the cut. As the general depth of excavation advances from one level to another, wales and struts are inserted in the same manner as for steel sheeting.

If the width of a deep excavation is too great to permit the economical use of struts across the entire excavation, inclined bracing may be used, provided there is adequate support for the inclined members. In some instances, it is possible to excavate the central portion of the site to its maximum depth and to place part of the permanent foundation. The completed part of the foundation then serves as support for the inclined bracing or rakers required when the remainder of the site is excavated. This procedure is shown in Fig. 8.4.

As an alternative to cross-lot bracing or inclined struts, *tiebacks* are often used. According to one system, shown in Fig. 8.5, inclined holes are drilled into the soil outside the sheeting or H piles; in favorable ground, an enlargement or bell is formed at the end of the hole. Tensile reinforcement is then inserted and concreted into the hole.

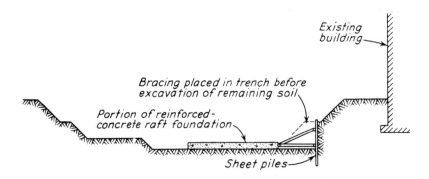

FIGURE 8.4. Typical bracing in deep, wide excavation.

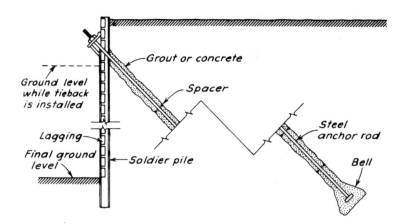

FIGURE 8.5. One of several tieback systems for supporting vertical sides of open cut. Several sets of anchors may be used, at different elevations.

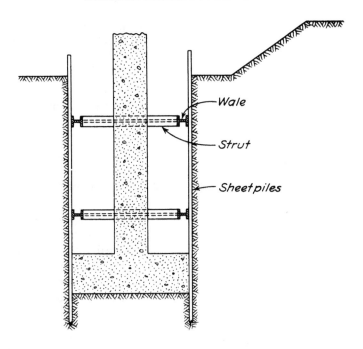

FIGURE 8.6. Method of constructing basement walls in trenches before general excavation.

Each tieback is usually prestressed before the depth of excavation is increased. The equipment and methods for forming the hole are similar to those used for drilling piers (Art. 13.2).

Sometimes it is preferable to complete the exterior walls of a structure before removing the material in the space occupied by the basements. The walls are constructed in narrow braced trenches, as shown in Fig. 8.6. After the walls have been completed and the floor system established at the top, the block of soil between the walls may be excavated. The floor provides the bracing for the upper part of the walls and additional bracing may be inserted as needed while excavation progresses.

Occasionally the exterior walls are constructed in a trench filled with a slurry or heavy fluid consisting of a clay suspension similar to drilling mud (Art. 5.2). The slurry stabilizes the walls of the trench and permits excavation without the need for sheeting or bracing. Cages of reinforcement are lowered into the slurry; the slurry is then displaced by tremie concrete. Special equipment is needed for the various operations, and occasional imperfections must be anticipated and repaired. Although more common in Europe, the method has been used on a number of projects in North America including the World Trade Center in New York City, where the walls constructed in the slurry trenches were supported by a tieback system.

8.5. Movements Associated with Excavation

Excavation always involves removal of material and, consequently, causes a change in the state of stress in the rock or soil beneath and beside the excavated space. Such a change occurs whether or not the sides of the cut are supported by bracing. Inasmuch as no material can experience a change in stress without corresponding deformations, excavation is always associated with movements of the adjacent ground surface. Such movements usually have the character of settlements, but under certain rather rare conditions the ground surface may rise.

On the other hand, properly designed and carefully placed bracing can materially reduce the change in lateral pressure in the material adjacent to the excavation and is thus capable of reducing the settlements to a value that may be regarded as the practicable minimum for a particular job. Where the settlements might damage adjacent structures it is one of the duties of the foundation engineer to make sure that the method proposed for bracing the sides is capable of reducing the settlement to a tolerable value. If this condition is satisfied and the settlements become excessive, the damage may properly be ascribed to lack of good workmanship on the part of the construction forces. These matters are discussed in greater detail in Chap. 16.

A common and desirable precaution for reducing movements of the adjacent ground to the practicable minimum is to prestress each strut as it is inserted. One method is illustrated in Fig. 8.7. Before a strut is inserted, two auxiliary struts equipped with hydraulic jacks are used to increase the distance between the wales against which the strut is to bear. The strut is then inserted and wedged tightly, so that a stress of several tons remains in the strut when the jacks are released.

On many projects, excessive movements and even failures have occurred even though the bracing system was well designed. The difficulties arose because the steps in the sequence of excavating and bracing were not clearly specified or followed, and excavation was allowed to advance too far before the next set of supports was installed (Art. 27.3). Not infrequently, excavation and bracing are done on the same job by different contractors. Since bracing interferes with excavation, the tendency of the excavation contractor to proceed as far as possible before bracing is placed must be counteracted by rigorously stipulated and enforced procedures.

Where the consequences of settlements of adjacent property or of a failure would be serious, field observations are often undertaken to provide early warning of unfavorable developments (Art. 16.5). These may include measurement of lateral and vertical movements of the sheeting and bracing, of forces in rakers or struts, of settlements of adjacent structures, and of piezometric levels below and adjacent to the excavation.

SUGGESTED READING

A. B. Carson (1961), *General Excavation Methods*, New York, F. W. Dodge Corp., 392 pp. A general description, intended primarily for contractors, of equipment and procedures for excavation, including methods of support and dewatering. Contains much background information useful to the designer concerning field practice.

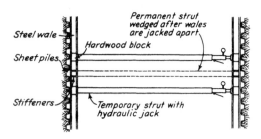

FIGURE 8.7. Method of prestressing struts for bracing sides of open cut.

Robert Stephenson (1803–1859)

Engineer of construction for some of the earliest English railroads, including the Birmingham and London Railway, first to enter London (1838). This new form of transportation demanded the construction of deep cuts and tunnels, some of which had to be driven through cohesionless sand below the ground-water table. Stephenson successfully met these unprecedented problems and developed many techniques in the art of drainage. He was a pioneer in structural engineering as well, serving as engineer for such famous early structures as the Menai Straits tubular girder and the first railroad bridge across the St. Lawrence River at Montreal.

PLATE 9.

CHAPTER 9

Drainage and Stabilization

9.1. Introduction

When the depth of excavation is greater than the distance to the free water surface in a pervious soil having a coefficient of permeability greater than about 10^{-3} cm/sec, the soil must be drained to permit construction of foundations in the dry. If the coefficient of permeability of the soil is within the range 10^{-3} to 10^{-5} cm/sec, the quantity of water that seeps into the excavation may be inconsequential but drainage may still be required to maintain the stability of the sides and bottom of the excavation. If the coefficient of permeability is smaller than about 10^{-7} cm/sec, the soil is likely to possess sufficient cohesion to overcome the influence of the seepage forces and drainage may not be required even if the excavation extends for a considerable depth below the water table.

After completion of structures with basements, it is often necessary to maintain the water level in a lowered position. This requires the installation of permanent drains.

The physical processes of drainage have been outlined in Art. 2.5. The possibility of inducing settlements by drainage is considered in Art. 16.3. The requirements for the drainage of various types of foundations in different subsurface formations will be described in Part C. The present chapter is concerned primarily with the general techniques of drainage and the equipment necessary to accomplish it.

9.2. Ditches and Sumps

Where space permits, ditches may be used to lower the water table in sand or in other materials made pervious by cracks or joints. In silty or fine sands, the side slopes must ordinarily be relatively flat on account of the seepage pressures exerted by the entering water.

The relatively flat slopes required for open ditches in sand generally preclude the use of ditches for lowering the water table more than a very few feet. However, open ditches are commonly used in the bottom of an excavation to collect the water that seeps into the hole. Such ditches lead to sumps from which the water is pumped.

A *sump* is a pit with its bottom below the level of the ditches that enter it. Considerable care must often be exercised to prevent sand and silt beside and beneath the sump from washing in and being pumped out with the water. To reduce the loss of sand by pumpage and to prevent the consequent instability, it is often desirable to line the sides of the sump and to cover the bottom with coarse-grained material that acts as a filter (Art. 2.5). Such a sump for an open

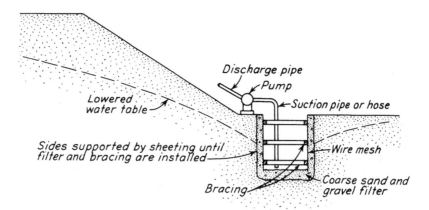

FIGURE 9.1. Filter-protected sump for open cut in sand.

cut in sand is shown in Fig. 9.1. A large-diameter pipe, set vertically, with filter material in the lower part, is often satisfactory.

Drainage for either temporary or permanent construction may also be accomplished by excavating trenches rather than ditches, placing drain tiles or perforated pipes in the trenches, and filling over the drains with pervious material. To prevent washing the fine material out of the backfill or surrounding soil, it may be necessary to surround the drains with granular material satisfying the requirements for a filter. The width of the openings in the drain pipe should preferably be equal to about the 60 per cent size (D_{60}) of the surrounding material.

9.3. Well Points

The water table in granular materials may be lowered by means of well points. A *well point* is a perforated pipe about 3 ft long and $1\frac{1}{2}$ in. in diameter, covered by a cylindrical screen to prevent the entrance of fine particles. It is attached to the bottom of a $1\frac{1}{2}$- or 2-in. riser pipe inserted vertically in the ground. The point can usually be jetted into the ground without driving, although stiff strata sometimes require the use of a punching device or of an auger. On the job, a line of well points spaced at 2 to 5 ft is connected to a 6-, 8-, or 10-in. header pipe laid on the surface of the ground. The header, in turn, is connected to a suction pump. The various parts of the assembly are shown in Fig. 9.2.

If the depth of excavation below the water table is greater than about 15 ft, several stages of well points are required. The first excavation is made to a depth of about 15 ft, whereupon additional well points are jetted into the ground before the next 15 ft are excavated. The points are generally arranged in such a manner that the sides of the excavation consist of slopes connected by flat berms containing drainage ditches. This arrangement, known as a *multiple-stage setup*, is shown in Fig. 9.3.

When the quantity to be pumped per well point is small, a *jet-eductor system* may be

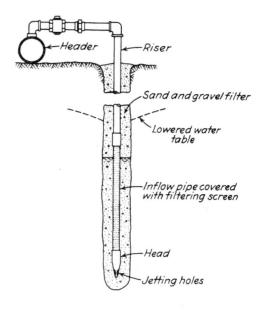

FIGURE 9.2. Details of well-point assembly.

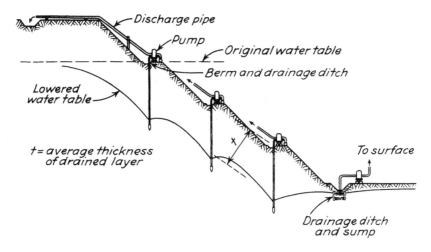

FIGURE 9.3. Multiple-stage well-point setup.

used in place of a multiple-stage setup. Each well point is installed in the bottom of a cased hole. The well point is attached to the under side of a jet-eductor pump, which in turn is connected to the surface by two pipes, one for incoming high-pressure water that operates the pump, and the other for the return water including that furnished by the well point. The efficiency is low because most of the water taken from the system was previously injected to operate the pumps. However, a single-stage system can lower the water table as much as 100 ft. When limitations of space preclude the use of a multiple-stage system, jet eductors may prove economical.

If the permeability is less than about 10^{-4} cm/sec, drainage cannot be accomplished simply by pumping from well points because the capillary forces prevent the flow of water from the pores of the soil. However, drainage can be accomplished by consolidation. This may be done by means of the vacuum method of well-point operation (Fig. 9.4). In this method, the well points are set in holes about 8 in. in diameter made by either augering or jetting. A filter of medium to coarse sand is then placed around the point and pipe to within about 2 or 3 ft of the surface. Above the filter an impervious material such as clay is tamped to form a seal. Special techniques may be required in holes that fail to stand open.

The pumps for such an installation must be capable of maintaining a vacuum in the well points and the surrounding filter. The pressure around the well points is thereby reduced to a small fraction of atmospheric pressure whereas the surface of the ground is acted upon by the weight of the atmosphere. Thus, the soil becomes consolidated under a pressure of about 1 ton/sq ft.

The vacuum process is highly effective in silts and organic silts, but the time required to achieve consolidation and stability is likely to be several weeks.

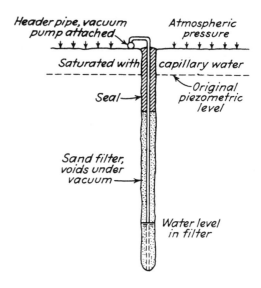

FIGURE 9.4. Vacuum well-point installation.

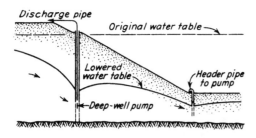

FIGURE 9.5. Drainage by means of deep-well pumps.

9.4. Deep-Well Pumps

For very deep excavations the multiple-stage well-point setup has the disadvantage that the water level is pulled down rather abruptly at the edges of the excavation. As a consequence, the hydraulic gradient near the excavation is quite large and the resulting seepage pressures may lead to instability of the side slopes. Under these circumstances, it is safer and sometimes more economical to install large-diameter drainage wells equipped with deep-well pumps. A typical arrangement for such wells is shown in Fig. 9.5. The spacing, which commonly varies from 20 to 200 ft, depends on several factors, including the permeability of the soil and the depth of the permeable stratum.

The drainage wells for such an installation consist of cased holes with diameters commonly ranging from 6 to 24 in. The casing is perforated in the pervious zones. The pumping unit consists of a submersible multistage turbine pump and motor mounted on a common vertical shaft. A 10-in. pump of this type is capable of discharging about 1000 gpm against a head of 80 ft and requires about a 30 hp motor.

9.5. Sand Drains

In many instances, it is necessary to build a structure or to construct an embankment on fine-grained soils with low shearing resistance. The initial strength of the soils may be too low to support the weight of the structure without failure. However, if the weak soils can be drained rapidly enough to let consolidation occur at nearly the same rate as the load is applied, the strength of the material may increase sufficiently to permit safe construction.

For the purpose of accelerating drainage in relatively impervious deposits, vertical drains may be installed. In the United States, these drains commonly consist of columns of sand about 2 ft in diameter arranged in a pattern of squares or triangles at a spacing of 10 to 15 ft. The ground surface at the top of the drains is covered by a pervious blanket, and the structure or embankment is constructed on top of the blanket (Fig. 9.6). As the weight increases, the water is squeezed out of the subsoil into the drains from which it escapes through the drainage blanket to ditches. The rate at which consolidation will occur can be controlled by varying the spacing and diameter of the drains.

The act of installing sand drains may considerably disturb the structure of the soil, whereupon the permeability and strength may be decreased and the compressibility

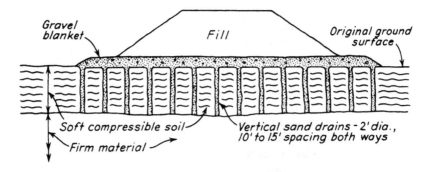

FIGURE 9.6. Sand-drain installation.

increased. The disturbance is especially great if the drains are formed with the aid of a mandrel that displaces the soil. Failure to minimize or to take account of these unfavorable effects has led to many unsatisfactory installations.

9.6. Miscellaneous Methods of Drainage and Stabilization

The stabilization of sands and silts by drainage is not always practicable. Hence, various other methods have been devised, most of which involve the injection of slurries or solutions into the voids of the soil. These materials harden to varying degrees and impart cohesion to the injected soil. Since they partly fill the voids, they also reduce the permeability.

The injection of *cement grout* has been attempted on many occasions. Experience has demonstrated that the method may lead to very satisfactory results, but only if the soil is relatively homogeneous and unstratified, and if the grain size is not too small. The grout will not penetrate the voids of a loose soil with an effective size D_{10} less than about 0.5 mm or a dense soil with an effective size less than about 1.5 mm. Thus, cement grouting is not appropriate for soils much finer than coarse sands.

The grain size of cement particles limits the fineness of a sand that is suitable for cement grouting. Clay suspensions of any desired fineness can be obtained, however, by removing the coarser fractions from natural clays. This has led to attempts to grout soils with *clay slurries*. In practice, it is found that the penetration of the slurry is impeded by the formation of a filter skin that seals the voids. The development of the filter skin appears to be strongly influenced by the electrolytes present in the groundwater; this influence introduces an element of considerable uncertainty regarding the effectiveness of the procedure. As a matter of fact, it appears that the materials that can be successfully injected with clay slurries have roughly the same characteristics as those suitable for cement grouting. Moreover, although clay grouting may greatly reduce the permeability of a sand, it does not significantly increase the strength.

Solidification of soils by *injection of chemicals* has been widely practiced. One common procedure consists of the successive injection of solutions of water glass and calcium chloride, which react in the soil to form a cohesive binder. In a variation of this method a single solution is injected containing a buffer that delays and controls the time of setting. These methods are highly successful in clean relatively homogeneous sands having an effective size greater than 0.1 mm, but the efficacy of the procedures decreases rapidly as the grain size or the homogeneity of the sand decreases. Furthermore, it depends greatly on the chemical composition of the groundwater.

Considerable use is also made of polymers that are mixed with catalysts and retarders before injection and that react after a lapse of time to form an almost impervious gel. Before the reaction the viscosity of the mixture is only about twice that of water. Moreover, the reaction time is not significantly affected by the chemical composition of the groundwater. The reaction time can be designed to occur after a few seconds or at a later time up to many minutes; with such control it is sometimes possible to stabilize materials through which water is flowing rather rapidly.

All the injection procedures are expensive and, even under favorable conditions, are uncertain. Although numerous successful applications have been made, many other attempts have resulted in disappointing failures. Hence, stabilization of this type should usually be considered only under exceptional circumstances where the risk of an unsuccessful attempt is worth taking in view of the possible benefits from a successful application. In any event, injection should not be considered a routine matter and should not be undertaken without the advice of competent and experienced specialists.

If the coefficient of permeability of the soil lies between about 10^{-4} and 10^{-6} cm/sec, no injection procedure is satisfactory. In such soils, stabilization may be accomplished by *electro-osmosis*, an electrical method for creating seepage pressures and producing consolidation. Careful studies are required to determine the likelihood of success and to estimate the power requirements.

In several instances soils have been rendered impervious and stable by *freezing* the water in the voids. The method has most frequently been used in sinking shafts or in tunneling, but it has also been successfully employed for the temporary stabilization of at least one large landslide in silt. Freezing is accomplished by circulating a refrigerant through a series of double pipes inserted in the soil to be stabilized. Each double pipe consists of an outer casing into which the cold liquid is pumped and an inner pipe through which it returns. Several weeks or months may be required to freeze a block of soil with a volume of a few hundred cubic yards. On account of the time required as well as the cost of the refrigerating equipment, the method is expensive.

SUGGESTED READING

W. F. Swiger (1960), "Control of Ground Water in Excavations," *ASCE J. Constr. Div.*, 86, CO1, 41–53.

D. A. Werblin (1960), "Installation and Operation of Dewatering Systems," *ASCE J. Soil Mech.*, 86, SM1, 47–66.

C. I. Mansur and R. I. Kaufman (1962), "Dewatering," Chapter 3 in *Foundation Engineering*, G. A. Leonards, ed., New York, McGraw-Hill, pp. 241–350. Technical details of the design of dewatering systems.

Useful information is also contained in brochures of the various dewatering contractors; their advertisements may be found in current engineering and construction periodicals.

A most interesting two-part paper on the history and application of injection methods of soil stabilization, with numerous examples, is R. Glossop (1960), "The Invention and Development of Injection Processes, Part I: 1802–1850," *Géotechnique*, 10, 3, 91–100; and "Part II: 1850–1960," *Géotechnique*, 11, 4 (1961), 255–279. Extensive bibliographies have been prepared by the ASCE Committee on Grouting, including "Bibliography on Chemical Grouting," (1966), *ASCE J. Soil Mech.*, 92, SM6, 39–66; and "Chemical Grouting" (1957), *ASCE J. Soil Mech.*, 83, SM4, Paper No. 1426, 106 pp. A series of papers on "Cement and Clay Grouting of Foundations" is found in *ASCE J. Soil Mech.*, 84, SM1, Papers No. 1544–1552. Electro-osmosis is treated in L. Casagrande (1952), "Electro-osmotic Stabilization of Soils," *J. Boston Soc. Civ. Eng.*, 39, 1, 51–83; reprinted as Harvard Soil Mechanics Series No. 38. The freezing method is discussed in detail in F. J. Sanger (1968), "Ground Freezing in Construction," *ASCE J. Soil Mech.*, 94, SM1, 131–158; the article contains a bibliography with practical examples.

A comprehensive treatise on sand drains was prepared at the request of the Bureau of Yards and Docks, U. S. Navy, by Moran, Proctor, Mueser, and Rutledge (1958), "Study of Deep Soil Stabilization by Vertical Sand Drains," U. S. Dept. of Commerce, Office Tech. Serv., Wash., D. C., 192 pp.

A state-of-the-art paper, by one of the principal authors of the foregoing report, is S. J. Johnson (1970b), "Foundation Precompression with Vertical Sand Drains," *ASCE J. Soil Mech.*, 96, SM1, 145–175; extensive bibliography. Factors influencing the performance of sand drains are considered, with appropriate case histories, in L. Casagrande and S. Poulos (1969), "On the Effectiveness of Sand Drains," *Canadian Geot. Jour.*, 6, 3, 287–326.

Frederick Baumann (1826–1921)

Prominent early Chicago architect, and author of a pamphlet entitled "The Art of Preparing Foundations for All Kinds of Buildings, with Particular Illustration of the 'Method of Isolated Piers' as Followed in Chicago." This pamphlet, which appeared in 1873, contains the first statements published in the United States that the areas of footings should be in proportion to the column loads, and that the centroid of the area of the base of each footing should coincide with the center of gravity of the loads acting on the footing. It also contains values of soil pressure recommended for the Chicago clays. (Photo courtesy of Chicago Historical Society.)

PLATE 10.

CHAPTER 10

Footing and Raft Foundations

10.1. Types of Footings

A footing is an enlargement of the base of a column or wall for the purpose of transmitting the load to the subsoil at a pressure suited to the properties of the soil. A footing that supports a single column is known as an *individual column footing*, an *isolated footing* or a *spread footing*. The footing beneath a wall is known as a *wall footing* or a *continuous footing*. If a footing supports several columns, it is called a *combined footing*. A particular form of combined footing commonly used if one of the columns supports an exterior wall is a *cantilever footing*. The various types are illustrated in Fig. 10.1.

10.2. Historical Development

Footings undoubtedly represent the oldest form of foundation. Until the middle of the nineteenth century, most footings consisted of masonry. If they were constructed of stone cut and dressed to specific sizes, they were known as *dimension-stone footings*. In contrast, *rubble-stone footings* were constructed of pieces of random size joined by mortar. Masonry footings were adequate for most structures until the development of tall buildings with heavy column loads. Such loads required large and heavy footings that occupied valuable basement space.

In the earliest attempts to enlarge the areas of footings without increasing weight, timber grillages were constructed and conventional masonry footings built on them. In 1891 a grillage consisting of steel railroad rails embedded in concrete was devised as an improvement over the timber grillage (John Wellborn Root, Montauk Block, Chicago). The rail grillage was an important forward step because it saved much weight and increased space in the basement. Within the following decade, railroad rails were superseded by steel I beams that occupied slightly more space but that were appreciably more economical of steel. Typical grillage foundations of timber, railroad rails, and steel I beams are shown in Fig. 10.2.

The steel I beam proved admirably suited for the construction of cantilever footings. These were introduced in 1887 almost simultaneously in two buildings in Chicago. One of these early footings is illustrated in Fig. 10.3.

With the advent of reinforced concrete shortly after 1900, grillage footings were almost entirely superseded by reinforced-concrete footings, which are still the dominant type.

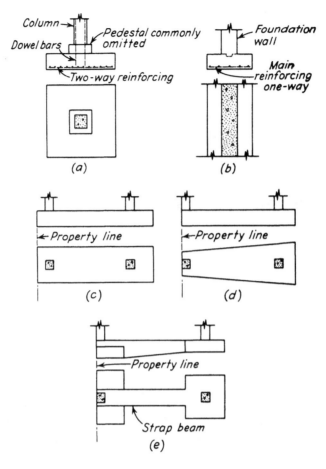

FIGURE 10.1. Types of footings. (*a*) Individual column footing. (*b*) Wall footing. (*c*) and (*d*) Combined footings. (*e*) Cantilever footing.

10.3. General Considerations

In temperate latitudes footings are commonly located at a depth not less than that of normal frost penetration. In warmer climates, and especially in semiarid regions, the minimum depth of footings may be governed by the greatest depth at which seasonal changes in moisture cause appreciable shrinkage and swelling of the soil.

The elevation at which a footing is established depends on the character of the subsoil, the load to be supported, and the cost of the foundation. Ordinarily the footing is located at the highest level where adequate supporting material may be found. In some instances, if an especially firm layer is encountered at greater depth, it may be more economical to establish the footing at a lower elevation because the area required for the footing is smaller.

The excavation for a reinforced-concrete footing should be kept dry so that the reinforcement can be set and held in its proper position while the concrete is being placed. To do this in waterbearing soil it may be necessary to pump either from sumps or from a previously installed system of drains. Forms are usually required around the sides of the footing. The necessity for pumping and for supporting the sides of the excavations in which the footings are placed may add appreciably to the cost of a footing foundation.

10.4. Allowable Soil Pressures

In the earliest days of foundation engineering the area of a footing was selected

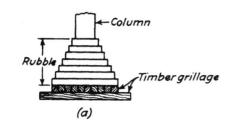

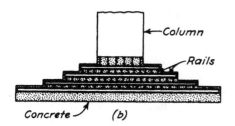

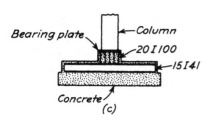

FIGURE 10.2. Historical development of grillage foundations of (a) timber, (b) railroad rails, (c) steel I-beams.

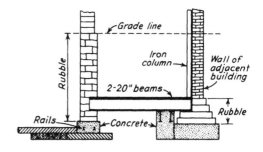

FIGURE 10.3. Cantilever footing supporting exterior column of Auditorium Building, Chicago, 1887.

according to the judgment of the engineer on the basis of his experience. In most localities simple rules of thumb developed. For example, in some parts of the United States the width of a continuous footing in feet was made equal to the number of stories in the structure. No attempt was made to provide larger footings for the support of heavier loads.

In the early 1870s the proportioning of footings was placed on a more rational basis. Progressive engineers of that day recommended that the areas of footings on a given site be made proportional to the loads on the footings and that the center of gravity of the load on each footing be made to coincide with the centroid of the footing. It was believed that the settlements of all footings beneath a structure would be equal and that no footing would tilt if these recommendations were conscientiously followed. Furthermore, it was believed that for each soil there existed a specific pressure under which the settlements of the various footings would not exceed reasonable values. This pressure, known as the *allowable soil pressure*, was generally specified in the building code or ordinances of the city in which the construction was to be located.

With the development of soil mechanics, it became evident that the safety or the settlement of a footing depended on many factors besides the pressure exerted on the subsoil. Nevertheless, the concept of an allowable soil pressure is so convenient that it has been retained in modern foundation engineering, but with modifications and limitations dictated by the improved state of our knowledge. These modifications and limitations constitute a large part of the information contained in Part C.

10.5. Combined Footings

If the loads from several columns are transmitted to the same footing, the footing should be proportioned so that its centroid coincides with the center of gravity of the column loads under normal conditions and so that the maximum pressure beneath the footing does not exceed the safe soil pressure under the most severe loading. Combined footings are customarily used along the walls of buildings at property lines where the footing for a wall column cannot extend outside the limits of the structure, Figs. 10.1c, 10.1d, and 10.1e. Under these circumstances, the wall footing is usually combined with an interior footing by one of the three methods shown.

10.6. Raft Foundations

A *raft* or *mat* foundation is a combined footing that covers the entire area beneath a structure and supports all the walls and columns. Wherever the building loads are so heavy or the allowable soil pressure so small that individual footings would cover more than about half the building area, a raft foundation is likely to be more economical than footings.

Ordinarily, rafts are designed as reinforced-concrete flat slabs. The downward loads on the raft are the loads from the individual columns or walls. If the center of gravity of the loads coincides with the centroid of the raft, the upward load is regarded as a uniform pressure equal to the sum of the downward loads divided by the area of the raft. The weight of the raft is not considered in the structural design because it is assumed to be carried directly by the subsoil. Since this method of analysis does not take into account the moments and shears caused by differential settlement, it is customary to reinforce the raft more heavily than required according to the analysis.

Raft foundations are also used to reduce the settlement of structures located above highly compressible deposits. Under these conditions, the depth at which the raft is established is sometimes made so great that the weight of the structure plus that of the raft is wholly compensated by the weight of the excavated soil. The settlement of the structure is then likely to be insignificant. Where complete compensation is impracticable, a shallower raft may be acceptable if the net increase in load is small enough to lead to tolerable settlements.

If the column loads are not more or less uniformly distributed or if the subsoil is such that large differential settlements would tend to develop, large rafts must be stiffened to prevent excessive deformations. This stiffening has been accomplished by using partitions as the stems of T beams connected to the raft (Fig. 10.4a), by constructing a cellular or rigid-frame foundation (Fig. 10.4b), or in some instances by utilizing the stiffness of a reinforced-concrete superstructure. The larger the foundation, the more expensive these procedures become; often pile or pier foundations are preferable.

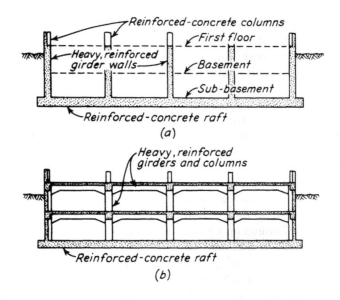

FIGURE 10.4. Methods of stiffening large raft foundations. (a) Use of ribs or walls as T beams. (b) Rigid-frame construction.

10.7. Drainage, Waterproofing, and Dampproofing

Distress to buildings can often be traced to foundation movements caused by inadequate removal of surface water. Hence, the grading of the site around a building should be carefully designed. If the ground surface next to a building is to be unpaved, its slope should be not less than 1 in./ft.

The nature of the subsoil and the groundwater conditions at a site should always be considered in the choice of elevations for basements and floor levels. If the basement must be established below normal groundwater level, special precautions must be taken to prevent seepage into the structure. Two general methods are in common use: *drainage*, whereby the water is prevented from reaching the exterior of the structure, and *waterproofing*, whereby entrance of the water adjacent to the structure is prevented by some sort of impervious barrier. The two methods are often combined.

Drainage may be suitable where the seepage is small enough to permit removal of the water at little expense, usually by gravity flow into sewers or ditches. The most common installations for this purpose are footing drains (Figs. 10.5a and 10.5b), and floor drains, (Fig. 10.5c). The footing drains may consist of short cylindrical sections of clay tile, laid with a gap of about $\frac{1}{4}$ in. between sections and with a strip of roofing felt over the joint to exclude soil. Increasingly they are likely to consist of corrugated perforated metal or plastic pipes because of economy and ease of installation. The drains are laid in trenches backfilled to within about a foot of the ground surface with free-draining material. The upper foot of backfill is preferably of less pervious soil that will prevent inflow of surface waters. Where the presence of water in the drains might reduce the strength of the soil beneath the footings, type b is more suitable than type a. If the soil being drained is likely to clog the backfill, the backfill material should satisfy the requirements for a filter (Art. 2.5).

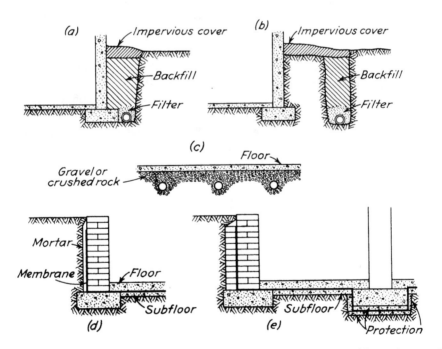

FIGURE 10.5. Methods of draining and waterproofing basements. (a) Footing drain. (b) Intercepting drain. (c) Floor drain. (d) Membrane waterproofing where exterior wall is accessible from outside. (e) Membrane waterproofing where outside of wall is not accessible (after Huntington, 1963).

Floor drains (Fig. 10.5c) are not usually necessary if the footing drains are effective. In some instances, however, there may be slow seepage from beneath the structure and the drainage may be advisable. Floor or footing drains should not be connected to downspouts or surface drains.

If the quantity of water that would be collected in a system of drains would be excessive, it may be necessary to waterproof the basement and permit the structure to be subjected to the full water pressure. The most positive procedure is the *membrane method*. In this method, a membrane consisting of alternate layers of fabric and bituminous material is constructed on or near the exterior of the building. The bituminous material is applied hot. Such a membrane is sufficiently flexible and extensible to maintain its integrity even if a moderate amount of cracking occurs in the walls or floors.

Membrane waterproofing to be fully effective must be continuous over the entire surface of the structure that extends below water level. This requires the placement of a concrete subfloor upon which the membrane is built before the structural floor is constructed, and requires special details at walls and footings (Figs. 10.5d and 10.5e). Figure 10.5e also shows the arrangement necessary if the exterior wall is not accessible for covering with the membrane.

Since the walls and floors inside the membrane are acted upon by full water pressure, they must be designed for this force as well as for earth pressure.

The seepage of capillary moisture through basement walls and floors can be reduced by placing pervious material outside the walls and beneath the floors. Ordinarily, the exterior of the wall is painted with a bituminous material, and the floor is cast on a sheet of polyvinyl chloride resting on the pervious material. These measures commonly prevent entrance of moisture into the wall or floor. Various admixtures are available for decreasing the permeability of the concrete itself and for reducing the flow of moisture by application on the inside of the structure. The effectiveness of these substances is extremely variable.

SUGGESTED READING

The evolution of footing foundations and the transition to deep foundations in Chicago, where much of the development of American foundation practice took place, is described in R. B. Peck (1948), "History of Building Foundations in Chicago," *Univ. of Ill. Eng. Exp. Sta. Bull. 373*, 64 pp. A similar historical account of one of the earliest examples (1783–1786) of an industrial building on a partly compensated raft foundation is A. W. Skempton (1971), "The Albion Mill Foundations," *Géotechnique, 21*, 3, 203–210.

Details of waterproofing are described in G. W. Gill (1959), "Waterproofing Buildings Below Grade," *Civ. Eng., ASCE, 29*, 1, 3–5.

Consolidation by Surcharging

Marine terminal for transfer from water to land at Elizabeth, New Jersey, being constructed on swampland underlain by 10 to 20 ft of peat and organic silt. A surcharge loading of about 20 ft of sand was used to induce settlement which amounted to 3 or 4 ft. The berms were required to avoid mud waves. In the background, next to the channel, are shown surcharge fills which have been moved progressively from left to right; the areas already stabilized are being covered by pavement which will become the floors of the soil-supported transit sheds and warehouses. (Photo courtesy of Port Authority of New York and New Jersey.)

PLATE 11.

CHAPTER 11

Foundations on Compacted Fill

11.1. Historical Development

Until recently, filled ground has generally been regarded as unsuitable for supporting the foundations for structures, whether residential, commercial, or industrial. Many fills were merely wasted material, often mixtures of various types, dumped without compaction and without treatment of the surficial soils on which the fill rested. Structures established on such fills usually experienced severe differential settlement and cracking. Consequently, good practice required extending the footings, piles, or piers through the fill and into the natural ground until adequate support was found. The ground floors either were supported structurally by the building foundations or were placed on the fill with the realization that repairs, overlays, or replacements would be needed from time to time.

In contrast to the random uncontrolled fills of the past, many fills today are placed by carefully controlled procedures on natural ground from which weak and compressible surficial materials have first been stripped. Such fills may provide better support for structures than natural deposits and are widely used in developing residential subdivisions and industrial areas. In some instances, soils that would give rise to large settlements are subjected to special treatment to reduce their compressibility and are left in place beneath the fill.

11.2. Design Considerations

The principal considerations in designing a structure supported by fill are illustrated by Fig. 11.1. A one-story light manufacturing building is presumed to be located in a lowland underlain by a deep deposit of compressible soil. Before construction is started on the building proper, the grade of the entire area is raised several feet by a controlled compacted fill. Additional fill is placed to support the floor, located about 5 ft above the surrounding ground level, at a height convenient for direct unloading of goods from railroad freight cars or truck transports. The columns are established on footings supported by the fill.

The fill under these conditions provides the immediate local support for the footings, retaining walls, and floors of the building. If the filling and compaction are properly controlled, the fill is likely to be stronger and less compressible than most natural deposits. In this sense, it constitutes an excellent foundation material. On the other hand, the fill itself imposes a substantial load on the underlying compressible soil. The 5 ft of fill within the retaining walls, for example, add a load of about 600 lb/sq ft

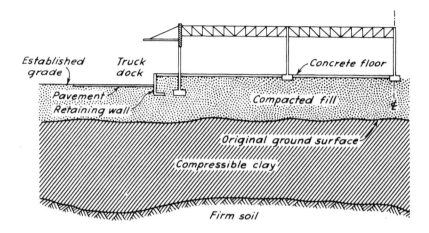

FIGURE 11.1. Cross section through typical one-story light manufacturing building on compacted fill.

over the entire area of the structure, a load roughly equal to that added by a 5-story office building with no basement, or by a 15-story office building with a basement. Moreover, because the load is distributed over a large area, a corresponding increase in stress occurs throughout the depth of the compressible deposit. Thus, although the fill provides excellent support for the footings and floor loads, it may cause detrimental subsidence of the entire area, including the superposed building. Under these circumstances, no amount of refinement in the layout or design of the footings themselves can appreciably improve the unsatisfactory performance of the structure.

The design of a fill-supported structure, therefore, requires two steps. The first is to determine whether the weight of the fill and building will lead to excessive deep-seated settlement. If it will, measures must be considered to avoid the settlement or its consequences, or else the site must be judged unsuitable for the purpose. Only if the unfavorable consequences of deep-seated settlement can be accepted or eliminated should the second stage of design be undertaken. The settlement of the underlying soft material can then be ignored and the foundations proportioned to suit the characteristics of the fill with due regard to the strength of the soil beneath the fill. The appropriate methods are discussed in Part C.

11.3. Deep-Seated Settlement

The preceding section has demonstrated that design of a structure resting on compacted fill must be preceded by a forecast of the settlement that will be caused by compression of the underlying deposits. The procedures for making such forecasts, with respect to both magnitude and rate, are discussed in Art. 3.4 and in Chaps. 18 and 21.

If the compressible material is comparatively thin and is located just below the original ground surface, it can in some instances be removed economically by excavation. If it is also very weak, it is sometimes displaced by advancing the fill from one direction so that a mud wave is progressively swept across the site. Displacement, even with careful supervision, can hardly be recommended because it is rarely possible to avoid entrapment of soft pockets of remolded material within the fill. Serious settlements may then develop.

In many instances a large part of the settlement due to the weight of the fill occurs during and shortly after filling. If the additional weight of the building and its contents is likely to be a small fraction of the weight of the fill, the amount of deep-seated settlement that occurs after construction may be small enough to be tolerable, and no special measures may be needed. If the postconstruction settlement would be excessive, the top of the fill may be brought temporarily several feet above final grade. The weight of additional fill, known as the *surcharge*, increases the rate and magnitude of settlement of the compressible deposit. When the settlement becomes equal to that which would ultimately be attained by the deposit beneath only the normal fill plus the weight of the occupied building, the surcharge can be removed. In this manner the construction settlement can be accelerated whereas that remaining to occur during the lifetime of the structure can be reduced substantially or almost eliminated. As the function of the surcharge fill is only to provide weight, compaction of the surcharge is unnecessary except for the portion that will settle below final grade.

When the compressible deposit is highly organic, much of the settlement may arise from secondary consolidation (Art. 3.2) and the results achieved by surcharging may not be as favorable as anticipated. Moreover, if the deposit consists of a thick clay bed without drainage layers, the time required for settlement either during filling or under the surcharge may be far too great to be acceptable. Under favorable conditions, sand drains (Art. 9.5) may prove beneficial, but in many instances a fill-supported structure may not be a suitable type of construction.

Predictions of the rate of settlement of a compressible deposit under the influence of a surcharge are likely to be rather unreliable, largely because of the difficulty of judging the lengths of the drainage paths for the escaping water and, if sand drains are used, because of the differences in coefficient of consolidation in vertical and horizontal directions. To provide assurance that the time allowed for consolidation is neither too short nor unnecessarily long, control observations (Fig. 11.2) are customarily carried out. Settlement plates (Fig. 11.2b) are set at the base of the fill and observed periodically during and after filling. Piezometers are also commonly installed (Fig. 11.2c) at various depths in the fill for observing the excess pore pressure produced by the fill, and the rate at which the pressure dissipates. If the fill is of limited extent, surface reference stakes are also set beyond the toe of the fill to provide warning of lateral movement or heave indicative of failure of the soft material under the weight of the surcharge.

Occasionally, compacted fills are placed above deposits of submerged sand so loose that an earthquake shock or other dynamic disturbance would transform the sand into an unstable mass responsible for large settlements or even outright loss of support and collapse of structures founded on the fills. Such sands require densification by pile driving (Art. 19.5) or by other means such as vibroflotation (Art. 19.7).

11.4. Placement and Compaction of Fill

The most suitable materials for filling at building sites are well-graded sands and gravels, possibly containing a small percentage of clayey fines. Unfortunately, economic considerations usually dictate the use of less satisfactory materials that may be available close to the project. Most inorganic soils are acceptable, with the exception of highly plastic swelling clays (Art. 3.11) and clays at natural moisture contents well above the Standard Proctor optimum (Art. 1.6) in localities where the climatic conditions preclude drying by manipulation and exposure to the atmosphere. Cohesionless silts and very fine uniform sands are also undesirable because they are difficult to compact.

Modern practice calls for placement of the fill in layers, usually not thicker than 12 in. after compaction, and compaction by equipment suited to the type of soil. The placement moisture content should be close to the optimum value corresponding to the type of soil and compaction procedures being used (Art. 1.6).

Wherever practicable, placement and compaction of the fill should be done when the area is still unobstructed by footings, utilities, and other construction. The fill is normally distributed in windrows, spread by bulldozers or graders, and compacted by tractor-drawn equipment. Pneumatic-tired rollers (Fig. 11.3a) exerting 25 tons or

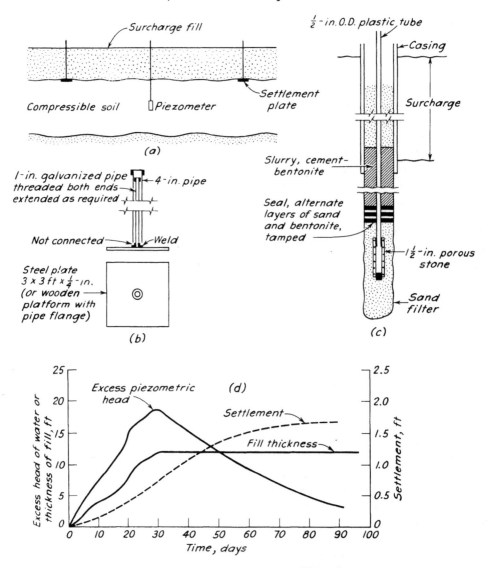

FIGURE 11.2. Control observations during precompression of foundation material by surcharging. (a) Typical arrangement. (b) Details of settlement plate. (c) Details of one form of open-standpipe porous-tube piezometer. (d) Typical results of observations.

more on a group of four wheels abreast, with a tire pressure of about 60 lb/sq in, usually achieve adequate compaction after about six coverages of most slightly cohesive granular materials at or near their optimum moisture content. Sheepsfoot rollers (Fig. 11.3b) are more effective for silty or clayey soils; they consist of steel drums about 4 ft long, with diameters of about 3.5 ft, to which are affixed radial protrusions or feet extending about 10 in. from the face of the drum and having bearing areas at the outer ends of about 4 to 8 sq in. The drums are usually ballasted to weigh about 4000 lb and exert foot pressures of about 100 lb/sq in. Six to eight passes are usually needed to develop the required compaction. For purely cohesionless sands and gravels, vibratory rollers are preferable. They consist of steel drums exerting a weight of about 3 to 5 tons plus a vertical oscillatory force of slightly smaller magnitude applied at a

FIGURE 11.3. Commonly used compaction equipment. (a) Pneumatic-tired roller. (b) Sheepsfoot roller. (c) Vibrating plate for hand operation [(a) and (b) courtesy Bros Div. American Hoist; (c) courtesy Stow Mfg. Co.]

frequency of about 20 Hz. Two coverages are usually sufficient.

Where practicable, excavations for footings, retaining walls, and utilities are made in the fill after compaction. In many instances, however, footings and walls are in place before the final lifts of the fill are placed. The fill cannot then be compacted by large-scale equipment. If the fill is cohesive, it can be compacted in local corners and pockets by hand-held pneumatic tampers; the layer thickness should not exceed 4 in. Cohesionless sands and gravels can be compacted by small vibrators acting on rollers or plates (Fig. 11.3c). Great care is required to avoid overcompaction behind retaining walls; otherwise the walls may be displaced or even cracked.

Backfill in utility trenches and other spaces later to be covered by floors or pavements should preferably consist of the best available well-graded sand and gravel mixtures. Materials for this purpose are often dumped into place loosely and then flooded in an attempt to compact them. This procedure, although still widely used, should not be permitted. In cohesive backfills it inevitably leads to weakening and softening of the soil and to future loss of support and subsidence. In uniform or fine sands it can do no more than cause the collapse of the extremely loose unstable zones associated with bulking (Art. 2.5) and leave the sand at a density index close to zero. If the backfill is a well-graded sand and gravel, the effect of bulking is negligible and no benefit is derived from flooding.

11.5. Control of Compaction

The placement of a compacted fill requires careful control. Two different approaches for achieving the control are in common use; circumstances govern which approach should be given the preference on a particular project.

If the materials to be used as fill have been designated and investigated by the engineer, and if there is ample experience in the locality in compacting the materials with the equipment generally available, the engineer may prefer to specify the details of the compaction procedure, such as the placement moisture content, layer thickness, type of equipment, and number of passes. Control then consists primarily of enforcing the procedure and making sure that the fill has the engineering properties required by the design. If the inspector judges that a change such as additional rolling is required, he gives the order and the contractor is paid accordingly. This approach recognizes that uniformity of application of an appropriate construction procedure is perhaps the best single guarantee of satisfactory performance of a fill. It is best suited to large projects on which extensive preliminary investigations have been carried out; indeed, it is the approach adopted by such organizations as the Corps of Engineers in connection with large dams and airfields.

On many smaller jobs neither the source of the fill nor the type of compaction equipment may be known before the contract is awarded. Under these conditions the engineer usually specifies general criteria for acceptability of fill material and requires that the fill be placed within a specified range of water content close to the optimum value and that a minimum per cent compaction be achieved. He also specifies the standard of compaction applicable; on most fills for buildings the Standard Proctor test is designated. The field control procedure consists of determining the Standard Proctor moisture-density curve for each variant of fill material, and of performing check tests in the field to determine whether the placement moisture content is within the specified range and whether the required dry density has been achieved. The details of the compaction procedure are left to the contractor. The approach has two principal shortcomings; in variable soils much testing may be needed to establish the identity of the materials on which the field check tests have been made with those for which the moisture-density curves have been obtained, and considerable fill may have to be removed if check tests disclose inadequacies in soil already covered with additional fill.

Use of the Standard Proctor test (Art 1.5) for control of compaction of fills for buildings is more appropriate than more rigorous standards such as the modified AASHO standard because the equipment used for compacting fills beneath buildings is generally much lighter than that adopted for highways, earth dams or airfields. For most projects the placement moisture content should be within the range $(w_{opt} \pm 2)$ per cent for ML soils, $(w_{opt} \pm 3)$ per cent for CL and MH soils, and $(w_{opt} \pm 5)$ per cent for CH soils of low to moderate swelling potential. For such cohesive soils, a dry density corresponding to 95 per cent of the Standard Proctor density is usually sufficient. For purely cohesionless soils, a minimum density index (Art. 1.5) is sometimes specified in place of a minimum percentage of the maximum Standard Proctor dry density. However, since reliable determination of the density index is difficult under field conditions, indirect methods such as standard load tests (Art. 5.5), standard penetration tests (Art. 5.4), or Dutch cone tests (Art. 5.4) may be preferable. A relative density corresponding to an N-value of at least 30 should be attained.

Clays with high swelling potential should be avoided as fill beneath either foundations or soil-supported floors of structures. The long-time conditions of moisture equilibrium in the fill beneath the structure rarely correspond to those at the time of placement and are hardly possible to predict. If the soil dries, it is likely to shrink differentially and deprive portions of the floor or structure of support; irregular subsidence then develops. If the moisture content increases, the structure and especially the floor may heave irregularly and crack, and large lateral forces may develop on foundation walls. If there is no practical alternative to the use of a swelling clay as fill, it is generally preferable to place and compact the material somewhat wetter than at the optimum moisture content because the effects of swelling are usually more detrimental than those of shrinkage. If the moisture content of the fill can alternately increase and decrease, serious damage can eventually be anticipated.

Swelling clays can in many instances be permanently transformed into soils of much lower plasticity and swelling potential by mixing into them a small percentage of hydrated lime, $Ca(OH)_2$. Such treatment should be undertaken only after suitable investigations to determine the amount of lime required and the degree of benefit that may be expected (Eades and Grim, 1966).

The addition of lime may also be beneficial in improving the workability of silts and clays having moisture contents in the field higher than the optimum for compaction. The principal effect of the lime is to reduce the free water in the soil by hydration; it also reduces the plasticity. The compacted soil is likely to develop additional strength and stiffness with time. Portland cement is often used for the same purpose; it is generally less efficient in reducing the free water but may lead to greater long-time strength. Premixing with lime followed by the addition of cement may be most economical where the additional strength is required.

11.6. Proportioning and Details of Foundation Elements

The floor of the structure illustrated in Fig. 11.1 is located about 5 ft above the established grade surrounding the building. The differential in height is created by the compacted fill, which supports not only the floor but also the interior footings. The lower part of the exterior wall of the structure serves to retain this portion of the fill. If the possibility of deep-seated settlement has been eliminated, all these structural units may now be proportioned without regard to deep-seated settlements.

The ground floor of such a structure consists almost invariably of a concrete slab (Fig. 11.1). In warehouses or manufacturing buildings the floor may be subjected to heavy distributed loads and to concentrated wheel loads from forklift trucks. On the other hand, if the structure has only one story, the footings support columns subject to little more than the weight of the roof and possibly occasional snow loads. Thus, the floor may be regarded as the principal load-carrying element. The footings can be

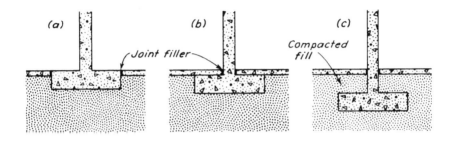

FIGURE 11.4. Common arrangements of interior footing with respect to ground-floor slab.

established in shallow excavations such that the upper surfaces of the footings constitute part of the floor, as shown in Fig. 11.4a. Alternatively, the floor slab may be cast directly on top of the footings (Fig. 11.4b). Footings established several feet below floor level (Fig. 11.4c) require compacted backfill under the floor. As the degree of compaction above the footing is likely to differ from that of the adjacent fill, particularly close to the columns, cracking of the floor is difficult to avoid; consequently, the arrangements shown in Figs. 11.4a and 11.4b are preferable. Exterior footings must, of course, be carried below frost level and placing the adjacent floor on backfill cannot be avoided. The backfill requires careful compaction to provide adequate support for the floor slab near the exterior wall.

SUGGESTED READING

S. J. Johnson (1970a), "Precompression for Improving Foundation Soils," *ASCE J. Soil Mech.*, 96, SM1, 111–144. A state-of-the-art paper including examples and an extensive bibliography.

P. C. Kotzias and A. C. Stamatopoulos (1969), "Preloading for Heavy Industrial Installations," *ASCE J. Soil Mech.*, 95, SM6, 1335–1355. Several case histories.

J. K. Mitchell (1970), "In-place Treatment of Foundation Soils." *ASCE J. Soil Mech.*, 96, SM1, 73–110. State-of-the-art paper considering vibroflotation, densification by blasting, use of compaction piles, grouting, electro-osmosis, thermal treatment, and use of additives.

D. J. D'Appolonia, R. V. Whitman, and E. D'Appolonia (1969), "Sand Compaction with Vibratory Rollers," *ASCE J. Soil Mech.*, 95, SM1, 263–284.

W. J. Turnbull and C. R. Foster (1958), "Stabilization of Materials by Compaction," *Trans. ASCE*, 123, pp. 1–15. Fundamentals of compaction of cohesive soils by sheepsfoot and rubber-tired rollers.

M. R. Thompson (1968). "Lime-Treated Soils for Pavement Construction," *ASCE J. Highway Div.*, 94, HW2, 191–217. Although oriented toward highway design, contains much information applicable to fills for foundations and soil-supported floors; extensive bibliography.

Albert Edward Cummings (1894–1955)

For 40 years an employee of the Raymond Concrete Pile Company, he rose successively from field clerk to field superintendent, district manager, and director of research. His vocation was the pile industry, his avocation the theory of foundations. His exposition of the fundamental fallacies of pile formulas is considered a classic of soil mechanics, and his recognition of the significance of the wave equation in pile-driving dynamics preceded the widespread use of that approach by more than two decades.

PLATE 12.

CHAPTER 12

Pile Foundations

12.1. Functions of Piles

When the soil beneath the level at which a footing or raft would normally be established is too weak or too compressible to provide adequate support, the loads are transferred to more suitable material at a greater depth by means of *piles* or *piers*. The distinction between the two is somewhat arbitrary. Piles are structural members of small cross-sectional area compared to their length, and are usually installed by a driver consisting of a hammer or a vibrator. They are often grouped into clusters or rows, each containing enough piles to support the load delivered by a single column or wall. Piers, on the other hand, are usually of larger cross section, each capable of transferring the entire load from a single column to the supporting stratum. Piers are considered in Chap. 13.

A lightly loaded column may, in some instances, require only a single pile. However, since under field conditions the actual position of a pile may be as much as several inches from its planned location, an eccentric loading can hardly be avoided. Consequently, the heads of single piles are usually braced in two directions by grade beams (Fig. 12.1a). If only two piles are needed, their heads may be connected by a concrete cap braced by grade beams in only one direction, perpendicular to the line joining the two piles (Fig. 12.1b). Clusters containing three of more piles are provided with reinforced concrete caps, as shown in Fig. 12.1c, and are considered stable without support by grade beams.

Vertical piles may also be used to resist lateral loads as, for example, beneath a tall chimney subject to wind. Compared to their axial capacity, the lateral capacity is usually small. Where large lateral loads are to be resisted, inclined or batter piles may be used (Fig. 12.1d). Batters of 4 horizontal : 12 vertical represent about the greatest inclination that can be achieved with ordinary driving equipment. Economy usually favors smaller inclinations even if more piles have to be battered.

12.2. Types of Piles

Classification. Piles display a remarkable variety of sizes, shapes, and materials to suit many special requirements, including economic competition. Although their variety defies simple classification, they may be discussed from the point of view of the principal materials of which they consist. These include timber, concrete, and steel.

Timber Piles. The use of the trunks of trees for piling was well established by Roman times; details of pile foundations were de-

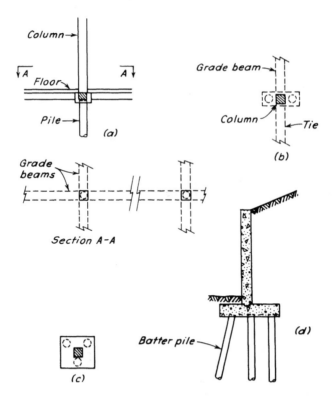

FIGURE 12.1. (a) Single pile supported by grade beams in two directions. (b) Two-pile group supported by grade beams in weak direction. (c) Laterally unsupported pile cap for group of three piles. (d) Use of batter piles in foundation for retaining walls.

scribed by Vitruvius in 58 A.D. Timber piles are still probably the most commonly used type throughout most of the world. Under many circumstances, they provide dependable, economical foundations. Their length is limited by the height of available trees; piles 40 to 60 ft long are common, whereas 85-ft piles cannot be obtained economically in all areas.

Timber piles cannot withstand the stresses due to hard driving that may be required to penetrate highly resistant layers. Damage to the points may be reduced somewhat by the use of steel shoes but, for a given type of hammer, the danger of breaking the piles can be significantly reduced only by limiting the stress induced in the head of the pile and the number of blows of the hammer. Timber piles cannot be driven against high soil resistances without damage; therefore, they are rarely used for loads in excess of 30 tons; in many localities the working load is restricted to 25 tons or less.

Although timber piles may last indefinitely when permanently surrounded by saturated soil, they are subject to decay above the zone of saturation. In some localities they may be damaged or destroyed by insects such as termites. The life of timber piles above the water table can be considerably increased by treatment with creosote under pressure. The effective duration of the treatment has not yet been fully established, but is known to exceed 40 years.

Timber piles in brackish or salt water are also subject to attack by various marine organisms such as *teredo* and *limnoria*. The deterioration may be complete within a few years or, under extremely unfavorable conditions, within a few months. Chemical treatment does not appear to be fully effective. Hence, timber piles should not be used where they will be exposed to open salt waters unless full investigations demonstrate that the destructive organisms are not present.

Concrete Piles. Shortly before 1900, several types of concrete piles were devised. Since that time, numerous variations have been introduced, and today a wide variety of piles is available from which the engineer may choose those best suited to a particular project. Concrete piles may be divided into two principal categories, *cast-in-place* and *precast* piles. The cast-in-place piles may be further divided into *cased* and *uncased* piles.

The concrete of a cased pile is cast inside a form that usually consists of a metal shell or thin pipe left in the ground. The shell may be so thin that its strength is disregarded in evaluating the structural capacity of the pile, but it must still have adequate resistance against collapse under the pressure of the surrounding ground before it is filled with concrete. Very thin shells and pipes cannot be driven without being supported internally by a mandrel, which is itself a source of expense and, at least occasionally, of construction difficulty. Of the several types of thin-shell mandrel driven concrete piles used in North America, three are illustrated in Figs. 12.2a to 12.2c. The pile shown in Fig. 12.2d is a modification of an uncased pile and is described in the next paragraph. The shell of the pile shown in Fig. 12.2e is heavy enough to permit driving without a mandrel.

Elimination of the casing or shell reduces the cost of materials incorporated in the pile; hence, there have been economic incentives for developing uncased piles. Several early types were formed by driving an open-ended pipe into the ground, cleaning it out, and filling the hole with concrete as

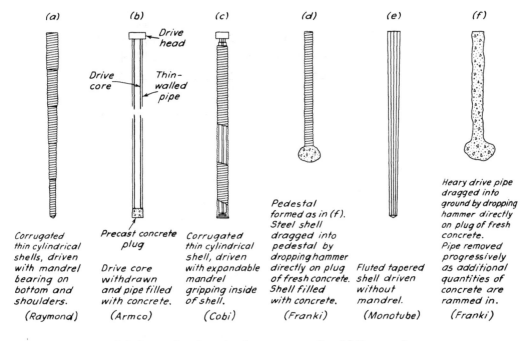

FIGURE 12.2. Examples of cast-in-place concrete piles. (*a*) Raymond step taper mandrel-driven pile. (*b*) Armco core-driven thin-walled pipe pile with end closed by precast concrete point. (*c*) Cobi-type corrugated cylindrical thin-walled pile driven by mandrel when expanded against shell by compressed air. (*d*) Cased Franki pile driven by drop hammer falling on fresh concrete within shell. (*e*) Union Metal Monotube driven without mandrel. (*f*) Uncased Franki pile.

the pipe was withdrawn. Such piles often contained imperfections or even discontinuities, and more positive measures are now universally required to assure continuity of the concrete. For example, in forming the uncased Franki pile (Fig. 12.2f), a drop hammer is allowed to fall directly on a mass of concrete in the lower part of a drive pipe; friction between the concrete and the pipe drags the pipe into the ground. When the required depth is reached, the drive pipe is lifted slightly and held against further penetration as concrete is fed into the pipe alongside the hammer and rammed into the soil to form a pedestal. The pipe is then withdrawn progressively as additional quantities of concrete are rammed in to form the shaft, which exhibits a rough exterior where it is in contact with the soil. The cased variant (Fig. 12.2d) is formed in the same fashion until the pedestal is created. A corrugated steel shell is then inserted into the drive tube, a plug of concrete is placed in the bottom of the shell on the pedestal, and the plug is driven to drag the shell into the upper part of the still plastic pedestal. The drive pipe is then removed and the remainder of the shell filled with concrete.

The installation of piles by casting them in bored holes instead of by driving is similar to that of piers and is discussed in Chap. 13. Several hybrid types are available, moreover, such as those formed by pumping concrete under continuous positive pressure through the hollow shaft of an auger, by which the hole is drilled, as the auger is pulled from the ground.

In choosing among the wide variety of cast-in-place piles, the engineer needs a detailed knowledge, which can be obtained in part from current brochures of pile contractors, of the characteristics and dimensions of the available piles, and a skeptical

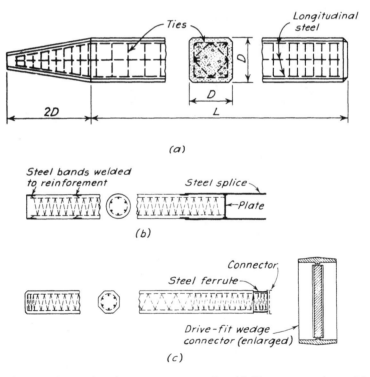

FIGURE 12.3. Examples of precast concrete piles. (a) Type commonly used for bridge trestles. (b) Sectional (Fuentes) pile with connections made by welding sleeve to steel band after inserting next section (c) Prestressed sectional pile (Brunspile) with connection made by driving fit of steel ferrule into slightly tapered connector.

attitude toward all operations that occur under conditions where no direct inspection is possible.

Precast concrete piles are manufactured in a variety of shapes. A type commonly used for bridge trestles, and occasionally for buildings, is illustrated in Fig. 12.3a. Such piles must be reinforced to withstand handling until they are ready to be driven and must also be reinforced to resist the stresses caused by driving. If the required length has been underestimated, they can be extended only with considerable difficulty. If the length has been overestimated, they are expensive to cut off. Sectional precast piles (Fig. 12.3b), on the other hand, can readily be varied in length.

Precast piles may also be prestressed. Prestressing tends to reduce tension cracking during handling and driving, and provides efficiently for bending stresses. Full-length prestressed piles have the same disadvantages as ordinary reinforced concrete piles if their lengths are misjudged. Prestressed sectional piles (Fig. 12.3c) overcome this difficulty. Hollow cylindrical prestressed spun-concrete piles with diameters up to 5 ft or more and wall thicknesses of 4 to 6 in. have been developed for high capacities and have been used extensively for bridge piles.

Since most varieties of concrete piles can be driven to high resistance without damage, they are usually assigned greater allowable loads than timber piles (Table 12.1). Under ordinary conditions they are not subject to deterioration and they can be used above the water table. High concentrations of magnesium or sodium sulphate salts (over 1000 ppm of SO_3 in the porewater) may cause deterioration and require special precautions or selection of a different material. The salts in sea water and spray attack the reinforcement in piles through cracks in the concrete; as rust develops it causes the concrete cover to spall off. The best protection against attack is dense concrete of high quality. Prestressed piles are not so rapidly susceptible because tension cracks are minimized.

Steel Piles. Steel pipes, usually filled with

Table 12.1 Customary Range of Working Loads on Driven Piles[a]

Type	Load (tons)
Timber (8-in. tip diameter)	10–30
Concrete, precast, or prestressed	
10-in. diameter	20–60
18-in. square	70–200
Steel pipe or shell, concrete-filled, not mandrel-driven	
$10\frac{3}{4} \times 0.188$ pipe	30–50
$10\frac{3}{4} \times 0.250$ pipe	40–70
$12\frac{3}{4} \times 0.250$ pipe	50–80
14×0.312 pipe	60–90
16×0.375 pipe	100–120
Monotube, 7-gage	30–50
Steel pipe or shell, concrete-filled, mandrel-driven[b]	
Raymond Step-taper with $10\frac{1}{4}$-in. point	30–50
Raymond Step-taper with $12\frac{1}{8}$-in. point	40–70
12-in. corrugated, 16-gage	30–60
10-in. $\times$ 0.125 pipe	30–50
Steel H-section[c]	
HP 10 $\times$ 42	50–75
HP 12 $\times$ 53	50–95
HP 14 $\times$ 89	100–160
HP 14 $\times$ 117	150–200

[a] Indicated maximum loads can be exceeded if freeze or setup (Art. 12.5) occurs after pile has been driven to resistance corresponding to tabulated value.

[b] Use of the mandrel permits driving these piles to a resistance great enough to warrant working loads based on the full structural capacity of the pile.

[c] When driven with adequate hammer to resistance indicated by wave equation (Art. 12.5), H-piles may be stressed to as much as 12,600 lb/sq in. under working loads; in soils likely to deform the tips, the same stress may be allowed if the piles are equipped with drive points.

concrete after being driven, and steel H-sections are widely used as piles, especially if conditions call for hard driving, unusually great lengths, or high working loads per pile.

Steel H-piles penetrate into the ground more readily than other types, partly because they displace relatively little material. Consequently, they are often used to reach a strong bearing stratum at great depth. If driving is hard, and especially if the overburden contains obstructions or heavy gravels, the flanges are likely to be damaged and the piles may twist or bend. Serious defects may produce few if any symptoms that can be noticed during driving. Whenever conditions suggest the possibility of such damage, the points of the piles should be reinforced by measures such as the point shown in Fig. 12.4a. Furthermore, since damage to H-piles occurs only if the driving stresses in the pile become excessive, particular attention should be paid to dynamic stress analyses carried out by means of the wave equation (Art. 12.5). By such analyses the hammer, cushion, and pile can be selected to form a compatible system in which, with proper field supervision, the driving stresses can be kept below the value that will lead to damage.

Pipe piles usually range in diameter from 10 to 30 in. The wall thickness is rarely less than 0.10 in., and pipes with thicknesses up to 0.179 in. are usually driven with a mandrel. If the wall thickness exceeds 0.10 to 0.125 in., the steel is usually considered to participate with the concrete in carrying the structural load. If the piles are driven open-ended they must be cleaned out before they are filled with concrete. Ordinarily they are closed at the lower end, usually with a plate, as shown in Fig. 12.4b. More elaborate closures such as conical points rarely display any significant advantages. In a few soils, such as stiff plastic clays, the overhang of the plate should be eliminated. Since pipe piles can be inspected after driving, damaged piles can be identified and, if not capable of repair, can be rejected.

Steel piles are subject to corrosion. The deterioration is usually insignificant if the entire pile is buried in a natural soil formation, but may be severe in some fills on account of the entrapped oxygen. If the piles extend upward to or above the ground surface, the zones just above and below the ground line are particularly vulnerable. Moreover, severe attack may be expected between tide levels in sea water, and above high tide where the pile may be subjected to salt spray. Expert advice from corrosion specialists may be needed to assess the probability of damage and to select suitable defenses. Factory-applied epoxy coatings are effective and not easily damaged by pile driving. Concrete encasement of the most vulnerable zones may provide adequate protection.

Specifications usually include tolerances

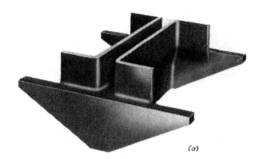
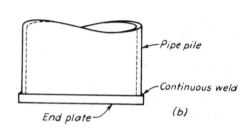

FIGURE 12.4. (a) Point reinforcement for H-pile (Associated Pile and Fitting Corp.). (b) Plate for closed-ended steel pipe pile.

on the verticality and straightness of driven piles but, as a practical matter, inspection to judge compliance is possible only if the piles are hollow. Thus, pipe piles and mandrel-driven steel shell piles can be inspected but timber piles, H-piles, and most precast-concrete piles cannot. There is no evidence that deviation from verticality of as much as 5 to 10 per cent of the length of the pile is harmful, or that a considerable curvature free from sharp bends is detrimental, even in soft soil, if the point of the pile is within the limits stated above for verticality. Since many piles that cannot be inspected undoubtedly deviate substantially from their theoretical position and are nevertheless accepted, it is unreasonable to penalize by too severe restrictions those piles that can be examined.

Composite Piles. Occasionally, piles are formed by joining upper and lower sections of dissimilar materials, such as concrete above the water table and untreated timber below. The cost and difficulty of forming a suitable joint have led to the virtual abandonment of this type of construction in the United States and Canada. On the other hand, a wide variety of piles is available consisting of various combinations of shells, pipes, and other components.

12.3. Installation of Piles

Pile Driving Equipment. Piles are commonly driven by means of a hammer or, occasionally, by a vibratory force generator. The hammer operates between a pair of parallel guides or *leads* suspended from a standard lifting crane. At the bottom the leads are connected to the base of the crane boom by a horizontal member, known as the *spotter*. The spotter can be extended or retracted to permit driving piles on a batter and also to plumb the leads over the location of a vertical pile. The hammer is guided axially by rails incorporated in the leads.

Impact Hammers. Originally, pile drivers were equipped with hammers that fell from the top of the leads to the top of the pile. Occasionally devices of this type, known as *drop hammers*, are still used, but most impact hammers are of the steam or diesel type. Drop hammers falling on the fresh concrete, however, are used in the installation of Franki piles (Figs. 12.2d and 12.2f).

Steam hammers contain a ram lifted by steam pressure and allowed to fall by gravity with or without the assistance of steam pressure. Compressed air may be substituted for steam. If the fall is due to gravity alone, the hammer is called *single acting* (Fig. 12.5a). If steam or air pressure adds to the

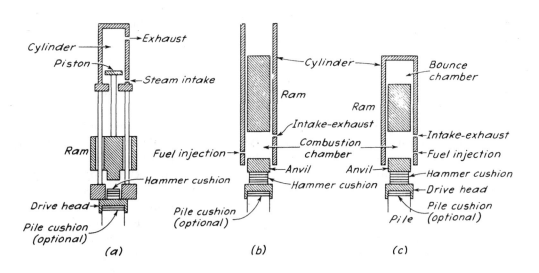

FIGURE 12.5. (a) Diagrammatic sketch of a single-acting steam pile hammer. (b) Open-ended diesel hammer. (c) Closed ended diesel hammer.

Table 12.2 Properties of Selected Impact Pile Hammers

Rated Energy (ft-lb)	Make	Model	Type[a]	Blows per Minute[b]	Stroke at Rated Energy (in.)	Weight Striking Parts (lb)
7,260	Vulcan	2	S	70	29	3,000
8,750	MKT[c]	9B3	DB	145	17	1,600
13,100	MKT	10B3	DB	105	19	3,000
15,000	Vulcan	1	S	60	36	5,000
15,100	Vulcan	50C	DF	120	15½	5,000
16,000	MKT	DE–20	DE	48	96	2,000
18,200	Link-Belt	440	DE	86–90	36⅞	4,000
19,150	MKT	11B3	DB	95	19	5,000
19,500	Raymond	65C	DF	100–110	16	6,500
19,500	Vulcan	06	S	60	36	6,500
22,400	MKT	DE–30	DE	48	96	2,800
22,500	Delmag	D–12	DE	42–60		2,750
24,375	Vulcan	0	S	50	39	7,500
24,400	Kobe	K13	DE	45–60	102	2,870
24,450	Vulcan	80C	DF	111	16	8,000
26,000	Vulcan	08	S	50	39	8,000
26,300	Link-Belt	520	DE	80–84	43⅙	5,070
32,000	MKT	DE–40	DE	48	96	4,000
32,500	MKT	S10	S	55	39	10,000
32,500	Vulcan	010	S	50	39	10,000
32,500	Raymond	00	S	50	39	10,000
36,000	Vulcan	140C	DF	103	15½	14,000
39,700	Delmag	D–22	DE	42–60		4,850
40,600	Raymond	000	S	50	39	12,500
41,300	Kobe	K–22	DE	45–60	102	4,850
42,000	Vulcan	014	S	60	36	14,000
48,750	Vulcan	016	S	60	36	16,250

[a] S = single-acting steam; DB = double-acting steam; DF = differential-acting steam; DE = diesel.
[b] After development of significant driving resistance.
[c] For many years known as McKiernan–Terry.

downward energy, the hammer, depending on the details of its construction, is called *double acting* or *differential*. Hydraulically actuated differential hammers are also used.

The ram of hammers of the Vulcan type (Table 12.2), such as that of the single-acting hammer shown in Fig. 12.5a, strikes a *hammer cushion* positioned in the base of the hammer.

The original purpose of the cushion was to prolong the life of the hammer by reducing the impact stresses. In other hammers, notably the MKT type, the ram strikes directly on the base or anvil. The top of the pile itself is protected by a *drive head* suspended from the base of the hammer and fitted to the dimensions of the pile. Between the drive head and the pile may also be inserted a *pile cushion*. These various elements not only protect the top of the pile from local overstress, but have a significant influence on the stress waves developed in the pile during driving. The selection of suitable cushions affects the driving characteristics of a pile, the depth to which it can be driven and, to some extent, its load-carrying capacity. These matters will be discussed in Art. 12.5.

The rating of a hammer is based on the gross energy per blow; for a drop hammer the rated energy is the product of the weight W_H of the ram and the height of fall H. Energy is lost because of friction in the ram guides. The *efficiency* of the hammer is defined as the energy delivered at impact divided by the gross rated energy. Steam hammers for greatest efficiency must be operated at the pressure for which they were designed. Their efficiency decreases markedly at lower pressures. On the other hand, too high a pressure causes the hammer to jump off the pile, a behavior known as racking, and damages the equipment. Even at the appropriate pressure, the efficiency of well-maintained steam hammers as devices for developing and transmitting energy is on the order of 70 per cent.

The weights of rams, heights of fall, and other pertinent information for commonly used single-acting steam hammers are shown in Table 12.2, together with data concerning double-acting and differential hammers delivering comparable energy. Hammers delivering the greater energies are usually used for the heavier and longer piles. For special purposes, such as driving piles of great length and large diameters for offshore drilling platforms, hammers delivering energies from 50,000 to more than 180,000 ft-lb are manufactured but are rarely used for ordinary foundations.

Diesel hammers are of two general types, open-ended (Fig. 12.5b) or closed-ended (Fig. 12.5c). The ram of either type is lifted by explosion of fuel and compressed gas in a chamber between the bottom of the ram and an anvil block in the base of the housing. In the open-ended hammers the ram falls by gravity and delivers energy to the anvil by direct impact. As it descends, however, fuel is injected into the space, known as the combustion chamber, between the ram and the anvil. At approximately the instant of impact the fuel ignites and again lifts the ram. For a significant time the pressure of the burning gases also acts on the anvil and increases the magnitude and duration of the driving force. In this respect the driving characteristics of diesel hammers differ appreciably from those of drop or steam hammers.

In the closed-ended hammers the housing extends over the cylinder to form a bounce chamber in which air is compressed by the rising ram. The compressed air acts as a spring that limits the rise of the ram and thereby shortens the stroke. It returns its stored energy to the ram on the downstroke. On account of the shorter stroke the number of blows per minute is increased with respect to that for open-ended hammers.

The energy delivered by several commonly used diesel hammers is listed in Table 12.2.

Vibratory Drivers. Piles are also driven by force generators consisting of a static weight and a pair of counterrotating eccentric weights (Fig. 12.6), arranged so that the horizontal components of the centrifugal force cancel whereas the vertical components are additive. The vibrating part of the machine is positively attached to the head of the pile to be driven but the remainder of the machine is isolated from the vibrator by springs so that it does not participate in the vibratory motions. The pulsating force facilitates penetration of the pile under the influence of the constantly acting downward weight.

Vibratory drivers differ from each other in the type of motive power and in the fre-

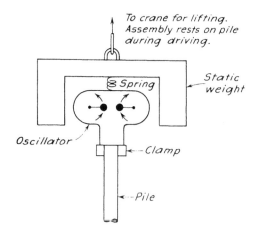

FIGURE 12.6. Diagrammatic sketch of vibratory pile driver.

quency of the pulsating force. Low-frequency drivers ordinarily operate at a constant frequency in the range of 10 to 30 Hz. If the frequency can be made equal to the natural frequency of the system consisting of the pile, the driver, and the soil, the device is known as a *resonant driver*. The frequency of a resonant driver must, therefore, be adjustable. The resonant frequency is often within the range of 50 to 150 Hz. When the system is at resonance, the pile exhibits energetic up-and-down displacements limited only by the damping furnished by the surrounding soil. Almost the full downward weight of the driver and pile are then effective in causing the point of the pile to penetrate the underlying material. Penetration may be very rapid unless the resistance to point penetration exceeds the weight of the pile driving assembly. Because the pull of the crane (Fig. 12.6) can exceed the downward weight and there is, of course, no point resistance, vibratory drivers are effective extractors.

Driving Resistance. Piles installed by impact hammers are ordinarily driven to a resistance measured by the number of blows required for the last inch of penetration. For wood piles driven by steam hammers delivering 15,000 ft-lb of energy, the final number of blows per inch should not be more than 3 or 4 to reduce the danger of breaking or brooming the piles. Damage is, moreover, likely to be less if the piles are driven with a diesel hammer even at the same rated energy. Resistances of 6 to 8 blows/in. are commonly specified for concrete and steel piles. Resistances of more than 10 blows/in. usually constitute excessively severe requirements.

Jetting, Drilling, and Spudding. If piles must penetrate through dense layers of sand or gravel in order to continue through underlying soft deposits, the sand or gravel may be loosened by jetting. In this procedure, a stream of water is discharged near the point or along the sides of the pile through a pipe 2 to 3 in. in diameter. The jetting pump must be capable of delivering about 500 gpm at a pressure of 150 to 200 lb/sq in. The water loosens the sand and creates a quick condition so that the pile can easily penetrate through the sand to the underlying material.

Rotary drilling (Art. 5.2) with specially adapted equipment is capable of penetrating similar deposits as well as stiff or hard cohesive soils to considerable depth.

Relatively thin deposits of stiff clays or soft rock at shallow depths may sometimes be penetrated with spuds consisting of hard metal points driven into the ground and withdrawn before the piles are inserted. Jetting and drilling are relatively common practices whereas spudding is not.

Predrilling. When piles are driven through saturated plastic clay, they displace a volume of soil that may be as great as that of the piles. This displacement usually takes the form of a heave of the ground that may lift adjacent structures or piles already driven. If the piles depend for their support on a firm layer beneath the plastic clay, the heave may cause loss of bearing capacity; in many instances, the heaved piles may be redriven to firm bearing.

The displacements may cause not only heave but also undesirable lateral movements, especially if there are adjacent excavations extending below the level from which the piles are driven. If the piles are of the composite type, the heave may cause separation at the joint.

Where the displacements are likely to be objectionable, they can be considerably reduced by removing part of the soil in the space to be occupied by each pile. This is usually done by predrilling with an auger or by the use of a rotary cutting tool combined with water jets which transform the clay where the pile is to be located into a heavy slurry. These procedures are known as *predrilling* or *coring*.

12.4. Action of Piles Under Downward Loads

Behavior of Single Piles. Piles are commonly classified as either *point-bearing* or *friction*. A point-bearing pile derives practically all its support from the rock or soil near the point, and very little from the soil surrounding the upper part. On the other hand, a friction pile derives its support principally from the surrounding soil through the development of shearing resistance between the soil and the pile. A very small percentage of the load is carried by the soil near the lower end of the pile.

In contrast to the relatively simple conditions of support implied by the classification of piles into these two categories, the manner in which piles are actually supported is by no means simple. For example, under many conditions encountered in nature, the stiffness or relative density of the subsoil increases in a general way with depth. Piles may be driven through the soft upper layers and through progressively stiffer layers until the required capacity is achieved. Such piles derive an appreciable part of their support from the surrounding material, especially in the lower portion of their length, but they may also derive considerable support by direct bearing at their points.

One of the most important engineering decisions in connection with any pile job is the choice of the most appropriate type of pile to suit the particular circumstances. Many factors enter into the final decision, including the behavior during driving (Art. 12.5). Hence, it appears unlikely that simple, definite rules can be formulated for the sure guidance of the inexperienced engineer. Nevertheless, a sound conception of the manner in which piles of different characteristics transmit their load to the soil under working conditions is an invaluable asset and can serve as a basis for the development of good judgment as the engineer's experience accumulates.

A point-bearing pile surrounded by soil is sometimes erroneously regarded as a free-standing column unsupported laterally by the soil in which it is embedded. However, both experience and theory have amply demonstrated that there is no danger of buckling an axially loaded point-bearing pile of conventional dimensions because of inadequate lateral support, provided it is surrounded by even the very softest soils. Hence, the stresses in such piles under working loads may be taken as those appropriate for the materials of the pile under direct compression. For piles that decrease in size with depth, the critical section is at the tip. Working stresses in cast-in-place concrete should not exceed 0.25 to $0.33f_c'$, where f_c' is the 28-day unconfined compressive strength of the concrete in test cylinders. The lower part of the range of values should be used for piles in which placement of the concrete would be difficult; that is, where the tip is of small diameter, the side walls of the shell contain irregularities, reinforcement is required, or the pile is battered. In prestressed piles an allowance must be made for the amount of prestress. Working stresses at the tips of steel piles are ordinarily limited to 12,600 lb/sq in.

To assure that the required concrete strength of cast-in-place piles is obtained, the slump should be controlled between the limits of 3 to 6 in. or more. For a straight, smooth pipe driven vertically, a $3\frac{1}{2}$-in. slump is suitable. On the other hand, in a batter pile, or in a vertical pile with a corrugated shell or reinforcement, the slump should be at least 6 in.

Once the requirement is satisfied that the material at the tip is not overstressed, the capacity of a point-bearing pile depends entirely on the capacity of the material on which the point finds its bearing and on the degree to which the point has a satisfactory seat on or in the bearing material. If the supporting stratum does not consist of extremely resistant material such as sound rock or firmly cemented hardpan but, in contrast, is a fairly thick but only relatively firm deposit, two different types of piles deserve consideration. A pile of a small point diameter, but of a type capable of transmitting the driving stresses to the point without excessive loss of energy, is likely to penetrate for several feet into the firm stratum and develop high capacity by a combination of direct bearing at the point and intense skin friction within the zone of embedment in the firm stratum. The contribution of the skin friction would be markedly increased if the lower part of the pile were given a small uniform taper.

The other alternative if the bearing stratum is not exceptionally firm is a pile with a very large point. The point may consist of a plate or precast concrete tip, or it may have the form of an enlargement or pedestal made by expanding fresh concrete into a soft deposit directly above the bearing stratum. The capacity of such a pile is not enhanced by any wedging action because of penetration of the pile into the bearing stratum but is governed entirely by the capacity and compressibility of the soil on which the bottom of the pile rests. Piles with enlarged points are most beneficial in loose granular materials; they are also appropriate if the bearing stratum is very firm but is so thin that piles of small diameter could punch through it.

The term *friction pile* is somewhat of a misnomer, inasmuch as it implies that the shearing forces between pile and soil are necessarily derived from friction; they may also consist of adhesion. In any event, the capacity of friction piles depends on the characteristics of the material surrounding the pile. Hence, as a general rule the structural strength of an axially loaded friction pile does not govern its design. If a friction pile has parallel sides, the load is transferred from the pile to the soil exclusively through shear. If the pile has a taper, a portion of the load is transmitted by direct bearing, but still the greater part is transferred by shear. If the piles are driven in soft clay, the shear consists primarily of adhesion and the difference in supporting capacity between parallel-sided and tapered piles is relatively small. However, for soils possessing appreciable frictional resistance, such as sands, silts, and clays containing air, the wedge action of a tapered pile increases the lateral pressure and increases the shearing resistance correspondingly. Hence, a uniformly tapered pile may be advantageous under such circumstances. For example, light loads may often be supported efficiently on short timber piles in loose sand.

The foregoing discussion serves as a basis for a general understanding of the manner in which individual piles support their loads and gives some indication concerning the most appropriate types of piles for given conditions.

Pile Load Tests. The many variables influencing the behavior of a pile under load and the complex nature of the phenomena involved have led to the practice of carrying out load tests on one or more piles at the site of important projects. The test piles are preferably of the same type and are driven by the same equipment to the same requirements as anticipated for the job. In some instances, several alternatives are investigated to permit refinement in design.

All pertinent details of the equipment and procedure are recorded during the driving of a test pile, including the blows per inch of penetration, preferably throughout the entire embedded length of the pile. Any pause in driving, as for breakdown of equipment or for splicing the pile, is noted.

The load is usually applied in increments by means of a hydraulic jack reacting against a dead weight or against a yoke fastened to a pair of anchor piles, Fig. 12.7. Under each increment the settlement of the head of the pile is observed as a function of time until the rate of settlement becomes very small. A new increment is then applied. As the capacity of the pile is ap-

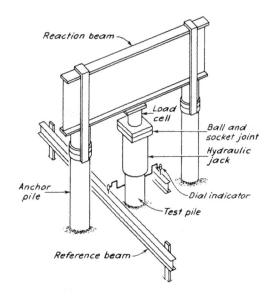

FIGURE 12.7. Arrangement for load test on pile by use of anchor piles.

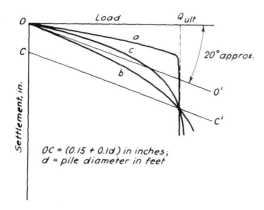

FIGURE 12.8. Typical results of load test on (a) friction pile, (b) end-bearing pile and (c) pile deriving support from both end-bearing and friction.

proached, the size of the increments is decreased in order to permit better definition of the load at which the capacity of the pile is reached. The rise of the head of the pile is measured when the load is removed.

The results of typical load tests are shown in Fig. 12.8 in which the total load is plotted as a function of the settlement of the pile head. Curve a represents a pile that slipped or plunged suddenly when the load reached a definite value termed the *ultimate pile load* or *pile capacity*. Curves b and c, on the other hand, show no well-defined breaks; consequently, the determination of the ultimate pile load is to some degree a matter of interpretation. A reasonable procedure (Davisson, 1973) that takes account of the significant variables is illustrated in the figure. The elastic deflection of the pile is computed by means of the expression PL/AE and plotted on the load-settlement diagram as line OO'; for the best interpretation the scales of the diagram should be chosen so that the slope of OO' is about 20°. The line CC' is drawn parallel to line OO' with an intercept on the settlement axis equal to $(0.15 + 0.1d)$ in., where d is the diameter of the pile in feet. The intercept is a measure of the tip settlement required to develop the capacity. The ultimate load is defined as the load at which the line CC' intersects the load-settlement curve. This criterion is applicable to load-settlement records obtained from tests in which each load increment is held for periods not exceeding 1 hr. It is overly conservative, however, for load-settlement records based on holding periods of 24 hr or longer because creep or consolidation settlements may then become significant portions of the total.

Detailed procedures for conducting load tests and for deriving from them the ultimate capacities or working loads have been standardized by various organizations (ASTM D-1143) and are incorporated into various building codes. Some of the procedures are quite elaborate. The cost of a load test depends greatly on its duration, especially if the job is delayed while the results are awaited. Specifications that require maintaining the final load (usually twice the design load) for several days are rarely justifiable. In most instances the performance under twice the design load in 24 hr or more can be judged by applying 2.25 times the design load for 1 hr.

A load test can furnish data concerning the load-settlement characteristics and capacity of a pile only at the time and under the conditions of the test. Numerous factors may lead to different behavior of a similar pile beneath the actual structure. These will be given consideration under the next subheading and, in greater detail, in Part C. In addition to the information that a load test furnishes about the validity of the assumptions made in the design, it also provides a useful and often necessary check on the adequacy of the equipment and field procedures used during construction.

Pile Groups. In the foregoing discussion only the behavior of single piles has been considered. Nevertheless, piles are almost never used singly, but are combined in groups or clusters. The changes in stress conditions as well as alterations in the consistency and relative density associated with driving previous piles may have a marked influence on the behavior of the remaining piles not only during driving but also during their support of the loads to which they will be subjected. The behavior of a group of piles may not be directly related to the behavior of single piles subjected to the same load per pile in the same deposit.

Of particular importance are the relative contributions of skin friction and point bearing to the support of a single pile reaching a firm stratum as compared to their corresponding contributions in a group of piles reaching the same stratum. When a single pile is loaded, as in a load test, a large part of its support may be furnished by the soil alongside the pile through skin friction, even if the soil is rather weak and compressible. If the same pile has many neighbors, to each of which the surrounding soil furnishes support, the stress accumulated in the entire block of soil in which the group is embedded may tend gradually to compress the soil and correspondingly to allow the piles to settle at least slightly, whereupon a larger portion of the load is transferred directly from the pile to the firm stratum. In large groups, most of the load may sooner or later be carried by point bearing, irrespective of the magnitude of skin friction that may have been developed at higher elevations around a single pile in a load test.

Under other subsoil conditions, other differences arise between the action of single piles and pile groups or entire pile foundations. Such differences are discussed in Part C in connection with foundations in various types of deposits.

Settlements of Pile Foundations. Not only may the capacity of a single pile when loaded individually be different from its capacity when all the piles in a group are loaded, but the entire load-settlement relationship may be strikingly different. Consequently, the settlements of pile groups cannot, in general, be predicted on the basis of pile load tests. Moreover, if the pile foundation is underlain, even at considerable depth, by a compressible deposit, the entire foundation may settle because of consolidation of the underlying strata, even if the individual piles hardly move at all with respect to the soil in which they are embedded. Failure to recognize this possibility has led to spectacular examples of excessive and unexpected settlements.

12.5. Dynamics of Pile Driving

Significance of Behavior During Driving. In a foundation, the capacity of a pile of given length and dimensions is determined by the ability of the surrounding and underlying soil to provide skin friction and point resistance. Except for the effects of alteration of the character of the soil, which may sometimes be appreciable, the manner in which the pile arrives at its final position in the subsoil is immaterial. The capacity of a pure friction pile is equal to the adhesion or friction per unit of contact area between soil and pile, multiplied by the contact area itself; it is hardly influenced by slight differences in the embedded length or by the ease or difficulty with which the penetration is accomplished under a given hammer. On the other hand, unless a point-bearing pile can achieve adequate contact with or sufficient penetration into a bearing layer, it cannot develop the required capacity. An increase of a few inches in penetration may greatly increase the capacity, but whether the needed penetration can be achieved depends on the effectiveness of the pile-driving procedure. Since the ability to drive the pile until suitable resistance is developed is an essential requirement for assuring the required capacity, the engineer cannot avoid considering the implications of the dynamics of pile driving.

Pile Formulas. It seems obvious that the greater the resistance of a pile to driving, the greater should be the pile's capacity to support load. With this apparent truism as a starting point, many engineers have concluded that it should be possible to calculate the capacity of a pile from a knowledge of the energy delivered by the hammer and the penetration of the pile under a hammer blow. The resulting expressions for pile capacity are known as *pile formulas*. Their variety and number are matched only by their shortcomings.

All the common pile formulas equate the energy delivered by the hammer to the work done by the pile as its tip penetrates a distance s against a resistance R, with various allowances for the losses of energy associated with the procedure. The time-dependent aspects of the phenomena of stress transmission are ignored; as the next subsection will demonstrate, these are of fundamental importance. For example, in the *Engineering News formula* all the energy losses associated with each blow of a single-acting steam hammer are assumed to be equivalent to the work that would have been done by a penetration of 0.1 in. against the resistance R. All the work accomplished, during the useful penetration s and the assumed lost penetration, is

$$W_H H = R(s + 0.1)$$

where W_H is the weight of the ram expressed in the same units as R, and s is in inches. If H is expressed in feet, and if a factor of safety of 6 is assumed, the resistance under working conditions is

$$R = \frac{2 W_H H}{s + 0.1} \qquad 12.1$$

For other than single-acting steam hammers, the numerator is replaced by $2E$, where E is the hammer energy per blow.

Because of its simplicity, the *Engineering News* formula (eq. 12.1) has been widely used, but statistical comparisons with the results of load tests on driven piles have shown such poor correlation and such wide scatter that further use of the formula cannot be justified.

Comparisons of computed and measured capacities have shown a few of the many pile formulas to be statistically superior to the others. All such formulas, however, are fundamentally unsound because of their neglect of the time-dependent aspects of the dynamic phenomena. Hence, except where well-supported empirical correlations under a given set of physical and geological conditions are available, the use of formulas apparently superior to the *Engineering News* formula is also not justifiable.

Pile formulas have remained in vogue in foundation engineering for many decades because of their great convenience. If the engineer, on the basis of only scanty information about the subsurface conditions, can reach the conclusion that a pile foundation is necessary and can arrive at a reasonable working load to be assigned to the piles, he can proceed with the design of the remainder of the structure with little further regard for the details of the foundation. He needs only to specify that the required capacity be attained by driving the piles to the resistance determined for that capacity by whatever pile formula he selects. The responsibility for achieving the foundation is thus transferred to the pile-driving contractor. Unfortunately, some of the errors of even the best of the current formulas may, under certain conditions, be large and misleading. For example, most of the more complex formulas would indicate that increasing the weight of the pile with respect to that of the ram of the hammer should decrease the capacity corresponding to a given driving resistance; in reality, the opposite effect is often experienced.

Most of the defects of pile formulas can be eliminated by a more realistic analysis of the dynamics of pile driving in which the pile is considered to be a long elastic bar subjected to transient stress waves set up by the impact of the hammer. The implications of this approach are considered in the following section, although the detailed information needed for routine application to specific design problems is still being developed.

The validity of any dynamic analysis depends in part on the assumption that R, the dynamic resistance to penetration, is equal to or is at least related to the static capacity of the pile after driving. In fine-grained soft saturated soils, pore pressures are likely to build up during driving and significantly influence the resistance to penetration, whereas under a static load of long duration the pore pressures dissipate and the effective stresses in the soil correspondingly change. Under these conditions, no correlation between dynamic and static resistance can be expected. In free-draining soils such as medium to dense sands, and in stiff to hard clays, dynamic and static resistances, on the other hand, are more closely related.

Transmission of Stresses During Driving. A realistic representation of the dynamics of pile driving must consider the complex chain of events initiated by a single blow of a pile hammer. The energy delivered by the hammer sets up time-dependent stresses and displacements in the pile-head assembly, in the pile, and in the surrounding ground. As the length of a pile is always large compared to its diameter, a pile does not behave as a concentrated mass but, as previously mentioned, more nearly as an elastic bar in which the stresses travel longitudinally as waves. When the waves are compressive, as at the tip of a pile being driven into a hard material, they cause the pile to penetrate into the ground. Yet, if the compressive stresses are too great, they may damage the pile. On the other hand, when the soil at the tip is soft and driving is easy, the compressive wave may be reflected upward from the end of the pile as a tensile wave. If at some point in the pile the tensile stress is not canceled by other stresses that are compressive, a net tension may develop at least for an instant. The stress may be sufficient to crack a precast concrete pile. Thus, the behavior of the pile, with respect both to its ability to penetrate into the soil and to its structural integrity during driving, is intimately related to the mechanics of stress-wave transmission within the pile. To the extent that the dynamic force developed at the tip during driving is related to the static bearing capacity of the pile, a knowledge of this force is useful in estimating the static capacity.

The theory of wave transmission in a prismatic elastic bar struck longitudinally by a rigid object was developed more than a century ago and numerical solutions were obtained for several simple boundary conditions. The conditions were so far removed from the complexities of actual pile driving, however, that the solutions had little practical value. The introduction of more realistic conditions became possible only with the development of suitable theoretical models and electronic computation.

According to the theory of longitudinal impact of a prismatic elastic bar, the stress waves travel axially at a velocity

$$c = \sqrt{\frac{E}{\rho}} \qquad 12.2$$

where E is the modulus of elasticity and ρ the mass density of the material in the bar. The mass density is defined as

$$\rho = \frac{\gamma_P}{g} \qquad 12.3$$

where γ_P is the unit weight of the material constituting the bar and g is the acceleration of gravity. The velocity c of the stress wave is not to be confused with the velocity v at which a particular point in the bar actually moves. The former, known as the *velocity of longitudinal wave propagation*, or sometimes as the *seismic velocity*, is a constant for a given solid, elastic material. During the passage of a single wave, the longitudinal direct stress in the bar at any point is related to the particle velocity at that point by the simple expression

$$p = \frac{E}{c} v = \rho c v \qquad 12.4$$

The force transmitted across a section of the bar is then

$$P = pA = \rho c A \cdot v \qquad 12.5$$

Since v is a function of position and time, P and p are similarly functions of these quantities. The capability of the bar to transmit longitudinal force is measured by the product $\rho c A$, which is designated as the *impedance* of the pile.

The capacity of a pile at a given depth is the force that can be exerted by the surrounding soil against downward displacement. At least this much force must have been transmitted to the soil by the pile during driving in order for the point to have penetrated to a given position under the last blow of the hammer. In particular, enough force must have been transmitted to the pile to overcome the side and point resistances. Conversely, no matter how much energy may be applied to the head of the pile, the force that can be transmitted down the pile is limited by the impedance.

Since the impedance $\rho c A$ determines the maximum force that can be transmitted along the pile as long as the material remains elastic, it is, therefore, a measure of the ability of the pile to develop the required capacity as a consequence of its being driven into the ground. If the impedance of a pile is increased, the potential for obtaining a greater capacity with a particular hammer is also increased, provided the pile is not of such unusually large dimensions with respect to the hammer that its action resembles that of a large mass instead of that of a bar.

Relative magnitudes of impedance are listed for several common pile sections in Table 12.3. They indicate that piles of roughly the same exterior dimensions but of different materials have widely different impedances. For example, piles of 10-in. diameter or width may be arranged in order of increasing impedance with respect to that of wood as follows: wood (1.0); steel pipe with 0.279-in. wall (1.9); steel pipe with 0.365-in. wall (2.3); concrete (3.1); HP10 × 57 (3.3); concrete-filled steel pipe with 0.279-in. wall (4.6). Qualities tending to increase impedance are increased density, higher modulus of elasticity, and larger cross-sectional area. For a given material the impedance depends only on the area. The influence of the concrete filling in the pipe pile is notable.

The force actually developed at the tip of a pile depends, however, not only on $\rho c A$ but also on the energy that can be furnished by the hammer and on a wide variety of other factors including the nature of the impulse delivered by the hammer, the stress-transmission characteristics of the cushions and pile-head assembly, the general pattern of distribution of the resistance exerted by the soil along the pile, and the proportion of the total resistance developed

Table 12.3 Stress-Transmission Characteristics of Typical Piles

	γP (lb/ft^3)	$\rho = \dfrac{\gamma P}{g}$ (lb sec^2/ft^4)	c (ft/sec)	ρc (lb sec/ft^3)	A (in.^{2})	$\rho c A$ (lb sec/in.)	Ratio[a]
Wood							
10-in. diameter kiln dry	40	1.24	13,600	16,900	78.5	768	1
10-in. diameter treated southern pine	60	1.86	10,600	19,750	78.5	898	1.2
Concrete	150	4.66	11,100	51,800			
10-in. diameter					78.5	2360	3.1
20-in. diameter					314.2	9410	12.3
Steel	490	15.2	16,900	257,000			
HP 10 × 57					16.76	2500	3.3
HP 12 × 53					15.58	2430	3.2
HP 14 × 117					34.44	5370	7.0
10¾ × 0.188 pipe					6.24	928	1.2
10¾ × 0.279 pipe					9.18	1440	1.9
10¾ × 0.365 pipe					11.91	1770	2.3
10¾ × 0.188 pipe[b]					53.30	7930	10.3
Steel/concrete							
10¾ × 0.279 pipe filled with concrete	185	5.76	12,100	69,800	87.9	3550	4.6

[a] Ratio of $\rho c A$ to that for 10-in. wood pile.
[b] With steel driving mandrel weighing 160 lb/ft.

along the sides of the pile in comparison to that beneath the point.

Hammers differ greatly in the manner in which they deliver energy to the anvil or hammer cushion. Diesel hammers exert forces of relatively great duration in comparison with those exerted by steam hammers. The total energy actually transmitted by a hammer can best be ascertained by continuous measurement of the velocity of the ram as it approaches the pile, reverses direction, and rises again. Such measurements have shown that the efficiency of well-maintained hammers has little relation to the efficiency factors quoted by the manufacturers, and that the efficiency of poorly maintained or improperly operated hammers may be extremely low.

Pile cushion blocks may be described as soft or hard. For a given hammer and pile, the stress wave induced if the cushion is soft is longer and exhibits a lower peak stress than if the cushion is hard. The lower stress leads to longer hammer life and less potential for damaging the pile. If, however, the peak force generated with the soft cushion is not sufficient to produce the desired ultimate capacity for the pile, a hard cushion may be required.

Hammer and pile cushions may be regarded as springs having a modulus equal to their column stiffness AE/L, where A and L are respectively the cross-sectional area and the height of the cushion and E is the modulus of elasticity of the material. Both the dimensions and the modulus of elasticity are important in determining the spring modulus. Relatively few materials have been found satisfactory as cushioning material. For soft cushions, wood and asbestos are most common. Hard cushions usually consist of alternating disks of aluminum and micarta, although other materials similar to micarta are being developed and used. These materials are either inexpensive or possess long life relative to their cost. Other materials, such as wood chips or coiled steel cable, are often utilized but are undesirable because their properties cannot be controlled. Since cushions absorb much energy, it is not uncommon for wood cushions to catch fire.

Experience indicates that cushions should be available with stiffnesses covering the range from soft to hard, but only cushions of known characteristics should be used. Lack of control over cushioning materials permits a degree of subterfuge. According to the pile formulas, a small penetration corresponds to a high bearing capacity; thus, an inadequate pile can, by manipulation of cushion block materials, be made to appear acceptable to the unwary inspector.

For the most satisfactory results, the type and dimensions of the cushion block and the characteristics of the hammer should be chosen to satisfy two criteria: (1) to assure development of a peak driving force in the pile at least equal to the desired ultimate capacity of the pile being considered, without overstressing the pile; and, (2) to transmit as much of the available hammer energy as possible to the pile. The second requirement leads to economy in driving; it sometimes must be sacrificed to achieve the first. The significance of the requirements is illustrated in Fig. 12.9, which refers to long piles driven by Vulcan single-acting hammers operating at 75 per cent efficiency. If an 800-kip ultimate capacity is required, it is apparent that a pile with an impedance of at least 6400 lb sec/in. would be needed, that the 010 hammer would be required if that impedance were chosen, and that an aluminum–micarta cushion would be effective whereas the capacity could not even remotely be achieved with the softer pine plywood cushion. A lighter hammer could be used if the pile impedance were increased. On the other hand, if an ultimate capacity of only 100 kips were needed, a pile with much lower impedance could be driven, and a soft cushion would be more efficient in transmitting the energy. The figure is strictly applicable only if the stresses at the head of the pile are not modified by reflected waves during the period of impact. For short piles this condition is not likely to be satisfied and more complex analyses are required; short piles can often be driven to the required capacities with

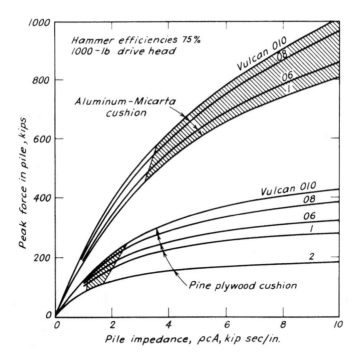

FIGURE 12.9. Relation between peak driving force in pile and pile impedance ρcA for Vulcan single-acting hammers of various energies, and for stiff (aluminum-micarta) and soft (pine plywood) cushion blocks. Conditions associated with maximum transmission of driving energy from hammer to pile are indicated by shaded areas (after Parola, 1970).

somewhat smaller hammers than indicated by Fig. 12.9 because of compressive reflections at the pile tip. The figure also implicitly assumes that the pile is structurally capable of withstanding the driving stresses.

The properties of the soil also play a decisive role in the behavior of the pile during driving. In addition to the nature of the point resistance, discussed in the preceding paragraphs, the frictional forces along the sides of the pile may have significant effects. The distribution of the side forces from top to tip of the pile and the relative magnitudes of side and point resistance have been determined by measurement in only a few instances. The results suggest that at least a crude approximation to both can be made on the basis of borings and soil tests.

The foregoing factors influencing the stresses developed in a pile under actual driving conditions can be taken reasonably but approximately into account in the theoretical model (Fig. 12.10), by means of which the wave analysis has been extended from unrealistically simple cases to those of interest in practice. The pile is assumed to consist of a series of elements, each possessing a weight W_n, connected by springs with stiffnesses K associated with the elastic properties of the pile material. The resistances on the side of the pile, which damp the vibrations, are assumed to be of viscoelastic nature and to possess spring constants K' representative of the soil material. The point resistance is also represented by a viscoelastic element. The weights and stiffnesses of the ram and drive head and the properties of the cushion blocks are represented by appropriate elements.

In the investigation of a specific problem, the velocity of the ram at impact must be introduced, as well as numerical values for all the appropriate weights, stiffnesses, damping factors, and resistances. It is not usually necessary, however, to assign spe-

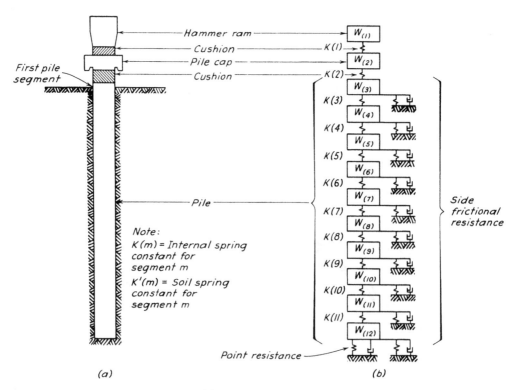

FIGURE 12.10. Actual pile (a) and idealized mechanical model (b) considered as basis for dynamic analysis of pile driving (after Davisson, 1970b).

cific numerical values to the point resistance and the side frictional resistance; it is necessary only to estimate the fraction of the total pile resistance likely to be developed by point resistance, and to assume the manner in which the side resistance is distributed along the pile. The results of a calculation for a given pile under particular soil conditions, driven by a specified hammer, may be expressed in the same form as those derived from the ordinary pile-driving formulas. That is, the ultimate static resistance may be plotted as a function of the resistance to penetration in blows per inch. A second quantity of significance, the maximum stress in the pile, may also be computed as a function of the resistance to penetration.

The results of such a calculation are shown in Fig. 12.11 for a steel pipe pile of 18-in. diameter and 0.375-in. wall thickness. The pile is 75 ft long, but because it extends through water for a wharf, it is embedded only 35 ft. It is assumed that 50 per cent of the pile resistance is carried by the tip and that the remaining 50 per cent, carried by skin friction, is distributed uniformly along the length of the pile. It is further assumed that the pile is driven by a No. 1 Vulcan hammer operating at 70 per cent efficiency. A cushion block consisting of alternate disks of aluminum and micarta is used.

The solid curve in Fig. 12.11 represents the ultimate resistance as a function of the blows per inch during driving. The curve applies strictly only to the specific length and embedment for which the calculation was made, but the results are rather insensitive to changes in length and only small errors are introduced even at substantially different lengths. On the assumption that the dynamic driving resistance is related to the static resistance, the curve represents the ultimate capacity of the pile if driven to a given penetration resistance. At 10 blows/in., for example, the ultimate resistance is indicated as 150 tons.

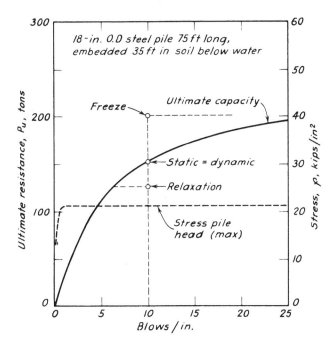

FIGURE 12.11. Resistance curve from dynamic analysis of steel pipe pile by application of wave equation. Soil reaction assumed to be 50 per cent at tip, 50 per cent side friction distributed uniformly along 35-ft embedded length. Driven by No. 1 Vulcan hammer operating at 70 per cent efficiency, with aluminum-micarta cushion block (after Davisson, 1970b).

If a load test is made on the pile, if the theory is correct, and if the static resistance is properly related to the dynamic resistance, the test load should correspond to the computed load of 150 tons. In some types of ground there is a tendency after driving for the capacity to increase. This phenomenon is known as *freeze*; its effect might be to increase the capacity, as shown in the figure. On the other hand, in some materials stress *relaxation* occurs and the capacity after driving decreases. The occurrence of freeze or relaxation is significant in design and can be investigated by means of a calculation such as that indicated in Fig. 12.11, together with the results of a load test (Art. 12.4). After the magnitude of freeze or relaxation has been ascertained in this manner, the information can be used to modify the computations that might be carried out for piles of other dimensions, materials, or driving conditions at the site.

During driving, the resistance of the soil along the sides of the pile is likely, because of the continued disturbance, to be a minimum. However, it is not, in general, equal to zero. The side resistance during driving plus the subsequent freeze constitute the total skin friction on the single pile, provided the static point resistance has not changed. Not infrequently, the point resistance during driving exceeds that which can be developed later. If relaxation decreases the point resistance, the increase of load ascribed to freeze in Fig. 12.11 is the difference between the increase in side resistance and the decrease of the point resistance after driving. Thus, freeze and skin friction are not synonymous; freeze may be a combination of changes in friction and point resistance.

The calculation in Fig. 12.11 also shows that for any driving resistance the stress in the pile is on the order of 22 kips/sq in., a value well below the yield point of the material. Therefore, in this instance, driving should not damage the pile.

The significance of the wave equation to

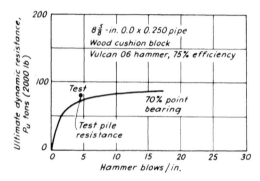

FIGURE 12.12. Result of dynamic analysis of pile indicating impossibility of achieving ultimate capacity of 100 tons by means of selected hammer and cushion block (after Davisson, 1970b).

the designer is further indicated in Fig. 12.12. This diagram gives the results of calculations for a cylindrical pipe pile with a diameter of $8\frac{5}{8}$ in. and a wall thickness of 0.25 in. The pile extended through a deposit consisting of 107 ft of medium to stiff clay underlain by sand. It had a length of 110 ft. It was driven closed-end to a resistance of 4.5 blows/in. with a Vulcan 06 hammer and a wood cushion block. According to the specifications for the project, the pile was to carry a design load of 50 tons and was to be load-tested to demonstrate a 100-ton ultimate capacity. The load test actually failed at 80 tons. According to the wave analysis, the pile at a resistance of 5 blows/in. should have had a capacity of about 70 tons. The results of the load test suggest that a small amount of freeze occurred. Nevertheless, failure of the pile under the load test could have been predicted and, according to the wave equation, should have been expected. Furthermore, the shape of the curve in Fig. 12.12 indicated without any doubt that the required test load of 100 tons could not have been obtained for the given pile, hammer, and cushioning, no matter to how high a resistance the pile might have been driven. The dynamics of the problem were such that the energy delivered by the hammer could not be transmitted through the pile to develop sufficient force resistances along the sides and at the point to meet the designer's load-test requirement. With the availability of the wave equation and rapid methods of calculation, the designer has the ability and, hence, the obligation to check whether his proposed requirements are compatible with the type of pile and other conditions he may have specified.

Studies with the wave equation demonstrate clearly that, except for the influence of freeze and relaxation, the attainment of a specified capacity by a particular pile in a given soil depends partly on the impedance of the pile and partly on the success with which pile, cushioning, and hammer are matched. The system must effectively transmit the driving energy to the point, and it must also keep the driving stresses within safe limits with respect to failure or crippling of the pile.

For a particular combination of pile, hammer, and cushioning, the relation between ultimate driving resistance or capacity and blows per inch of penetration, as exemplified by the resistance diagrams (Figs. 12.11 and 12.12), is practically independent of the soil conditions, inasmuch as the influence of the soil profile is reflected only in the minor effects of damping and of the ratio of point-to-side resistance. The soil conditions, on the other hand, are of paramount importance in determining the actual penetration of the pile per blow and are thus implicitly contained in the driving record for the pile. If the penetration per blow is not as small as that corresponding to the required capacity, the soil conditions are not adequate for support of the pile as driven. It is evident that the pile represented in Fig. 12.12 was not embedded deeply enough in the sand to develop the required capacity. A pile of the same external dimensions but of greater impedance would have penetrated somewhat deeper under the same hammer to achieve the same number of blows per inch, and its resistance diagram would have indicated a higher capacity. As a matter of fact, the pile was successfully redriven to the required capacity after being filled with concrete.

The influence of freeze and relaxation must be investigated experimentally, by comparing the results of wave analyses with field load tests (Art. 12.4). Information can also be obtained by returning to a driven pile after several days and determining the increase or decrease in blows per inch required to start the pile with the same hammer and driving assembly. The procedure may be misleading, however, because the first few blows of a hammer on which the comparison must be based are likely to be struck with hammer efficiencies far below normal. This source of error may be avoided by warming up the hammer by driving on an adjacent pile and immediately repositioning the hammer on the test pile; the interruption of use of the hammer is kept as brief as possible. Piles extending only partly through deep soft deposits may develop little dynamic resistance. Most of their capacity may ultimately be a consequence of freeze. Under these circumstances, which will be discussed in Part C, estimates of capacity are usually based on static rather than dynamic calculations, supplemented by load tests.

12.6. Choice of Type of Pile

The final choice of the type of pile for any one job is dictated by the subsurface conditions, the driving characteristics of the piles, the probable performance of the foundation, and also by economy. Economic comparisons should be based on the cost of the entire foundation instead of on the cost of the piles alone. For example, the cost of twelve 20-ton wood piles might be less than that of four 60-ton concrete piles, but the larger pile cap required to transfer the column load to the wood piles might increase the cost of the wood-pile foundation above that of the concrete-pile foundation.

12.7. Lateral and Upward Loads on Pile Foundations

Many types of pile-supported structures are subjected to lateral loads applied at an elevation considerably higher than that of the base of the foundation. Therefore, the foundation must resist not only lateral forces but also moment. Beneath such structures as lock walls, retaining walls, and ordinary buildings, the downward pile loads due to the weight of the structure are usually greater than the upward loads due to the moment caused by the lateral forces, and none of the piles is required to resist uplift. On the other hand, the piles beneath the windward side of tall steel towers or tall piston-type gas holders must often be counted on to provide the reaction against uplift forces. Finally, piles may be required to resist the uplift due to the buoyancy of tanks and similar structures established below the groundwater level.

When lateral loads on a structure must be transmitted to the subsoil by a pile foundation, one of the principal decisions to be made by the designer is whether or not some of the piles must be installed on a batter. This decision requires an estimate of the ability of vertical piles to withstand horizontal loads. If the soil immediately below the pile caps consists of sand, irrespective of its relative density, or of silt or clay having an N-value greater than about 5, it is reasonable to allow a horizontal load at the top of each pile of as much as 1500 lb. In denser or stiffer soils greater loads can be resisted, but no decision to use higher values should be made without careful study of the subsurface conditions and structural requirements. Under some conditions there may be a tendency for the soil in which the piles are embedded to move laterally under the influence of lateral forces other than those originating in the superstructure. The soil surrounding the piles may, for example, be involved in a slope failure. If such a tendency may exist, the vertical piles should not be counted upon to resist the movement, but should be expected to move together with the surrounding soil. Attention must then be turned to the causes of the lateral instability of the soil itself.

When the horizontal load per pile exceeds that which can be withstood by vertical piles alone, batter piles must be used in combination with vertical piles. Batter piles are common beneath retaining walls (Fig. 12.1d), bridge piers, and abutments. They are also used to provide lateral stability in the transverse rows of piles, known as *bents*,

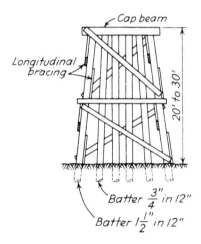

FIGURE 12.13. Use of batter piles in transverse bent of bridge trestle.

that constitute the vertical supports for trestles (Fig. 12.13). If both the batter and vertical piles beneath a structure are point-bearing and if all are driven to the same stratum, the axial capacity of each of the piles is generally assumed to be the same. The horizontal load per batter pile is then considered to be the horizontal component of the axial load. If the piles depend only on skin friction for their support and are all of equal length, the same assumption is usually made.

The uplift resistance of piles depends on many factors, including the type, dimensions, and tensile strength of the piles, and on the subsoil conditions. It is obvious that only skin friction can be effective in resisting the upward loads. The uplift resistance for various conditions will be discussed in Part C.

12.8. Negative Skin Friction

If a mass of soil is experiencing settlement because of incomplete consolidation under its own weight or under the weight of a fill or surcharge, the insertion of piles into the mass interferes with the settlement. As the soil tends to slip downward with respect to the piles, it exerts a downward drag known as *negative skin friction*. Similar downdrag may be produced by consolidation due to lowering the water table and other causes. In some instances, the loads on piles from negative skin friction approach or exceed those derived from the superstructure. Methods for evaluating negative skin friction and for reducing its effects are considered in Part C.

PROBLEMS

1. Two types of piles are under consideration to support a large working load in a dense sand layer beneath a deposit of soft clay and loose silt 100 ft thick. The piles of one type are $12\frac{3}{4}$ O.D. $\times$ 0.250-in. steel pipes closed by a steel plate at the bottom. The others are of 12×12-in. precast prestressed concrete. The same hammer and cushioning are considered appropriate for both types. If driven to the same penetration per blow, which pile will have the greater capacity? Which will penetrate the sand more deeply?

   Ans. The value of $\rho c A$ for the steel pile is 1.46 kip sec/in., and that for the concrete pile 4.32 kip sec/in. Hence, the concrete pile will transmit the greater force and develop the greater capacity. To do so, it will penetrate more deeply.

2. A long HP12 $\times$ 53 is to be driven to a working load of 100 tons with a factor of safety of 2. A Vulcan No. 1 hammer is specified. Can the required capacity be obtained without recourse to freeze?

   Ans. The value of $\rho c A$ is 2.43 kip sec/in. Reference to Fig. 12.9 would indicate that the required peak force of 400 kips could not be obtained with a Vulcan hammer delivering less energy than the No. 010, irrespective of the cushioning used or the required penetration per blow. A complete analysis by the wave equation would show, however, that a 20-ft pile could be driven to an ultimate load capacity of 200 tons at 14 blows/in. with an 06 hammer and an aluminum–micarta cushion; on the other hand, a No. 1 hammer could not satisfy the requirements.

SUGGESTED READING

Details of the various types of piles can be learned from brochures and catalogs furnished by the manufacturers or pile-driving companies, as listed in such periodicals as *Engineering News-Record* or *Civil Engineering—ASCE*. The latest information about pile hammers can be obtained similarly. A storehouse of such data is R. D. Chellis (1961), *Pile Foundations*, 2nd ed., New York, McGraw-Hill, 704 pp., but in a commercial industry as competitive as providing pile foundations, no book of this type can long remain fully up to date.

The classic paper describing the shortcomings of dynamic pile-driving formulas and suggesting development of the wave equation is A. E. Cummings (1940), "Dynamic Pile Driving Formulas," *J. Boston Soc. Civ. Eng.*, 27, 6–27; reprinted in *Contributions to Soil Mechanics 1925–1940*, Boston Soc. Civ. Eng., pp. 392–413. An introduction to the wave equation and its implications can be found in T. J. Hirsch, L. L. Lowery, H. M. Coyle, and C. H. Samson, Jr. (1970), "Pile-Driving Analysis by One-Dimensional Wave Theory: State of the Art," *Hwy. Res. Rec.*, 333, 33–54.

The requirements for load tests to be used for basic studies of pile-soil behavior or for determination of the capacity of heavily loaded piles under complex soil conditions are set out by M. T. Davisson (1970a), "Static Measurements of Pile Behavior," *Proc. Conf. on Design and Installation of Pile Foundations and Cellular Structures*, Lehigh Univ., Lehigh Valley, Pa., Envo Publ. Co., pp. 159–164.

The literature of piles and pile driving is vast and, to a large extent, unrewarding. Much of it is concerned with the relative merits of pile-driving formulas, results of miscellaneous load tests without complete corollary data, speculation about group effects, and descriptions of designs unaccompanied by performance records. Some of the better references are listed in appropriate chapters in Part C. A few of general interest include:

B. C. Gerwick, Jr. (1970), "Current Construction Practices in the Installation of High-Capacity Piling," *Hwy. Res. Rec.*, 333, 113–122.

J. H. Thornley (1959), *Foundation Design and Practice*, New York, Columbia Univ. Press, 298 pp. Somewhat dated, but sets out the practical engineering approach to choice of type of foundation and choice of types, lengths, and capacities of piles.

M. T. Davisson (1970b), "Design Pile Capacity," *Proc. Conf. on Design and Installation of Pile Foundations and Cellular Structures*, Lehigh Univ., Envo Publ. Co., pp. 75–85. Brief outline of designer's steps in selecting appropriate pile for a particular situation.

The lateral resistance of piles and methods of analysis are summarized by M. T. Davisson (1970c), "Lateral Load Capacity of Piles," *Hwy. Res. Rec.*, 333, 104–112.

Daniel E. Moran (1864–1937)

One of the foremost foundation builders and engineers of recent times. His responsibilities included the substructures of the Philadelphia-Camden, Ambassador, George Washington, and San Francisco-Oakland Bay bridges, and of many of the tallest buildings in the financial district of New York. In 1894 he invented improvements of the air lock that greatly facilitated the use of the pneumatic method of pier construction. In 1936, he devised techniques for controlling the sinking of the large open-dredged caisson that formed the center pier or anchorage of the San Francisco-Oakland Bay bridge. This pier was 92 by 197 ft in plan and was estalished at a depth of 220 ft below the water level. (Photo courtesy of Wm. H. Mueser.)

PLATE 13.

CHAPTER 13

Pier Foundations

13.1. Definitions

In foundation engineering the term *pier* has two different meanings. According to one usage, a pier is an underground structural member that serves the same purpose as a footing, namely, to transmit the load to a stratum capable of supporting it without danger of failure or excessive settlement. In contrast to a footing, however, the ratio of the depth of foundation to the base width of piers is usually greater than 4, whereas for footings this ratio is commonly less than unity.

According to the second usage, a pier is the support, usually of concrete or masonry, for the superstructure of a bridge. The pier usually rises above the ground surface, and commonly also extends through a body of water to a level above maximum high water. According to this definition, a pier may be regarded as a structure in itself, which must in turn be supported on an adequate foundation. To avoid confusion, the term *pier shaft* will be used with reference to the part above the foundation. The base of a pier shaft may rest directly on a firm stratum, or it may be supported on piles or on several piers as defined in the preceding paragraph. A pier shaft located at the end of a bridge and subjected to lateral earth pressure is known as an *abutment*.

No sharp distinction can always be drawn between piers and piles. Large-diameter steel pipes, driven open-ended and subsequently cleaned out and filled with concrete, may properly be considered either as piers or piles. The pipes themselves might be regarded as the linings of shafts or as caissons. The terminology in these respects differs considerably in different localities.

13.2. Methods of Construction

General. The methods of constructing piers are divided into two principal groups. In one, a hole is excavated to the base level of the foundation and the pier built inside the hole. Usually, the sides of the excavation must be lined and braced to prevent caving. Such a hole is known as a *sheeted pit* or a *cased hole*, depending on whether the lining consists of individual boards or sheet piles, or of cylindrical metal shells. The hole is sometimes stabilized by a heavy fluid instead of casing. If the ground surface is below water, the area to be occupied by the pier may be surrounded by a relatively watertight enclosure known as a *cofferdam*. Within the protection of the cofferdam the excavation is carried to the desired level and the pier constructed.

The other principal method for constructing piers is the use of caissons. A *caisson* is a hollow shaft or box that is sunk into position and becomes the outer part of the finished pier. To facilitate sinking, the lower part of the caisson is provided with a cutting edge. The material inside the caisson is removed by dredging through openings in the top, or by hand excavation. The lower part of the caisson may be sealed from the atmosphere and filled with air under pressure to exclude water from a space in which men can work. This procedure, known as the *compressed-air method*, permits removal of obstructions beneath the cutting edge and facilitates cleaning the bottom of the excavation. It constitutes a hazard to the health of the workmen, however, and is avoided insofar as possible.

Piers in Sheeted Pits and Cylindrical Holes. Hand-excavated pits with timber sheeting were once common and may still often be used to advantage, especially in underpinning (Chap. 15). The best known method originated in Chicago in 1892 (Gen'l Wm. SooySmith, Stock Exchange). It is particularly appropriate for clays in which waterbearing inclusions are not present. In the Chicago method, a circular hole at least 3.5 ft in diameter is excavated by hand for a depth varying from 2 to 6 ft, depending on the consistency of the clay. The walls of the hole are then lined with vertical boards, known as *lagging*. The lagging is held in place by two circular steel rings (Fig. 13.1). Excavation is then continued until another set of lagging and rings can be installed. When the hole reaches the stratum on which the foundation is to be supported, the bottom may be enlarged or *belled out* to increase the bearing area. The rings and lagging are left in place as the hole is filled with concrete.

Piers over 50 ft deep and 4½ ft in diameter, extending through fill and clay to limestone, were constructed in cased holes drilled by a power-driven auger in Kansas City as early as 1890 (L. Curtis, City Hall). Few similar attempts were made in the next half century, but today most piers passing through or into cohesive soils are excavated by truck or crawler-mounted drilling machines (Fig.

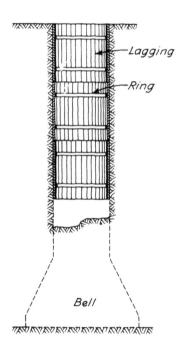

FIGURE 13.1. Chicago method for excavating shafts for piers.

13.2) equipped with rotating augers or buckets provided with cutting vanes. Holes ranging in diameter from 12 in. to 12 ft, and with depths exceeding 100 ft, have been made by this procedure. Various attachments are available for cutting bells in stiff soils at the bottom, or for drilling into or through rock. The holes, when filled with concrete, are known as *drilled piers* (Art. 13.3).

If the subsurface conditions are too unfavorable to permit installation of foundations by any of the methods described in the preceding paragraphs, large-diameter steel pipes may be installed by a procedure that combines the methods of pile driving and open excavation. The pipes are driven open-ended, a few feet at a time, and are cleaned out by means of air jets, water jets, or cable tools similar to those used for percussion drilling (Art. 5.2). Such shafts are almost always carried to rock. They are commonly continued into the rock for several feet by percussion or rotary drilling before the shaft is finally cleaned and concreted. Piers formed by this method are ex-

FIGURE 13.2. Crawler-mounted drilling machine spinning soil from auger after removal from shaft being excavated to diameter of 4 ft.

pensive but they can usually be assigned very high loads, and they can be installed under almost any subsurface conditions. The development of the large-diameter pipe pier has greatly reduced the circumstances under which compressed-air methods must be employed.

Cofferdams. Where piers are to be constructed in water and the depth of water does not exceed 8 or 10 ft, they may be built in cofferdams made of wood sheet piles. The piles may have one of the various forms shown in Fig. 13.3. They are driven around the area to be unwatered and are braced close to the water line by wales and struts. The lower part of the sheet piles is supported by the soil in which the piles are embedded.

For greater depths, wood sheet piles are inadequate, and the enclosure is usually made of steel sheet piles. One of the simplest types of cofferdams consists of a sheet-pile enclosure with internal bracing (Fig. 13.4). The sheet piles are driven until their lower ends are embedded and sealed in the underlying soil. Ordinarily, they extend to at least the full depth of the proposed pier. Before the cofferdam is pumped out, one set of bracing is installed just above the water line. The water level is then lowered to the elevation at which another set of bracing must be installed. Successive lowering of the water level and installation of bracing continue until the cofferdam is pumped out, whereupon the rest of the excavation is completed in the dry. Frequently several sets of bracing are prefabricated, placed simultaneously in the cofferdam, and set in posi-

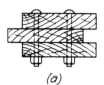 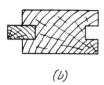 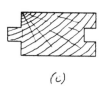

(a) (b) (c)

FIGURE 13.3. Types of wood sheet piles used for cofferdams in shallow water. (a) Wakefield. (b) Splined. (c) Tongue and groove.

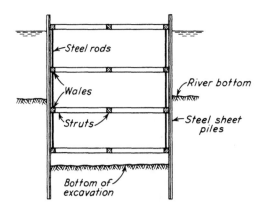

FIGURE 13.4. Braced single-wall sheet-pile cofferdam.

tion with the aid of divers before the cofferdam is pumped out. One of the principal difficulties with single-wall cofferdams is the rather great leakage through the interlocks of the sheet piling, especially at the beginning of dewatering. Cinders, manure, and various other materials are often dumped outside the cofferdam to plug the leaks. As the water level is lowered, the sheet piles bend inward and the interlocks become tighter. If the depth of the cofferdam becomes too great, it may be impracticable to lower the water table enough to expose bottom without running the risk of a blowup. A much deeper excavation can be

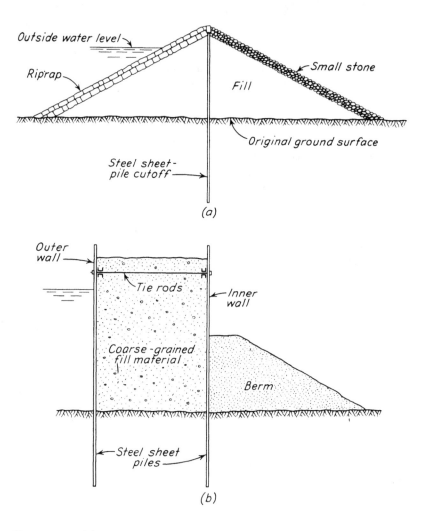

FIGURE 13.5. (a) Single-wall sheet-pile cofferdam protected by embankments. (b) Double-wall sheet-pile cofferdam.

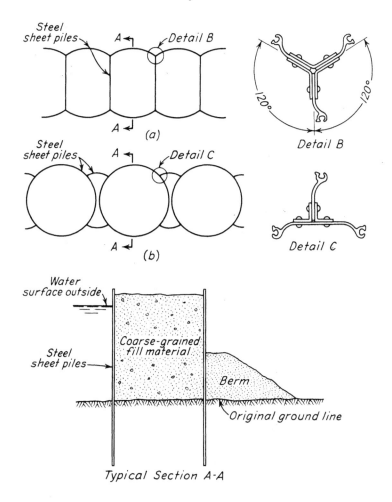

FIGURE 13.6. Cellular cofferdams. (a) Diaphragm type. (b) Circular type.

made by dredging under water. A concrete seal, heavy enough to resist the uplift, is then placed on the bottom before the cofferdam is dewatered.

For larger excavations in not over about 20 ft of water, a single wall of sheet piles is often driven and braced with berms of earth (Fig. 13.5a). A cofferdam of this type is easily attacked by currents in the water and must usually be protected on the outside by riprap. The double-wall sheet-pile cofferdam (Fig. 13.5b) is suitable for somewhat greater depths of water. It consists of two lines of sheet piles connected by tie rods. The space between the piles is filled with rock or soil.

Where the depth of water is very great, cellular sheet-pile cofferdams are used. The two principal arrangements are shown in Fig. 13.6. Each cell is filled with rock or gravel. The circular type has the advantage that each cell is independently stable. Hence, during construction the cofferdam is quite invulnerable to sudden floods or storms. For small bridge piers, single cells are sometimes made large enough to surround the entire foundation. Since such cells cannot be filled, the sheet piles must be supported by circular rings.

Caissons. If the ground surface is above water, construction of a caisson may be started on the ground directly above the area where its base will be located. If the ground level is below water, the lower part of the caisson may be constructed elsewhere,

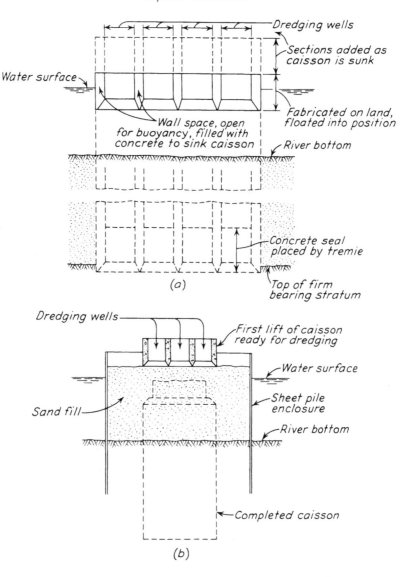

FIGURE 13.7. Cross section through typical open caissons. (a) Floating type. (b) Sand-island construction.

floated into place, and sunk (Fig. 13.7a). As an alternative, a ring of sheet piles may be driven to form an enclosure in which an island of sand is constructed. The caisson is then started on the sand and sunk through it as if the ground surface were above water table (Fig. 13.7b).

Open caissons are usually sunk by dredging. Hence, they must be provided with a number of wells extending from top to bottom through which the dredging can be done. The wells should be large enough to provide easy passage of the excavating buckets but small enough to leave substantial walls to provide some of the weight necessary for sinking the caisson. In small caissons a single central well is provided. In very large caissons several may be used. One of the piers for the Tagus River Bridge in Portugal extends to the record depth of 260 ft below water level. It was constructed by means of an open-dredged caisson that

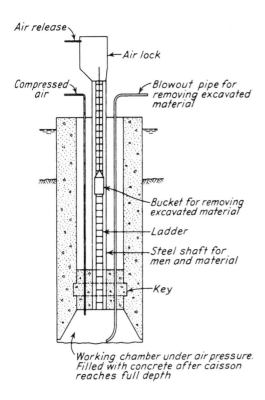

FIGURE 13.8. Section through typical pneumatic caisson.

contained 28 wells. The weight of a caisson must at all times be sufficient to overcome the side friction against the structure. In some instances, loads must be added and jetting facilities used to force the caisson down. After the caisson has reached its final position, the bottom is sealed with concrete deposited under water by means of a tremie.

The pneumatic caisson (Fig. 13.8) is used for depths between about 40 and 110 ft. Compressed air prevents water and mud from entering the working chamber. The working chamber is usually about 6 ft high and is provided at the top with openings for passage of men and materials. The openings are protected by air locks. The use of pneumatic caissons permits the removal of sunken logs, large boulders, etc., that may be encountered by the cutting edge. The foundation bed can be prepared carefully and the concrete placed in the dry. On the other hand, the cost of construction is relatively great compared to open dredging. Furthermore, the depth is limited by the air pressure under which men can work. As the air pressure increases, the time of exposure of the workmen must be shortened. Little detrimental physiological effect is felt at pressures above atmospheric up to about 15 lb/sq in. and normal working hours are customary. At pressures as high as 45 lb/sq in. the working periods may have to be reduced to 1 or 2 hr and long decompression periods are required.

Figure 13.9 shows a typical pier for a railroad bridge. The foundation is a caisson 96 ft high. The caisson was sunk by open dredging to the elevation of the shale, whereupon it was converted to the pneumatic type. It was then sunk into the shale to provide a good seat for the bottom. This procedure combined the economy of open dredging with the reliability and certainty of the pneumatic process.

13.3. Drilled Piers

Importance of Subsurface Conditions. The decision to use drilled piers, to a far greater extent than that to use spread footings or rafts, requires careful consideration of the construction conditions likely to be encountered. The behavior of drilled piers is determined at least as much by the success with which the construction operations are carried out as by the load-settlement characteristics of the surrounding and underlying soils. Such details as the presence of cobbles or boulders that interfere with drilling, the presence or lack of the slight cohesion necessary to prevent caving of the walls of the shaft or of the bell, or concentration of small amounts of seepage in occasional pervious zones may have a decisive effect on the practicability of forming a satisfactory and economical drilled pier.

Groundwater is particularly influential in determining the difficulty and hence the cost of drilled pier construction. Inflow, even in small quantities, may require the use of drilling mud or casing to permit advancing the hole without caving; it may interfere with preparation of the bottom; it may cause difficulty in concreting; and it

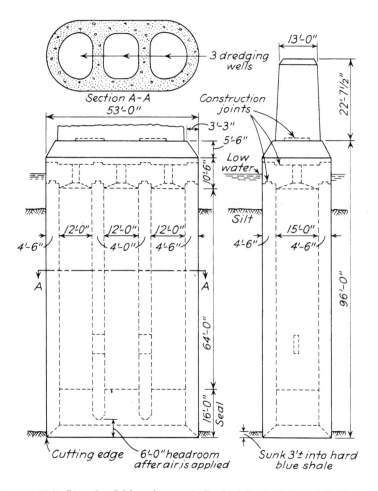

FIGURE 13.9. Pier for Milwaukee and Rock Island Railroad Bridge at Kansas City (courtesy of Howard, Needles, Tammen, and Bergendoff).

may lead to damage to the fresh concrete if the casing is removed.

The type of soil is hardly less significant. In stiff clays, in cemented sands above the water table, and in soft rock, shafts can be drilled rapidly and the walls can be expected to stand without support until the concrete is placed. Bells can usually be formed readily. Under most other conditions, however, provisions must be made to stabilize the walls. These conditions include, roughly in order of increasing difficulty of construction, relatively impermeable soils containing layers or lenses of cohesionless waterbearing material that tends to flow into the shafts; soft and very soft clays or silts that may squeeze into the hole; moist granular materials with enough apparent cohesion to provide general support for the shaft walls but with pockets of coarse less cohesive materials; perfectly dry cohesionless soils above the water table; and cohesionless soils below the water table. In all types of soils, the conditions for stability of bells are less favorable than those for the walls of the shaft.

The subsurface exploration program may require several stages (Art. 7.1) before the final design of a drilled-pier foundation is completed. The initial borings should permit a detailed study of the stratigraphy with special attention to the presence of cohesionless layers or lenses and of cobbles, boulders, or cemented zones, and to the position of the groundwater table. The entrance of groundwater at any level into a previously dry boring, or caving of the hole, deserve particular consideration. The preliminary borings should provide enough information for concluding whether drilled piers will be suitable and for determining their probable depth and dimensions. If a drilled-pier design is tentatively adopted, further borings may then be needed to define the construction problems that may be encountered. The groundwater conditions near the base of the shaft or in the vicinity of proposed bells should be explored carefully by such techniques as bailing the holes when they have reached the critical depths and observing the rate of inflow. At least some of the borings should be converted to observation wells. Finally, on large or potentially difficult jobs, it may be advisable to excavate one or more full-sized shafts with equipment of the type likely to be used on the job. The mobility of modern drilling equipment often makes this procedure feasible and economical. The test shafts facilitate the preparation of realistic specifications and give the bidders a clear perception of the work to be performed. If, even at this late stage in design, the results should prove to be unfavorable, the type of foundation can be changed before the contract documents have been made final.

Excavation. The equipment for drilling has been described briefly in Art. 13.2. If the holes stand open and remain dry until concreting can be completed, the foundation can be constructed rapidly and economically. Otherwise, steps must be taken to keep the holes open. In some instances the ground can be converted from potentially unstable to stable material by dewatering or grouting. If, for example, the instability of the side walls is associated with permeable cohesionless zones, it may be possible to drain the entire site by lowering the water table to a level below that of the bottom of the drilled holes. The shafts may then be excavated in the dry and support may not be necessary. If there are only a few such zones and their location is well defined, they may sometimes be stabilized by grouting (Art. 9.6) before drilling, although the possibility of incomplete grouting introduces considerable uncertainty into the procedure. By far the most widely used method for advancing drilled shafts that would otherwise become unstable is the introduction of a slurry similar to drilling mud (Art. 5.2) into the hole. The heavy fluid prevents the inflow of water and surrounding materials.

Under differing circumstances, drilling mud may be used alone or in combination with casing. Conditions often encountered are illustrated in Fig. 13.10, which represents steps in drilling a shaft through cohesive soil suitable for dry augering except for a zone of submerged cohesionless soil that would cave into the hole. The hole is advanced to the caving zone by augering in the dry (Fig. 13.10a). Before it penetrates the zone, however, it is filled with soil, bentonite, and water, proportioned to form a heavy, viscous fluid, and mixed by simultaneously rotating and lifting and lowering the auger. When the fluid attains the proper consistency, the hole is advanced through the cohesionless zone (Fig. 13.10b) in the usual way by means of the auger. The slurry stabilizes the walls of the hole, prevents inflow of groundwater, and imparts enough cohesion to the soil being drilled to permit it to be lifted by the auger. As the hole deepens, slurry is added to maintain its surface near ground level. When the caving layer has been passed, a casing is inserted (Fig. 13.10c). The casing usually consists of a single length of pipe with an internal diameter slightly larger than the diameter of the auger. It is seated in the underlying cohesive soil by turning it with the kelly (Fig. 5.4), which normally rotates the auger, and by simultaneously forcing the kelly downward. The slurry is then bailed from the hole and drilling resumed in the dry (Fig. 13.10d).

Under these soil conditions the entire hole could equally as well be drilled with slurry and without casing. Since the hole would collapse, however, if the slurry were pumped out and no casing were inserted, the concrete would necessarily be placed in the slurry-filled hole by underwater methods described under the next subheading.

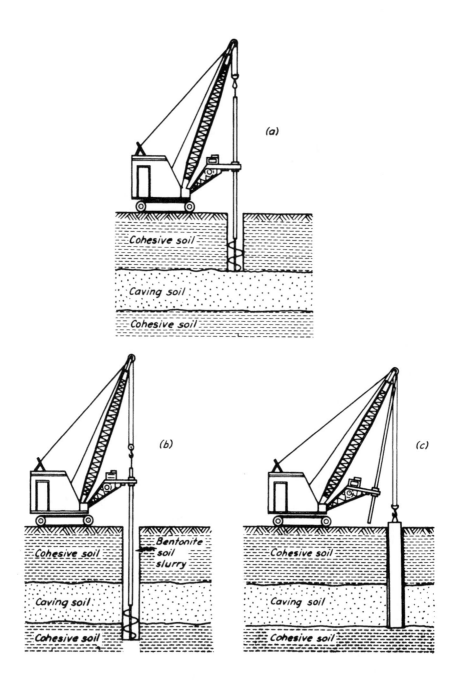

FIGURE 13.10. Steps in constructing typical drilled pier. (a) Dry augering through self-supporting cohesive soil. (b) Augering through waterbearing cohesionless soil with aid of slurry. (c) Setting casing. (d) Dry augering into cohesive soil after sealing casing. (e) Forming bell (after O'Neill and Reese, 1970).

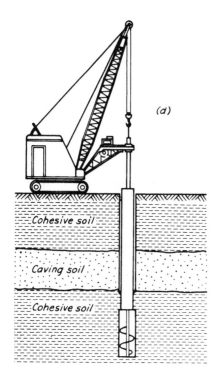
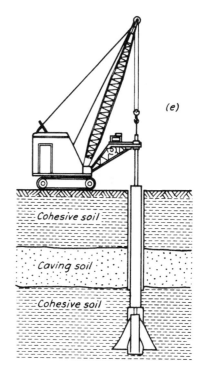

This procedure is often used in Europe. In North America, on the other hand, after the shaft has been drilled while filled with slurry, it is more usual to insert a casing with a diameter a few inches less than that of the hole, to seat it on the bottom, and to bail out the slurry. The bottom can then be completed in the dry. The thin annulus of slurry that remains between the casing and the walls of the drilled hole prevents or reduces the movement of the surrounding material toward the shaft. If the shaft terminates on rock overlain by pervious material or containing pervious joints near the rock surface, a heavy casing with hardened steel teeth is sometimes rotated and ground into the rock for several inches or a few feet until a seal is formed that cuts off the flow of water. The lower part of the casing is left in place. If the seepage is not stopped, the bit may sometimes be sealed in by pressure grouting.

Belling. Bells should not be attempted unless the ground is cohesive enough to allow the roof to stand without collapse during the time between excavation, cleanup of the bottom, and placement of the concrete in the bell. Because of the difficulty of satisfying this condition, it is preferable in many localities to extend a straight-shafted pier into the firm materials far enough to permit carrying the load by side friction.

Bells can be excavated by hand, as in the Chicago method (Fig. 13.1), but they are generally formed by attaching a *belling bucket* to the kelly in place of the drilling bucket or auger. The bucket illustrated in Fig. 13.10e consists of a cylinder with two cutting blades, pivoted at the top, that fold inside the cylinder while the bucket is lowered into the hole. At the bottom of the hole the blades are spread outward through vertical slots in the bucket, the bucket is rotated, and the soil cut by the blades drops into the bucket. After a few revolutions the blades are retracted and the bucket is raised and emptied. The procedure is then repeated. In comparison to the time required for drilling the shaft, that for forming a bell is quite long. Since the tendency of any unsupported excavation to slough or collapse increases with time, the soil conditions for forming a shaft with a bell must be generally more favorable than those for forming a shaft alone.

If a bell collapses, the shaft must be carried deeper to a level where another bell can be formed, or to a depth great enough for support of the loads by a shaft alone. If the soil conditions do not permit these alternatives, the foundation must be redesigned. In some instances two piers are drilled as replacements and are capped by a girder to carry the column load. Occasionally it is necessary to resort to piles.

Concreting. If the hole is dry, the concrete is normally allowed to fall freely from the ground surface. Undesirable segregation of the cement and aggregate may occur if the concrete falls against the sides of the shaft; hence, if the diameter is small a short vertical guide tube is usually located at the center of the top of the shaft where the concrete is introduced. Vibration is usually necessary only in the top 5 to 10 ft of the shaft where the impact of the falling concrete is ineffective. Reinforcement may be introduced as a circular cage inside which the concrete can fall freely. The slump of the concrete depends on the dimensions of the pier, on whether casing is needed and will be pulled, and on the reinforcement. A slump of about 6 in. is suitable under most conditions but higher slumps are used in heavily reinforced piers and in small-diameter piers from which casing will be removed.

The presence of an inch or two of water in the bottom of the shaft or bell, except if localized in a small sump, may appreciably reduce the strength of the concrete. Bags of cement are sometimes laid on the bottom to absorb the excess water before the concrete is placed. More than about 2 in. of water may cause segregation of the concrete. The water rises on top of the rising concrete and all additional concrete must drop through it. Almost complete separation of cement and aggregates is likely to occur where the depth of water is 6 in. or more. Hence, sources of incoming water should be shut off if possible.

If the inflow cannot be stopped but if the rise of the water level does not exceed about $\frac{1}{4}$ in./min, a sump with a small cross section compared to that of the base may be excavated near the center and the water lowered by a pump. With the concrete in readiness the pump is removed as rapidly as possible, and a substantial charge of concrete is then dropped into the shaft.

If the inflow is too great to permit concreting in the dry, the water level may be allowed to rise freely until it reaches equilibrium, provided the inflowing water does not transport an objectionable amount of loose material into the hole. After the inflow has stopped, concrete should be placed by means of a tremie with its bottom initially within 1 ft of the bottom of the hole. The tremie should be withdrawn as the concrete is placed, but the depth of burial of the bottom of the tremie should at no time be less than 1 ft.

Tremie concrete can also be placed in an uncased slurry-filled hole, but refined techniques and experienced specialist contractors are mandatory. The slurry must be slowly circulated and cleaned of coarse fragments until the entire hole is filled with a smooth thixotropic suspension that prevents accumulation of coarse material at the bottom before the tremie concrete can be introduced.

Removal of Casing. Because casing is costly it is often pulled as the concrete is placed. This procedure may, unless controlled with the greatest care, lead to serious defects in the concreted shafts. It should not be attempted if the concrete has been placed by tremie.

The operation is usually carried out by lifting the casing slowly as the concrete is deposited, while the bottom of the casing is kept at least 5 ft below the top of the concrete in the pier at all times. A greater distance will be required if the pressure of the concrete will be less than that exerted by the surrounding soil or drilling fluid. Otherwise, the surrounding material will invade the fresh concrete or reduce the diameter of the pier, or the concrete will fail to press tightly against the soil. The casing must be kept vertical during the withdrawal to avoid displacement of the reinforcement and to prevent mixing of soil with the concrete along the sides of the shaft. After concreting starts, the withdrawal must be accomplished within about an hour, before the concrete experiences its initial set. If a delay occurs beyond this period, the casing should not be pulled further and the unburied portion should be cut off.

Many factors enter into the decision whether to pull the casing. If casing has been installed in a hole drilled with the aid of drilling mud, it can usually be pulled rather easily. On the other hand, if mud has not been used the casing may have to be twisted free; the adhesion between casing and soil may cause delays and may also not permit a uniform rate of withdrawal. It may then be preferable to insert a thin corrugated-steel liner in the hole, to cast the concrete, to fill the space between the liner and the casing with grout or sand, and finally to pull the casing. If the pier extends through soil to rock into which casing has been extended, it is preferable to leave the casing in the rock and in at least the lower portion of the soil. According to some building codes the strength of the steel casing can be included in calculating the capacity of the shaft. The diameter of the shaft and possibly the amount of reinforcement can then be reduced. If the rock is capable of supporting the greater intensity of pressure associated with the smaller bearing area, the economic advantage of recovering the casing is considerably decreased.

Some of the most serious foundation defects of recent years have been associated with attempts to recover casings during concreting, especially in deep piers. Unless conditions are at least reasonably favorable and unless the best supervision and control can be assured, the practice should be avoided. The cost of the casing can be negligible in relation to the cost of the damage that may be done by a defective pier.

Inspection. Because details of construction have a decisive influence on the performance of drilled piers, to the extent that apparently minor deviations from the best practice may invalidate a well-conceived design, inspection has an extraordinarily important role, particularly if the piers are deep or cannot be drilled and concreted in the dry.

During excavation the inspector should assure that the shafts are within the tolerances for verticality and dimensions, that the walls of the shaft have not collapsed, that the bottom has been adequately cleaned, that the bell is properly formed and is intact, that the nature of the materials at the base of the pier agrees with that on which the design is based, and that groundwater is properly controlled.

If drilling mud is used, the inspector should check that the pumping equipment is adequate, that proper screens are provided to remove the coarse material from the circulating fluid, and that the initial consistency of the slurry is suitable. During drilling, he should continually check the consistency and observe the nature of the materials brought up by the fluid.

Inspection of the bell and bottom materials requires that the inspector descend to the bottom of the hole. Even in a dry hole, this should not be done without protection. If the hole is not already cased, a casing is usually suspended from a crane and lowered in the hole. A proper harness and a supplementary safety rope should also be provided for the inspector.

Inspection of the bottom is potentially dangerous. Explosive or poisonous gases frequently accumulate in the bottom of drilled shafts. Suitable detection devices and ventilation should be insisted on by any person descending into the shaft. Minimum precautions are specified by law and penalties for violations may be severe, especially if accidents occur.

Water pumped from the hole may be passed through a filter or a settling basin to permit the inspector to judge whether or to what extent fines are being eroded from the waterbearing soil.

Concrete is usually delivered by transit-mix trucks. Because of the influence of the properties of the fresh concrete on the final quality of the foundation, especially if the casing is to be pulled, the inspector should give particular attention to the slump and to making certain that the mixing time is not excessive. He should form at least one cylinder from each truckload. He should make sure that the concrete falls freely without striking the sides of the hole or the reinforcement. If the concrete is placed by tremie, he should see that the bottom of the tremie is kept adequately below the surface of the concrete.

The greatest vigilance is required during removal of casing. Direct observation of the elevation of the top of the concrete is difficult. Telltales and sounding devices can be devised to suit the job conditions; in large shafts it is sometimes possible for the inspector, with suitable protection, to enter the casing. If the surface of the concrete rises even momentarily as the casing is being withdrawn, it is virtually certain that foreign material such as slurry, water, or soil has invaded the shaft and created a defect. The appearance of a sinkhole or depression around the top of the casing is almost a sure indication of invasion of foreign matter. To supplement direct observation, a comparison of the volume of concrete and the volume of the drilled hole should always be made. If any upward movement of the concrete inside the casing is noted or if there is any other indication of defective conditions, the casing should not be pulled further and the part of the casing still in the ground should be left in place. The concrete of such a pier should be cored to indicate the nature of any defects and to provide information for determining remedial measures.

Loss of Ground. The loss of ground associated with excavation is discussed in Chap. 16. In the construction of drilled piers, loss of ground is not likely to be a serious matter unless the surrounding soil consists of relatively soft plastic clay or silt that may squeeze into the holes or of cohesionless sands or silts that may flow into the holes under the influence of groundwater.

If the shafts are drilled through soft clays with the aid of a drilling fluid, the fluid restrains the inward movement of the clay. Casing is often placed in the hole to permit removal of the fluid and concreting in the dry. If the casing is left in place, the surrounding soil may gradually invade the slurry-filled annulus around the casing and the accumulated settlement associated with the construction of a large number of piers may be excessive. It may be necessary to grout behind the casing to fill the space and reduce the loss of ground.

Pervious cohesionless zones beneath the water table may be encountered in an otherwise dry hole and substantial inflows may occur before the water is brought under control. Silts and sands flow freely into the hole under these conditions. Often the material may migrate from a considerable distance. If it is drawn from beneath existing structures, serious settlements can take place. Furthermore, if the material is drawn from beneath other piers in the foundation being constructed, settlements of the new building may develop as loads are applied to the piers. Control of the groundwater by filter wells is far preferable to pumping from within the piers being excavated. In some instances, cement and chemical grouting have been utilized to stabilize the soils before the shaft is drilled into the mobile materials. Although this procedure has sometimes met with notable success, it has frequently failed. Its use is generally reserved for emergencies.

Conditions for Success. The foregoing paragraphs have discussed a number of conditions under which construction difficulties and their attendant defects can be anticipated. Many of these are associated with the particular soil conditions at a site. The more the likelihood of such difficulties, the less justification there is for the use of drilled piers.

On the other hand, where such difficulties are unlikely, pier foundations may prove extremely satisfactory and economical. In addition to the nature of the soil conditions, the qualifications of the contractor and the kind of equipment have a strong influence on the cost and success of the work. Good techniques lead to more rapid excavation and less likelihood of defects. Poor techniques can lead to unsatisfactory results even under relatively good subsurface conditions. In contrast to adequate equipment, equipment too light or insufficiently powered to cope with the soil conditions can lead to slower excavation, increased difficulty with groundwater and caving, greater cost, and defective piers.

SUGGESTED READING

A thorough review of caissons, cofferdams, and deep piers is contained in R. E. White (1962), "Caissons and Cofferdams," Chap. 10 in *Foundation Engineering*, G. A. Leonards, ed., New York, McGraw-Hill, pp. 894–964. An extensive bibliography is included.

Piers for bridges occupy a prominent place in the literature. Perhaps the finest account of sinking a compressed-air caisson, written for a nontechnical audience but completely authoritative, is to be found in D. B. Steinman's *The Builders of the Bridge* (1945), New York, Harcourt, 457 pp., Chap. XVIII, "Down in the Caissons." The chapter is part of the story of the Roeblings and such famous structures as the Brooklyn Bridge.

The use of open-dredged caissons, supplemented by compressed-air domes first introduced on the San Francisco Bay Bridge, is described in L. W. Riggs (1966), "Tagus River Bridge-Tower Piers," *Civ. Eng. ASCE*, 36, 2, 41–45.

The various types of foundations considered and adopted for a single project are discussed in A. Hedefine and L. G. Silano (1968), "Newport Bridge Foundations," *Civ. Eng. ASCE*, 38, 10, 37–43.

The most comprehensive treatise on drilled piers is *Drilled Pier Foundations*, by R. J. Woodward, W. S. Gardner, and D. M. Greer, New York, McGraw-Hill, 1973. An excellent description of current practice in the construction of drilled piers, with a discussion of many practical problems and their consequences, is contained in Chapters I and II of *Behavior of Axially Loaded Drilled Shafts in Beaumont Clay*, by M. W. O'Neill and L. C. Reese, Research Report 89-8, Part One–State of the Art, Center for Highway Research, The University of Texas at Austin, Dec. 1970. Modern specifications dealing with many details of design and construction are contained in "Suggested Design and Construction Procedures for Pier Foundations," by I. Schousboe, *ACI Journal*, August 1972, No. 8, Proc. Vol. 69, pp. 461–480. Typical problems in soft clays, particularly in the Chicago area, are described by C. N. Baker, Jr., and F. Khan (1971), "Caisson Construction Problems and Correction in Chicago," *ASCE J. Soil Mech.*, 97, SM2, 417–440.

Pier Construction

Barge-mounted rig driving Z-piling for single-wall braced cofferdam within which bridge pier will be constructed. Completed twin double-pier shafts in background. (Photo courtesy of Illinois Department of Transportation.)

PLATE 14.

CHAPTER 14

Pier Shafts, Retaining Walls, and Abutments

14.1. Pier Shafts

The dimensions of the top of a pier shaft for a bridge are determined by practical considerations such as the magnitude of the bridge-shoe reactions, the distance required to provide for expansion of the superstructure, and the distance between trusses or girders. Frequently, the top of the shaft is provided with a coping or overhang that extends about 6 in. beyond the edges of the top of the shaft proper. If the shaft extends through a body of water, its shape may be streamlined below high water to prevent eddy currents and scour.

In northern latitudes the upstream edge may be provided with an inclined cutting edge to lift and break cakes of ice. For the sake of appearance, a slight batter is sometimes given to the shaft as a whole.

Solid shafts (Fig. 14.1a) are commonly used for railroad bridges. The double shaft (Figs. 14.1b and 14.1c) is often adopted for highway bridges, although type b is also common for railroads. The hammer-headed pier (Fig. 14.1d) is one of several types often used to avoid skew spans in passing over existing railroad tracks or highways.

Most modern bridge piers are of reinforced concrete. For protection against the elements, stone facing is sometimes used especially near the water line.

Although pier shafts are commonly regarded as part of the substructure of a bridge, they are not part of the foundation in the sense that their design requires consideration of the properties of the subsur-

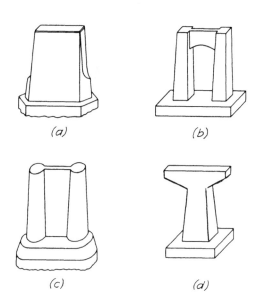

FIGURE 14.1. Typical pier shafts for railway and highway bridges. (a) Solid shaft. (b) and (c) Double shafts. (d) Hammer-headed shaft.

face materials. Therefore, they will not be considered further in this text.

14.2. Retaining Walls

A retaining wall is a structure that provides lateral support for a mass of soil and that owes its stability primarily to its own weight and to the weight of any soil located directly above its base. Retaining walls constitute inherent parts of many foundations and their design is one of the functions of the foundation engineer.

Before about 1900 retaining walls were usually constructed of stone masonry. Since that time concrete, either plain or reinforced, has been the predominant material. The most common types in current use are gravity, semigravity, cantilever, counterfort, and crib walls.

The *gravity wall* (Fig. 14.2a) depends for its stability entirely on the weight of the stone or concrete masonry and of any soil resting on the masonry. No reinforcement is provided except in concrete walls, where a nominal amount of steel is placed near the exposed faces to prevent surface cracking due to temperature changes.

The *semigravity wall* (Fig. 14.2b) is somewhat more slender than a gravity wall and requires reinforcement consisting of vertical bars along the inner face and dowels continuing into the footing. It likewise is provided with temperature steel near the exposed face.

The *cantilever wall* (Fig. 14.2c) consists of a concrete stem and a concrete base slab, both relatively thin and fully reinforced to resist the moments and shears to which they are subjected.

The *counterfort wall* (Fig. 14.2d) consists of a thin concrete face slab, usually vertical, supported at intervals on the inner side by vertical slabs or counterforts that meet the face slab at right angles. Both the face slab and the counterforts are connected to a base slab, and the space above the base slab and between the counterforts is backfilled with soil. All the slabs are fully reinforced.

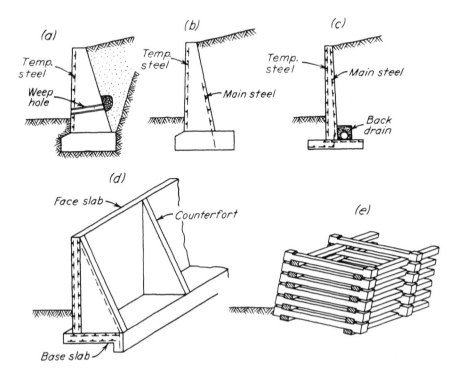

FIGURE 14.2. Types of retaining walls. (a) Gravity section. (b) Semigravity section. (c) Cantilever section. (d) counterfort wall. (e) Crib wall.

The preceding four types are known as *monolithic walls*, in contrast to *crib walls* (Fig. 14.2e), which consist of individual structural units assembled at the site into a series of hollow bottomless cells known as cribs. The cribs are filled with soil, and their stability depends not only on the weight of the units and their filling, but also on the strength of the soil used for the filling. The units themselves may consist of reinforced concrete, fabricated metal, or timber.

Of the monolithic types, those most commonly constructed today are the cantilever and the semigravity. Cantilever walls generally have the advantage of lowest first cost and are widely used in connection with buildings and highways. However, because of the relatively small thickness of the concrete sections they may be vulnerable to the effects of freezing and thawing, expansion and contraction, and concrete deterioration. Therefore, where permanence and low maintenance costs are primary considerations, as for railroad structures, the thicker semigravity walls are sometimes considered preferable.

All retaining walls are expected to withstand the pressure of the earth that they support, but they are not usually designed to resist water pressure in addition to the earth pressure. Therefore, well-designed retaining walls are provided with means for draining the water that would otherwise accumulate in the backfill. The drains commonly consist of pipes known as weepholes (Fig. 14.2a), which have a diameter of 6 or 8 in. and which extend through the stem of the wall and are protected against clogging by pockets of gravel in the backfill. The drains should be spaced at about 10 ft both vertically and horizontally; in counterfort walls there should be at least one drain in each pocket between adjacent counterforts.

Weepholes are not highly efficient in draining semipervious backfills. Unless the pockets of gravel satisfy the requirements for a filter (Art. 2.5), they are likely to become clogged. In freezing weather the outlets may become obstructed with ice. For these reasons, a *continuous back drain* (Fig. 14.2c) is considered preferable when the physical conditions at the site permit its use. The drain consists of a perforated pipe having a diameter not less than 6 in. The pipe, which must be surrounded by a filter, usually terminates at an open ditch where it should be accessible for cleaning.

The character of the material used for backfill has an important influence on the forces acting against the inner face of a retaining wall. Clean sands and gravels are considered superior to all other soils because they are free draining, are not susceptible to frost action, and do not become less stable with the passing of time. Silty sands, silts, or granular soils containing a small percentage of clay are less desirable because they cannot be drained readily, are likely to be subject to frost action, and may experience a decrease of shearing strength with accumulation of moisture. Clays are undesirable as backfill because they can hardly be drained, are likely to experience alternate swelling and shrinking with the seasons, and may lose much of their shearing strength if moisture accumulates. If shrinkage cracks in a clay backfill become filled with rainwater, the wall may be subjected to full hydrostatic pressure as well as earth pressure, even if drains have been provided. Wherever possible, it is considered good practice to insert a wedge-shaped body of free-draining material between the wall and a clay backfill, as shown in Fig. 14.2a.

14.3. Abutments

An abutment serves two principal functions. It supports the end of a bridge span, and it provides at least some lateral support for the soil or rock on which the roadway rests immediately adjacent to the bridge. Hence, an abutment combines the functions of a pier shaft and a retaining wall.

One of the most common types of abutment is shown in Fig. 14.3a. It consists of a central pier supporting the bridge seat, and two wing walls to retain the fill. All three elements rest on a single footing. If the wing walls are at right angles to the pier, the structure is known as a U abutment (Fig. 14.3b). The wing walls of a U abutment are

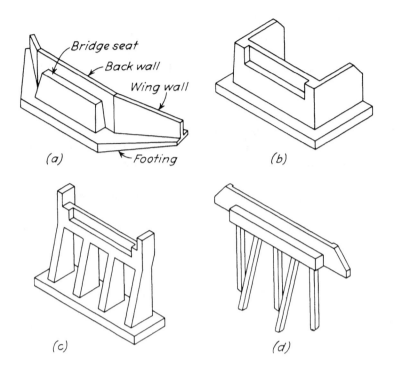

FIGURE 14.3. Types of abutments. (*a*) Typical gravity abutment with wing walls. (*b*) U abutment. (*c*) Spill-through abutment. (*d*) Pile-bent abutment with stub wings.

sometimes tied together to reduce their tendency to overturn.

The spill-through or open abutment (Fig. 14.3c) is also widely used. It consists of two or more vertical columns carrying a beam that supports the bridge seat. The fill extends on its natural slope from the bottom of the beam through the openings between the columns. In its extreme form a spill-through abutment is no more than a row of piles driven through the fill and supporting a bridge seat (Fig. 14.3d). The bridge seat is usually provided with small wings to keep the bridge shoes free of soil. Another common variation is a simple pier with small wings near the top. The fill in this case spills around the abutment.

SUGGESTED READING

A survey of factors leading to the failure or excessive movements of retaining walls and abutments is reported by R. B. Peck, H. O. Ireland, and C. Y. Teng (1948), "A Study of Retaining Wall Failures," *Proc. 2nd Int. Conf. Soil Mech.*, Rotterdam, *3*, 296–299. The importance of drainage and proper backfilling procedures is emphasized in a nontechnical note intended for field inspectors: R. B. Peck and H. O. Ireland (1957), "Backfill Guide." *ASCE J. Struct. Div.*, *83*, ST4, 10 pp.

Lazurus White (1874–1953)

Foundation constructor especially noted for his skill in executing difficult shoring and underpinning operations. He was one of the first engineers to realize that the behavior of the subsoil of a foundation under load is comparable to that of an elastic solid. In accordance with this conception he introduced the practice of prestressing the soil beneath newly installed temporary or permanent supports for a structure in order to prevent settlement during the transfer of load to the new supports. He vigorously advocated adequate drainage of granular materials during the excavation of cuts below water table and during construction operations within cofferdams. The records of his observations of construction difficulties are an invaluable source of information to the engineering profession. (Photo courtesy of Robert E. White.)

PLATE 15.

CHAPTER 15

Shoring and Underpinning

15.1. Shoring

Excavation below the foundation level of an adjacent structure usually leads to the necessity for supporting the structure temporarily. The provision of temporary supports is known as *shoring*.

In one method of shoring, notches are cut in the walls of the neighboring building and inclined posts inserted in these notches to carry the weight of that part of the wall above the supports (Fig. 15.1a). The lower ends of all the posts must be established on foundation pads similar to those provided for rakes (Fig. 8.1). This method is adequate only for relatively unimportant structures.

Better support is provided by cutting the walls or columns at their bases and inserting jacks between the foundations and the walls or columns that they support. As excavation progresses and causes subsidence of the surrounding soil, the jacks may be adjusted to keep the walls or columns at their original level.

15.2. Underpinning

In some instances it is necessary to replace or strengthen the foundation of an existing structure. The operation of providing new permanent foundations is known as *underpinning*. Underpinning is an art practiced by specialists, but an acquaintance with underpinning methods and procedures is an essential part of the knowledge of every foundation engineer. One of the simplest forms of underpinning is the replacement of

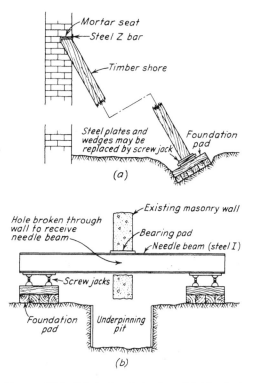

FIGURE 15.1. Methods of shoring. (a) Notched wall. (b) Needle beam.

251

a wall footing by another footing at greater depth. Masonry or concrete walls are capable of temporarily bridging gaps in their foundation. Hence, it is possible to excavate a pit beside the existing wall to a depth corresponding to the base of a new foundation, to excavate beneath the existing footings for a length of 4 to 8 ft, and to construct a short section of the new footing in this space. After this section is complete, another adjacent section can be done.

If the wall is incapable of supporting itself, holes can be broken through the wall and horizontal beams inserted as shown in Fig. 15.1b. These supports, known as *needle beams*, rest at both ends on temporary footings and jacks. Thus, the needle beam is a form of shore. While the wall is supported on the needle beam, its new foundation can be constructed. Similarly, needle beams can be used to support columns that are being underpinned.

In many instances the foundations of existing structures have proved inadequate, and it has been necessary to provide new support at a much greater depth than the original footings. The new support usually consists of hand-excavated piers or of piles. Steel H beams or pipes may be jacked into the ground in short sections against the reaction of the weight of the footing.

The transfer of load from shores or from existing supports to new foundations inevitably causes a deformation or settlement of the new foundations. This settlement may crack the structure. The cracking may be reduced or eliminated by prestressing the new foundations, usually by jacking them down against the reaction of the existing structure. While the load is still on the jacks, wedges are inserted to maintain the units in their stressed condition. The jacks are then removed, and the space that they occupied is filled with concrete.

Methods of underpinning are often elaborate and extremely expensive. The possibility that a structure may have to be underpinned as a result of adjacent construction often leads initially to the use of deep foundations for a structure that could be supported adequately at a higher level if it would not be disturbed.

SUGGESTED READING

E. E. White (1962), "Underpinning," Chap. 9 in *Foundation Engineering*, G. A. Leonards, ed., New York, McGraw-Hill, pp. 826–893.

Damage Due to Adjacent Construction

Two-story structure distorted by differential settlement. The settlements had its seat in deposits of highly compressible clay which consolidated partly because of the weight of the building on the right and partly because of a general lowering of the groundwater level. Serious differential movements did not occur, however, until the steel-frame structure on the left was built. The pile foundation of this structure arrested the normal settlement of the edge of the two-story building and caused ruinous distortions. Movements of this magnitude are rare except in Mexico City, where the photograph was taken, but smaller movements of the same character have often occurred elsewhere. (Photo courtesy of Professor Leonardo Zeevaert.)

PLATE 16.

CHAPTER 16

Damage Due to Construction Operations

16.1. Settlement Due to Excavation

Introduction. Every process of excavation is associated with a change in the state of stress in the soil. The change in stress is inevitably accompanied by deformations. These commonly take the form of a subsidence of the area surrounding the excavation, an inward movement of the soil at the sides, and an upward movement of the soil beneath the bottom. Structures supported by foundations resting on the material that deforms experience corresponding movements. They commonly settle and move toward the excavation.

Inasmuch as no excavation can be made without altering the state of stress to some extent, some movement of surrounding structures must be regarded as inevitable. However, to the inevitable movements associated with a given type of construction there may be added others due to poor workmanship. If the bracing of the sides of an excavation is done haphazardly, for example, large displacements may occur. The portion of the movement associated with poor workmanship must be regarded as unnecessary and should be avoided.

Different methods of construction involve different amounts of inevitable movements. In some instances, a given procedure cannot be used because damage to adjacent structures would unquestionably be too great, even if the workmanship were excellent. Therefore, the engineer should be aware of the consequences of using different construction procedures and should be careful not to specify methods of excavating and bracing that cannot be carried out without causing damage to adjacent property.

In order to avoid this danger, the engineer should be familiar with the various procedures for excavating and bracing the sides of excavations. On a given job, he should make use of such observations as are necessary for determining whether the movements are greater than those that must be considered inevitable and for determining how best to reduce any preventable movements.

Excavation in Sand. Sand above the water table is commonly moist and possesses enough cohesion to facilitate excavation. In properly braced large cuts, the settlement of the adjacent ground surface does not usually exceed about 0.5 per cent of the depth of the cut, and the influence of the settlement does not extend farther from the sides of the cut than a distance equal to the depth.

When large cuts are made in sand below the water table, it is advisable to lower the water table before construction. The excavation can then be carried out with no more subsidence than would be associated with a similar cut in moist sand. However, the process of lowering the water table may itself lead to subsidence under some conditions. These will be pointed out subsequently.

Piers are often established in sand below the water table by sinking a caisson and dredging from the inside. If the water level inside the caisson falls lower than that corresponding to the pressure at the cutting edge, sand is likely to flow into the caisson. The volume of sand removed may be several times that of the caisson. Such a procedure results in subsidence in the neighborhood of the excavation. In many instances it may be prevented by keeping the water level inside the caisson higher than that outside. If the removal of the sand by dredging is not practicable, it may be necessary to balance the water pressure by means of a clay slurry or compressed air to keep the sand from flowing in.

Excavations in Clay. When large cuts are excavated in soft clay, the weight of the soil surrounding the edge of the excavation acts like a surcharge on the soil at the level of the bottom of the cut and develops lateral forces in the subsoil. If the depth of the cut becomes so great that the bearing capacity of the soil beneath the sides is approached, large movements are inevitable, irrespective of the care with which the sides of the cut may be braced.

If the bottom of the excavation is underlain at a shallow depth by a firm stratum, the tendency for such a bearing-capacity failure is greatly reduced. The movements can also be decreased by driving piles around the boundary of the cut until they are firmly embedded in the underlying stiff stratum. These piles are braced by struts as excavation progresses.

Field measurements have demonstrated that the volume of soft clay that moves inward at the sides of a cut plus that which moves upward beneath its base is represented by an equivalent volume of subsidence surrounding the excavation. Therefore, any measures that can be taken to prevent lateral movements or heave will be reflected in a decrease in settlement. Hence, it is advisable to keep such cuts braced tightly at all times and to insert braces as soon as possible as the excavation progresses.

The excavation of open shafts or the sinking of open caissons in clay is likewise associated with loss of ground. The material beneath the bottom of the excavation at any stage rises toward the excavation. If the sides remain unsupported even for a short time, lateral movements also take place. Such movements may not produce appreciable settlement around a single shaft, but the cumulative settlement due to the excavation of many shafts beneath a small area may amount to several inches or feet.

The inevitable settlements may be reduced by choosing construction procedures that permit smaller loss of ground. These include the use of sheet piles or cylindrical shells that eliminate exposed vertical faces. If the movements are still too great, the hole may be filled with a heavy fluid and excavation carried out by a mechanical auger or rotating bit. Still further reductions can be accomplished, but at great expense, by the use of compressed air. Similar results can be achieved by driving heavy steel pipes of large diameter and removing the soil by dredging, by jetting, by a process of washing similar to that employed in making test borings, or by means of a blast of compressed air.

Excavation in Stratified Deposits. In some localities, deposits of clay or other impervious materials are underlain by pervious silts, sands, and gravels. Excavation in an open shaft through the impervious soils may be accomplished without difficulty. However, if the voids of the underlying granular materials are occupied by water or gas under pressure, the materials will tend to break into the shaft and flood at least its lower part. In this process, the clay strata may be undermined because of the removal of the underlying sand. This may cause subsidence even at a considerable distance from the point at which the excavation is being made.

16.2. Settlement Due to Vibrations

Vibrations have relatively little effect on cohesive soils, but they may greatly increase the relative density of cohesionless materials. The moisture contained in many sands above the water table provides sufficient cohesion to retard or prevent a rearrangement of the grain structure. However, relatively dry sands and sands beneath the water table are readily compacted. Any operations that produce vibrations in such materials may result in appreciable subsidence of the ground surface.

The principal sources of vibration due to construction operations are pile driving and blasting. Since both operations have been used for the purpose of compacting loose sands, it is obvious that both are likely to produce subsidence in connection with construction activities.

In one instance about 100 piles were driven to a depth of 50 ft in a loose deposit of sand and gravel. Settlement of the ground surface within the area occupied by the piles was about 6 in. It decreased to about $\frac{1}{8}$ in. at a distance of 50 ft (Cummings, 1949). If the driving of piles in such sands is likely to damage adjacent structures, it may be necessary to resort to some other type of construction.

16.3. Settlement Due to Lowering the Water Table

Cause of Settlement. Whenever the water table is lowered, the effective weight of the material between the original and final position of the water table is increased from that of submerged to that of moist or saturated soil. This causes corresponding increase in pressure between the grains of all the soil beneath the original position of the water table, and it is accompanied by strains in accordance with the stress–strain relationships for the material in question. The resulting displacements produce a settlement of the ground surface that is roughly proportional to the descent of the water table.

Effect of Lowering Water Table in Sand. A single increase of effective pressure in a mass of sand does not ordinarily produce significant settlement because even a loose sand is relatively incompressible. Only if a sand is extremely loose, so that its structure may actually collapse, is there likelihood of important subsidence. On the other hand, fluctuations of the water table may ultimately produce large settlements because the deformation of sand increases perceptibly with each application of load as indicated in Fig. 4.13. During normal construction operations, the water table is usually lowered only once or twice and the cumulative effect is not important.

In some instances, large subsidences have occurred near drained excavations in sand because springs have been permitted to form at or near the bottom of the excavations. Water flowing into springs from the surrounding sand may carry the material into the excavation grain by grain, and a tunnel may be formed beneath some slightly cohesive stratum. When the tunnel becomes sufficiently large, its roof may collapse and the ground surface above it subside. The subsidence may take the form of a sink hole that may be located at a considerable distance from the edge of the excavation.

Effect of Lowering Water Table in Clay Strata. Lowering the water table within or above a stratum of clay ultimately increases the intergranular pressure precisely as in sand. Moreover, on account of the high compressibility of clay and various organic soils, the resulting settlements may become very large. However, the process of consolidation initiated by lowering the water table may require considerable time for the development of the ultimate amount of settlement. The time depends primarily on the permeability of the material. The settlement of peat and organic silt may occur very rapidly and may amount to several feet. The settlement of more impervious clays may not be excessive during the relatively short construction period. The rate of settlement and its magnitude can be estimated in accordance with the principles discussed in Chap. 3.

If the compressible layers are underlain by a layer of sand in which the water table is lowered, the piezometric levels may be decreased for a very large distance from the excavation. The corresponding settlements of the overlying materials may extend far from the site of the construction and may lead to serious structural defects. During the construction of a set of locks in Holland, lowering of the water table through a distance of 21 ft in soil consisting of about 20 ft of clay and peat underlain by sand produced settlements noticeable as far as 2500 ft. Even at a distance of 130 ft, the settlement amounted to 24 in. (Brinkhorst, 1936).

16.4. Displacement Due to Pile Driving

When piles are driven into loose cohesionless materials, the ground surface is likely to subside, even though the volume of the piles may represent an appreciable fraction of the volume of the subsoil. In most other materials, the driving of a pile is likely to be associated with the displacement of a volume of material equal to or somewhat less than that of the pile. When many piles are driven in a foundation, especially if the spacing is close, the ground surface may heave as much as several feet. The heave may extend to a considerable distance from the boundaries of the pile group. Structures located within this distance are displaced upward.

In some instances, especially if the soils are silty, the high stresses set up in the mass of soil due to the introduction of the piles cause an intense and rapid consolidation, whereupon the heave is followed within a few days by a settlement. In less pervious materials such as clays, a considerable amount of the heave is likely to remain.

The displacement associated with the driving of piles is also accompanied by lateral movements. Structures located near the piles are likely to be displaced away from the operations. In some instances, the displacement may be very important. For example, a large number of foundation piles may be driven behind a dock next to a river with a relatively deep channel. The resistance of a dock to outward movement is usually small. Most of the displacement due to pile driving is then likely to be represented by lateral movements of the dock. Encroachment of dock structures into river channels for this reason have in many instances amounted to a foot or more.

16.5. Importance of Field Observations for Control of Construction Operations

Legal Importance of Observations. In the past the inevitability of a certain amount of settlement or heave adjacent to construction operations was often ignored. Contractors and engineers alike considered it unwise to admit the influence of their operations on neighboring structures. Numerous specifications have been written containing the statement that excavation shall be carried out in such a manner that no settlement of the surrounding soil shall occur. If no complaints were received from adjacent property owners, it was generally conceded that these provisions had been complied with. On the other hand, if complaints arose, there were always wide grounds for differences of opinion concerning the magnitude of the settlement and the amount of damage resulting therefrom.

More recently it has become the practice for the engineer responsible for a construction project or for the contractor, or both, to establish reference points not only within the limits of their own property but also on structures belonging to adjacent property owners. Observations are made on these reference points often enough to indicate the trend of any movements. Likewise, the adjacent property owners themselves commonly employ independent engineers to observe the movements of their structures. If damage occurs, the facts of the case can be established readily, and a fair settlement can generally be agreed upon without extensive litigation.

Observations for Improving Technique. Observations of settlement, heave, and lateral movement serve another vital purpose, especially when combined with a careful record of the details of the construction procedure. They serve as an indication of unsatisfactory techniques of excavating, bracing, or pile driving. If settlements are consistently larger than have been observed in connection with similar projects under similar soil conditions, it may be concluded that the method of construction is not being carried out in the most workmanlike manner. Variations in technique may be tried and their merit judged on the basis of any changes in the trend of the settlements. By adopting those techniques that lead to reductions in settlement and discarding those that lead to increases, the settlement can be reduced to the smallest amount compatible with the general method of construction.

In many instances, measurement of the deflections of the members of a system of bracing, or measurement of loads in these members, has led to a substantial reduction in the cost of construction, together with a decrease in the accompanying settlements. Hence, systematic observations of this type are almost always economical. They have the further advantage that unforeseen conditions that might lead to disaster are certain to be preceded by apparently unmotivated increases in deflection or pressure. With adequate warning, the cause of the increases may be investigated and remedial measures taken before difficulties arise.

16.6. Influence of Construction Procedures on Design

In the following chapters, methods will be established for determining the adequacy of various kinds of foundations. In general, the foundation will be considered satisfactory if it does not transmit pressures to the subsoil that exceed the safe load or that will cause excessive settlement. However, certain types of foundations that may be entirely acceptable from these two points of view may also be extremely difficult or impossible to install. Furthermore, some installations may lead to excessive subsidence of surrounding property. Hence, the practicability of constructing each type of foundation that may be considered is an important matter. In many instances, it is a decisive factor in the final choice.

The choice of the type of foundation for a given structure may also be influenced by the possibility of damage due to adjacent construction operations at a subsequent date by other parties. For example, a satisfactory foundation for a given building may consist of piers excavated through soft clay to a stratum of very hard clay resting on waterbearing dense sand, which in turn rests on bedrock. Nevertheless, if a new building is constructed on adjacent property and is founded on piers extending through the hard clay and waterbearing sand to bedrock, there is great likelihood that the sand may flow into the excavations for the new piers, and that the support for the hard clay stratum beneath the older building may be lost. In view of this possibility, it may be preferable to establish the first building on piers extending to rock, even at extra expense. Hence, the possibility of damage due to future construction in the neighborhood may also be a significant factor in the final choice of the type of foundation.

SUGGESTED READING

Examples of the various causes of damage due to construction operations and similar disturbances are contained in the following articles.

D. J. D'Appolonia (1971), "Effects of Foundation Construction on Nearby Structures," *Proc. 4th Panamerican Conf. Soil Mech.*, Puerto Rico, 1, 189–236. State-of-the-art report.

R. B. Peck (1969b), "Deep Excavations and Tunneling in Soft Ground," *Proc. 7th Int. Conf. Soil Mech.*, Mexico. State-of-the-art volume, pp. 225–290. Movements associated with braced cuts and tunnels.

J. Feld (1968), *Construction Failure*, New York, Wiley, 399 pp., Chaps. 2 and 3 deal with foundations and subsurface work.

M. G. Lockwood (1954), "Ground Subsides in Houston Area," *Civ. Eng. ASCE*, 24, 6, 48–50. Effect of groundwater lowering.

R. E. Gray and J. F. Meyers (1970), "Mine Subsidence and Support Methods in Pittsburgh Area," *ASCE J. Soil Mech.*, 96, SM4, 1267–1287.

D. D. Barkan (1962), *Dynamics of Bases and Foundations* (translated from the Russian by L. Drashevska; translation edited by G. P. Tschebotarioff), New York, McGraw-Hill, Chap. IX, "Effect on Structures of Waves from Industrial Sources," pp. 407–424. General discussion with brief case histories.

PART C

Selection of Foundation Type and Basis for Design

Since all structures are supported by foundations and ultimately by soil or rock, the success of a project in which structures are included weighs heavily on the foundation engineer. Yet, the overall planning, design, and construction of most projects require the integrated efforts of various disciplines. The foundation engineer then finds himself to be a member of a design team; he also finds that types of foundations and methods of construction may be compromises resulting from many requirements other than subsurface conditions.

On certain work, such as waterfront construction and dam and levee projects, the foundation engineer may appropriately be designated the prime professional. On other work, including most buildings, the structural engineer or the architect is likely to be the leader of the team and may assume many, if not all, of the responsibilities of the foundation engineer. The complexity of foundation conditions, however, is not necessarily related to the size of the project. Hence, the demands for expertise in foundation engineering may be greater on small than on large projects.

No matter what the arrangement of professional responsibilities or the size and type of project may be, the engineering procedures remain the same for choosing the most appropriate type of foundation for given soil conditions and for selecting suitable values of the allowable soil pressure or allowable load per pile. Part C deals with these procedures.

Gen. Wm. SooySmith (1830–1916)

Eminent American foundation engineer during the era of the westward expansion of the railways and the development of the skyscraper. He employed a wide variety of foundations, depending upon the requirements of his projects. He was first in this country to use pneumatic caissons (Wagoshance Lighthouse, Straits of Mackinac, 1867); he was a leader of the transition in Chicago from shallow footings to long piles; and he was the inventor of the "Chicago method" of pier construction. He also conducted one of the earliest tests on a group of piles (Chicago Public Library, 4-pile group, 1893). (Photo courtesy of Chicago Historical Society.)

PLATE 17.

CHAPTER 17

Factors Determining Type of Foundation

17.1. Steps in Choosing Type of Foundation

The type of foundation most appropriate for a given structure depends on several factors: the function of the structure and the loads it must carry, the subsurface conditions, and the cost of the foundation in comparison with the cost of the superstructure. Other considerations may enter into the selection, but these three are commonly the principal ones.

Because of the interplay of these several factors, there are usually several acceptable solutions to every foundation problem. Faced with a given situation, different engineers of long experience may arrive at somewhat different conclusions. Thus, judgment plays a primary part in foundation engineering. It is doubtful that a strictly scientific procedure can ever be established for foundation design, although scientific developments have contributed greatly to the advancement of the art.

When an experienced engineer begins to study a new project, he almost instinctively discards the most unsuitable types of foundation and concentrates on the most promising ones. When he has narrowed his choice to a few alternatives well fitted to the subsurface conditions and to the function of the structure, he studies the relative economy of these selections before making his final decision.

Engineers of less experience can follow a similar procedure without danger of serious error if they avail themselves of the results of scientific studies and the collective experience of others. To be useful, however, this information must be organized in a logical fashion. Part C of this book represents a digest of experience with various types of foundations and subsurface conditions, arranged so that the reader can become familiar with the mental processes used by men who have achieved success in practice. By doing so, the reader can use similar techniques and can expect to arrive at reasonable solutions for all but the most complex foundation problems.

In choosing the type of foundation, the engineer must perform five successive steps:

1. Obtain at least approximate information concerning the nature of the superstructure and the loads to be transmitted to the foundations.
2. Determine the subsurface conditions in a general way.
3. Consider briefly each of the customary

types of foundation to judge whether they could be constructed under the existing conditions, whether they would probably be capable of carrying the required loads, and whether they might experience detrimental settlements. Eliminate, in this preliminary way, obviously unsuitable types.

4. Make more detailed studies and even tentative designs of the most promising types. These studies may require additional information concerning the loads and subsurface conditions and generally must be carried far enough to determine the approximate size of footings or piers or the approximate length and number of piles required. It may also be necessary to make more refined estimates of settlement in order to predict the behavior of the structure.
5. Prepare an estimate of the cost of each promising type of foundation, and choose the type that represents the most acceptable compromise between performance and cost.

Steps 3 and 4 require a knowledge of the probable behavior of each type of foundation for each type of subsurface condition. The presentation of such information is the principal object of Part C.

17.2. Bearing Capacity and Settlement

On the assumption that it is practicable to construct a given type of foundation under the conditions prevailing at a site, the probable performance of the foundation must be judged with respect to two types of unsatisfactory behavior. On the one hand, the entire foundation or any of the elements of which it is composed may break into the ground because the soil or rock is incapable of supporting the load without failure. On the other hand, the supporting soil or rock may not fail, but the settlement of the structure may be so great or so uneven that the superstructure may become cracked and damaged. Misbehavior of the first type is related to the strength of the supporting soil or rock and is known as a *bearing-capacity failure*. That of the second is associated with the stress-deformation characteristics of the soil or rock, and is known as *detrimental settlement*. In reality the two types of unsatisfactory behavior are often so closely related that the distinction is entirely arbitrary. For example, a footing on loose sand settles by greater and greater increments, out of proportion to the increases in load, until even the settlements under very small increments are intolerable; yet no outright plunging of the footing into the ground occurs. In other instances the distinction is clear; a footing on stiff clay underlain by a layer of soft clay may be entirely safe against breaking into the ground, but the settlement due to consolidation of the soft clay may be excessive. In many practical problems the two types of unsatisfactory behavior can be investigated separately, as if they had independent causes. Such separation considerably simplifies the engineer's approach.

In each of the following chapters, one of the principal kinds of natural deposits is considered. For each kind, the various types of foundations are listed, and methods are described for determining the load that can safely and without excessive settlement be transferred to the soil by the foundation. Finally, for each kind of natural deposit, a summary is given of construction difficulties that may have a bearing on the suitability of each type of foundation.

17.3. Design Loads

The selection of the loads on which the design of a foundation is to be based influences not only the economy but sometimes even the type of the foundation. Moreover, the soil conditions themselves have a bearing on the loads that should be considered.

Every foundation unit should be capable of supporting, with a reasonable margin of safety, the maximum load to which it is ever likely to be subjected, even if this load may act only briefly or once in the lifetime of the structure. If an overload or a misjudgment of the soil conditions would result merely in an excessive increase in settlement but not an outright failure of the subsoil, a smaller factor of safety might be justified than if the overload would lead to a sudden and catastrophic bearing-capacity failure.

The maximum loads and corresponding soil pressures or pile loads are often specified by building codes; these requirements are legal restraints on the design that must be satisfied. However, since they may not take into account all eventualities, the foundation engineer must assure himself that the foundations are safe, even though they satisfy the code. Furthermore, the loadings required for investigations of safety or to satisfy the legal requirements may not be appropriate for assuring the most satisfactory performance of the structure with respect to settlement.

For example, since sands deform quickly under change in stress, the settlements of footings on sand reflect the actual maximum load to which they are subjected. The actual live load may never approach the value prescribed by the building code, whereas the actual and computed dead loads should be practically equal. Hence, a column for which the code ratio of live to dead load is large is likely to settle less than one for which it is small. Thus, to proportion footings on sand for equal settlement, the engineer should use the most realistic possible estimate of the maximum live loads instead of arbitrarily inflated ones.

On the other hand, the settlement of a structure supported by footings underlain by a saturated clay is virtually unaffected by a short application of even a fairly high load to one or more footings, provided a bearing-capacity failure is not approached. Because of the slow response of clay to applied loads, the settlement should be estimated on the basis of the dead load plus the best possible estimate of the long-term average instead of the maximum live load.

In the following chapters, appropriate loads and factors of safety are suggested for various conditions. Each design, however, requires a careful appraisal with respect to the conditions peculiar to the particular site and structure. As a general rule, a factor of safety of 3 should be provided against the loads specified by the building code if the subsoil is not of an unusual type and if its properties have been investigated in a customary, competent manner. The factor of safety should not ordinarily be less than 2 even if the maximum loads are known with a high degree of certainty and the soil properties are known exceptionally well.

The permissible settlement depends on the type of structure and on its function. The load transmitted to the soil at the base of a bridge pier may be due largely to the dead weight of the pier itself, and the corresponding settlement may be several inches. If the settlement occurs during construction of the pier, it is of no practical significance. If it takes place over a long time, it may be of no consequence provided the superstructure consists of simple trusses or cantilevers, but it may be of serious consequence if the superstructure is a continuous girder or truss. Few concrete building frames can stand a differential settlement between adjacent columns of more than about $\frac{3}{4}$ in. without showing some signs of distress. A steel structure can stand somewhat more, and a brick masonry structure can stand three to four times this amount without serious damage. Irregular or erratic settlement is more harmful to a structure of any type than uniformly distributed settlement.

Inasmuch as the cost of a foundation is appreciably influenced by the amount of differential settlement considered tolerable, the engineer should not underestimate the settlement that his structure can stand.

The foregoing considerations pertain to the loads that influence the behavior of the soil or rock supporting the foundation. In addition, the structural design of reinforced-concrete foundation elements such as footings, pile caps, or rafts, as presently carried out by ultimate-strength procedures, requires the assignment of load factors that take account of the nature of the loading and the probability of its occurrence. This aspect of the evaluation of the design loads is discussed in Part D.

SUGGESTED READING

The classic statement concerning the influence of soil mechanics on foundation engineering is K. Terzaghi's Presidential Address at the First International Conference on Soil Mechanics and Foundation Engineering, held at Harvard University in June 1936. It deserves to be read repeatedly by every foundation engineer. It is to be found in Vol. III of that conference, and is reprinted on pp. 62–67 of the volume *From Theory to Practice in Soil Mechanics—Selections from the Writings of Karl Terzaghi*, New York, Wiley, 1960. Three other of Terzaghi's writings give to foundation engineers the perspective of the originator of soil mechanics:

"The Influence of Modern Soil Studies on the Design and Construction of Foundations," *Building Research Congress*, London, 1951, Div. 1, Part III, pp. 139–145. Also reprinted in *From Theory to Practice in Soil Mechanics*, pp. 68–74.

"Origin and Functions of Soil Mechanics," *Trans. ASCE*, Vol. CT (1953), pp. 666–696. (A more general paper dealing with all aspects of applied soil mechanics.)

Address at Opening Session of Fourth International Conference on Soil Mechanics and Foundation Engineering, *Proc. 4th Int. Conf. Soil Mech.*, London, 1957, *3*, 55–58. Also reprinted in *From Theory to Practice in Soil Mechanics*, pp. 75–78.

The following papers are concerned with the effects of settlements on structures and criteria for allowable settlements. The paper by Feld summarizes worldwide experience and contains an extensive bibliography.

J. Feld (1965), "Tolerance of Structures to Settlement," *ASCE J. Soil Mech.*, *91*, SM3, 63–77.

J. R. Salley and R. B. Peck (1969), "Tolerable Settlements of Steam Turbine-Generators," *ASCE J. Power Div.*, *95*, PO2, 227–252.

A. W. Skempton and D. H. MacDonald (1956), "The Allowable Settlement of Buildings," *Proc. Inst. Civ. Eng.*, *5*, 3, Pt. 3, 727–784.

Discussions of safety factors in relation to foundation design are presented in:

G. G. Meyerhof (1970), "Safety Factors in Soil Mechanics," *Canadian Geot. Jour.*, *7*, 4, 349–355. The paper contains a useful bibliography.

G. F. Sowers (1969), "The Safety Factor in Excavations and Foundations," *Hwy. Res. Rec.*, *269*, 23–34.

Laurits Bjerrum (1918–1973)

Director of the Norwegian Geotechnical Institute since its establishment in 1951. While under his leadership the Institute devoted its efforts primarily to problems associated with the soft sensitive clays of Norway, but the knowledge derived from its studies, ranging from investigations of the most fundamental aspects of soil behavior to the most elaborately instrumented full-scale field observations, has exerted a remarkable influence on geotechnical engineering throughout the world.

PLATE 18.

CHAPTER 18

Foundations on Clay and Plastic Silt

18.1. Significant Characteristics of Deposits of Clay and Plastic Silt

Clays are encountered in soft normally loaded to very stiff preloaded states. Their consolidation and strength characteristics have been described in Arts. 3.2 and 4.8. Preloaded clays frequently contain secondary structural defects such as cracks and slickensides, which influence their strength and compressibility and may require special investigation before design.

Plastic silts may derive their plasticity either from a high percentage of plate-shaped particles or from an appreciable organic content. Deep beds of silt, often more or less organic, are encountered near the present or former shores of oceans and lakes and in beds of present or ancient rivers. When these deposits are below the water table and have never been exposed to drying, they are likely to be as soft and compressible as normally loaded clays near the liquid limit. Hence, plastic silts have many of the characteristics of soft to medium clays, and the design of foundations on plastic silt is based on considerations similar to those that govern structures on clay. For this reason, clay and plastic silt are considered synonymous in this chapter.

Under various circumstances, footings, rafts, piers, and piles may be used to support structures on deposits of clay and plastic silt. For each type of foundation, independent investigations are required to determine the factor of safety against a bearing-capacity failure and the amount of settlement to be expected.

The bearing capacity depends primarily on the shearing resistance of the soil. When the load is first applied to footings on saturated clay, it produces excess pore pressure which, if the clay is at least fairly impermeable, does not quickly dissipate. Hence, for at least a short time after loading, undrained conditions prevail and the $\phi = 0$ analysis (Art. 4.8) is applicable. The strength may then be taken as the undrained shear strength (eqs. 4.7), or one half the unconfined compressive strength. To the extent that consolidation does occur, the results of analyses based on the premise that $\phi = 0$ are on the safe side.

Unless the clay is very sensitive, the undrained shear strength can usually be investigated economically and with sufficient accuracy by means of tests on 2-in. samples taken continuously from drill holes in thin-walled tube samplers. Where only a minimal expenditure for exploration can be justified, the compressive strength can be estimated on the basis of examination of split-spoon samples (Table 1.5), and the correlation between strength and N-values from the standard penetration test (Table 5.3). However, the standard penetration test is at best a rather unreliable indicator of the compressive strength of clays, and tests on tube samples are preferable. In-situ vane tests are often an economical and technically superior alternative to the use of tube samplers. If the clay contains secondary structural defects, it may be necessary to resort to load tests (Art. 18.2).

The settlement depends primarily on the compressibility of the clay, which is intimately related to its history of loading. The compressibility can be evaluated most reliably by means of consolidation tests on representative undisturbed samples of 4- to 6-in. diameter, but the cost of obtaining such samples sometimes precludes their use. Smaller samples, if taken by the best techniques, may prove satisfactory. In some instances, the compressibility of soft clays can be estimated with sufficient accuracy from the Atterberg limits and natural water content, or from consolidation tests on completely remolded samples (Art. 3.3). Estimates based on such procedures are, however, inapplicable to preloaded or extra-sensitive clays (Arts. 3.5 and 3.6).

In the articles that follow, the bearing capacity and the determination of safe loads are discussed for each of the various types of foundations. Then the general procedures for making settlement forecasts are described. Finally, attention is given to excavation and stability of slopes in clay and to the lateral movement of structures founded on clay.

18.2. Footings on Clay

Ultimate Bearing Capacity. Figure 18.1a represents a cross section through a long footing with width B, resting at depth D_f below the ground surface. The quantity D_f is known as the *depth of foundation* or *surcharge depth*. Its meaning for a footing with different depths on the two sides is shown in Fig. 18.1b. The soil beneath the base of the footing consists of intact clay without structural defects such as slickensides or cracks, and the degree of saturation is practically unity.

If the footing illustrated in Fig. 18.1a fails, a wedge of soil $Ocbde$, as shown on the left side of the figure, must be displaced upward and to the left. The weight of the wedge and the shearing strength of the soil along $Ocbd$ tend to resist failure. No completely rigorous theory exists for calculating the ultimate bearing capacity under these circumstances, but satisfactory approximate

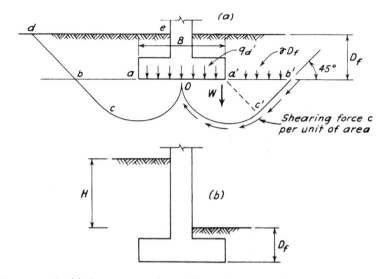

FIGURE 18.1. (a) Cross section through long footing on clay, showing basis for computation of ultimate bearing capacity. (b) Section showing D_f for footing with surcharge of different depth on each side.

solutions are available. It is assumed, as illustrated on the right side of Fig. 18.1a, that the influence of the soil above the base level of the footing can be replaced by a uniform surcharge γD_f. Theory and experiment then indicate that the surface of sliding consists of a circular section Oc' and a straight section $c'b'$, which rises at 45° to the horizontal. All the forces acting on the sliding mass $Oc'b'a'$ are shown in the figure. The condition that these forces must be in equilibrium can be used to evaluate the ultimate bearing capacity q_d'. The normal forces across the surface of sliding can produce no frictional shearing resistance on account of the postulated condition that $\phi = 0$. On the other hand, a shearing resistance c (Fig. 4.9) per unit of area acts along the surface to oppose the sliding. The results of the evaluation lead to the expression

$$q_d' = cN_c + \gamma D_f \qquad 18.1$$

The *net ultimate bearing capacity* q_d is defined as the pressure that can be supported at the base of the footing in excess of that at the same level due to the surrounding surcharge; hence

$$q_d = q_d' - \gamma D_f$$

and

$$q_d = cN_c \qquad 18.2$$

To emphasize the distinction between q_d and q_d', the latter is known as the gross ultimate bearing capacity. In eq. 18.2, N_c is a dimensionless *bearing-capacity factor* having, for a long continuous footing on the surface of the clay deposit, the value 5.14.

The value of N_c for a footing varies somewhat with the ratio of the width B to the length L, and with the depth of surcharge D_f, as indicated in Fig. 18.2 (Skempton, 1951). For any given value of D_f/B, Fig. 18.2 shows that the bearing capacity factor for circular and square footings is approximately 1.2 times the corresponding value for a long continuous footing. A straight-line interpolation may be used for rectangular footings having intermediate values of B/L.

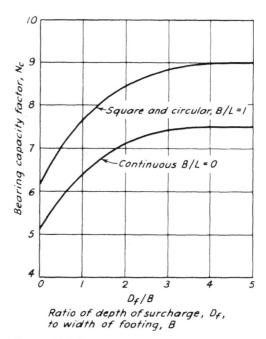

FIGURE 18.2. Bearing capacity factors for foundations on clay under $\phi = 0$ conditions (after Skempton, 1951).

Safe Soil Pressure. Under dead load plus the maximum live loads that can normally be expected, the factor of safety against a bearing-capacity failure should be on the order of 3. The allowable soil pressure q_a (Art. 10.4) may, therefore, be taken as one third the net ultimate soil pressure (eq. 18.2)

$$q_a = \frac{cN_c}{3} \qquad 18.3$$

or

$$q_a = \frac{q_u N_c}{6} \qquad 18.4$$

Equation 18.4 and values of N_c from Fig. 18.2 provide the basis for the curves shown in Fig. 18.3. For given values of soil strength and D_f/B, the net allowable soil pressure q_a for continuous footings may be obtained directly from the chart. For rectangular footings the chart values are multiplied by $(1 + 0.2B/L)$; hence, for square or circular footings the increase is 20 per cent. Appropriate adjustments may be made to the chart values for factors of safety other than 3.

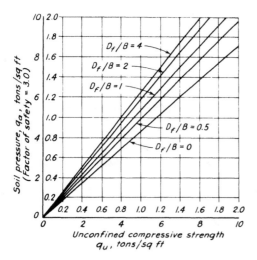

FIGURE 18.3. Net allowable soil pressure for footings on clay and plastic silt, determined for a factor of safety of 3 against bearing capacity failure ($\phi = 0$ conditions). Chart values are for continuous footings ($B/L = 0$); for rectangular footings, multiply values by $1 + 0.2\ B/L$; for square and circular footings, multiply values by 1.2.

The ratio of depth of surcharge to width of footing for most footings is commonly less than 1. Figure 18.3 reveals that, under these conditions, the net allowable pressure is roughly equal to the unconfined compressive strength of the clay.

Because of the variations that normally occur even in relatively uniform clay deposits, the value of q_u in eq. 18.4 or Fig. 18.3 should represent the average for a depth B below the base of the footing. The unconfined compressive strength should be determined at 6-in. intervals in the vertical direction. The strength of some of the samples may be estimated if the technician is experienced, but at least one out of every five samples should be tested. If, on the other hand, the clay is not fairly uniform but instead a soft layer is located within a depth B below the base of the footing, the strength of the soft layer is likely to determine the factor of safety of the footings. This condition is shown in Fig. 21.2, and its significant effects on footing design are discussed in Art. 21.3.

If the strength of the soil decreases with depth, the safety of a large footing may be much less than that of a small one. Moreover, under these conditions eq. 18.4 and the curves of Fig. 18.3 can be used for selecting the safe load on a single footing only if that footing is so far from its neighbors that the stresses in the subsoil beneath it are not significantly influenced by the surrounding footings. If one or more soft layers are located in the subsoil at a depth even greater than B below the base of the footing, a computation should be made to ascertain whether the pressure at the top of any of the soft layers exceeds the safe value for that layer. If it does, the design should be revised.

The pressure at the top of a deep layer of clay can be estimated by means of the influence chart described in Art. 18.6. However, it is often more expedient and sufficiently accurate to assume that the pressure at the base of the footing spreads out uniformly within the confines of a truncated prism with sides sloping outward from the edges of the footing at 2 vertical to 1 horizontal.

If the clay contains a network of cracks or slickensides, the procedure based on the undrained shear strength as determined from unconfined compression or field vane tests should not be used because the strength of the clay depends on the spacing and nature of the defects instead of the shearing resistance of the intact fragments themselves. Under these circumstances it may be necessary to resort to load tests.

The details of the load-test arrangement depend on the spacing of the cracks, the size of the footing, and the degree of uniformity of the clay. To make sure that a representative number of cracks is included, the test plate (Fig. 18.4) should be at least 2 feet square. The effect of surcharge around the plate should be eliminated. Hence, the test should be conducted in a pit with a width at least three times that of the plate. The load should be applied in increments, and after each increment the deformation should

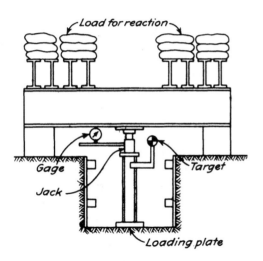

FIGURE 18.4. Arrangement for load test on fissured clay.

be observed until its rate becomes very small. The load-settlement relationship should be determined either to the point of failure or to a value of load three times that to be used for design. One load test should be made at the level of the base of the footings, and enough additional tests should be made at greater depth to establish the average value of the bearing capacity within the depth B. If the clay weakens consistently with depth, the safe load should be based on the lowest values of bearing capacity. On the other hand, if the variations are erratic, the average value may be used.

Design Procedure. If the footings supporting a structure are designed for a factor of safety of 3, the differential settlements having their origin in compression of the clay immediately beneath the footings are not likely to exceed 0.75 in., providing the footings are far enough apart that the action of each is independent of the others, and provided the soil directly beneath the footings is not a soft or very soft normally loaded clay. The loads considered in selecting the sizes of the footings to provide a factor of safety of 3 should be those that will normally act for a large proportion of the time. Exceptional values of live load due to improbable combinations of loading should not be included, nor should wind load, snow load in temperate climates, or earthquake forces. On the other hand, the factor of safety should not be less than 2 under the most severe conditions and probable combinations of loading. Hence, the maximum soil pressure under the most unfavorable loading combination should not exceed 1.5 times the values of q_a obtained from eq. 18.4 and Fig. 18.3.

As previously discussed in this article, the value of q_u to be used in obtaining the allowable soil pressure q_a for a given footing by means of eq. 18.4 and Fig. 18.3 should represent the average for a depth B below the base of the footing. However, the sizes, shapes, and depths of footings commonly vary considerably within any given project. Moreover, soil borings are not ordinarily made at every footing. For these reasons, and because of the inherent variations of most natural deposits of clay, the use of different average values of q_a for different footings is seldom justified. The allowable pressure is commonly based on data from the boring showing the least favorable conditions. The determination of its magnitude and the decision whether it should be varied for different footings and for different elevations require the exercise of judgment by the foundation engineer.

Continuous footings beneath exterior basement walls are subjected to unbalanced loading including lateral earth pressure (Fig. 18.1b). If the difference γH in surcharge on the two sides of the footing does not exceed about $0.5q_u$, the unbalanced loading may be neglected. Otherwise, the footing and basement wall should be designed as a retaining wall (Chap. 26).

In many instances, except where stiff or hard clays are involved, the action of any one footing in a group is significantly influenced by the presence of the neighboring footings or other loaded areas. In this event the differential settlement of a footing foundation may be much greater than 0.75 in. Consequently, an investigation of the settlement of the entire foundation must always supplement the determination of the allowable soil pressure for the individual footings. These matters are considered in Art. 18.6.

Footings with Eccentric Loading. Beneath retaining walls and other types of structures subjected to lateral loads, the calculated pressure on the base may not be uniform but may vary linearly from one toe to the other. The width of such footings on clay should be determined as if the entire base were subjected to an average pressure equal to the maximum toe pressure.

ILLUSTRATIVE DESIGN. DP 18–1. FOOTINGS ON CLAY

DP 18–1 illustrates the use of various principles discussed in Art. 18.2. The dimensions of the footings are determined exclusively on the basis of the strength of the clay. The subject of settlement of foundations above clay is discussed in Art. 18.6 and in DP 18–4 and DP 18–5.

The least favorable generalized data obtained from several borings at the building site are given on Sh. 1. Also listed are the

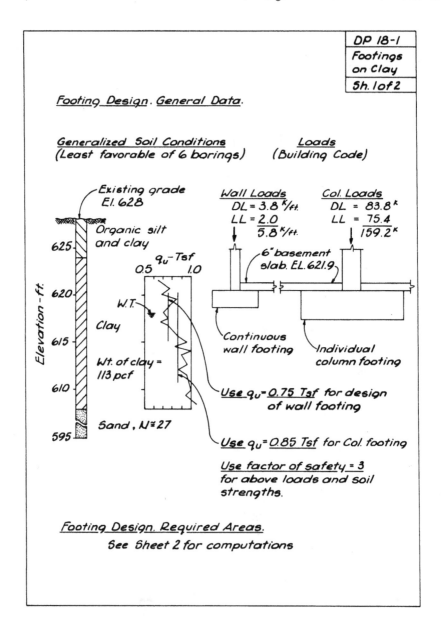

footing loads for which a factor of safety of 3 is to be provided. It is assumed that these loads govern the design. If, because of more severe short-time combinations, the total loading on any footing might exceed these values by more than 50 per cent during the life of the building, the footing for which this condition would exist should be proportioned for the most severe combination of loads at a factor of safety of 2.

The allowable net soil pressure is determined from Fig. 18.3 in which the value of q_u is taken as the average unconfined compressive strength of the clay for depths below the base equal to the widths of the footings. For illustrative purposes, although the procedure might be considered an unnecessary refinement in practice, a higher value of q_a is used for the square than for the continuous footing because of the shape factor $(1 + 0.2B/L)$.

Initially, of course, the width of the footing is not known. Hence, the depth over which the unconfined compressive strengths

---

DP 18-1
Footings on Clay
Sh. 2 of 2

Footing Design. Required Areas.

<u>Continuous Wall Footing</u>:

Total load = 5.8 ^{k}/ft. = 2.9 ^{T}/ft.
    ÷ say 0.70 = 4.14'

<u>Try B = 4'-3"</u>, assume footing depth = 1'-6"
    therefore D_f/B = 0.47

<u>Total wall load</u> = 2.9 ^{T}/ft.
    ÷ 4.25 = 0.68 Tsf

Additional load
 Floor slab = 0.5 × $\frac{0.15}{2}$ ⎫
                                    ⎬ = $\frac{0.15}{0.83}$ Tsf
 Footing   = 1.5 × $\frac{0.15}{2}$ ⎭

Surcharge load
 Floor slab = 0.5 × $\frac{0.15}{2}$ ⎫
                                    ⎬ = $\frac{0.12}{0.71}$ Tsf Net
 Soil   = 1.5 × $\frac{0.11}{2}$ ⎭

Fig. 18.3, q_u = 0.75, q_a = 0.73 Tsf <u>Use 4'-3" wall footing</u>

<u>Individual Column Footing</u>:

Total Col. load = 159.2^k = 79.6^T
    ÷ say 0.95 = 83.8 sq. ft.

<u>Try B = 9'-3" × 9'-3"</u>, A = 85.6 sq. ft.
    assume depth = 2'-3" and D_f/B = 0.30

<u>Total Col. load</u> = 79.6^T
    ÷ 85.6   = 0.93 Tsf

Additional load ⎫
minus surcharge ⎬ 2.25 $\left(\frac{0.15 - 0.11}{2}\right)$ = $\frac{0.05}{0.98}$ Tsf

Fig. 18.3, q_u = 0.85, q_a = 0.79 × 1.2 = 0.95 Tsf, say ok

<u>Use 9'-3" × 9'-3" Col. footing</u>

should be averaged is also not known, and a trial-and-correction procedure is required. The design plate shows only the final trial for each footing. The initial trial is based on a rough average for q_u, approximated by eye from the data plotted on Sh. 1. Convergence to the final value of B is rapid; the third trial is usually satisfactory.

The trial-and-error procedures exemplified in this and the subsequent design plates, and the general form of the computations, are representative of usual practice in design offices. In many instances equations could be written and solved more directly for a desired quantity such as the width of a footing. The equations, however, tend to be complex and are sometimes a source of error. More seriously, they often obscure the influence of the principal variables. The trial-and-correction procedures, on the other hand, lead quickly to approximately correct values of the desired quantities and permit the designer, at each step of the calculation, to judge the reasonableness of the results. Moreover, the form of calculation facilitates the independent checking required in every design office.

Since Fig. 18.3 furnishes the value of the bearing capacity in excess of the weight of the surcharge adjacent to the footing, the weights of the floor slab and soil above the elevation of the base are subtracted from the load on the footing. However, this subtraction very nearly offsets the weights of the floor slab and footing that were previously added to the pressure produced by the load of the wall. Therefore, it is usually sufficiently accurate to divide the load from the wall or column by the area of the footing and to compare this pressure with the safe value determined from Fig. 18.3. If, however, the computations for additional load minus the surcharge are not omitted, they may be simplified as indicated for the column footing on Sh. 2 of DP 18-1, where the depth of the footing is multiplied by the difference between the unit weights of concrete and soil. It should be noted that the surcharge on the outer projection of the wall footing, although greater than γD_f, is ignored in the calculations. It is assumed that this load is fully resisted by the strength of the adjacent soil.

The area of the column footing in DP 18-1 is such that the allowable net soil pressure is exceeded slightly under the dead and live load used for design. It is common practice to allow such small excesses because footings are usually proportioned only to the nearest 3 in. In view of the probable limits of accuracy of the loads, soil strength, and methods of analysis, there is seldom justification for greater refinement of the design.

18.3 Rafts on Clay

Ultimate Bearing Capacity. The net ultimate pressure that can be sustained by the soil at the base of a raft on a deep deposit of clay or plastic silt may be obtained in the same manner as for footings on clay (Art. 18.2). The quantity q_d in eq. 18.2 is the pressure at the elevation of the base of the raft in excess of that exerted by the surrounding surcharge. Likewise, in eq. 18.4 and in Fig. 18.3, q_a is a net soil pressure. By increasing the depth of excavation, the pressure that can safely be exerted by the building is correspondingly increased. This can be accomplished by increasing the number or depth of basements. On the other hand, the area of a raft cannot usually be enlarged appreciably in an attempt to reduce the soil pressure because it is not feasible to extend a raft more than a few feet beyond the building proper. Therefore, if a raft foundation is to be constructed at a site underlain by clay too soft to provide support at the normal basement level, the only practical method to provide the required factor of safety is to lower the elevation of the raft.

Safe Soil Pressure for Rafts on Clay. In proportioning footings on clay (Art. 18.2), the net ultimate bearing capacity is divided by the factor of safety to obtain a net allowable soil pressure. For a factor of safety equal to 3, this procedure results in eqs. 18.3 and 18.4 and in Fig. 18.3. The same principles are applicable to rafts on clay. Accordingly, the factor of safety, in terms of net soil pressures, may be expressed as

$$F = \frac{cN_c}{q_b - \gamma D_f} \quad 18.5$$

where q_b is the *gross soil pressure* or *contact pressure* produced at the base of the raft by the weight of the building and live load, and which in the denominator of eq. 18.5 is reduced to a net pressure by subtracting the weight of the surrounding surcharge, γD_f.

According to eq. 18.5, the factor of safety is very large for rafts established at such depths that γD_f is nearly equal to q_b. When these terms are equal the raft is said to be a *fully compensated foundation*. The theoretical factor of safety against failure of the subsoil under these circumstances is infinite, provided no uncertainty is involved in the estimate of loads or in the action of γD_f. However, even if γD_f is fully effective, an increase in the gross soil pressure q_b, possibly caused by unanticipated additional loads on the raft, reduces the degree of compensation; the decrease in the factor of safety, moreover, is out of proportion to the increase in loads. Equation 18.5 also shows that an error in estimating the weight of the structure or the live load has a greater influence on the factor of safety for a weak clay than for a strong one.

As for footings on clay (Art. 18.2), the factor of safety against failure of the soil beneath a raft on clay should not be less than 3 under normal loads, or less than 2 under the most extreme loads. Therefore, Fig. 18.3 may be used to obtain the allowable net soil pressure for rafts on clay. The values from Fig. 18.3 may be multiplied by appropriate ratios to convert the pressures to those corresponding to factors of safety other than 3. Since N_c and, consequently, the allowable soil pressure are somewhat influenced by the depth of surcharge, the determination of D_f for the partial compensation required to attain a desired factor of safety is, strictly speaking, a trial procedure. However, the first trial based on an assumed D_f/B is ordinarily sufficiently accurate.

ILLUSTRATIVE DESIGN. DP 18-2. RAFT ON CLAY

Various principles regarding full and partial compensation and the relationships to the factor of safety of rafts on clay are illustrated in DP 18-2. The combination of loads chosen for design is the weight of the building plus normal live load. For this combination, a depth D_f of 20.5 ft is required for full compensation, whereas a depth of 15.5 ft provides a factor of safety of 3.

The computations in DP 18-2 show that, for a depth of 15.5 ft, an unforeseen increase in loads of only 25 per cent would reduce the factor of safety of the foundation approximately 50 per cent. At the same depth of foundation, the factor of safety would be further disproportionately reduced to slightly less than unity if the loads were increased 50 per cent.

The importance of the surrounding surcharge D_f is also illustrated in DP 18-2. It is seen that the factor of safety decreases to unity when the surcharge is reduced from 15.5 to 5.4 ft. If D_f were reduced to zero, the factor of safety for the design loads would equal 0.74, and failure would occur.

For any change in the strength of the clay, all values of the factor of safety computed in DP 18-2 change proportionately. Such changes are illustrated by the curves on Sh. 3 for different values of q_u.

The computations in the design plate are appropriate if the depth of the soft clay is equal to at least 60 ft. Deposits of less thickness and those of nonuniform character (Chap. 21) require special consideration.

18.4. Piers on Clay

The net ultimate and safe soil pressures that can be sustained by a clay deposit at the base of a pier may be determined in a manner similar to that described for footings (Art. 18.2). For piers, however, the ratio of depth of foundation to base width is large; usually it exceeds 4. According to Fig. 10.2, if the ratio D_f/B is equal to or greater than 4, it has no influence on the bearing capacity factor N_c. For piers, therefore, the net ultimate pressure q_d (eq. 18.2) may be expressed as

$$q_d = 7.5c\left(1 + 0.2\frac{B}{L}\right) \qquad 18.6a$$

If the pier is square or circular, if c is taken as the undrained shear strength equal to $q_u/2$, and if a factor of safety of 3 is desired, eq. 18.6a becomes

$$q_a = 1.5q_u \qquad 18.6b$$

The values q_a obtained from eq. 18.6b are equal to 1.2 times those obtained from Fig. 18.3 along the line $D_f/B = 4$.

A stratum suitable for the support of piers is often found only at a considerable depth, beneath soils too weak or compressible to participate in carrying the permanent loads on the pier. Nevertheless, eqs. 18.6a and 18.6b are still applicable, provided that the pier is established a distance of at least $4B$ below the upper surface of the bearing stratum. For shallower depths of

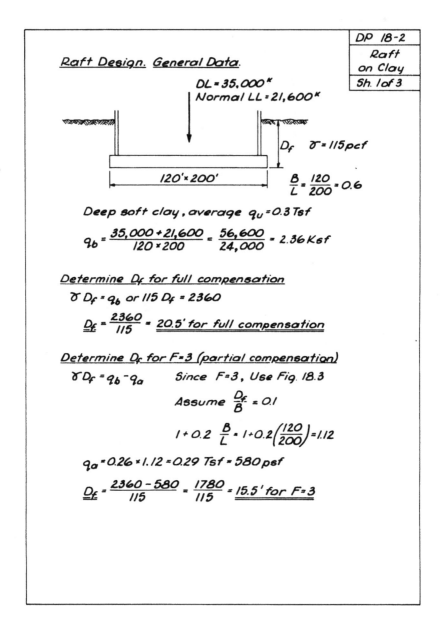

penetration, Fig. 18.3 may be used, but D_f should be considered the depth of penetration of the base of the pier into the bearing stratum.

On the other hand, the foundation material may consist of relatively stiff clay to great depth. Straight-shafted piers may be established in bored holes deep enough to develop sufficient skin friction to support a large portion of the load. The amount of skin friction that can be relied on depends on several factors, including the shearing strength of the undisturbed clay; the extent to which the clay becomes softened by absorbing water from the atmosphere, from the drilling operations, and from the concrete cast in the hole; the degree of remolding associated with drilling; and the presence or absence of a skin or smear zone of remolded clay on the walls of the hole. The relatively few full-scale tests available (Skempton, 1959; Woodward et al., 1961)

DP 18-2
Raft on Clay
Sh. 2 of 3

<u>Raft Design.</u>

<u>Determine Factor of Safety ($D_f = 15.5'$)</u>

(a) <u>If q_b increases 25%</u> (2360 × 1.25 = 2950 psf)

Use eq. 18.5 $\underline{\underline{F}} = \dfrac{3 \times 580}{2950 - 1780} = \dfrac{1740}{1170} = \underline{\underline{1.49}}$

(b) <u>If q_b increases 50%</u> (2360 × 1.5 = 3540 psf)

$\underline{\underline{F}} = \dfrac{1740}{3540 - 1780} = \dfrac{1740}{1760} = \underline{\underline{0.99}}$

<u>Determine D_f for Factor of Safety = 1.0</u>

Assume no change in $q_b = 2360$ psf

Use eq. 18.5 $1.0 = \dfrac{1740}{2360 - 115 D_f}$

$\underline{\underline{D_f}} = \dfrac{2360 - 1740}{115} = \underline{\underline{5.4'}}$

<u>Determine Factor of Safety ($D_f = 15.5$ & $q_b = 2360$ psf)</u>

If q_u decreases to 0.25 Tsf
Fig. 18.3 $q_a = 493$ psf

Use eq. 18.5 $\underline{\underline{F}} = \dfrac{3 \times 493}{2360 - 1780} = \dfrac{1479}{580} = \underline{\underline{2.55}}$

Check by direct proportion, $\underline{\underline{F}} = 3.0 \times \dfrac{0.25}{0.30} = \underline{\underline{2.50}}$

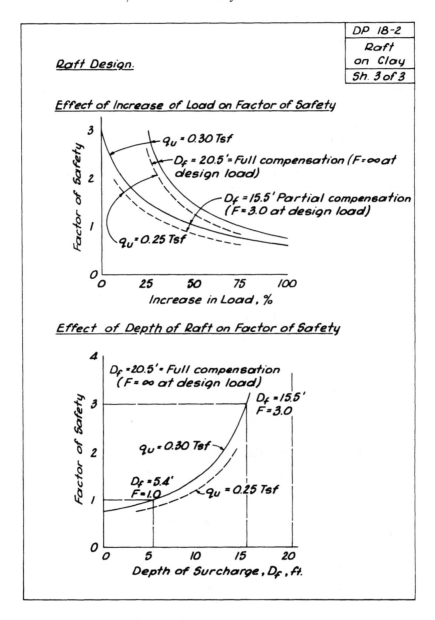

suggest that the ultimate skin friction c_a per unit of area may be taken as

$$c_a = \alpha_1 c \qquad 18.7$$

where α_1 ranges between about 0.3 and 0.5 and may generally be considered about 0.45, and c is the undrained shear strength of the undisturbed clay averaged over the height of the shaft. Because of the possibility of softening, the value of c_a should not be allowed to exceed about 1 ton/sq ft. The total load resulting from skin friction may be computed by multiplying c_a (eq. 18.7) by the surface area of the pier shaft.

The total ultimate capacity of a pier (Fig. 18.5) may be obtained by adding the total skin friction to the ultimate resistance of the clay at the base of the pier. The base resistance may be determined from the ultimate pressure (eq. 18.2) multiplied by the area of the base. When the skin friction along the shaft is utilized to support even a small portion of the load, D_f should be taken as zero in the determination of N_c

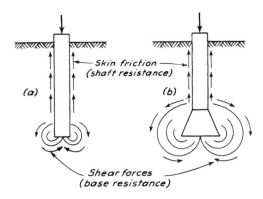

FIGURE 18.5. Forces acting on deep piers in clay. (a) straight shaft; (b) belled shaft.

(eq. 18.2 and Fig. 18.2), because the clay surrounding the pier cannot furnish upward skin friction and at the same time be effective as a downward surcharge (Fig. 18.5). It is common practice to divide both c_a (eq. 18.7), and q_d (eq. 18.2), by the desired factor of safety and to obtain the safe capacities of shaft and base separately. As a reasonable approximation, the sum of the capacities is taken as the total allowable load for the pier.

If the pier is provided with a bell, only the straight part of the shaft should be considered in computing skin-friction capacity. The width B is taken as the diameter of the bell in all determinations of the ultimate and safe soil pressures and for the computation of the area and total capacity of the base of the pier. The undrained shear strength of the clay used for design should be the average strength within the depth B below the bottom of the bell. However, the influence of any soft zones below the bell should be considered (Art. 21.3).

The factor of safety against the failure of piers on clay should be equal to 3 under the maximum conditions of loading that may normally be expected, and should not be less than 2 under the most severe combinations of loading.

Since the net pressure is a function of the total weight of the pier, a hollow pier is often advantageous.

When piers are constructed by sinking caissons through soft clay deposits, the skin friction must be known in order to judge whether the caisson may become stuck at some point in its descent. Theoretical methods for estimating the skin friction have not been found sufficiently reliable for practical use. Most of the available information has been obtained from a knowledge of the loads required to start caissons that had become stuck. The unit skin friction ranges from 150 to 600 lb/sq ft in silt and soft clay to values between 100 and 4000 lb/sq ft in very stiff clay. Even when data are available from experience on nearby jobs, care should be taken in applying the information to a new project because the total skin friction depends on a number of factors including the shape and diameter of the lower part of the caisson and on the method of excavation.

18.5. Piles in Clay

Conditions for Use. If the subsoil consists of clay too weak or too compressible to support footings or a raft, the weight of the structure may be transferred to piles.

The manner in which a pile driven into a clay deposit can be expected to carry its load may be inferred from the characteristics of the curve showing the resistance of the pile to penetration at different depths. Typical curves are shown in Fig. 18.6. Curve *a* is characteristic for friction piles. The number of blows per inch is practically constant with depth and remains relatively small. On the other hand, diagram *b*, for point-bearing piles, indicates relatively small resistance until the point encounters the firm layer, whereupon the resistance increases greatly. A typical diagram for a pile driven into a deposit of clay in which the strength increases with depth is shown in *c*.

Safe Load on Friction Piles. The load that can be carried by a single friction pile in a deposit of saturated clay can best be determined by means of a static load test. The shape of the load-settlement curve is characterized by Fig. 12.8a. The curve approaches a vertical asymptote corresponding to a sudden plunge of the pile into the ground or to a steady penetration of the pile under the applied load. The load that causes failure is represented by the abscissa

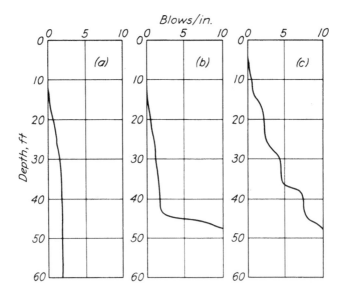

FIGURE 18.6. Relation between blows per inch of penetration and total depth or penetration for wood piles driven into (a) soft clay of great depth, (b) soft strata to firm bearing in very stiff or hard clay, and (c) clay strata increasing in stiffness with depth.

of the vertical tangent to the load-settlement curve. When the ultimate capacity of a pile is so well defined, the allowable load is commonly taken as one third of the failure load for normal design loadings or as one half of the failure load for extreme combinations of loading.

The load test may be supplemented by a pulling test. The corresponding pull-rise curve has the same characteristics as the load-settlement curve for a friction pile. If the pile is cylindrical, the difference between the failure loads determined from the load-settlement and pull-rise branches is a rough measure of the point resistance.

Since load tests on piles are relatively expensive, numerous efforts have been made to compute the safe load on friction piles on the basis of pile-driving records or laboratory tests. Experience has demonstrated that the driving resistance of friction piles in clay is likely to be low because of the disturbance of the structure of the clay, whereas after driving the strength may increase markedly over a period of time. The gain in strength or freeze (Art. 12.5) may be caused partly by thixotropic processes and partly by consolidation of the highly stressed clay immediately surrounding the piles. As the freeze is likely to constitute the greatest part of the capacity of the pile, and as it is not related to the phenomena of stress transmission in the pile during driving, dynamic pile driving formulas or analyses based on the wave equation are likely to give an erroneous conception of the capacity. This conclusion has been amply demonstrated by experience.

In contrast, static calculations based on the undrained shearing resistance of the clay have been found somewhat more reliable, although exceptions are not uncommon. For cylindrical piles the ultimate capacity is approximately

$$Q = \alpha_2 c \pi d L \qquad 18.8$$

where Q = ultimate pile capacity, tons
 α_2 = reduction coefficient
 c = undrained shear strength
  = $q_u/2$, tons/sq ft
 d = diameter of pile, ft
 L = length of embedment, ft

The reduction factor α_2 accounts for the disturbance of the clay due to pile driving

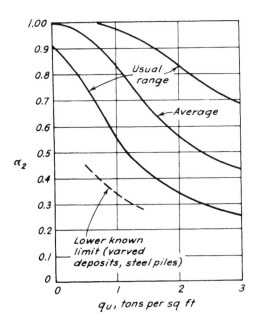

FIGURE 18.7. Values of reduction factor α_2 for calculation of static capacity of friction piles in clays of different unconfined compressive strengths q_u.

and for several other factors. Values can be approximated from Fig. 18.7. From the figure it may be seen that α_2 decreases with increasing stiffness of the clays into which the piles are driven (Tomlinson, 1957; Peck, 1958).

In some instances, the results of static computations have been found to be considerably too low. The most reliable information can be obtained by means of full-scale load tests (Art. 12.4). The load tests should not be made for at least several days after driving, in order to permit the freeze to develop. Usually an interval of three days is sufficient, but in some clays the increase in strength may continue for a much longer time. The results of a series of load tests on one pile are shown in Fig. 18.8. They represent a rather abnormal increase, but they accentuate the importance of the time effects in connection with clays.

Friction piles are commonly driven in groups or clusters beneath individual footings or as single large groups beneath mats or rafts. The bearing capacity of a pile cluster may be equal to the number of piles multiplied by the bearing capacity per pile, but it may also be much smaller. Therefore, a group of piles may fail when the load per pile is less than the safe load determined on the basis of load tests on a single pile. This possibility must be investigated.

Figure 18.9 shows a cluster of friction piles in a deep clay subsoil. The piles are connected at the top by a rigid cap. If they fail, they must fail as a group, and the failure is likely to occur along a surface such as that indicated by the dashed line. The area of the surface corresponds approximately to the product of the perimeter of the pile group and the length of the piles. The ultimate load that can be supported by the group is derived from two sources. The total shearing resistance of the soil around the boundary of the pile group is equal to PLc, where P is the perimeter of the group and the other quantities are defined as in eq. 18.8. To this must be added the capacity of the base; that is, the bearing capacity of the soil on the plan area of the group at the level of the tips of the piles. If the subsoil is uniform, this may be evaluated approximately by means of eq. 18.2 and Fig. 18.2 for the condition that D_f/B equals zero. For long piles in small groups, the base capacity is small compared to the shearing strength of the soil surrounding the pile group. On the other hand, as the num-

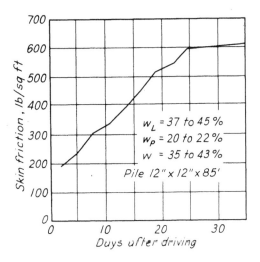

FIGURE 18.8. Increase of ultimate bearing capacity of friction pile with time.

ber of piles increases, the bearing area under the group increases much faster than the shearing surface around the group. The pile spacing that utilizes the full capacity of each pile can be found easily by trial. A spacing of three times the diameter of the piles is commonly selected as the trial spacing and checked against the criterion that the sum of the shearing and bearing capacities of the group of piles must be at least equal to the capacity of a single pile multiplied by the number of piles in the group.

According to the preceding paragraph, no benefits are derived from driving additional piles within the boundary of a given group of friction piles if the spacing is equal to or less than that determined by the criterion stated above. On the other hand, the greatest benefits from a friction pile foundation are obtained if the length of the piles is as great as possible within the limits of economy. In connection with the settlement of such foundations, it is found (Art. 18.6) that the effectiveness of a friction pile foundation also increases with increasing length of piles.

The freeze per pile in a group of piles may differ from that of a single pile. On large jobs it may be worthwhile to investigate this difference by making a load test on a pile in a test group driven before the design is made final.

Safe Load on Point-Bearing Piles. If a deposit of weak or compressible material is located above a stratum of very stiff or hard clay, the loads may be transferred to the firm stratum by means of piles. The total capacity of a single pile in a load test is distributed between point resistance in the firm bearing stratum and skin friction derived from the upper weak materials. If the weak materials contain a few thin stiff layers or a few sand layers, the skin friction of the upper materials may carry a considerable part of the load during a test. Beneath a completed structure, however, the friction of the upper materials soon disappears as the soft materials consolidate, and all the load is transferred to the lower end of the pile. Under these conditions the resistance developed by the pile in the bearing stratum must be adequate (Art. 12.4). Analyses by means of the wave equation may serve as a useful guide to estimating the point resistance, especially since the side resistance during driving is likely to be quite small. The results of load tests must be corrected for the skin friction. This may be done by performing the load tests in pairs. One pile should be driven through the soft material into the bearing stratum and tested in the usual manner. The other should be driven into the soft material and stopped when its point is 3 or 4 ft above the top of the layer upon which the piles are expected to bear. A test to failure of the latter pile will indicate the amount of load carried by skin friction in the upper layers. The load-settlement curve of the first pile should not be used for determination of the safe pile load until a correction of this amount has been made.

Estimates of the skin friction correction can be made, although less reliably, by means of the static formula (eq. 18.8) or by evaluating the freeze by means of the wave equation (Art. 12.5) on the basis of the redriving resistance.

If a pile cluster is surrounded by a fresh fill after the piles are driven, it is likely that

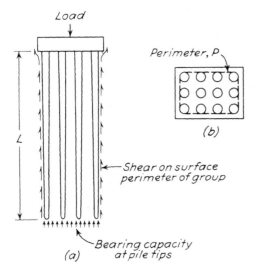

FIGURE 18.9. (*a*) Section and (*b*) plan of group of friction piles in clay.

the compressible materials above the bearing stratum will settle progressively for a considerable time because of the weight of the fill. Under these conditions, the piles may be acted on by the additional force due to the skin friction of the subsiding material, known as the *negative skin friction* or *drag* (Art. 12.8).

The magnitude of the drag per unit of area cannot exceed the shearing strength of the compressible soil which may usually be considered as one half the unconfined compressive strength. The area on which the drag acts is the vertical surface that surrounds the entire group of piles or the entire pile foundation.

Although it is not possible to estimate with great accuracy the additional pile load due to drag, a rough computation can be made to indicate whether or not the added load will be of serious consequence, and appropriate measures can be taken. Several examples of unexpected settlement of large magnitude have been attributed to neglect of negative skin friction.

If the piles are driven through a recently placed fill that has initiated a process of consolidation in the underlying compressible layers, the weight of the fill within the pile cluster is also likely to be transferred to the piles. The load per pile would then be increased still further, by an additional amount approximately equal to $A\gamma H/n$, where A is the area of soil within a horizontal section bounded by the perimeter of the cluster, γ and H are the unit weight and thickness of the fill, and n is the number of piles in the cluster.

The safe load on a group of piles driven through compressible layers to firm material is equal to the number of piles in the group multiplied by the safe load per pile. No reduction need be made because of close spacing of the piles. In fact, it may be preferable to keep the spacing as small as 2.5 times the diameter of the pile if there is a likelihood that negative skin friction may develop because of consolidation of the soft deposit. The additional pile load due to the negative skin friction should be taken into consideration.

Safe Load on Piles in Clay of Increasing Stiffness with Depth. In many localities stratified deposits are found in which each stratum has approximately uniform characteristics but in which the stiffness of the strata increases in a general way with depth. Piles are commonly used to transfer the load to a level below the most compressible upper layers. The depth to which the piles are driven depends primarily on the settlement that may be contributed by the materials below the pile tips. This aspect of the problem will be considered later. In general, it is economical to drive the longest piles compatible with the equipment readily available, in order to take advantage of both the increased strength and stiffness of the lower layers.

The load carried by a pile driven into such a deposit is supported partly by the material directly beneath the point of the pile and partly by skin friction. The skin friction developed in the firmer materials along the lower part of the pile may be permanently effective in supporting load, whereas that along the upper part, like the skin friction in soft materials above the bearing layer of point-bearing piles, may decrease with time, particularly if the pile is one of a large group.

The depth to which a particular pile can be driven under these circumstances and the capacity that can be developed at a given penetration per blow with equipment of certain characteristics can be investigated by means of the wave equation. From a study of the soil profile, the engineer selects one or more depths at which it seems likely the piles would develop the required capacity. For piles having lengths corresponding to each depth, wave analyses are performed. Resistance curves, such as that in Fig. 12.11, will indicate whether the chosen combination of pile, hammer, and cushioning is capable of developing the required capacity at reasonable driving resistance. In the interpretation of the curves, an allowance should be made for freeze in addition to the capacity associated with the driving resistance. This allowance may be estimated on the basis of experience with other piles in the locality or by conducting a redriving test (Art. 12.5), but preferably it should be determined by conducting one or more load tests at different times after driving a test pile.

In large groups the portion of the freeze developed along the upper parts of the piles may gradually become less effective and load may be transferred to the lower parts. The mechanism is the same as that described in connection with piles that pass through compressible strata to reach firm point bearing. However, the portion of the freeze along the lower parts of the piles is likely to be appreciable and to be of permanent load-carrying value. Determination of the freeze by conducting load tests on piles analyzed by means of the wave equation, as illustrated by Fig. 12.11, furnishes the sum of both components; hence, use of the freeze so determined may be somewhat unconservative. Nevertheless, the error is likely to be small.

Since the major part of the pile capacity has its origin in the firm strata beside the lower part of the pile and near the point, failure of the piles as a group is less likely than as individuals. Hence, the safe load on a group of piles driven into progressively stiffer materials is practically equal to the number of piles times the safe load per pile.

Uplift Resistance of Piles in Clay. The resistance to uplift of a single pile or of a group of piles in plastic clay is governed by the same considerations that apply to the safe downward load on friction piles; the bearing capacity of the soil at the points of the piles, of course, does not contribute to uplift resistance.

ILLUSTRATIVE DESIGN. DP 18-3. PILE FOUNDATION IN CLAY

The computations presented in DP 18-3 demonstrate the application of the criterion for pile spacing in the design of groups of friction piles in clay. In the first step, the number of piles has been determined by assuming that the individual capacity of each pile can be developed in the group. Second, the capacity of the group has been investigated and found to exceed the total load on the group with the appropriate factor of safety. Hence, a satisfactory design is indicated. In fact, the spacing of the piles could be decreased to $2\frac{1}{2}$ ft and the capacity of the group would still be greater than 190 tons. However, whether this reduced spacing should be used is a decision that can be made only if one is able to anticipate the behavior of the clay during the driving of the piles.

It should be noted that when individual action of the piles is considered, the shearing strength of the clay is multiplied by the reduction factor α_2 (Fig. 18.7). In the investigation of group action, on the other hand, no reduction is necessary.

This design plate is concerned only with the strength of the clay. The problem of settlement of friction piles in clay caused by the compressibility of the clay is discussed in Art. 18.6 and illustrated in DP 18-5.

18.6. Settlement of Foundations Underlain by Clay

Introduction. For all types of foundations on clays and plastic silts, the factors of safety must be adequate against a bearing capacity failure. It is equally important, however, to make a reliable estimate of the amount of differential settlement that may be experienced by the structure even if the soil pressure does not exceed the safe load. If the estimated differential settlement is excessive, it may be necessary to change the layout or type of foundation under consideration. In most instances a high degree of accuracy is not warranted. The proper choice of foundation can usually be made if it is known whether the differential settlement will be of the order of $\frac{1}{2}$, 2, or 20 in.

The general method for computing the settlement of a point on the base of a structure due to the consolidation of a thin layer of clay involves four steps. The first of these is the computation of the original effective pressure p_0 at midheight of the layer. The second is the computation of the increase in pressure Δp, directly below the given point and at midheight of the clay layer, caused

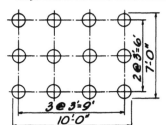

```
                                                    DP 18-3
                                                    Pile
  Pile Footing Design. Required Number              Footing
  and Spacing                                       Sh. 1 of 1

  General Data:
       Total load on pile group = 190ᵀ including wt.
  of pile cap.
       Borings indicate fairly uniform clay to a
  depth of 100 ft. The clay has an average
  qᵤ = 0.9 T/sq.ft. The sensitivity of the clay is
  low, and its water content is considerably
  below the liquid limit. From Fig. 18.7, α₂ = 0.87.
       A factor of safety of 3 is desired, using
  40 ft. piles, having an average diameter of
  12 in.

  Individual Action:
       Surface perimeter = 40 × 3.14 × 1 = 125.6 sq.ft./pile
       Reduced shearing strength of clay = (0.9/2)(0.87) =
                      0.39 T/sq.ft. ÷ 3 = 0.13 T/sq.ft.
       Safe pile load = 125.6 × 0.13 = 16.3ᵀ
       Piles required = 190/16.3 = 11.7  Use 12 piles

  Group Action:
       Try 3 rows of 4 per row @ 3'-0" ctrs. both ways

                            Surface perimeter =
                              40 × 34 = 1360 sq.ft.
                            Total shear = × 0.15 = 204ᵀ
                            Plan area of group =
                              7 × 10 = 70 sq.ft.
                              × 0.77 (1 + 0.2 × 7/10) = *
                            Safe base load = 61ᵀ

       Total safe group load = 204 + 61 = 265ᵀ > 190ᵀ ok
       Use 12 piles spaced as shown
                    * Since F = 3, use Fig. 18.3
                      (D_f/B = 0, B/L = 0.7)
```

by the construction of the building. The third is the determination of the compressibility of the clay.

Finally, these quantities are utilized in a computation of the decrease in thickness of the stratum at the point in question. If there is only one thin layer, the surface subsidence is assumed equal to this decrease in thickness. If there are several layers, similar computations are performed for each layer, and the decreases in thickness of the various layers below the point are added.

The computation of the original effective pressure p_0 and the determination of the compressibility have been described in Art. 2.4 and Chap. 3. In the following paragraphs, methods are developed for determining the increase in stress Δp and for computing the decrease in thickness of the clay layer.

Computation of Pressure. Since the stresses in the subsoil of a structure having an adequate factor of safety against a bearing capacity

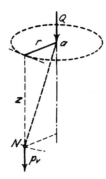

FIGURE 18.10. Intensity of vertical pressure at point N in interior of semiinfinite soil acted on by point load Q.

failure are relatively small in comparison with the ultimate strength of the material, a reasonable approximation to the manner in which the vertical load is distributed through the subsoil can be obtained on the assumption that the soil is elastic in its behavior.

If the subsoil is also assumed to be homogeneous, the vertical pressure p_v produced at any point N (Fig. 18.10) in the soil mass by the application of a load Q on the surface of the ground may be computed as

$$p_v = \frac{3Q}{2\pi z^2}\left[\frac{1}{1+(r/z)^2}\right]^{5/2} \qquad 18.9$$

Equation 18.9 is one of a set of stress equations developed by Boussinesq (1885) on the basis of the theory of elasticity. It is very likely that the real pressures in the soil mass differ appreciably from those computed by the means of the theory. However, because of the simplicity of the equations and the comparative reasonableness of the assumptions, Boussinesq's method is almost universally used, and the results are justified by the general agreement between observed settlements and those computed with the aid of the theory.

Since the load is usually applied to the subsoil through footings, piers, or rafts that are assumed to exert a uniform pressure on the soil, the engineer usually wishes to know the vertical pressure produced at some point on a horizontal plane in the interior of the soil mass by a load uniformly distributed over a portion of the ground surface. To solve this problem, a graphical representation of Boussinesq's equation was developed by N. M. Newmark (1942). It consists of an influence chart (Fig. 18.11). The chart represents a map of the ground surface drawn to such a scale that the distance AB on the map is equal to the depth below the surface at which the value of the pressure is to be determined. When this condition is satisfied, a uniform load on any one of the "square" areas bounded by two adjacent circular arcs and two adjacent radii produces a pressure at a point directly below the center of the chart equal to 0.005 times the intensity of the uniform load. In determining the depth corresponding to the distance AB, the ground surface is always considered to be the plane through the base level of the foundation.

If, for example, the chart is to be used for computing the increased pressure at a depth of 40 ft below footing level due to construction of a building on footing foundations, the footing plan is drawn on a sheet of tracing paper to such a scale that the distance AB equals 40 ft. The point on the building beneath which the pressure is to be determined is placed over the center of the chart. The number of squares covered by loaded footings is counted. This number multiplied by $0.005q$ equals the increase in pressure Δp due to the uniform load q exerted by all the footings.

To obtain the pressure Δp at various other points in the subsoil, the tracing paper is moved so that each of these points is successively below the center of the chart, and the counting operation is repeated. In this manner, the distribution of the pressure Δp at a given elevation below the footings can be determined readily. To obtain the pressures at a different depth, a new tracing must be prepared of the foundation plan drawn to such a scale that the new depth is equal to the distance AB.

The reduction in pressure at any point caused by the excavation of the basement may be computed in the same manner on the assumption that the weight of the exca-

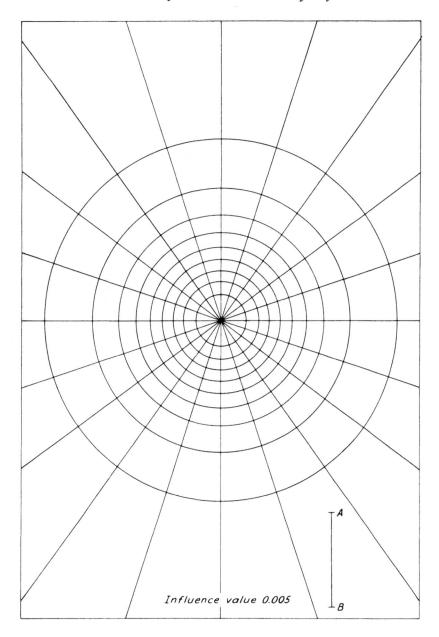

FIGURE 18.11. Influence chart for computation of vertical pressure (after Newmark, 1942).

vated soil is a uniformly distributed upward load acting at the level of the bottom of the excavation.

The increase or decrease of pressure Δp at midheight of a thin clay layer causes consolidation or swelling of the clay. Considerable time is usually required for these processes. Hence, temporarily applied loads have little influence on the settlement of a structure above clay strata. The loads that should be considered in a settlement computation commonly include the upward pressure due to the weight of the removed soil, the dead weight of the structure and any backfill within it, and the average live load that can be expected over fairly long periods during the life of the structure. If construction of the building permanently changes the water table, the change in effective pressure due to this cause must also be incorporated into the quantity Δp.

Computation of Settlement. Procedures for the computation of the settlement of a deposit of clay are described in Art. 3.4. The decrease in thickness of the clay layer, or the settlement S above point A (Fig. 3.5) is given by

$$S = H \frac{\Delta e}{1 + e_0} \qquad 3.3$$

which, for normally loaded clays, may be expressed according to eq. 3.1 as

$$S = \frac{C_c}{1 + e_0} H \log_{10} \frac{p_0 + \Delta p}{p_0} \qquad 3.4$$

Whereas eq. 3.4 is applicable only to normally loaded clays, eq. 3.3 is general and may be used whenever the original void ratio and change in void ratio are known. Regardless of the type of clay, e_0 is readily obtained; Δe can be obtained from the results of a consolidation test. Hence, even for normally loaded clays, eq. 3.3 may be preferred to eq. 3.4 if consolidation tests have been performed. If the clay is overconsolidated and if $p_0 + \Delta p$ does not exceed the preconsolidation pressure, eq. 3.3 must be used and Δe must be determined by the procedure described in Art. 3.5.

If the pressure Δp is negative, the clay would be expected to swell. However, if the clay is soft, the swelling due to relief in pressure caused by excavation is so small that the value of S may usually be taken as zero.

Reliability of Settlement Forecast. The ultimate differential and total settlement can be predicted with a high degree of reliability if the subsoil consists of normally loaded clay of ordinary sensitivity and if the clay is practically homogeneous or consists of homogeneous layers. The compression index may be determined by means of consolidation tests on undisturbed or remolded samples, or by means of eq. 3.2.

If the deposit is preloaded (Art 3.2) but this fact is not recognized, the computed settlements are likely to be substantially more than the real ones. The discrepancy decreases as the ratio $\Delta p/(p_0' - p_0)$ approaches unity. However, even if the computation is based on the e-log p curves for good undisturbed samples, the computed settlements are likely to exceed the real ones, and the amount of the discrepancy cannot often be predicted reliably.

Since the principal source of error in connection with settlement computations for a homogeneous clay deposit is misjudgment of the degree of preloading, all available evidence should be utilized. In many instances it is enlightening to estimate the values of Δp at critical points in the deposit beneath an older, neighboring structure of known behavior. If such a structure has not experienced detrimental settlement and if it produces stresses in the subsoil not less than those expected beneath the proposed structure, the satisfactory behavior of the new structure may usually be inferred. Such a conclusion should be verified by investigations to make sure that the subsurface conditions at the two sites are actually comparable.

If a clay deposit contains various elements of compressible soil arranged in an erratic manner, a settlement forecast can do no more than indicate the maximum probable differential settlement without furnishing a conception of the distribution of the settlement.

Highly sensitive clays call for the best possible samples. Settlement computations for nonerratic deposits of this type, however, can usually be made with reasonable accuracy.

Settlement of Footing and Raft Foundations. Footings located above soft normally loaded clay deposits and deposits of plastic silts commonly settle excessively even if the soil pressures are relatively low. The amount of settlement can be estimated by the procedures described in Art. 3.4 and in the preceding sections of this chapter. The computations are made on the assumption that the superstructure has no rigidity. Hence, the differential settlements indicated by the forecast are likely to be somewhat greater than those that actually occur. However, the forecast can be expected to be sufficiently reliable to permit the designer to estimate whether the deformations are too great for his structure. If they are, he must either combine his footings into a raft of adequate strength and stiffness or provide the structure with pile or pier support.

The settlement of individual footings is likely to exhibit a somewhat erratic pattern because of local soft or stiff spots that may exist beneath some of the footings. If a raft foundation exerts the same pressure per unit of area on the clay, the stresses beneath the foundation are increased to a considerably greater depth than beneath the footing foundation, and the soft and stiff spots largely compensate for each other. Consequently, the settlement pattern of a raft foundation is likely to be much more regular than that of a footing foundation on the same material and is likely to have a bowl-shaped character with the largest settlement near the middle.

The settlement of a raft foundation above clay can be reduced by lowering the elevation of the raft and thus increasing the amount of excavation. It is often possible to compensate for the entire weight of the structure by the weight of the soil excavated for the basements. In this manner structures have been built successfully on very soft clay deposits. In spite of the complete load compensation, they are likely to experience at least a small amount of settlement because of the heave and subsequent settlement of the clay during excavation and reloading, and because of the slight increase in compressibility caused by disturbing the structure of the clay during excavation.

In some instances, a layer of dense sand or stiff clay overlies a deep deposit of soft clay. A structure may be founded on footings resting on the sand or stiff clay. The safe load on the footings is determined by the characteristics of the material directly beneath them for a distance equal to the width B of the largest footing. However, the entire layer of sand or stiff clay acts as a natural raft, and the settlement of the structure as a whole depends on the increase of stress in the soft deposit and on the compressibility of the deposit. The forecast of settlement may be made by procedures similar to those used for rafts. If the differential movements will be too great for the structure, the situation can rarely be improved by changing the size of the footings. It is necessary to resort to a foundation on which the loads are compensated by the weight of the excavated soil, or to use piles or piers extending through the upper crust and the underlying compressible material to a firm base.

Settlement of Friction Piles in Clay. Friction piles may be used in clusters to support footings, or they may be placed beneath an entire raft foundation. Their function is to transmit the loads to a considerable depth in the subsoil where the values of the effective pressure p_0 are larger than near the surface. By so doing, the piles may reduce the settlement to less than that which would be experienced by footings without piles. Moreover, the load is distributed to the soil by skin friction throughout the length of the piles. In this manner, large concentrations of stress are avoided.

As a rough approximation, the settlement of a group of friction piles can be computed on the assumption that the clay contained between the top of the piles and their lower third point is incompressible and that the load is applied to the soil at the lower third point of the piles. The presence of the piles below this level is ignored.

It is apparent that friction piles are most beneficial if they are long compared to the width of the footing or the raft that they support. Long friction piles beneath a relatively small footing may greatly reduce the settlement, whereas a large number of short friction piles beneath a large raft may have practically no beneficial effect. Indeed, if pile driving increases the compressibility of the clay by remolding, short piles may be detrimental.

At the same load per pile the settlement

of a group of friction piles increases with increasing number of piles in the group, much as the settlement of a footing on clay increases with the size of the footing. Efforts have been made to compensate for the increase in settlement by reducing the allowable load per pile in some arbitrary manner as the number of piles in the group is increased. The expressions for accomplishing the reduction are known as efficiency equations. None of the existing equations, however, takes into account all the variables that are likely to be associated with clay deposits, and their use is not recommended. It seems preferable to make rough estimates of the settlement on the basis of the procedures described above and to proportion the pile groups with the aid of judgment.

Settlement of Piers and Point-Bearing Piles. The settlement of piers on normally loaded soft clays is likely to be large even if the net load at the base is small. Therefore, piers usually are not economical or satisfactory unless they either penetrate relatively stiff clay to a considerable depth or rest on stiff or hard preloaded clays. Even on fairly stiff clays, the progressive settlement of very large piers may become appreciable with the passage of time. Nevertheless, settlement computations based on the results of consolidation tests are likely to indicate greater settlement than will actually occur because of the difficulties associated with determining the compressibility of preloaded clays.

Although the ultimate bearing capacity of a drilled pier in stiff clays may be taken as the sum of the ultimate capacities of the shaft and base (Art. 18.4), the downward displacement required to develop the base capacity is much larger than that required to develop the skin friction. Hence, under working loads, the settlement may be governed primarily by the ratio of the actual to the ultimate skin friction.

The load on a pile driven through compressible strata into stiff or very stiff clay is ultimately carried by the point of the pile. This produces a large concentration of stress near the point. If a load test of customary duration is made on a single pile, the results may be encouraging because during the test most of the load is likely to be carried by skin friction in the upper layers and very little load may be transmitted to the point. Nevertheless, the settlement of the same pile beneath a building might eventually become quite large.

It is seldom possible to make a satisfactory estimate of the amount of settlement of point-bearing piles driven into very stiff clay. The best results are obtained by performing a long-time load test on a pile driven to suitable resistance in the stiff clay and separated by a casing from the surrounding soil to within a few feet of the point. The pile during the test should be subjected to the anticipated average working load. Such tests are expensive and are justifiable only on the most important jobs.

While point-bearing piles are being driven, other similar piles driven earlier may be lifted off the bearing stratum on account of the upward movement of the clay displaced by the additional piles. The lifted piles must be redriven to firm bearing in the supporting stratum or else the settlement may be excessive. To determine whether redriving is necessary, it is advisable to establish the elevation of the top of each pile when it is driven and to check it after all the piles have been driven. If the piles are not rigid units, the movement of the top may not be an indication of the movement of the point. This is the condition, for instance, when the shells for cast-in-place concrete piles are temporarily left unfilled. In these instances, the possibility of heave should be checked with the aid of pipes known as *telltales* placed within the shells and extending from the tips to an elevation near the ground surface where the movements can be measured.

If point-bearing piles are driven into a dense or stiff stratum that overlies a softer deposit, the bearing stratum should be considered as a natural raft and the settlement computed by the procedures described above.

Settlement of Structures on Erratic Deposits. In the preceding discussions, it has been assumed that the consolidating clay occurs in the form of well-defined layers or in fairly uniform masses that can arbitrarily be divided into layers. The properties of any layer have been assumed to vary only slightly from point to point in horizontal directions.

In many instances, deposits of soft clay and plastic silt are heterogeneous. Attempts to compute the distribution of settlement over the area to be occupied by a structure are very unlikely to be successful because the location of all the soft spots can hardly be learned from sampling and testing. If the deposit has an erratic structure, an accurate pattern of the differential settlements cannot usually be predicted. It may often be possible to estimate the maximum and minimum probable settlements, but not to determine the parts of the structure beneath which they will occur. This may be done by assigning to the clay the maximum compression index compatible with the results of the soil exploration within that part of the subsoil where Δp is likely to be a maximum, and the minimum compression index where Δp is likely to be a minimum. The real differential settlement is not likely to exceed that computed on this basis. Sometimes, a preferable procedure is to make a comprehensive series of subsurface soundings at the site to define the outlines of the soft pockets and to get a general conception of the variations in consistency and compressibility. A few borings are then made at the softest and stiffest places and rough estimates of settlement are based on these extreme conditions. These are generally sufficiently reliable to demonstrate whether or not the proposed type of foundation will be satisfactory.

ILLUSTRATIVE DESIGN. DP 18–4. SETTLEMENT OF RAFT FOUNDATION ABOVE CLAY

DP18-4 illustrates the application of some of the principles discussed in this article. The computations are for the most part self-explanatory. However, it should be noted that, since the water content of the clay is closer to the liquid than to the plastic limit, it is advisable to assume that the clay is normally loaded. The initial void ratio, the initial pressure at the middle of the clay layer, and the compression index are found in accordance with the discussions in Arts. 1.6, 2.4, and 3.3, respectively. It must be recognized that the differential settlement between the center and a corner of the raft may be considerably less than that indicated by the computation on Sh. 2 because of the stiffness of the superstructure.

ILLUSTRATIVE DESIGN. DP 18–5. SETTLEMENT OF FRICTION PILES IN CLAY

An estimate of the settlement of a group of friction piles in normally loaded clay is presented in DP 18–5. The total settlement of 6 in. (Sh. 2) should be regarded as only a rough approximation. Nevertheless, the magnitude of the settlement predicted in this manner usually gives the designer a satisfactory basis for deciding whether the foundation can be considered adequate or should be rejected in favor of one of some other type, such as longer piles driven to rock or a pier resting on rock.

It may be noted that the properties of the clay that affect its compressibility change significantly at elevations 308 and 300. For this reason, the clay is divided into three layers below the lower third point of the piles where the load is assumed to be delivered to the soil. The settlements contributed by these separate layers are added to obtain the total settlement. The increase in pressure at midthickness of each layer is computed on the assumption that the load is spread at an angle of 2 vertical to 1 horizontal, starting at the lower third point of the piles. The 2 : 1 spread of the load results in a reasonable approximation to the average increase in pressure at any level under the entire group of piles instead of to the increase in pressure under a specific point, as would be obtained by the use of Fig. 18.11.

Investigation of the adequacy of the pile

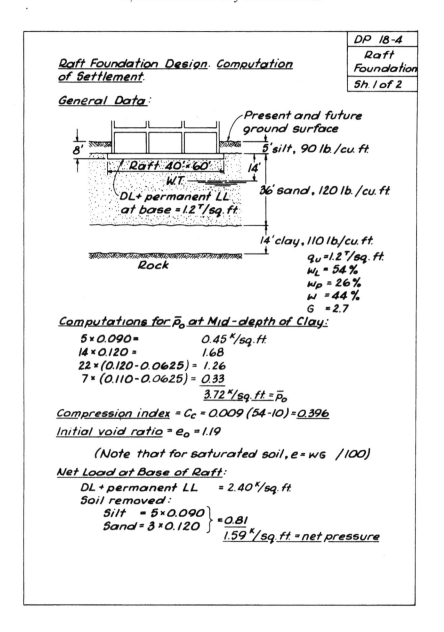

group as determined by the strength of the clay is omitted from this design plate because it would duplicate many of the computations given in DP 18-3.

Rate of Settlement. Equation 3.3 permits calculating the ultimate settlement due to the primary consolidation of a clay layer. The rate at which the settlement will occur can be predicted with considerable accuracy if the coefficient of consolidation c_v for the clay is known and if the distance between drainage layers is known. Many clay deposits contain a large number of sand or silt layers or lenses that drain the deposit to a greater or less degree. Unless the continuity of these layers can be established and unless their spacing can be learned, it is not possible to estimate the rate of settlement except within very wide limits. The estimate can be made by the procedures described in Art. 3.12.

In connection with foundations for buildings, the rate of settlement is not often a

> **DP 18-4**
> **Raft Foundation**
> **Sh. 2 of 2**
>
> ### Raft Foundation Design. Computation of Settlement.
>
> **Computations for Δp at Mid-depth of Clay:**
>
> Use sketch of 40'×60' raft to scale 1"=40'
> Use Fig. 18.11
>
> - 85.5 squares × 1.59 × 0.005 = **0.68 k/sq. ft.**
> - 38.3 squares × 1.59 × 0.005 = **0.30 k/sq. ft.**
>
> **Settlement Computations:**
>
> At center: $\dfrac{14 \times 12}{1+1.19} \times 0.396 \times \log \dfrac{3.72 + 0.68}{3.72} =$ **2.2"**
>
> At corner: $\dfrac{14 \times 12}{1+1.19} \times 0.396 \times \log \dfrac{3.72 + 0.30}{3.72} =$ **1.0"**

matter of primary concern because the damage will eventually occur if the final differential settlements are greater than those the structure can stand. Occasionally, however, an estimate of the rate of consolidation is necessary. Thus, if a structure has been damaged by increasing differential settlement, underpinning operations to arrest the movement may be contemplated. Before undertaking such a costly venture, it is desirable to learn the degree of consolidation that has occurred. In several instances, buildings have been underpinned at great expense at a time when practically all the primary consolidation had occurred. Knowledge of the degree of consolidation would have indicated that the underpinning was unnecessary.

The primary consolidation may be followed by secondary compression (Art. 3.12) that may occur at a nearly constant rate for many years. Ultimately, in some instances, it may become excessive. As yet, there are no fully satisfactory means for estimating

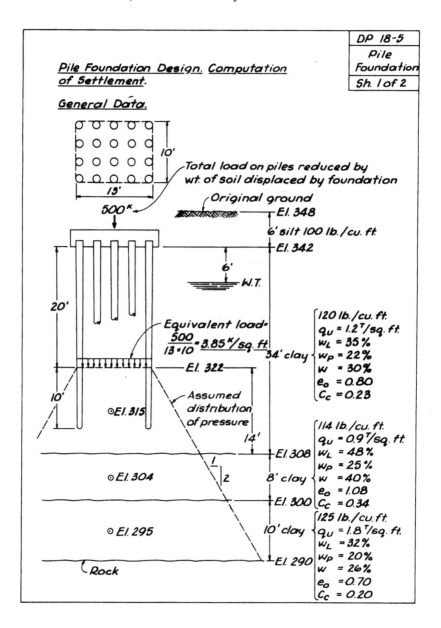

the rate of secondary consolidation. In normally loaded or slightly precompressed clays, a conservative upper limit may be judged on the basis of eq. 3.15 and Fig. 3.14, but for moderately or highly overconsolidated clays the actual rate may be much smaller. Calculations based on values of C_t obtained from consolidation tests on undisturbed samples may prove misleading because the rates in the field and in the laboratory appear to depend on factors not yet fully understood. Where the behavior of similar buildings on the same deposit is known, such information provides the best guide. For inorganic clays, the rate varies from less than 0.05 in./yr to over 0.5 in./yr in unusual instances. The primary consolidation of organic clays and silts may occur rapidly and may be small in comparison to the secondary compression. For this reason, secondary compression is given particular attention in such countries as Holland and Belgium where large organic deposits are found.

```
                                                    DP 18-5
                                                    Pile
    Pile Foundation Design. Computation             Foundation
    of Settlement.                                  Sh. 2 of 2

    Computations for Values of p̄₀:
        6(100)        =  600
        6(120)        =  720
        21(120-62.5)  = 1207
                        2527 lb./sq.ft. = p̄₀ at El. 315
        7(120-62.5)   =  403
        4(114-62.5)   =  206
                        3136 lb./sq.ft. = p̄₀ at El. 304
        4(114-62.5)   =  206
        5(125-62.5)   =  312
                        3654 lb./sq.ft. = p̄₀ at El. 295

    Computations for Values of Δp:
        A at El. 315 = [13+2(7)(0.5)][10+2(7)(0.5)] = 340 sq.ft.
        Δp at El. 315 = 3.85 × 130/340 = 1.47 ᵏ/sq.ft.

        A at El. 304 = [13+18][10+18] = 868 sq.ft.
        Δp at El. 304 = 3.85 × 130/868 = 0.58 ᵏ/sq.ft.

        A at El. 295 = [13+27][10+27] = 1480 sq.ft.
        Δp at El. 295 = 3.85 × 130/1480 = 0.34 ᵏ/sq.ft.

    Settlement Computations:
        El. 322 to 308   S = (14×12)/(1+0.80) × 0.23 × log (2.53+1.47)/2.53 = 4.3 in.
        El. 308 to 300   S = (8×12)/(1+1.08) × 0.34 × log (3.14+0.58)/3.14 = 1.2 in.
        El. 300 to 290   S = (10×12)/(1+0.70) × 0.20 × log (3.65+0.34)/3.65 = 0.5 in.
                                      Total settlement  = 6.0 in.
```

18.7. Excavation in Clay

Factors Affecting Stability of Cuts in Clay. Unless a deposit of clay either contains or is underlain by layers or lenses of waterbearing pervious soils, groundwater does not cause difficulties in making temporary excavations. On the other hand, the stability of temporary slopes in clay depends not only on the strength of the clay and on the slope of the cut, but also on the depth of the excavation and on the depth to a firm stratum if one exists not too far below the bottom of the excavation.

The stability of a slope during the construction period can be estimated quite reliably by theory and the results of soil tests if the clay is saturated, is of soft to medium consistency, and is protected from drying out and forming desiccation cracks at and beyond the crest of the slope. Stiffer clays usually contain joints, cracks, or slickensides that reduce their strength appreciably below that of intact samples taken from the deposit; the stability of slopes in such materials is difficult to estimate reliably. Clays containing pockets of water-bearing cohesionless material may stand up reasonably well when first exposed by excavation, but may deteriorate progressively.

If the excavation remains open and unprotected for a long time, the stability may be reduced by cracks that develop by desiccation and later become filled with water. Progressive deterioration may also result from seepage while the cut is open, particularly if the clay contains layers or pockets of cohesionless silts and fine sands below the water table. Since these factors are not under the complete control of the constructor, considerable judgment is required to select slopes that will achieve a proper balance between safety and economy.

Slopes in Homogeneous Soft to Medium Clays. Experience has indicated that the failure of a slope in a uniform clay takes place by slippage along a surface of nearly circular cross section. Sufficiently accurate estimates of the stability of such slopes can be made if the surface of sliding is assumed to be circular.

The procedure is illustrated in Figs. 18.12a and 18.12b, which represent respectively a fairly steep and a fairly flat slope in a homogeneous clay resting on a firm base. Since the period of construction is assumed to be short as compared to the time required for the water content of the clay to change, the $\phi = 0$ conditions are approximately satisfied. The analysis may then be based on total rather than effective stresses and on the undrained shear strength c determined from unconfined compression or vane tests of the clay (Art. 4.8).

To investigate the stability of a slope, a circular arc (Fig. 18.12a or 18.12b) is drawn with an arbitrary center O and an arbitrary radius r. The weight W of the mass of soil delineated by the arc is determined graphically or analytically. The moment Wl_w of this weight about O tends to cause sliding along the arc. The corresponding shear stresses t along the arc are determined by the requirement that the moment of the shear stresses about O must equal that of the weight if equilibrium exists. That is,

$$t\hat{l}r = Wl_w \qquad 18.10$$

where $\hat{l}$ is the length of the arc. Hence, the shear stress required for equilibrium is

$$t = \frac{Wl_w}{r\hat{l}} \qquad 18.11$$

The factor of safety against shear failure along the arbitrarily selected circular arc is the ratio of the available strength c to the required strength t. Thus

$$F = \frac{c}{t} = \frac{cr\hat{l}}{Wl_w} \qquad 18.12$$

The failure may not, of course, take place along the arbitrarily selected arc; it will occur along the arc for which the factor of safety is a minimum. Therefore, various locations for the center O and various radii must be selected until the minimum factor of safety of the slope is found.

Both theory and trial calculations have been utilized to reduce the number of trials necessary to locate the critical center and radius. It has been established, for example, that if the slope angle β is 53° or steeper, the critical circle passes through the toe of the slope. For flatter slopes the critical circle is generally tangent to the firm base and its center generally lies on a vertical line through the midpoint of the slope. However, if the firm stratum is at a very shallow depth below the toe of the slope, failure may take place along a circle, known as a slope circle, that intersects the slope above the toe.

Charts have been prepared (Taylor, 1937) for the solution of the problems illustrated in Fig. 18.12. A convenient form of these charts is shown in Fig. 18.13 (Terzaghi, 1943). For any slope angle and any depth n_dH to the firm base, a dimensionless stability number

$$N_s = \frac{\gamma H_c}{c} \qquad 18.13$$

may be determined, and the value H_c computed from eq. 18.13. The quantity H_c is

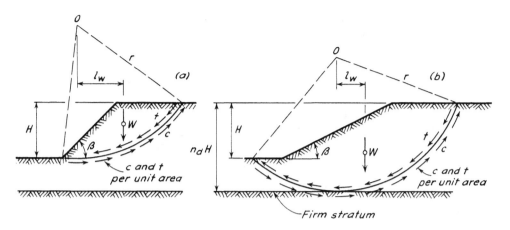

FIGURE 18.12. Position of failure surfaces. (a) Toe circle. (b) Base circle.

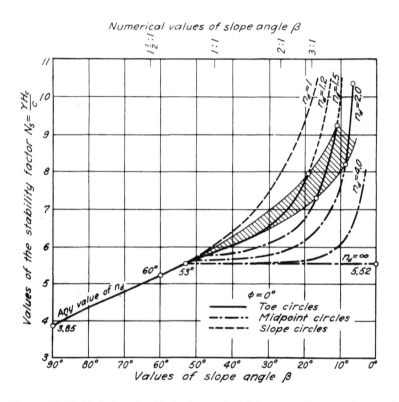

FIGURE 18.13. Relation for frictionless material between slope angle β and stability factor N_s, for different values of depth factor n_d (after Taylor, 1937).

the *critical height* of the slope; that is, the vertical height of the slope for a factor of safety of unity. The factor of safety of the proposed slope may be determined approximately from

$$F = \frac{H_c}{H} \qquad 18.14$$

where H is the height of the slope.

If the deposit consists of strata of soft and medium clays, similar analyses may be developed. These are discussed in Chap. 21.

Slopes in Stiff Clays. The maximum height to which slopes can be excavated in stiff, saturated or nearly saturated, intact clays can be estimated from Fig. 18.13. Ordinarily, even steep slopes will stand initially to considerable depths in such material. As joints open and tension cracks develop under the influence of the strains associated with excavation, especially if rainfall may fill the cracks or joints, slabs of the clay may break off or slide along surfaces having no resemblance to those assumed in the analysis. Experience with similar cuts in the same formation is the best guide under these circumstances, but even such experience may be misleading if the duration of the work and the weather are not properly evaluated.

ILLUSTRATIVE DESIGN. DP 18–6. SLOPE STABILITY IN CLAY

DP 18–6 illustrates the application of principles and procedures described in Art. 18.7 for determining the factor of safety of an excavation in clay by means of eq. 18.12. The factor of safety is then checked by an alternate procedure based on Fig. 18.13 and eq. 18.14.

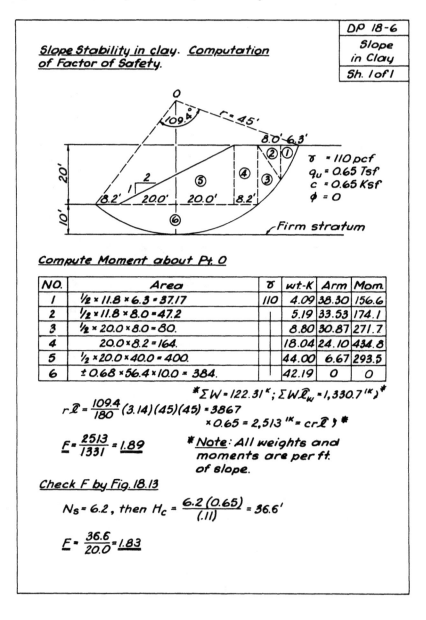

18.8. Lateral Displacements Due to Vertical Loads on Clay

Retaining walls, abutments, and wing walls are ordinarily designed to withstand the lateral pressure of the supported soil. The computation of the lateral earth pressure against such structures is a routine procedure and is discussed in Chap. 26. However, if the subsoil consists of clay or plastic silty materials, the structures may, under some circumstances, exhibit excessive horizontal movements even if adequate provision has been made for the earth pressure itself.

The fill behind a retaining structure constitutes a vertical load on the subsoil. The intensity of the loading, for a level fill placed above a level ground surface, is γH where H is the height of the fill. If the intensity of the loading exceeds the bearing capacity of the subsoil, the fill breaks into the ground, the retaining structure is displaced laterally, and the ground surface in front of the retaining wall heaves. As a rough approximation, the bearing capacity of the subsoil, if $\phi = 0$ conditions are applicable, may be calculated by means of eq. 18.2, where N_c is determined from Fig. 18.2 for B/L and D_f/B equal to zero. Hence

$$q_d \simeq 5c \qquad 18.15$$

If c is assumed to equal one half the unconfined compressive strength, the factor of safety against a bearing-capacity failure is then roughly

$$F \simeq \frac{2.5 q_u}{\gamma H} \qquad 18.16$$

The factor of safety by eq. 18.16 should not be allowed to become appreciably less than 2, or progressive lateral movements of the retaining structure are likely to develop. As the factor of safety decreases toward unity, the rate of progressive movement increases, and outright failure occurs at a factor of safety of unity.

The seat of lateral movements of this nature has not generally been recognized by engineers. As a consequence there have been many examples of progressive movements that have ultimately eliminated the clearance provided between the ends of bridge spans and the back walls of abutments (Peck, Ireland, and Teng, 1948). In several instances the movement has been rapid and of sufficient magnitude to be regarded as a failure. Figure 18.14 shows the pertinent data concerning a typical ex-

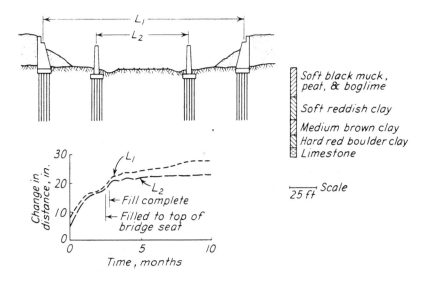

FIGURE 18.14. Example of progressive movement of bridge piers due to lateral forces caused by weight of backfill.

ample of progressive forward movement of large magnitude. The piles penetrated the soft material and rested on the medium clay at a depth of 40 ft. The two abutments and two intermediate piers for the bridge were completed and backfilling was in progress when movements of all four elements of the substructure were observed. Measurements yielded the results indicated on the figure. The distance L_2 between the intermediate piers decreased almost as much as the distance L_1 between abutments. Thus, the movement was deep-seated and involved the lateral squeeze or flow of clay toward the center of the bridge from each end. Under the weight of the fill the bearing capacity of the soft reddish clay and overlying material was approached. The lateral pressure against the abutments themselves was probably not excessive, because they were of the open type and the fill was allowed to extend through them.

The height of fill behind the abutments in Fig. 18.14 was about 27 ft and the unit weight of the fill material about 120 lb/cu ft. According to eq. 18.16, a factor of safety of unity against a bearing-capacity failure would have required an unconfined compressive strength of the underlying clay on the order of

$$q_u = \frac{\gamma H}{2.5}$$
$$= \frac{120 \times 27}{2.5}$$
$$= 1300 \text{ lb/sq ft}$$
$$= 0.65 \text{ ton/sq ft}$$

This value corresponds to the consistency of a medium clay, whereas the subsoil was described as soft. Therefore, it is not surprising that substantial displacements occurred before the fill was completed.

Vertical piles beneath the wall or abutment can provide little restraint against horizontal movements of this type because the bending strength of piles is relatively small and can be developed only if the lower part of the piles is firmly embedded in highly resistant material. Batter piles may be somewhat more efficient, but their effectiveness is likely to be considerably less than is generally believed. As the wall begins to move forward, the load on the batter piles increases greatly and the points penetrate further into the ground. Even a small penetration at the pile points permits appreciable lateral movement of the base of the structure. Hence, if the batter piles do not reach firm bearing in an underlying dense stratum, they are of little or no value.

Numerous abutments have experienced progressive movement in spite of the presence of batter piles. If the subsoil is too weak to support the weight of the backfill without progressive movement, it may be necessary to remove the unsuitable foundation material or to support the approach to the structure on a viaduct or trestle and to eliminate the backfill.

PROBLEMS

1. A continuous wall footing will rest at a depth of 3 ft on a saturated clay that possesses an unconfined compressive strength of 1.1 tons/sq ft. At a load of 4.3 tons/ft of wall a factor of safety of 3 is required; in addition, the factor of safety should not be less than 2 if the footing should be subjected to a load of 5.8 tons/ft. Ignore the difference in the unit weights of concrete and clay, and determine the width of the footing.
   Ans. 3.8 ft.

2. Proportion a square footing to carry a column load of 160 tons at a factor of safety of 2.5. The base of the footing will be 4 ft below the level of the surrounding ground. The clay beneath the footing has an unconfined compressive strength of 1.4 tons/sq ft.
   Ans. 9.1 x 9.1 ft, say 9 x 9 ft.

3. A footing 10 ft square rests at a depth of 3 ft on clay that has an unconfined compressive strength of 1.2 tons/sq ft. If the factor of safety is not to be less than 2.5, what is the maximum column load that can be supported by the footing?
   Ans. 165 tons.

4. A building is to be supported on a rein-

forced-concrete raft covering an area of 46 × 69 ft. The subsoil is clay with an unconfined compressive strength of 0.8 ton/sq ft. The pressure on the soil, due to the weight of the building and the loads it will carry, will be 1.4 tons/sq ft at the base of the raft. If the unit weight of the excavated soil is 120 lb/cu ft, at what depth should the bottom of the raft be placed to provide a factor of safety of 3?

Ans. 9.6 ft, say 10 ft.

5. A raft 60 × 72 ft in plan has its base 10 ft below the surface of a deposit of clay with a unit weight of 120 lb/cu ft. The unconfined compressive strength of the clay is 0.75 ton/sq ft. The factor of safety against a bearing-capacity failure must be 3. What total weight of building plus foundation can safely be supported by the raft?

Ans. 5980 tons.

6. A factor of safety of 3 is required for the group of friction piles shown below. Find the maximum load P as determined by
   a. The piles acting as individuals. (Use the average relation for obtaining α_2, Fig. 18.7.)
   b. The piles acting as a group.

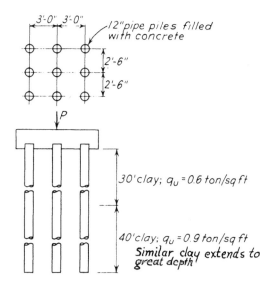

Ans. (a) 229 tons; (b) 271 tons.

7. A test pile with a diameter of 12 in. penetrates 30 ft of clay to a stratum of dense sand. The clay has an average unconfined compressive strength of 0.4 ton/sq ft. Estimate how much of the ultimate test load may be the result of skin friction in the clay deposit.

Ans. 18 tons.

8. Estimate the settlement of the center of the tank, shown in the sketch, due to consolidation of the 20-ft layer of clay.

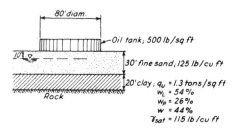

Ans. 1.85 in.

9. Estimate the settlement at the center and at a corner of the raft in DP 18–4 if the thickness of the layer of sand is 46 ft instead of 36 ft. All other data are the same as those given on Sh. 1 of DP 18–4.

Ans. 1.4 in.; 0.8 in.

10. What is the approximate factor of safety against failure for a 30-ft fill, weighing 125 lb/cu ft, if it rests on clay that possesses an unconfined compressive strength of 1.5 tons/sq ft?

Ans. 2.0.

11. A cut is to be made in soft clay to a depth of 40 ft. The material has a unit weight of 120 lb/cu ft and an unconfined compressive strength of 0.9 ton/sq ft. A hard layer underlies the soft layer at a depth of 50 ft below the original ground surface. If the slopes make an angle of 45° with the horizontal, what is the factor of safety against failure of the cut?

Ans. 1.08.

12. A wide cut was made in a stratum of soft clay that had a level surface. The sides of the cut rose at 25° to the horizontal. Bedrock was located 50 ft below the original ground surface. When the cut reached a depth of 30 ft, failure occurred. If the unit weight of the clay was 120 lb/cu ft, what was its average cohesive strength? What was the character of the surface of sliding?

Ans. 580 lb/sq ft; midpoint circle.

SUGGESTED READING

The following references are arranged in the order in which the chapter is subdivided.

Footings on clay:

A. W. Skempton (1951), "The Bearing Capacity of Clays," *Proc. British Bldg. Research Congress*, 1, 180–189. Contains development of data in Figs. 18.2 and 18.3.

A. W. Skempton (1942), "An Investigation of the Bearing Capacity of a Soft Clay Soil," *J. Inst. Civil Engrs.*, London, 18, 307–321; discussions pp. 567–576. Full-scale verification of bearing-capacity equations.

R. B. Peck (1948), "History of Building Foundations in Chicago," *U. of Ill. Eng. Exp. Sta. Bull. 373*, 64 pp.

Rafts on clay:

L. S. White (1953), "Transcona Elevator Failure: Eye-Witness Account," *Géotechnique*, 3, 5, 209–214. Details of bearing-capacity failure of large raft described in the following paper:

R. B. Peck and F. G. Bryant (1953), "The Bearing-Capacity Failure of the Transcona Elevator," *Géotechnique*, 3, 5, 201–208.

R. L. Nordlund and D. U. Deere (1970), "Collapse of Fargo Grain Elevator," *ASCE J. Soil Mech.*, 96, SM2, 585–607.

A. Casagrande and R. E. Fadum (1944), "Application of Soil Mechanics in Designing Building Foundations," *Trans. ASCE*, 109, 383–416. Partly compensated raft stiffened by deep concrete beams. Of primary interest are pp. 399–415 and the portion of the closing discussion on pp. 485–488.

Piers on clay:

G. C. Gauntt (1962), "Marina City—Foundations," *Civ. Eng. ASCE*, 32, December, 61–63. Construction methods for deep drilled piers in soft clays.

A. W. Skempton (1959), "Cast-in-situ Bored Piles in London Clay," *Géotechnique*, 9, 153–173.

C. N. Baker, Jr. and F. Khan (1971), "Caisson Construction Problems and Correction in Chicago," *ASCE J. Soil Mech.*, 97, SM2, 417–440. Influence of construction procedure on integrity of pier foundations in soft clays with waterbearing inclusions.

T. Whitaker and R. W. Cooke (1965), "Bored Piles with Enlarged Bases in London Clay," *Proc. 6 Int. Conf. Soil Mech.*, Montreal, 2, 342–346.

Piles in clay:

Some of the effects of driving piles through clay deposits are illustrated in the following:

E. J. Klohn (1963), "Pile Heave and Redriving," *Trans. ASCE*, 128, Part I, 557–577. Discussion by Olko, pp. 578–587.

A. E. Cummings, G. O. Kerkhoff, and R. B. Peck (1950), "Effect of Driving Piles into Soft Clay," *Trans. ASCE*, 115, 275–285. Discussions by Avery and Wilson, pp. 322–331; Rutledge, pp. 301–304; Zeevaert, pp. 286–292.

W. G. Holtz and C. A. Lowitz (1965), "Effects of Driving Displacement Piles in Lean Clay," *ASCE J. Soil Mech.*, 91, SM5, 1–13.

T. W. Lambe and H. M. Horn (1965), "The Influence on an Adjacent Building of Pile Driving for the M.I.T. Materials Center," *Proc. 6 Int. Conf. Soil Mech.*, Montreal, 2, 280–284.

D. J. Hagerty and R. B. Peck (1971), "Heave and Lateral Movements due to Pile Driving," *ASCE J. Soil Mech.*, 97, SM 11, 1513–1532.

O. Orrje and B. Broms (1967), "Effects of Pile Driving on Soil Properties," *ASCE J. Soil Mech.*, 93, SM5, 59–73.

The capacity of single piles and pile groups is considered in the following publications, which also contain useful bibliographies.

R. B. Peck (1958), "A Study of the Comparative Behavior of Friction Piles," *Hwy. Res. Bd. Special Report 36*, 72 pp.

M. J. Tomlinson (1957), "The Adhesion of Piles Driven in Clay Soils," *Proc. 4 Int. Conf. Soil Mech.*, London, 2, 66–71.

A. G. Stermac, K. G. Selby, and M. Devata (1969), "Behavior of Various Types of Piles in a Stiff Clay," *Proc. 7 Int. Conf. Soil Mech.*, Mexico, 2, 239–245.

T. Whitaker (1957), "Experiments with Model Piles in Groups," *Géotechnique*, 7, 147–167. Laboratory investigations of efficiency of pile groups with respect to ultimate capacity.

G. F. Sowers, C. B. Martin, L. L. Wilson and M. Fausold, Jr. (1961), "The Bearing Capacity of Friction Pile Groups in Homogeneous Clay from Model Studies," *Proc. 5 Int. Conf. Soil Mech.*, Paris, 2, 155–159.

Data concerning negative skin friction and its effects:

I. J. Johannessen and L. Bjerrum (1965), "Measurement of the Compression of a Steel Pile to Rock Due to Settlement of the Surrounding Clay," *Proc. 6 Int. Conf. Soil Mech.*, Montreal, 2, 261–264. One of the few quantitative full-scale observations.

L. Bjerrum, I. J. Johannessen, and O. Eide (1969), "Reduction of Negative Skin Friction on Steel Piles to Rock," *Proc. 7 Int. Conf. Soil Mech.*, Mexico, 2, 27–34.

M. Endo, A. Minou, T. Kawasaki, and T. Shibata (1969), "Negative Skin Friction Acting on Steel Pipe Pile in Clay," *Proc. 7 Int. Conf. Soil Mech.*, Mexico, 2, 85–92.

B. H. Fellenius and B. B. Broms (1969), "Negative Skin Friction for Long Piles Driven in Clay," *Proc. 7 Int. Conf. Soil Mech.*, Mexico, 2, 93–98.

Of the many publications concerning settlements of structures above clay, the following represent a variety of consistencies of clay and types of foundations:

L. Bjerrum (1967), "Engineering Geology of Norwegian Normally-Consolidated Marine Clays as Related to Settlements of Buildings," *Géotechnique*, 17, 83–117. The Seventh Rankine Lecture: a classic study.

A. W. Skempton, R. B. Peck, and D. H. MacDonald (1955), "Settlement Analyses of Six Structures in Chicago and London," *Proc. Inst. C. E.*, 4, Part I, July, 525–544.

C. B. Crawford and J. G. Sutherland (1971), "The Empress Hotel, Victoria, British Columbia. Sixty-five Years of Foundation Settlements," *Canadian Geot. Jour.*, 8, 1, 77–93.

D. J. D'Appolonia and T. W. Lambe (1971), "Floating Foundations for Control of Settlement," *ASCE J. Soil Mech.*, 97, SM6, 899–915.

S. J. Johnson (1970a), "Precompression for Improving Foundation Soils," *ASCE J. Soil Mech.*, 96, SM1, 111–144.

L. S. Brzezinski (1969), "Behavior of an Overpass Carried on Footings and Friction Piles," *Canadian Geot. Jour.*, 6, 4, 369–382.

L. Zeevaert (1957c), "Compensated Friction-Pile Foundation to Reduce the Settlement of Buildings on the Highly Compressible Volcanic Clay of Mexico City," *Proc. 4 Int. Conf. Soil Mech.*, London, 2, 81–86.

H. G. Poulos (1968), "Analysis of the Settlement of Pile Groups," *Géotechnique*, 18, 4, 449–471. Theoretical studies based on elastic medium surrounding piles; results compared with the few available data, largely from model tests.

The standard reference for routine analysis of the stability of slopes is A. W. Bishop (1955), "The Use of the Slip Circle in the Stability Analysis of Slopes," *Géotechnique*, 5, 1, 7–17.

The following papers describe foundation projects in clay soils and illustrate the application to practical problems of some of the principles in this and preceding chapters:

L. Casagrande (1966), "Subsoils and Foundation Design in Richmond, Va.," *ASCE J. Soil Mech.*, 92, SM5, 109–126.

L. Zeevaert (1957b), "Foundation Design and Behaviour of Tower Latino Americana in Mexico City," *Géotechnique*, 7, 3, 115–133.

G. G. Meyerhof and G. Y. Sebastyan (1970), "Settlement Studies on Air Terminal Building and Apron, Vancouver International Airport, British Columbia," *Canadian Geot. Jour.*, 7, 4, 433–456.

A. Casagrande (1947), "The Pile Foundation for the New John Hancock Building in Boston," *J. Boston Soc. Civ. Eng.*, 34, 4, 297–315. Reprinted in *Contributions to Soil Mechanics, 1941–1953*, Boston Soc. Civ. Eng. (1953), pp. 147–165; also as Harvard Soil Mech. Series No. 30.

Dewatering Excavation in Sand

Two-stage wellpoint system dewatering gravelly sand for construction of deep foundation for steel manufacturing plant, Hennepin, Illinois. Illinois River in background. Despite the relatively high permeability of the glacio-alluvial deposit, the excavation is completely dry. (Photo courtesy of Professor H. O. Ireland.)

PLATE 19.

CHAPTER 19

Foundations on Sand and Nonplastic Silt

19.1. Significant Characteristics of Sand and Silt Deposits

If the site is underlain by sand, the foundation may consist of footings, rafts, piers, or piles. The choice depends primarily on the relative density of the sand and on the position of the water table. The relative density determines the bearing capacity and settlement of footings, rafts, or piers, and it also establishes the resistance of piles. The position of the water table is important chiefly because excavation below water level requires drainage and increases the cost of the foundation. However, it also has an appreciable influence on the bearing capacity and settlement.

Nonplastic cohesionless silt exhibits most of the characteristics of fine sand. In this chapter, the two materials are usually considered synonomous; exceptions, however, are noted.

19.2. Footings on Sand

Basis for Design. Typical load-settlement relationships for footings of different widths on the surface of a homogeneous sand deposit are shown in Fig. 19.1a. The wider the footing, the greater the ultimate capacity per unit of area. However, for a given settlement S_1, such as 1 in., the soil pressure is greater for a footing of intermediate width B_b than for a large footing with a width B_c, or for a narrow footing with width B_a. The pressures corresponding to the three widths are indicated by points b, c, and a, respectively.

The same data may be used to plot Fig.

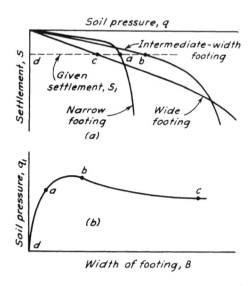

FIGURE 19.1. Relationships among soil pressure, width of footing, and settlements for footings of constant D_f/B ratio on sand of uniform relative density. (*a*) Load-settlement curves for footings of increasing widths B_a, B_b, and B_c. (*b*) Variation of soil pressure with width of footing for given settlement S_1.

307

19.1b, which shows the load q_1 per unit of area corresponding to a given settlement S_1 as a function of the width of the footing. The soil pressure for settlement S_1 increases with increasing width of footing if the footings are relatively small, reaches a maximum at an intermediate width, and then decreases gradually with increasing width.

Although the relation shown in Fig. 19.1b is generally valid for the behavior of footings on sand, it is influenced, nevertheless, by several factors including the relative density of the sand, the depth at which the foundation is established, and the position of the water table. Furthermore, the shape of the curve suggests that, for narrow footings, small variations in the actual soil pressure may lead to large variations in settlement and in some instances to settlements so large that the movement would be considered a bearing-capacity failure. The reason is apparent from the shape of the load-settlement curve for a narrow footing (Fig. 19.1a). On the other hand, a small change in pressure on a wide footing has little influence on settlements as small as S_1; moreover, the value of q_1 corresponding to S_1 is far below that which would produce a bearing-capacity failure of the wide footing.

The simple procedure described in the remainder of this article for proportioning footings on sand is based on the approximations illustrated in Fig. 19.2. The concave-upward portion of the curves similar to the right-hand portion of Fig. 19.1b is replaced by a straight line fg (Fig. 19.2), according to which the soil pressure corresponding to a settlement S_1 is independent of the width of the footing. The error for footings of usual dimensions is ordinarily less than ±10 per cent. The steeply ascending left-hand portion of the curves is replaced by a straight line ef, which lies to the right of the actual curve and thus provides a margin of safety against a bearing capacity failure.

The position of the broken line efg differs for different sands. The procedures for establishing the lines for different sands and for using them in the design of footings are described under the following subheadings.

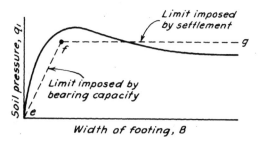

FIGURE 19.2. Actual relation (solid line) between soil pressure and width of footing on sand for given settlement S_1, and substitute relation (dashed lines) used as basis for design.

Considerations of Settlement. The soil pressure that produces a given settlement S_1 on a loose sand is obviously smaller than the soil pressure that produces the same settlement on a dense sand. Hence, in a rough way, there should be a relation between the soil pressure to produce the given settlement and the N-values from the standard penetration test (Art. 5.4). Such a relation was developed in 1948 (Terzaghi and Peck) on the basis of general knowledge of the loads, settlements, and N-values of various sand-supported footings; the value of S_1 was taken as 1 in. on the premise that if the maximum settlement were restricted to this amount the differential settlements among the footings of a given building would be within tolerable limits. The information then available was interpreted conservatively, so that in most instances the actual settlement of a footing proportioned on the basis of the relation would be less than 1 in. Experience has indicated that the relation was indeed conservative and sometimes excessively so; hence, various modifications have been suggested from time to time. The one proposed for present use is represented by the horizontal lines on the right side of the three parts of Fig. 19.3. Each line corresponds to a particular N-value and indicates the soil pressure corresponding to a settlement of 1 in. The lines are drawn for the condition that the water table is at great depth. The necessary correction for other positions is considered later.

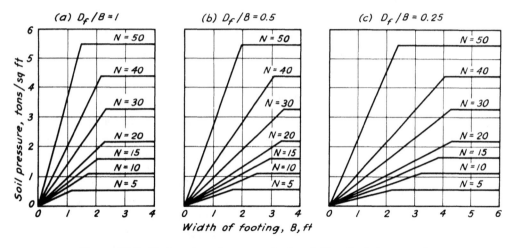

FIGURE 19.3. Design chart for proportioning shallow footings on sand.

The horizontal lines in Fig. 19.3 form part of a chart for design of footings on sand. The use and limitations of the chart are discussed after an investigation of the limitations imposed by the bearing capacity of the sand.

Considerations of Bearing Capacity. It has been pointed out that, for narrow footings, small increases in soil pressure may lead to such large increases in settlement that the movement would be considered a bearing-capacity failure. Hence, any acceptable procedure for proportioning footings on sand must provide assurance that, even if the settlement under the anticipated conditions would appear not to exceed 1 in., the margin against a bearing-capacity failure would be ample in spite of the inevitable differences between anticipated and real conditions. Moreover, under some circumstances even large settlements of wide footings or piers may be acceptable if the possibility of an outright failure of the supporting sand is excluded. Hence, a knowledge of the ultimate bearing capacity is essential for design. Reasonable estimates can be based on theoretical considerations.

Figure 19.4 represents a cross section through a long footing with width B, resting at depth D_f below the ground surface on a deep deposit of sand. If the footing fails by breaking into the ground, a zone $a0'a'$,

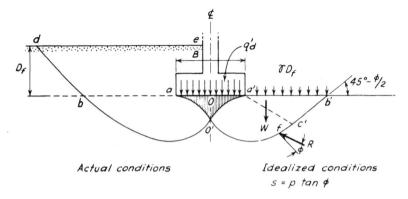

FIGURE 19.4. Cross section through long footing on sand showing (left side) pattern of displacements during bearing-capacity failure, and (right side) idealized conditions assumed for analysis.

within which the sand cannot slip with respect to the base of the footing because of the roughness of the base, moves downward as a unit. As it moves it displaces the adjacent material. Consequently, the sand in two symmetrical zones $a0'bde$, one of which is illustrated on the left side of Fig. 19.4, is subjected to severe shearing distortions and slides outward and upward along the boundaries $0'bd$. The movement is resisted by the shearing strength of the sand along $0'bd$ and the weight of the sand in the sliding masses.

No completely adequate rigorous theory exists for calculating the ultimate capacity of a footing under such circumstances, but satisfactory approximate solutions have been obtained on the basis of various simplifying assumptions (Terzaghi, 1943; Meyerhof, 1955). It is assumed, as illustrated on the right half of Fig. 19.4, that the influence of the soil above the base level of the footing can be replaced by a uniform surcharge γD_f. Theory and experiment then indicate that the surface of sliding consists of a curved portion $0'c'$ and a straight section $c'b'$ that rises at an angle of $45° - \phi/2$ with the horizontal. The load q_d' on the footing, the surcharge γD_f, and the weight W of the sliding mass all produce normal stresses across the surface of sliding $0'c'b'$, which, in turn, develop frictional shearing resistance along the surface of sliding. When the mass is on the verge of sliding the resultant R of the normal and shearing stresses at any point such as f on the surface of sliding is inclined at the angle ϕ to the normal to the surface of sliding. The wedge $0'c'b'a'$ may be considered as a free body and its equilibrium investigated to evaluate q_d'. Various trials must be made to find the surface of sliding corresponding to the least value of q_d' that can be developed. This least value is designated the *gross ultimate bearing capacity*.

The results of such studies indicate that the gross ultimate bearing capacity may be expressed as

$$q_d' = \tfrac{1}{2}B\gamma N_\gamma + \gamma D_f N_q \qquad 19.1$$

and the *net ultimate bearing capacity* as

$$q_d = q_d' - \gamma D_f$$
$$= \tfrac{1}{2}B\gamma N_\gamma + \gamma D_f(N_q - 1) \qquad 19.2$$

In these equations, N_γ and N_q are dimensionless *bearing-capacity factors* depending primarily on ϕ. They may be evaluated by means of the chart, Fig. 19.5.

Equation 19.2 demonstrates that the bearing capacity of a footing on sand is derived from two sources: the frictional resistance due to the weight of the sand below the level of the footing and the frictional resistance due to the weight of the surrounding surcharge or backfill.

The unit weights of most sands, whether dry, moist, or saturated, lie within a fairly narrow range. Therefore, the unit weight of the sand is in itself not an important variable in the determination of the bearing capacity of a footing. However, if the sand is located below the free water surface, only its submerged weight is effective in pro-

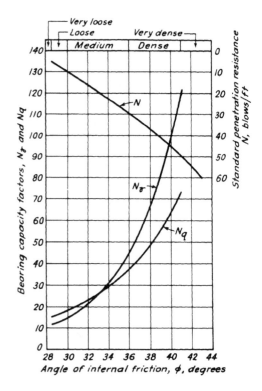

FIGURE 19.5. Curves showing the relationship between bearing-capacity factors and ϕ, as determined by theory, and rough empirical relationship between bearing capacity factors or ϕ and values of standard penetration resistance N.

ducing friction. The submerged weight is about one half of the moist, dry, or saturated weight. The value of ϕ is not appreciably changed by submergence. Hence, it may be concluded that a rise of the water table from a depth greater than about B below the base of the footing to the top of the surcharge would have the effect of reducing the bearing capacity to about one half of its value for moist, dry, or saturated sand. Thus, the position of the water table is of outstanding practical importance in establishing the bearing capacity of a footing on sand.

The values of N_γ and N_q increase rapidly as ϕ increases (see Fig. 19.5). Since ϕ depends to a considerable extent on the relative density of the sand but is practically independent of the grain size, it may be concluded that the bearing capacity is greatly influenced by the relative density but hardly at all by the grain size.

According to eq. 19.2, the portion of the bearing capacity due to the weight of the soil beneath the footing increases directly with the width of the footing. The portion due to the surcharge, however, is independent of the width of the footing.

In summary, eq. 19.2 shows that the ultimate bearing capacity of a footing on sand depends mainly on four variables: the position of the water table, the relative density of the sand, the width of the footing, and the depth of the surcharge surrounding the footing.

The bearing-capacity factors N_γ and N_q are functions of ϕ. The N-values as determined by means of the standard penetration test (Art. 5.4) can be correlated at least roughly with the values of ϕ and, therefore, with N_γ and N_q. The results of such correlations are shown in Fig. 19.5. Hence, if the N-values are known, eq. 19.2 can be evaluated with the aid of Fig. 19.5.

It should be understood that the relationship between N and ϕ (Fig. 19.5) is only approximate; the limitations of the standard penetration test have been pointed out in Art. 5.4. On the other hand, the relationship between ϕ and N_γ or N_q is based primarily on theory and is much more reliable. Hence, if ϕ is determined by a procedure more refined than use of the standard penetration test, Fig. 19.5 may still be utilized to evaluate N_γ or N_q.

Equation 19.2 can be expressed in the form

$$q_d = \left[\frac{\gamma N_\gamma}{2} + \gamma(N_q - 1)\frac{D_f}{B}\right]B \quad 19.3a$$

and, for a given factor of safety F against a bearing-capacity failure,

$$q_a = \frac{q_d}{F} = \left[\frac{\gamma N_\gamma}{2} + \gamma(N_q - 1)\frac{D_f}{B}\right]\frac{B}{F} \quad 19.3b$$

For a particular value of D_f/B and a given deposit of sand, the expression within the brackets is a constant. Thus, the relation between the width of footing and the net soil pressure q_a for a given factor of safety can be expressed in a plot such as Fig. 19.3 as a family of straight lines radiating from the origin. Each line corresponds to a sand having a different N-value. The initial branches of the curves in Fig. 19.3 have been drawn to provide a factor of safety of 2. If the soil pressures indicated by these lines are not exceeded, runaway settlement of a footing is precluded.

Design Chart for Footings on Sand. The considerations of settlement and bearing capacity discussed under the preceding subheadings define the right- and left-hand branches of the curves shown in Fig. 19.3. The curves constitute a convenient means for proportioning footings on sand.

The width B in Fig. 19.3 may be taken as the side of a square footing, the smaller dimension of a rectangular footing, the width of a long continuous footing, or the diameter of a circular footing.

The chart applies to shallow footings ($D_f \lesseqgtr B$) resting on a uniform sand for which $\gamma = 100$ lb/cu ft, and in which the water table is at too great a depth to influence the behavior of the footings. In view of the other approximations in the procedure, variations of γ from the assumed value of 100 lb/cu ft are inconsequential and may be neglected. On the other hand: (1) the N-values must sometimes be adjusted for the influence of the overburden pressure during the performance of the standard penetration test; (2) the variability of the deposit, as reflected in variation in the N-values from boring to boring, is usually appreciable and should be taken into account; and finally, (3) the influence of the water table, if shallow enough to affect the behavior of the footings, must be evaluated. Each of these considerations is discussed in detail in the following steps.

1. The chart (Fig. 19.3) is based on the behavior of shallow footings ($D_f \lesssim B$) of normal dimensions and depths below the ground surface. Most of the structures considered in developing the charts had basements 8 to 10 ft deep; consequently the bases of the footings were usually 10 to 15 ft below the original ground level at the time the borings and standard penetration tests were made. The groundwater level was usually at or below base level of the footings. Thus, the N-values governing the behavior of the footings usually corresponded to depths at the time of boring of 10 to 25 ft, or to effective overburden pressures between about 0.6 and 1.2 tons/sq ft. According to eq. 5.3, the correction factor C_N, by which the N-values are adjusted to those corresponding to an overburden pressure of 1 ton/sq ft, varies for this range from 1.20 to 0.93. Hence, the data on which the chart is based correspond to conditions under which no correction of N-values for the effect of overburden pressure is needed. Under such circumstances, the chart can be used directly for the design of footings for a proposed structure.

Conversely, when the overburden pressure corresponding to the N-values expected to govern the design of proposed footings differs greatly from 1 ton/sq ft, the N-values should be corrected. If, for example, the borings are made from a ground level to be lowered, say 20 ft, by grading the site before the basement is excavated, the uncorrected N-values may be appreciably too large. On the other hand, if the structure is to have no basement and the footings will be at a high level, the pertinent N-values will correspond to shallow depths and, if uncorrected, may be much too small.

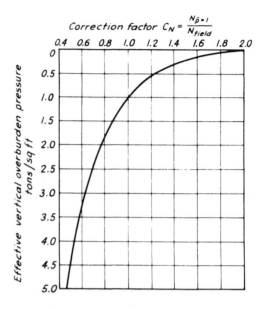

FIGURE 19.6. Chart for correction of N-values in sand for influence of overburden pressure (reference value of effective overburden pressure 1 ton/sq ft).

Hence, as a general procedure, the variation of effective overburden pressure with depth below the ground surface at the time of boring should be estimated, and each N-value corrected according to eq. 5.3. If it is apparent that the effective overburden pressures in the range of depths likely to govern the design of the footings would involve correction factors within the range 0.8 to 1.2, the corrections can be ignored without serious error. Otherwise, the appropriate corrections should be made. For convenience the chart, Fig. 19.6, may be used. It may be noted that, at very low effective overburden pressures, eq. 5.3 leads to unreasonably large correction factors, whereas in Fig. 19.6 the upper limiting value of C_N is 2.0. Hence, for values of effective overburden pressure less than about 0.25 ton/sq ft the correction factor should be taken from the chart.

2. No deposit of natural sand is perfectly uniform. To assure that the least favorable subsurface conditions are taken into account, the penetration resistance for the sand at a given site should be determined

by making borings with standard penetration tests at a number of points, preferably at least one boring for every 4 or 6 footings. Values of N should be determined at $2\frac{1}{2}$-ft vertical intervals between the level of the base of the footing and a depth B below this level. If the conditions described under Step 1 are applicable, the N-values should be corrected. The average of the N-values for each boring should be computed, and the smallest average value of N obtained in this manner should be used to determine the allowable soil pressure.

3. It has been shown that a rise of the water table from a depth greater than about B below the base of the footing up to the top of the surcharge has the effect of reducing the bearing capacity to about one half its value for moist, dry, or saturated sand. Such a rise, which reduces the effective pressures within the sand to roughly half their original values, also correspondingly reduces the stiffness of the sand (Fig. 4.5). Hence, the footing pressure required to produce a settlement of 1 in., if the water level is at the surface of the surrounding soil, is only about half that required to produce a 1-in. settlement if the water level is at a depth of B or more beneath the footing.

Therefore, if the water level occurs and will remain at or below a depth $D_f + B$ beneath the ground surface surrounding the footing, the footing may be proportioned for the soil pressure taken directly from the chart. If the water table is located at or may rise to the ground surface, the values from the chart should be multiplied by a correction factor $C_w = 0.5$. For a depth to groundwater level equal to D_w, measured from the surface of the surcharge surrounding the footing, the correction factor may be obtained with sufficient accuracy by linear interpolation, or

$$C_w = 0.5 + 0.5 \frac{D_w}{D_f + B} \qquad 19.4$$

Design Procedure Based on Chart for Footings on Sand. Since the permeability of sands is high enough to permit rapid adjustment to changes in load, the behavior of a foundation on sand is likely to reflect the influence of the maximum rather than the long-term average loads. This characteristic should be given consideration in choosing the loads for which the footings are to be proportioned.

In practical foundation design it is customary to proportion all footings for the same net soil pressure except for a few that may have to be treated as special cases. If the soil pressure for design is influenced by the presence of the water table, the largest footing is likely to require the greatest correction. Hence, the soil pressure should be determined for the footing to be subjected to the greatest design load. By selecting the soil pressure for this footing on the basis of the smallest average value of N, as described under Step 2 of the preceding section, the designer is assured that the largest footing, even if located above the loosest part of the deposit, will not settle more than 1 in. The differential settlement between this and any other footing then cannot exceed 1 in. and, in reality, will rarely exceed $\frac{3}{4}$ in. A differential settlement of this amount between adjacent columns in ordinary structures is generally considered tolerable (Art. 17.3) and, in fact, is the principal criterion on which the design procedure has been established.

After the dimensions of each of the smaller footings have been determined on the basis of the soil pressure selected for the largest footing, Fig. 19.3 is entered with the width B and the appropriate value of D_f/B of each of the smaller footings to check whether the allowable soil pressure for that footing may be governed by bearing capacity; i.e., whether the value of B corresponds to the left-hand rising branch of the design curve. If it does, the footing must be reproportioned for the smaller pressure. The water table correction appropriate to the new width of footing should be made. A footing having dimensions governed by bearing capacity may settle less than 1 in., but it is unlikely that the settlement will be smaller than $\frac{1}{4}$ in. Hence, the differential settlement of the building will still be within the customary tolerable limits.

The procedure just described inherently takes into account the normally erratic variations in the relative density of natural sand deposits. If there are no significant trends in the average N-values from one part of a site to another, but only erratic variations, the procedure may be used without modification. On the other hand, if the site is underlain by several depositional units of significantly different average relative density, it should first be subdivided into regions each of which may be regarded as having reasonably consistent properties, and the soil pressures for the footings in each region should be determined separately.

In all the foregoing discussion it has been assumed that the N-value from the standard penetration test gives at least a rough indication of the behavior of the sand, although the limitations are numerous and have been emphasized. Yet the greatest source of error in the use of N-values is that they may not actually represent the results of the standard penetration test, conducted with the standard equipment in accordance with the correct procedure. Because nonstandard techniques and equipment and careless procedures are not uncommon, the designer should not take for granted that reported boring logs represent the results of standard tests but should investigate the procedures carefully. He should, moreover, evaluate the logs for indications of doubtful or erroneous results such as those that might be caused by lowering the water level in the casing below the external water level when the drilling tools are removed from the hole (Art. 5.4).

Footings with Eccentric Loading. Beneath retaining walls and some other types of structures subjected to lateral loads, the calculated distribution of pressure against the base may be triangular (Fig. 19.7). The soil pressure q_t at the toe is then twice the average pressure. Therefore, the toe pressure is likely to govern the design. Theoretical and experimental information is available (Meyerhof, 1953) for estimating

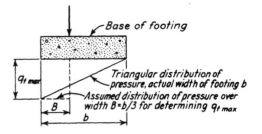

FIGURE 19.7. Diagram illustrating method for estimating allowable soil pressure beneath toe of eccentrically loaded footing.

the factor of safety against a bearing capacity failure under these circumstances, but no fully rational procedure has been developed for determining the allowable toe pressure to restrict the settlement. However, inasmuch as relatively high pressures act over only part of the base, the allowable soil pressure can be estimated on the assumption that the effective width B of the footing is less than the real width b.

It would appear conservative and reasonable to select the allowable soil pressure by means of the procedures given in this article for concentrically loaded footings, on the assumption that the footing has a width $B = b/3$. The toe pressure q_t should not exceed the allowable pressure determined in this manner.

ILLUSTRATIVE DESIGNS. DP 19–1 AND 19–2. FOOTINGS ON SAND

Two sets of design computations (DP 19–1 and DP 19–2) illustrate the application of the foregoing discussion of footings on sand.

Because the borings for the warehouse considered in DP 19–1 were made from the ground surface while the site was not yet graded, the effective overburden pressure will be considerably reduced before the foundations are constructed; hence, the N-values require correction. This is done for all penetration tests below footing level. For convenience, since the width B of the larger footing is not initially known, a cumulative average of the corrected N-values is

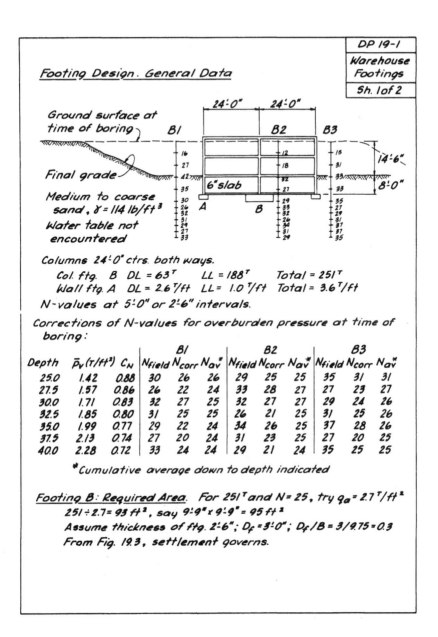

computed and tabulated as a function of depth for each boring. By inspection it is seen that the minimum average corrected N-value, irrespective of the width of the footing, is in the range 24 to 31; the corresponding soil pressures from the horizontal branches of the design curves (Fig. 19.3) are about 2.5 to 3.4 tons/sq ft. Hence, as a first trial, the column load of 251 tons is divided by about 2.5 tons/sq ft. The area of roughly 100 sq ft leads to a trial value for B of 10 ft. For B equal to 10 ft, the appropriate depths for calculating the minimum average corrected N-value are 25 to 35 ft inclusive, and N_{av} is found to be 24, from boring $B1$.

In DP 19–1, the width of footing B is such that q_a is determined by considerations of settlement alone. Use of the same soil pressure beneath the wall footing A would

```
                                                    ┌─────────────┐
                                                    │   DP 19-1   │
   Footing B (cont.)                                │  Warehouse  │
       Check soil pressure:                         │   Footings  │
          251^T / 95.0              = 2.64 T/ft²    │  Sh. 2 of 2 │
                                                    └─────────────┘
          Add'l load:
             Floor slab  0.5 × 0.15/2  ⎫
             Footing     2.5 × 0.15/2  ⎬ = +0.22   ⎫
                                                    ⎬ *
          Surcharge load:                           ⎪
             Floor slab  0.5 × 0.15/2  ⎫            ⎪
             Soil        2.5 × 0.114/2 ⎬ = -0.18   ⎭
                                        ─────────
                                        2.68 T/ft²

       From Fig. 19.3, q_a = 2.7 T/ft², ok  Use 9'-9" × 9'-9" footing B
```

*Note that difference between these two quantities is equal to thickness of footing multiplied by difference in unit weights of soil and concrete; that is, $2.5 \times (0.150 - 0.114)/2 = 0.04$ T/ft². In most instances the difference can be ignored.

Footing A: Required Area.
   Using soil pressure found for Footing B,
   3.6 T/ft / $2.68 = 1.34'$, say $1'-4"$.
   Assume thickness of ftg. $1'-0"$; $D_f = 1'-6"$; $D_f/B \cong 1.0$
   From Fig. 19.3a, for $N = 24$ and $B = 1.34$, $q_a = 1.6$ T/ft² < 2.7 T/ft²
                                    (therefore, bearing capacity
                                                        controls)
   By trial, say $B = 2.0$ ft

   Check soil pressure:
      3.6 T/ft / 2.0 = 1.80 T/ft²
      Add'l load - surcharge
         $1.0 \times (0.150 - 0.114)/2$ = 0.02
                                        ──────
                                        1.82 T/ft²

   Check min. average N-value for depth $B = 2$ ft below ftg.
      (consider values at depths 25.0 and 27.5 ft only). From
      Boring B1, $N_{av} = 24$
   Interpolate between Fig. 19.3a&b for $D_f/B = 0.75$;
      $N = 24$; $B = 2.0$; $q_a = 1.9$ T/ft² say ok
                              Use 2'-0" Footing A

---

require a width of only 1.34 ft but, according to Fig. 19.3a, the design of a footing of this width is governed by bearing capacity and the soil pressure must be reduced. Several trials may be needed to determine the proper combination of width and allowable soil pressure to satisfy the conditions of Fig. 19.3a. It is evident that the width will be on the order of 2 ft; therefore, only the N-values within about this depth below the footing should be included in the average. Those at depths 25.0 and 27.5 ft are appropriate; the minimum average corrected N-value for this range is 24 blows, from boring B1.

DP 19-1a illustrates the method of considering the effect of the water table. Since the presence of the water table affects the vertical effective overburden pressure for all standard penetration tests below water level, the corrected N-values at these levels differ from those in DP 19-1. The

		DP 19-1a
		Warehouse Footings
		Sh. 1 of 1

Footing Design. Required Areas. Effect of Water Table

All data are the same as on Sh.1 of DP 19-1, except that water table was encountered at 27 ft in all three borings and is expected to remain at that level after construction. Therefore, correct N-values for overburden pressure at time of boring.

$\bar{p}_v = [114 \times 27 + (D-27)(114-62)]/2000$ if below watertable

Depth	$\bar{p}_v$ (T/ft²)	C_N	N_{field}	N_{corr}	N_{av}*	N_{field}	N_{corr}	N_{av}*	N_{field}	N_{corr}	N_{av}*
25.0	1.42	0.88	30	26	26	29	25	25	35	31	31
27.5	1.55	0.86	26	22	24	33	28	26	27	23	27
30.0	1.62	0.84	32	27	25	32	27	27	29	24	26
32.5	1.68	0.83	31	26	25	26	22	25	31	26	26
35.0	1.75	0.82	29	24	25	34	28	26	37	30	27
37.5	1.81	0.80	27	22	24	31	25	26	27	22	26
40.0	1.88	0.78	33	26	25	29	23	25	35	27	26

*Cumulative average down to depth indicated

Footing B: Required Area. For 251ᵀ and N=24, try $q_a = 1.8$ T/ft²
251 ÷ 1.8 = 139 ft², say 12'-0" × 12'-0" = 144 ft²
Assume thickness of footing 3'-0"; $D_f = 3'-6"$; $D_f/B = 0.3$

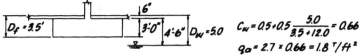

$C_w = 0.5 + 0.5 \dfrac{5.0}{3.5 + 12.0} = 0.66$

$q_a = 2.7 \times 0.66 = 1.8$ T/ft²

Check soil pressure:
251ᵀ/144 = 1.74 T/ft²
Add'l load - surch.
3 × (0.150 - 0.114)/2 = 0.05
1.79 T/ft² ok Use 12'-0" × 12'-0" Footing B

Footing A: Required Area.
Using soil pressure found for Footing B
3.6 T/ft /1.79 = 2.0 ft. Assume ftg. thickness = 1'-0"
$D_f = 1'-6"$; $D_f/B = 0.75$; Interpolate between Fig. 19.3 a & b
$D_w = 5'-0"$; $D_f + B = 3'-6"$; no water table correction req'd.
From Fig. 19.3 a & b, (N=24, B=2.0) $q_a = 1.95$ T/ft² ok < 1.79 T/ft²
Use 2'-0" Footing A

---

correction factor C_w (eq. 19.4) to be applied to values taken from Fig. 19.3 is found to be 0.66 and the value of q_a is found to be 1.79 tons/sq ft.

The wall footing, if proportioned for the same soil pressure, would have a width of 2.0 ft. The base of the footing, located about 1.5 ft below the basement floor level, would then be about 3.5 ft above water level. As this distance exceeds the width B of the footing, no correction for water table is needed for footing A; hence, it can be proportioned for the same pressure as the larger footing B.

DP 19-2 shows the influence of correcting the field N-values when the standard penetration tests are made at very shallow depths. The base for the structure was prepared by placing about 13 ft of sand fill above a natural sand deposit located beneath the surface of a body of water. The dense original ground is at a depth too

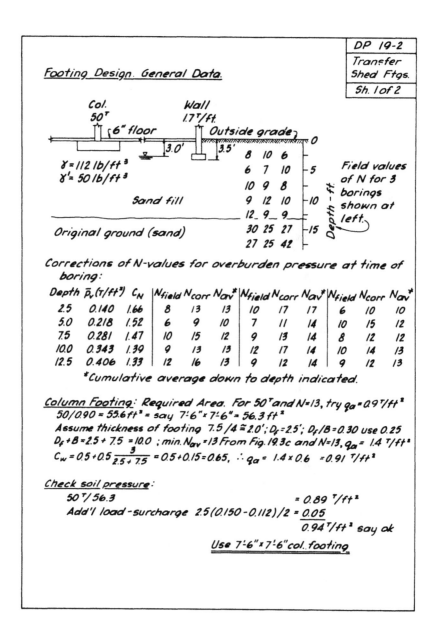

great to influence the behavior of the footings. The wall footing is located at a depth of 3.5 ft to avoid frost action.

19.3. Rafts on Sand

Soil Pressure. Because of the large size of rafts compared to that of footings, the factor of safety against a bearing-capacity failure of the underlying sand is always very great. This can be seen from eq. 19.2. If the width of the raft is only 20 ft, the depth of foundation only 10 ft, and the number of blows equal to 10 or more, the ultimate bearing capacity on submerged sand exceeds 7 tons/sq ft. With increasing width of the raft or increasing relative density of the sand, the ultimate bearing capacity increases rapidly. Hence, the danger that a large raft may break into a sand foundation is too remote to require consideration.

```
                                                    ┌─────────────┐
                                                    │   DP 19-2   │
                                                    │  Transfer   │
          Footing Design. (cont.)                   │ Shed Ftgs.  │
                                                    │  Sh. 2 of 2 │
                                                    └─────────────┘
```

<u>Wall Footing</u>: Required Area. Using soil pressure for col. ftg.
  $1.7/0.94 = 1.80$, say $2'\text{-}0"$
  Assume thickness $= 9" = 0.75'$; $D_f = 3.5'$ (for frost);
  $D_f/B = 3.5/2.0 = 1+$, use 1.0; $D_f + B = 3.5 + 2.0 = 5.5$; min. $N_{av} = 11$
  $C_W = 0.5 + 0.5 \dfrac{3}{3.5+2.0} = 0.5 + 0.27 = 0.77$ $q_a = 1.2 \times 0.77 = 0.92 \ ^T/ft^2$

<u>Check soil pressure:</u>
  $1.7^T/2.0$ $= 0.85 \ ^T/ft^2$
  Add'l. load - surcharge $1.25 \times (0.150 - 0.112)/2 = \underline{0.02}$
                                                          $0.87 \ ^T/ft^2 < 0.92$ ok
              <u>Use $2'\text{-}0"$ wall footing</u>

On account of the large size of rafts, the stresses in the underlying sand are likely to be relatively high to considerable depth. Therefore, the influence of local loose pockets distributed at random throughout the sand is likely to be about the same beneath all parts of the raft, and the differential settlement is likely to be smaller than that of a footing foundation designed for the same soil pressure. Although it is not improbable that a single footing may rest entirely above a pocket of loose sand and experience large settlement, a loose pocket beneath part of a raft has a much smaller influence.

Because the differential settlements of a raft foundation are less than those of a footing foundation designed for the same soil pressure, it is reasonable to permit larger allowable soil pressures on raft foundations. Experience has shown that a pressure approximately twice as great as that allowed for individual footings may be used because it does not lead to detrimental differential settlements. For a soil pressure that produces a differential settlement of $\frac{3}{4}$ in., however, the maximum settlement of a raft may be about 2 in. instead of 1 in. as for a footing foundation.

The shape of the curve in Fig. 19.1b shows that the net soil pressure corresponding to a given settlement is practically independent of the width of the footing or raft when the width becomes large. The allowable net soil pressure for design may with sufficient accuracy be taken as twice the pressure indicated by the horizontal lines in Fig. 19.3. The corresponding relation between allowable net soil pressure and N is

$$q_a \text{ (tons/sq ft)} = 0.22N \quad (5 \lesseqgtr N \lesseqgtr 50) \quad 19.5$$

The correction factor for the presence of the water table is given by eq. 19.4. For values of $N > 50$, the linear relation expressed by eq. 19.5 becomes somewhat unconservative. Moreover, N-values of this magnitude may be associated with the presence of gravel or boulders, or with cementation. Hence, they should be scrutinized carefully to permit judging whether the routine procedure described in this paragraph is applicable to the actual conditions.

The values of q_a from eq. 19.5, with appropriate corrections, serve as a rational basis for the design of a raft foundation on sand under most conditions encountered in the field. They may be increased somewhat if bedrock is encountered at a depth less than about one half the width of the raft.

If the average value of N after correction for the influence of overburden pressure is less than about 5, the sand is generally considered to be too loose for the successful use of a raft foundation. Either the sand should be compacted or else the foundation should be established on piles or piers.

The loads that should be considered in computing the gross soil pressure on the raft are the dead load of the structure including the raft, and the maximum live load that is really likely to be active. The surcharge due to the weight of the soil between the surrounding ground surface and the base level of the foundation is subtracted from the gross pressure to obtain the net soil pressure for comparison with the allowable soil pressure. That is, the net soil pressure at the base of the raft is

$$q_{\text{net}} = \frac{Q}{A} - \gamma D_f = q_b - \gamma D_f \quad 19.6$$

where Q = total weight of structure plus live load
A = base area of raft
q_b = gross soil pressure or contact pressure at base of raft

A raft-supported building with a basement extending below water table is acted on by hydrostatic uplift or buoyancy equal to $\gamma_w(D_f - D_w)$ per unit of area. The beneficial effect of the buoyancy is automatically taken into account in calculating the net pressure, provided the total weight of the surcharge γD_f is used in eq. 19.6. In many instances, however, the settlement is governed by conditions during construction rather than by those prevailing after completion.

During construction of the substructure the water table is usually drawn down below the base of the raft. If it then rises to a higher level, the gross soil pressure is reduced by the uplift of the full head of water on the base. Simultaneously, the effective weight of the surcharge is reduced by the same amount. Hence, the actual net pressure is not influenced by the buoyancy. The allowable net pressure, however, is a function of the water-table correction.

If the soil pressure is chosen in accordance with the foregoing procedures and if the corrected value of N is not less than about 5, the differential settlements between adjacent columns on a raft foundation on sand will not exceed about $\frac{3}{4}$ in., provided the base of the raft is located at least 8 ft below the surrounding ground surface. Experience has shown that, if the surcharge is less than this amount, the edges of the raft settle appreciably more than the interior because of the lack of confinement of the sand.

ILLUSTRATIVE PROBLEM

A reinforced concrete structure 100 ft square is to be supported by a raft with its base 16 ft below the surrounding ground surface. The subsoil consists of sand to great depth. Five borings have been made at the site; the average N-values, corrected for the influence of the overburden pressure, are respectively 36, 30, 32, 35, and 33. The average unit weight of the sand is 114 lb/cu ft. While test-boring was in progress, the water level was at a depth of 5 ft. During construction the water level will be lowered to 20 ft, but upon completion of the structure the level will return to its original position. What total load, including the weight of raft, structure, and contents, may be supported at a settlement not to exceed 2 in.; that is, at a differential settlement not to exceed $\frac{3}{4}$ in.?

Solution. According to eq. 19.5, the allowable net soil pressure for a value of $N = 30$ would be

$$q_a = 0.22 \times 30 = 6.6 \text{ tons/sq ft}$$

if the water table were at great depth. The water table correction is

$$C_w = 0.5 + 0.5 \frac{D_w}{D_f + B}$$
$$= 0.5 + 0.5 \times \frac{5}{16 + 100}$$
$$= 0.5 + 0.02$$
$$= 0.52$$

Hence, the allowable net pressure is $6.6 \times 0.52 = 3.4$ tons/sq ft. The surcharge $\gamma D_f = 114 \times 16/2000 = 0.91$ ton/sq ft. The corresponding gross pressure or contact pressure that would lead to a 2-in. settlement is, therefore, $3.4 + 0.9 = 4.3$ tons/sq ft, and the total weight that can be supported is $4.3 \times 100^2 = 43,000$ tons.

The total weight that will produce a settlement of 2 in. should, of course, be independent of whether the calculations are based on total or effective stresses. In the preceding paragraph a total-stress calculation is illustrated, in accordance with eq. 19.6. On the basis of effective stresses, the surcharge is only

$$\gamma D_w + \gamma'(D_f - D_w) = \frac{1}{2000}\left[114 \times 5 + (114 - 62.5)(16 - 5)\right]$$
$$= 0.57 \text{ ton/sq ft}$$

This value is $0.91 - 0.57 = 0.34$ ton/sq ft less than that calculated on the basis of the total weight of the surcharge, and the net capacity of the raft is decreased by the same amount. At the same time, however, the raft is buoyed up by the hydrostatic uplift on its base equal to

$$\gamma_w(D_f - D_w) = \frac{1}{2000} \times 62.5(16 - 5)$$
$$= 0.34 \text{ ton/sq ft}$$

and the downward load of the structure can be increased by the amount of the buoyancy. Hence, the total weight that can be supported remains 4.3 tons/sq ft.

19.4. Piers on Sand

Conditions for Use of Piers. Piers may be established on a bed of dense sand at a considerable depth below the ground surface if the overlying materials are too soft or compressible for the support of the structure or if they may be removed by scour. In these respects, the conditions under which pier foundations are appropriate are similar to those for piles (Art. 19.5), and the choice between piers or piles depends primarily on economy and on certain details that influence the construction procedure. For example, if the overlying material contains organic deposits that include trunks of trees, or if there are likely to be numerous large boulders above the bearing stratum, it may not be possible to drive piles to the necessary depth. Under these conditions, a pier foundation may be indicated. On the other hand, the excavation for a pier is likely to lead to some loosening of the sand deposit on which the structure is to rest, whereas driving piles into the sand tends to compact the bearing stratum. Because of these conditions, a pile foundation may be preferable.

It is also sometimes necessary to excavate piers through sand to reach bearing on rock or other firm material. The conditions to be considered under these circumstances are discussed in Chaps. 13, 21, and 22.

Bearing Capacity and Settlement of Piers. The ultimate bearing capacity of piers exceeds that calculated on the basis of eq. 19.2 because the shearing resistance along the surface *bd* through the surcharge (Fig. 19.4) is no longer negligible, as assumed in the derivation of eq. 19.2. However, if the material above the level of the base of the piers is weak or compressible, the increase in bearing capacity may be small. Moreover, if there is a possibility that the surrounding material may even occasionally be removed by scour, its beneficial influence must be neglected. Therefore, in general, it is conservative and justifiable to determine the safe bearing capacity by means of eq. 19.2 with an appropriate factor of safety.

The settlement of a pier under a given net soil pressure is less than that of a shallow footing on sand of comparable relative density because of the confining pressure due to the weight of the surrounding material. However, at the comparatively large depths associated with piers, the confining pressure also correspondingly increases the N-values in the standard penetration test, as indicated by eq. 5.3 or Fig. 19.6. Hence, unless the final level of the ground surface differs greatly from that at the time the standard penetration tests were made, the allowable soil pressure may be obtained by entering the horizontal portions of the design curves, Fig. 19.3, with N-values uncorrected for the confining pressure. The steep left-hand portions of the design curves may be ignored because the requirements for bearing capacity will have been checked separately by application of eq. 19.2. The values of q_a from Fig. 19.3 must, of course, be corrected for the position of the water table by means of eq. 19.4. The soil pressure determined in this manner should not be increased above that permissible for shallow footings if scour may remove most of the surcharge. Moreover, if the soil surrounding the upper portion of the piers is compressible, and if fresh fill is placed after construction of the piers, it may be necessary to take into account the negative skin friction or drag forces described in Art. 12.8.

In many instances the weight of a concrete pier itself is a large fraction of the total load transferred by the pier to the sand, but the settlement that occurs before the pier is completed may be of no significance. The bases of tall pier shafts supporting bridge spans, for example, may be allowed to settle appreciably, while the concrete of the shafts is being placed, with no detrimental effects. Under these circumstances the weight of the pier can be subtracted from the total net load before the base area is determined by the procedures described in the preceding paragraph. The requirements for an adequate factor of safety, however, must be satisfied for the total net load, including the weight of the pier.

19.5. Piles in Sand

Use of Piles in Sand. (1) Piles may be driven through soft or compressible materials into a layer of dense sand to which they may transfer the weight of the structure. (2) They may be driven into loose sand to compact it and increase its bearing capacity. (3) They may be driven into a bed of sand to establish the foundation below the greatest depth to which the sand may be removed by scour.

Piles Driven to Bearing in Dense Sand. When the material directly beneath foundation level is too compressible or unstable to support the foundation, the weight of the structure should, if possible, be transferred to a more suitable stratum at a lower level. If this stratum consists of sand, piles are often driven through the soft materials and far enough into the sand to develop adequate carrying capacity. If the sand is moderately loose, the piles may have to be driven into it for a considerable distance. If it is fairly dense, it may not be possible to obtain more than a few feet of penetration. In either event, the support provided to the piles by the sand is derived partly from the resistance of the point to further penetration and partly from skin friction between the lower part of the piles and the sand.

It is not usually possible on the basis of the results of laboratory soil tests to make an accurate estimate of the load that can be carried by a pile driven through soft materials into a deposit of sand. The most reliable information is obtained by means of load tests. In deposits of this type, the shape of the load-settlement curve for a single pile is similar to that shown in Fig. 12.8b. Unless the pile fails structurally, which is unlikely, the load-settlement curve approaches an inclined tangent. Hence, the pile does not fail by plunging into the ground, but merely continues to settle or to penetrate into the sand as the load is increased. The limiting load for design must therefore be based on the value of settlement that may be considered tolerable. The procedures for conducting and interpreting the load tests are discussed in Art. 12.4.

Prediction of the depth to which piles must extend into the sand to develop the required capacity is also fraught with uncertainties. The most satisfactory procedure, except for conducting a series of load tests on piles of different lengths, is to determine the necessary depth on the basis of the penetration record of the piles during driving. Since almost all the driving resistance is derived from the sand, and since the static and dynamic resistances are approximately equal for all but the finest sands and silts, use of the wave equation provides a reliable means for determining the resistance to penetration required for the postulated capacity of a given pile driven by given equipment. Studies can also be carried out to ascertain the most efficient combinations of pile and driving equipment to achieve the desired results.

Beneath several of the large cities of Holland, extremely soft materials such as organic silt or peat rest directly on sand. Under these circumstances, a satisfactory approximation to the ultimate point resistance per unit of cross-sectional area of the precast concrete piles widely used in that country has been found to be the static penetration resistance per unit of area of the Dutch cone (Art. 5.4). The cone values are averaged over a depth of about two pile diameters below the point of the pile (Van der Veen, 1953; Van der Veen and Boersma, 1957).

If the soil above the sand is capable of developing a significant amount of freeze, the influence of the freeze on group action, negative skin friction, and load-test techniques is comparable to that previously discussed in Art. 18.5 in connection with point-bearing piles passing through similar materials to bearing in stiff clays. In particular, if part of the dynamic resistance of a single pile is developed in the overlying soft layers, this resistance should not mislead the engineer into believing that the bearing capacity of the pile under static load in the foundation will be as great as the load-test value.

Compaction Piles. Driving piles into a bed of loose sand compacts the material, partly because of the decrease in void ratio necessary to compensate for the volume of the piles and partly because of the compacting effect of the vibrations produced by pile driving. Once the piles are driven, the settlements of the structure are approximately the same as those estimated according to the procedures described for footings and rafts on relatively dense sand.

Inasmuch as the purpose of compaction piles is merely to increase the density of the sand, the structural strength of the piles themselves is of little consequence. Indeed, it may be sufficient to make a hole by driving and removing a spud having the shape of a pile, and to fill the hole with compacted sand. By this procedure the density of the foundation material is increased but no piles remain in the ground. Bulbous uncased piles of the Franki type are also widely used for compacting loose sands.

Ordinarily, piles with a heavy taper are most effective and economical. If the structure is to rest on pile footings, the piles should be driven in groups and the innermost piles in each group driven first to achieve the maximum and most uniform compaction beneath the location of each footing. If the structure is to rest on a pile-supported raft, the piles should be spaced uniformly over the entire area.

The design load to be assigned to compaction piles is necessarily somewhat arbitrary. If the piles are driven to equal penetration per blow of the hammer, they are likely to be of progressively shorter lengths because each pile is driven into sand compacted somewhat more by the preceding pile. Moreover, if driving of the first pile in the cluster is discontinued at a relatively low driving resistance, resumption of driving after the neighboring piles are installed is likely to indicate an increased resistance because of the increase in relative density and confining pressure associated with driving the neighboring piles.

On small jobs, loads of 20 tons are usually assigned to compaction piles of timber and 30 tons to those of cast-in-place or precast concrete. The piles are usually driven to the corresponding capacities as indicated by the *Engineering News* formula (eq. 12.1). Use of the formula does not imply a knowledge of the actual carrying capacity of the piles, but merely provides a convenient criterion to indicate when adequate compaction has been achieved and to guard against overdriving. On larger jobs a test group of several piles should be driven. The center pile of the group should be driven first; it should be driven to a capacity, as indicated by the wave equation and the driving record for the pile, somewhat less than the capacity desired for the job. As the remaining piles in the test group are driven, their driving records should also be obtained. They should be driven to the same penetration per blow as the center pile. When the entire group has been installed, the center pile should be driven further by the same hammer and its capacity judged on the basis of the redriving record. The information will permit selection of the appropriate driving criteria for the project. After the center pile is redriven, it should be subjected to a load test to verify the equivalence of the actual capacity and that determined from the wave equation.

The length of compaction piles is also difficult to predict. It decreases markedly with increasing taper. Piles of 20- to 30-ton capacity, having a taper of 1 in. in 2.5 ft, rarely penetrate more than 25 ft even in loose sand.

Piles in Very Fine Sand and Silt. In the foregoing discussions it has been assumed that the sand into or through which the piles are driven is pervious enough to permit dissipation of the pore pressures due to driving almost as rapidly as they develop. This condition is not satisfied in fine sands and silts. The consequences depend on whether the soil is loose or dense. As successive piles are driven into such soils in a loose state, the pore pressures accumulate, reduce the effective stresses between the particles, and correspondingly reduce the shearing resistance of the soil. If the pore pressures become large enough, the soil is transformed into a viscous liquid in which previously driven piles are readily displaced laterally and upward. Timber or hollow-shell piles may in extreme cases actually float. Such behavior is especially undesirable if the piles are intended to derive their support from an underlying bearing stratum. To avoid such difficulties, it is advantageous to use types of piles that displace the smallest possible volume of soil. If the fine sand or silt is to be densified by compaction piles, it may be necessary to reduce the rate of progress of the job drastically to allow time for drainage; otherwise the pile driving merely converts the soil into a sea of mud. Nevertheless, if the piles can be driven, even with difficulty, dissipation of the pore pressures may be accompanied by substantial freeze and a satisfactory foundation may result.

If the fine sand or silt is dense, it may prove highly resistant to penetration of the piles because of the tendency for dilatancy and the development of negative pore pressures (Art. 4.7) during the shearing displacements associated with insertion of the piles. Analysis of the driving records by means of the wave equation may indicate high dynamic capacity, but instead of freeze, large relaxations may occur. An indication of the relaxation can be obtained from a wave analysis of redriving data obtained after the pore pressures have come to equilibrium, but it is preferable to base the evaluation on load tests.

Piles for Prevention of Failure Due to Scour. The bases of bridge piers located near the channels of rivers must be established below the level to which the river bottom is removed by scour during floods. In many streams, the depth of the river increases during floods at a rate greater than that at which the crest of the river rises. Moreover, since bridges are often located where the channel is narrow, the depth of scour is likely to be greater than average. Finally, the construction of the piers and abutments usually causes additional constriction of the channel and itself increases the depth of scour.

No universal rule can be given for estimating the maximum depth of scour in a stream with a bottom of sand or silt. Several records indicate depths of as much as 4 ft for each foot of rise of the water, and in some streams the ratio is known to be as great as 7. On the other hand, in the majority of rivers the depth is much less. Experience with a given stream is the best guide in estimating the depth of scour.

In many instances, the sands in the bottom of a river channel are relatively dense, and it is difficult or impossible to drive piles to adequate depth without the use of a water jet.

Uplift Resistance of Piles in Sand. The resistance to uplift of a pile driven into sand depends on the relative density of the sand, the length, diameter, and taper of the pile, the use or absence of jetting, and several other factors. For this reason reliable values of the uplift resistance can be obtained only by means of load tests.

As a rough guide, the resistance of a straight-shafted concrete pile driven without jetting is likely to be at least $\frac{1}{2} K \gamma L^2 \tan \phi$ times the perimeter of the pile, where L is the embedded length of the pile and ϕ the friction angle of the sand. The coefficient K varies from about 1.0 for a loose sand to over 3 for a dense sand. Somewhat smaller uplift resistances can be anticipated for steel piles, for piles with a taper, or for piles jetted before being finally driven (Ireland, 1957). The uplift resistance can be destroyed if the body of sand in which the piles are driven is saturated and is loose enough to lose its strength because of accumulated pore pressures during repeated applications of stress as in an earthquake (Art. 4.10). Under these conditions structures such as sewers may rise in spite of hold-down piles.

19.6. Excavation in Sand

Sand Above Water Table. No construction problems of consequence are encountered in making excavations for footing or raft foundations in sand above the water table. Slopes of 1 vertical to $1\frac{1}{2}$ horizontal are stable under all circumstances unless the sand deposit is underlain at shallow depth by soft clay, and steeper slopes are commonly used if the sand possesses a small amount of cohesion due to capillary moisture or cementing agents.

Large Excavations Below Water Table. If excavations for a basement or for the establishment of a foundation must extend below water table in sand, the water level must be lowered. This may be done either by pumping the water out of the excavation itself, or by predrainage of the site.

When the water is removed by pumping from an open excavation, ditches must be cut in the bottom to lead the water to a sump at a lower level than the rest of the excavation. The water level must be maintained in the sump at a low enough elevation to keep the free water surface in the surrounding sand below the bottom of the excavation at every stage. If this cannot be accomplished, the bottom of the excavation becomes *quick*. Water appears in the form of springs, the sand begins to boil, the slopes may begin to slough, and the entire base of the excavation may rise. The ditches around the edge of the hole must be kept clear to prevent water from emerging near the toe of the slopes and causing the banks to collapse.

The preceding statements indicate that pumping from sumps may be a hazardous procedure. In loose or fine sands, it sometimes cannot be accomplished at all. In coarse or dense sands, excavations can often be made with success, but, if conditions get out of control and the bottom becomes quick, the sand beneath the entire foundation may be loosened and its bearing capacity permanently decreased. Therefore, on large jobs, the sand is usually drained before excavation, either by means of single-stage or multiple-stage well points or by deep-well pumps (Chap. 9).

The quantity of water that must be pumped is a matter of importance because sufficient equipment should be provided at the beginning of the job to guarantee efficient removal of the water without the necessity for making additions and alterations. Inadequate pumping capacity may lead to sand boils and instability of the bottom of the excavation.

The quantity of water that must be pumped depends on the coefficient of permeability of the soil, the distance the water level must be lowered, and the dimensions of the area to be dewatered. If the excavation is fairly narrow and if the water table does not need to be lowered more than 15 ft, medium to fine sands can usually be drained by a series of well points spaced at about 3 ft with one 6-in. pump for every 500 ft of cut. Each pump requires a motor of about 20 hp, and drainage is likely to take from 2 to 6 days.

Estimates of the quantity of seepage under more severe conditions may require pumping tests to determine the coefficient of permeability and demand the exercise of considerable judgment based on experience.

The time required to accomplish drainage depends on the coefficient of permeability of the sand deposit. Only 2 or 3 days are necessary for relatively permeable soils (k greater than 10^{-3} cm/sec) whereas several weeks may be required for soils of moderate permeability ($k = 10^{-3}$ to 10^{-5} cm/sec). Silty sands and other materials having an effective size smaller than about 0.05 mm cannot be drained by ordinary well-point equipment, although somewhat finer soils can be drained if the pumps are capable of maintaining a vacuum in the system (Fig. 9.4).

Establishment of Piers Below Water Table. Piers may be carried below water level in open shafts or by means of caissons. The general techniques for making the excavations have been described in Chap. 13. Excavation of an open shaft through sand below water table may often be accomplished by pumping from sumps if the sand is of medium to high density. However, the sand is likely to be loosened and, consequently, its bearing capacity impaired. Use of well points or deep wells before excavation is usually preferable.

When shafts are excavated by hand and kept dry by the use of sumps, the lining is commonly made of pervious material so that the small amount of water that seeps to the excavation from the sides can enter the shaft and be collected in the sump. This prevents the development of a head of water outside the shaft that may lead to the formation of boils or springs at the bottom of the lining.

Large-diameter bore holes through sand may be stabilized by keeping them filled with a clay slurry similar to drilling mud. If the sand overlies an impervious soil, casing may be set in the holes and sealed into the impervious material, whereupon the slurry may be pumped out and the hole continued in the dry (Art. 13.3).

Open caissons are commonly used to establish bridge piers. The sand is removed from the caissons by dredging. To prevent the rise of sand in the bottom of the caissons, the water level inside should be maintained at a considerable height above the free water surface outside. By this procedure, the flow is maintained downward through the sand at the bottom of the caisson, and the material is not loosened by excavation.

If the excavation cannot be made by means of an open caisson, it may be carried out under compressed air. Since the pressure of the air in the working chamber is always greater than that of the water at a given depth, the sand is not loosened during the procedure, and the bearing capacity of the sand is not impaired.

19.7. Effect of Vibrations

Importance. Vibrations constitute the most effective means for compacting deposits of loose sand. By the same token, they constitute one of the most serious causes of excessive settlement of foundations located on loose sands.

Compaction. If the relative density of a deposit of sand is too low for the establishment of a raft or footing foundation, it may be increased by several means. One of these is the driving of compaction piles. Another is the use of a patented method known as the *vibroflotation* process. According to this method, a heavy steel capsule containing an internal vibrator is lowered into the deposit of sand. At the same time, powerful jets of water are forced into the sand beneath the capsule. Under the combined action of jetting and vibration the device sinks rapidly and creates a crater at the ground surface. As the crater develops, it is filled with sand. Compaction to the depth of penetration for a diameter of 6 to 8 ft is generally quite satisfactory. The procedure may, under some circumstances, be less expensive than driving piles. It is most effective in clean medium to coarse sands, but is not effective in silty sands or silts.

Somewhat similar results have been obtained without jetting by repeatedly inserting and removing a steel pipe attached to a low-frequency vibratory pile driver; the spacing of the points of insertion may have to be as close as 3 ft.

Settlement. Many types of machines subject the subsoil to periodic vibrations; these include air compressors, diesel engines for production of electric power, and turbogenerators. The amplitude of the vibrations and the accompanying settlement of the foundation for such a machine depend for a given installation on the frequency. Near a critical frequency the amplitudes may be greatly reinforced by a phenomenon related to resonance. Hence, especially if the subsoil consists of loose sand, a resonant condition must be avoided if possible. Unfortunately, the critical frequency is a function not only of the properties of the sand, but also of the weight, dimensions, and contact area of the exciting mass, including its foundation. The factors affecting the critical frequency and their relationships are complex and not yet fully investigated. Hence, the design of substructures to reduce settlement due to vibrations in sand calls for much judgment and experience. Reduction of the allowable soil pressure is relatively ineffective in decreasing the ultimate settlement. In some instances, machine bases are constructed with very large dimensions so that they are capable of absorbing much of the energy. In other instances, piles or piers may be preferable.

Repeated vibrations or impact at frequencies other than in the resonant range may also eventually produce large cumulative settlements, even on sands of moderately high relative density. Moreover, settlements are likely to be greater if the groundwater table is high. In one instance, impact loadings in a steel mill caused a cumulative settlement of 9 in. over a period of about 12 years. The water table was then lowered from a depth of 10 ft to 20 ft, whereupon the settlements were arrested. However, nine years later the groundwater rose to about the original level and the same rate of settlement was again observed.

Settlements caused by vibrations due to construction are discussed in Art. 16.2.

19 / Foundations on Sand and Nonplastic Silt

PROBLEMS

1. Select the width of the square footing shown in part *a* of the figure, below, to suit the subsoil conditions represented by the boring log (part *b* of the figure), and to satisfy the following conditions:
   a. Tolerable settlement 1 in. under full dead load and 50 per cent live load.
   b. Factor of safety equal to 2 for dead load plus 50 per cent live load.
   Ignore the correction of N-values due to the effect of overburden pressure. The unit weight of the cinder fill is 80 lb/cu ft and of the sand is 120 lb/cu ft.

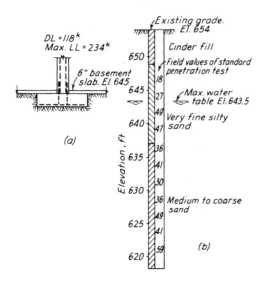

Ans. 6.5 ft.

2. Estimate the settlement of the footing shown below. Compute the factor of safety against a bearing-capacity failure. The unit weight of the sand is 120 lb/cu ft. The N-values have been corrected for overburden pressure.

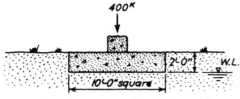

Ans. 1.3 in.; 4.4.

3. If the footing shown below is not to settle more than 1 in., what is the maximum load it can carry? The N-values have been corrected for overburden pressure.

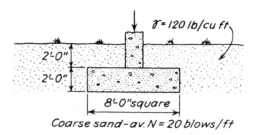

Ans. 139 tons.

4. A footing 5 ft square and 1.5 ft thick is to be located on sand. The surface of the ground is at the top of the footing and the water table is 10 ft below the base. What should be the minimum average N-value of the sand if the footing must support a column load of 200 kips without
   a. experiencing a settlement of more than 1 in. or
   b. possessing a factor of less than 2 against a bearing-capacity failure?
   Ans. 37 blows/ft corrected for effect of overburden pressure; 25 blows/ft uncorrected.

5. A footing 12 ft square and 2 ft thick is supported by sand with an average N value of 30 blows/ft. The surface of the ground is 3 ft above the top of the footing, and the water table is 4 ft below the base. Compute the maximum load that the footing can support if the settlement must not exceed 0.5 in.
   Ans. 181 tons.

6. A continuous wall footing is 2 ft wide and 9 in. thick. The ground surface and water table are at the top of the footing. The underlying sand has an N-value of 30 blows/ft, after correction for the influence of overburden pressure. What load will the footing support if the settlement is not to exceed 1 in. and the factor of safety against bearing capacity failure is not to be less than 2? Does

settlement or bearing capacity govern the load?

Ans. 1.75 tons/lin ft; bearing capacity.

7. A raft 30 ft wide and 50 ft long is to be supported 10 ft below the surface of the surrounding ground. The soil has been explored by means of the standard penetration test in four borings, each of which encountered rock at 40 ft. The borings indicate that the subsoil is a fairly uniform very fine sand. The minimum average penetration resistance in any boring between depths of 10 and 40 ft, after correction for the influence of overburden pressure, is 19 blows/ft. The water table is located 10 to 12 ft below the ground surface; hence, it can be assumed that the sand below the base of the raft is completely submerged. If the structure is to have a basement, and if the soil to be excavated has a unit weight γ of 100 lb/cu ft, what is the maximum soil pressure that should be allowed at the base of the raft?

Ans. 3.1 tons/sq ft.

8. A caisson for a bridge pier is 20 ft wide and 56 ft long. It extends 50 ft through soft clay and is supported on sand that has an N-value of 50 blows/ft. The water table is near the ground surface. If the settlement is limited to 0.5 in., what total load can be supported by the foundation? Neglect the difference between the weight of the caisson and the weight of the clay that is removed during construction.

Ans. 1540 tons.

SUGGESTED READING

An excellent article on the behavior of sand beneath loaded areas, fundamental to an understanding of the load-settlement relations, is A. Vesić (1963),"Bearing Capacity of Deep Foundations in Sand," *Hwy. Res. Rec.*, 39, 112-153.

Further information is contained in:

J. Feda (1961), "Research on the Bearing Capacity of Loose Soil," *Proc. 5 Int. Conf. Soil Mech.*, Paris, 1, 635-642.

A. Eggestad (1963), "Deformation Measurements Below a Model Footing on the Surface of Dry Sand," *Proc. European Conf. Soil Mech.*, Wiesbaden, 1, 233-239. Reprinted in *NGI Publ. 58*, pp. 29-35.

The relation between size of loaded area and settlement is summarized in:

L. Bjerrum and A. Eggestad (1963), "Interpretation of Loading Test on Sand," *Proc. European Conf. Soil Mech.*, Wiesbaden, 1, 199-203. Reprinted in *NGI Publ. 58*, pp. 23-27.

The evaluation of bearing-capacity factors occupies a large place in the literature. The standard references include:

G. G. Meyerhof (1951), "The Ultimate Bearing Capacity of Foundations," *Géotechnique*, 2, 4, 301-332. Revisions in some of the values in this paper are given in the following reference:

G. G. Meyerhof (1955), "Influence of Roughness of Base and Ground-Water Conditions on the Ultimate Bearing Capacity of Foundations," *Géotechnique*, 5, 227-242.

G. G. Meyerhof (1963), "Some Recent Research on the Bearing Capacity of Foundations," *Canadian Geot. Jour.*, 1, 1, 16-26.

The validity of semiempirical procedures for design of footings on sand is considered in D. J. D'Appolonia, E. D'Appolonia, and R. F. Brissette (1968), "Settlement of Spread Footings on Sand," *ASCE J. Soil Mech.*, 94, SM3, 735-760. Discussions appear in Vol. 95, No. SM3, pp. 900-916 and Vol. 96, No. SM2, pp. 754-762. The information in these references has been incorporated into the design rules given in this chapter.

A useful discussion of deep foundations, particularly piles, is included in the general report on that subject at the Montreal conference of the Int. Soc. of Soil Mech. and Found. Eng., by Á. Kézdi (1965), "Deep Foundations," *Proc. 6 Int. Conf. Soil Mech.*, Montreal, 3, 256-264. Empirical data relating the settlement of pile groups to the size of the group appears in a short discussion by A. W. Skempton (1953), *Proc. 3 Int. Conf. Soil Mech.*, Zurich, 3, 172. Further information on group behavior is found in A. Vesić (1969), "Experience with Instrumented Pile Groups in Sand, *ASTM Spec. Tech. Publ. No. 444*, pp. 177–222.

Attempts to calculate the capacity of piles in sand, taking into account the taper, roughness, and volume of the piles, are discussed and evaluated in R. L. Nordlund (1963), "Bearing Capacity of Piles in Cohesionless Soils," *ASCE J. Soil Mech.*, 89, SM3, 1–35. The references include many of the published load-test data on piles in sand. The relation between skin friction and point resistance for six piles in sand is described in A. H. Hunter and M. T. Davisson (1969), "Measurement of Pile Load Transfer," *ASTM Spec. Tech. Publ. No. 444*, pp. 106–117.

Instructive foundation projects in sands are described in the following papers:

J. D. Parsons (1959), "Foundation Installation Requiring Recharging of Ground water," *ASCE J. Constr. Div.*, 85, CO2, 1–21.

J. D. Parsons (1966), "Piling Difficulties in the New York Area," *ASCE J. Soil Mech.*, 92, SM1, 43–64. Includes experiences with inorganic silts.

C. I. Mansur and R. I. Kaufman (1960), "Dewatering the Port Allen Lock Excavation." *ASCE J. Soil Mech.*, 86, SM6, 35–55.

T. J. Lynch (1960), "Pile Driving Experiences at Port Everglades." *ASCE J. Soil Mech.*, 86, SM2, 41–62. Settlements of piles due to pile driving in calcareous sands.

Improvement of the properties of sand is discussed in:

C. E. Basore and J. D. Boitano (1969), "Sand Densification by Piles and Vibroflotation," *ASCE J. Soil Mech.*, 95, SM6, 1303–1323.

E. D'Appolonia, C. E. Miller, Jr., and T. M. Ware (1953), "Sand Compaction by Vibroflotation," *Proc. ASCE*, 79, Separate No. 200, 23 pp.

D. J. D'Appolonia, R. V. Whitman, and E. D'Appolonia (1969), "Sand Compaction with Vibratory Rollers," *ASCE J. Soil Mech.*, 95, SM1, 263–284. Extensive references.

D. C. Moorhouse and G. L. Baker (1969), "Sand Densification by Heavy Vibratory Compactor," *ASCE J. Soil Mech.*, 95, SM4, 985–994.

Two papers regarding scour at bridge piers are assembled in a single publication:

P. G. Hubbard (1955), "Field Measurement of Bridge-Pier Scour," *Proc. Hwy. Res. Bd.*, 34, 184–188.

E. M. Laursen (1955), "Model-Prototype Comparison of Bridge Pier Scour," *Proc. Hwy. Res. Bd.*, 34, 188–193.

References under appropriate chapters of Part B also contain information about foundations on sand and silt.

Damage Due to Swelling and Shrinking

Cracking of small church in Pierre, South Dakota. The Pierre shale, named for its typical occurrence in this locality, is found over a vast area in the northwestern great plains and has its counterpart in the Bearpaw shale that extends into the Canadian prairies. The two formations are notorious for their high swelling capacity and for the damage caused by volume changes of their weathered upper zones when foundations are supported above the depth of seasonal moisture variation.

PLATE 20.

CHAPTER 20

Foundations on Collapsing and Swelling Soils

20.1. General Considerations

Certain soils, even while under constant external load, exhibit large volume changes upon changes in moisture content. The possibilities are indicated in Fig. 20.1, which represents the results of a pair of tests in consolidation apparatus (Fig. 3.1) on identical undisturbed samples. Curve a represents the e-log p curve for a test started at the natural water content and to which no water is permitted access. Curves b and c, on the other hand, correspond to tests on samples to which water is allowed access under all loads until equilibrium is reached. If the resulting e-log p curve, such as curve b, lies entirely below curve a, the soil is said to have *collapsed* (Art. 3.8). Under field conditions, at pressure p_1 and void ratio e_0, the addition of water would, accordingly, cause the void ratio to decrease to e_1, with corresponding settlement $H\Delta e_1/(1 + e_0)$ (eq. 3.3). Soils exhibiting this behavior include true loess (Art. 6.3), clayey loose sands in which the clay serves merely as a binder, loose sands cemented by soluble salts, and certain residual soils such as those derived from granites under conditions of tropical weathering (Art. 6.7).

Conversely, if the addition of water to the second sample leads to curve c, located entirely above a, the soil is said to have *swelled* (Art. 3.2). At a given applied pressure p_1 the void ratio increases to e_1', and the corresponding rise of the ground is

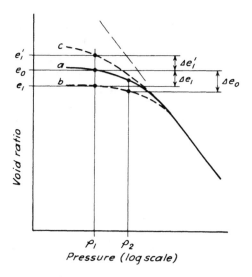

FIGURE 20.1. Behavior of soil in double oedometer or paired confined compression test. (*a*) Relation between void ratio and total pressure for sample to which no water is added. (*b*) Relation for identical sample to which water is allowed access and which experiences collapse. (*c*) Same as (*b*) for sample that exhibits swelling.

333

measured by $H\Delta e_1'/(1 + e_0)$. Soils exhibiting this behavior to a marked degree are usually montmorillonitic clays with high plasticity indices.

In the field, the external load on a foundation ordinarily increases the pressure in the soil from p_1 to p_2. The change in void ratio Δe_0 resulting from the increase in pressure combined with the addition of water can also be determined from the pair of tests, as illustrated by curve b for the collapsing soil in Fig. 20.1.

In practice, at least two factors complicate the direct application of paired consolidation tests (known as *double oedometer tests*) in estimating the amount of collapse or swell. The first of these is the virtual impossibility of obtaining two identical samples. It has been found satisfactory to perform the tests on two samples of similar properties, if both tests are carried to high enough pressures to define the virgin consolidation line (Art. 3.2), and to shift one of the curves vertically until the virgin branches coincide (Jennings and Knight, 1957).

The second complicating factor, fundamentally more serious, is the slow rate at which the water content may change in the field; in many instances equilibrium is not reached within the lifetime of the structure. Hence, the values of Δe_0 computed from the double oedometer test may be larger than the actual changes of void ratio beneath the structure.

Collapsing soils may exhibit large volume decrease upon application of load even without the addition of water. This behavior also requires consideration.

20.2. Foundations on Collapsing Soils

Identification of Collapsing Soils. All collapsing soils considered in this chapter contain an appreciable percentage of air in the voids. When sampled even in thin-walled samplers they are likely to compress significantly. Extrasensitive saturated clays such as the quick clays of Scandinavia and the St. Lawrence valley (Art. 1.7) are also sometimes said to have a collapsible structure, but do not decrease in volume during sampling because of their high degree of saturation and do not exhibit the behavior described in this chapter.

True loess is inherently collapsible, although the tendency decreases with increasing dry density. Most cohesive colluvial, fan, or windblown sands and silts in arid or semiarid regions are also suspect, particularly if the cohesion is imparted by precipitation of soluble compounds such as calcium carbonate, gypsum, or ferrous iron.

Collapsible soils usually slake upon immersion, but disintegration by slaking is not a definitive indicator because other types of soils also slake. Final judgment regarding the possibility of collapse should be based on consolidation tests such as those illustrated in Figs. 20.1 and 3.8, or on load tests conducted in pits into which water is introduced while the test plate is under load. The dimensions of the pit and test plate should correspond to those for the standard load test (Art. 5.5). Typical results on loess are shown in Fig. 6.21.

Performance of Structures on Collapsible Soils. In semiarid localities, the exterior walls of light structures such as dwellings and small commercial buildings are likely to settle excessively and unequally as soon as irrigation of landscaping or grass is started. Cracking of the structures often develops and becomes more pronounced as the moisture content of the soil rises above that formerly prevailing at the site. Behavior of this type occurs in many areas of the western United States. Large and sometimes catastrophic settlements of major structures located on collapsing soils have been caused by the escape of water from defective drains and sumps, from broken water pipes and leaky sewers, or even from reservoirs or swimming pools. Heavy structures such as grain elevators have tilted noticeably because surface waters were allowed to accumulate on one side.

Design of Foundations on Unwetted Collapsible Soils. The load-settlement relations obtained from plate-bearing tests on collapsible soils are illustrated by Fig. 6.21, which represents the results of standard load tests (Art.

5.5) on loess at five sites in the midwestern United States. The behavior differs for different natural water contents. For all tests, however, the relation is roughly linear to a critical pressure p_{cr}, at which the cohesive bonds between the particles begin to break down and the soil to crush. At pressures higher than p_{cr} the increase of settlement for a given increase of load becomes much greater. If the natural water content is relatively high a sudden shear failure may occur, as indicated by curves 1 and 2 in Fig. 6.21. If the natural water content is relatively low, no sudden failure occurs; rather, as exemplified by curves 5 and 6, the volume of the soil directly beneath the loaded area merely decreases as the air-filled pores crush. Loaded areas of greater width are likely to exhibit somewhat larger settlements than narrower ones at a given soil pressure, provided the pressure is less than p_{cr}, but the value of p_{cr} itself is not significantly influenced by the width of the loaded area.

If positive steps can be taken to insure against increase of the water content, footing or raft foundations can be proportioned on the basis of the results of field load tests or of confined compression tests (Fig. 3.1) in which no water is permitted to come into contact with the sample. The value of p_{cr} can be determined for each test by means of a plot similar to Fig. 6.21.

For most ordinary structures the soil pressure should not be allowed to exceed p_{cr}/F, where the value of the factor of safety F should be selected between 2 and 3 in accordance with the considerations discussed in Art. 17.3. The differential and total settlement of the foundations under these circumstances will not exceed those for properly proportioned footings or rafts on sand (Chap. 19). Although the allowable soil pressures determined in this manner may be relatively low, they usually lead to economical design.

On the other hand, grain elevators and other inherently heavily loaded structures often cannot be provided with rafts large enough to reduce the pressure to values as low as p_{cr}. Hence, in eastern Colorado and western Kansas, for example, where the load-settlement curves have the characteristics of curves 5 and 6 in Fig. 6.21, it has been customary to permit grain elevators to settle as much as a foot, providing tilting can be avoided. Since most thick loess deposits are quite uniform in horizontal directions, the settlement can be kept essentially vertical by avoiding eccentric loading. The total settlement can be estimated by computing the distribution of stress beneath the structure according to the procedures indicated in Art. 18.6, by determining values of Δe for the appropriate increases in pressure by means of confined compression tests, and by use of eq. 3.3. Inasmuch as the bonds between the particles of the soil are easily damaged, the samples for such tests should be carefully taken and handled. Because of the nature of the deformation under load, separate consideration of a bearing-capacity failure is unnecessary.

Residual collapsible soils or those having a colluvial origin are likely to be far more heterogeneous than loess. Hence, detailed investigations may be required to make certain that excessive tilting or differential settlement will not occur even if the loading is concentric with the raft.

If the settlements of a structure supported by footings or rafts would be too great, the loads may be transferred to a more suitable stratum or the soil may be densified. Since unwetted collapsible soils are slightly cohesive, piers can usually be drilled and belled out in them readily. The desirability of foundations of this type depends on the nature of the contact with the bearing stratum and, of course, on the character of the bearing stratum itself. Beneath loess the contact may be well defined and the properties and depth of the underlying material easily judged. On the other hand, residual soils may exhibit a gradual and irregular transition that requires evaluation, at the time of drilling, of the material encountered at the base of each pier. If the irregularities are too great, pier foundations may be impractical.

Piles may be driven through the collapsible soil to bearing in more resistant material, or they may be used to densify the soil and reduce its compressibility. In the first instance, the choice of type of pile and allowable load depends largely on the nature of the underlying resistant material; the general considerations that apply are similar to those discussed in connection with piles driven through weak or soft deposits to bearing on or in stiff clays (Art. 18.5), sands (Art. 19.5), or rock (Art. 22.2). If the transition is abrupt, as beneath loess on a firm base, penetration to satisfactory bearing is aided by the use of straight-sided piles of types that displace little volume. On the other hand, irregular and gradual transitions as in residual soils on feebly soluble rocks (Art. 6.7) call for piles preferably with sufficient taper to develop intense skin friction and bearing in their lower part, and lengths easily adjusted to the inevitably great variations in driving depth.

If the stresses below a given depth beneath the structure will not exceed p_{cr} for the soil, there is no need to extend the piles below this depth, provided the overlying soil is densified to the extent that it forms a natural raft. This can be accomplished by compaction piles. Like those in loose sand (Art. 19.5), such piles should displace a maximum volume of soil; heavily tapered piles may produce the desired densification to the given depth most economically.

Foundations on Collapsible Soils Subject to Wetting. If the possibility of wetting cannot be ruled out and if the ensuing settlement would be excessive, the foundation must be established below the zone of potential collapse, or else the collapse must be induced before the structure is built.

The design of piles or piers for support at depth is governed by the same considerations as in the preceding discussion, except that the subsequent wetting and collapse are likely to induce negative skin friction (Art. 18.5) on the foundation units. This added load must be taken into account.

Piles are subject to the further requirement that they must be driven against the resistance of the unwetted soil to adequate capacity in the bearing stratum. Since the unwetted soil may be quite strong, much of the driving energy may not reach the tip; to avoid misjudgment, the impedance of the pile should be given proper attention and the driving equipment suitably chosen (Art. 12.5). Predrilling holes with a diameter an inch or so less than that of the piles may prove advantageous and economical.

If the structure is supported on deep foundations, subsequent wetting may cause settlement of the ground around the structure, whereupon utility connections, sidewalks, and drains will be subjected to corresponding distortions. The types of construction specified should be capable of withstanding the distortions with minimum damage. Soil-supported floors may also experience settlements. Hence, floors should usually be provided with structural support.

Many attempts have been made to induce collapse by surrounding the site with low dikes and flooding the enclosed area. The procedure has been used successfully to treat the foundations for earth dams or dikes that fully load the soil during construction and that can usually tolerate moderate settlements. For several reasons it has not been so successful, however, in connection with foundations for buildings. In some instances collapse cannot be induced by flooding alone, but may require additional weight (curve 4, Fig. 6.21); furthermore, according to Fig. 20.1, even if collapse occurs upon flooding, further settlement must be expected as the load increases. The penetration of the water may occur too slowly to be complete within the time available for pretreatment. Moreover, the penetration may be so nonuniform before construction is complete that subsequent settlements may be extremely irregular. Hence, except under rather unusual circumstances, presettlement by flooding cannot be considered a suitable procedure.

Occasionally, special treatments may be appropriate. For example, in the very fine aeolian sands in parts of Denver, successful stabilization has been accomplished by flooding the foundation trenches for dwellings with a solution of sodium silicate and calcium chloride which, as it seeps through the subsoil, reacts to create a soft sandstone capable of resisting collapse by wetting (Art. 9.6). No such treatment is effective except under unusually favorable local conditions and a pessimistic view regarding the possible applicability of a simple method of stabilization is warranted.

20.3. Foundations on Swelling Soils

Identification of Swelling Soils. Swelling soils are widespread in North America. They are especially prevalent in a belt extending from Texas northward through Oklahoma, into the upper Missouri valley, and on through the western prairie provinces of Canada. In many parts of this belt, considerations of swelling dominate the design of foundations of structures.

A distinction must be made between soils that have the capacity to swell and those that actually exhibit the swelling characteristics in the field. Soils having little or no capacity to swell will not do so under any circumstances. On the other hand, soils with high swelling capacity may or may not swell; their behavior depends on the physical conditions of the material at the beginning of construction and the changes of stress and moisture content to which they are subjected.

The term *swelling soils* implies not only the tendency to increase in volume when water is available, but also to decrease in volume or *shrink* if water is removed. In this article the main emphasis is on swelling, but it should be kept in mind that shrinkage is merely the reverse process. A few specific comments about shrinkage are included at the end of the article.

The swelling potential of a soil is related in a general way to the plasticity index. Various degrees of swelling capacity and the corresponding ranges of plasticity index are indicated in Table 20.1.

Whether a soil with high swelling potential will actually exhibit swelling characteristics depends on several factors. That of greatest importance is the difference between the field moisture content at the time the construction is undertaken and

Table 20.1 Relation Between Swelling Potential of Soils and Plasticity Index

Swelling Potential	Plasticity Index
Low	0–15
Medium	10–35
High	20–55
Very high	35 and above

the equilibrium moisture content that will finally be achieved under the conditions associated with the completed structure. If the equilibrium moisture content is considerably higher than the field moisture content, and if the soil is of high swelling capacity, vigorous swelling may occur as evidenced by upward heaving of the soil or structure or by the development of large swelling pressures. If the equilibrium moisture content is lower than the field moisture, the soil will not swell but, on the contrary, will shrink. A second factor is the degree of compaction of the soil if in a fill, or the degree of overconsolidation if an undisturbed natural material. Relatively high compaction or high previous overburden pressures favor swelling as moisture becomes available. A third factor is the stress to which the material will be subjected after construction is completed. The less the imposed load, the greater the swelling.

The influence of these and several other factors introduces large uncertainties into the prediction of the swelling behavior of the soils on a given project. Local experience constitutes the best guide. Swelling tests conducted under conditions duplicating as closely as possible the anticipated field conditions may also provide useful information.

Swelling Tests. When it is practicable to obtain virtually identical pairs of undisturbed samples at the moisture content expected to prevail at the time construction is undertaken, the most reliable estimates of heave due to swelling can be obtained by means of the double oedometer test (Fig. 20.1). The reliability of predictions based on the test depends, however, not only on the degree to which the initial moisture content of each sample agrees with the actual moisture content at start of construction, but also on the extent to which the final or equilibrium moisture content in the field approaches that attained by the test specimen to which water has free access.

The practical difficulties in obtaining identical samples for double oedometer tests have led to the use of simpler tests giving results that can be interpreted only in qualitative rather than quantitative fashion. Two types of tests are commonly known as *unrestrained swelling tests* and *swelling-pressure tests*.

In tests of the first type, a sample at the moisture content expected to prevail at the time of construction is fitted as tightly as possible into a consolidation ring and subjected to a small vertical pressure such as 1 lb/sq in. The porous disks placed above and below the sample must be air-dry at the beginning of the test. Water is then admitted to the sample through the porous disks and the vertical expansion of the sample measured as a function of time until the expansion practically ceases. The increase in thickness expressed as a percentage of the original thickness is designated as the swell and is a measure of the maximum percentage increase in volume that the material could be expected to experience as a consequence of increase of moisture content. A volume change less than 1.5 per cent is regarded as low; between 1.5 and 5 per cent as medium; between 5 and 25 per cent as high; and above 25 per cent as very high.

The swelling-pressure test is conducted with similar equipment, but the vertical expansion of the sample upon access to water is prevented. The force required to prevent the expansion is determined as a function of time. The swelling pressure ultimately approached is a measure of the maximum force per unit area that can be exerted by the soil under extreme swelling conditions. Swelling pressures below about 0.2 ton/sq ft are regarded as low; pressures above 20 tons/sq ft are occasionally encountered.

The two types of swelling tests give useful indications of extreme behavior. In most instances, however, swelling is partly restrained. Consequently, the magnitudes of swell and of swelling pressure are likely to be intermediate between those determined by the two tests. Moreover, the results of the two types of tests are not always consistent, partly because of the inevitable variation in properties from sample to sample. If tests of either type indicate a high degree of swelling, a soil should be considered suspect. If the results of both tests indicate high swelling behavior, it may be taken for granted that extensive precautions are warranted.

Useful quantitative information can be obtained from a variation of the swelling-pressure test in which the sample, again initially at the moisture content expected to prevail at the time of construction, is subjected to an arbitrarily selected vertical pressure and allowed to come to equilibrium before moisture is given access to the sample. The pressure is usually taken as roughly twice the overburden pressure believed to have been acting on the sample before its removal from the ground, but if strong swelling tendencies are suspected, a higher initial pressure may be appropriate. Water is then introduced and the increase in height of the sample observed until equilibrium is approached. The vertical pressure is then reduced by a factor of about 2 and the swell again observed. The same procedure is followed for a second reduction of pressure, and finally under a reduction of vertical pressure to zero. The results are plotted as shown in Fig. 20.2. Curve *a* represents a soil exhibiting high swelling pressure at small expansion but which expands little; moreover, it exerts only a slight pressure after a moderate expansion. Curve *b* represents a more undesirable soil which, although it displays only moderate swelling pressure after moderate expansion, expands vigorously under reduction in pressure. Data obtained from such tests can be used to estimate the amount of surcharge or cover required to prevent the soil at a certain depth from swelling or to limit the swell to an acceptable amount. They can also be used to estimate the ultimate heave

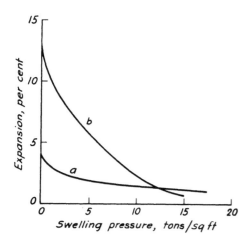

FIGURE 20.2. Typical results of modified swelling-pressure test.

of the surface corresponding to a given depth of excavation or fill.

The results of all swelling tests at best are rough approximations, partly because of inevitable changes in moisture content and structure of the soils during drilling, sampling, and handling in the laboratory. Disturbed samples taken by driving a thick-walled spoon into the ground may, for example, be denser than the soil in place and, consequently, may swell more than the undisturbed material. The greatest error associated with swelling tests, however, is likely to arise from a difference between the initial moisture content of the sample and the moisture content in the field at the critical early stage of construction after which swelling or shrinking will be detrimental to the structure.

Unsatisfactory Performance Associated with Swelling. The detrimental consequences of swelling are most apparent in arid and semi-arid localities because the water content of the soils near the surface is normally kept low by evaporation. Even in such localities, however, there are usually rainy seasons during which the rate of precipitation exceeds that of evaporation. Consequently, in a zone ranging in depth from a few feet to as much as possibly 20 ft, depending on the locality, the soil expands and contracts according to an annual cycle. Where there is no interference with the natural processes, the ground surface rises and falls but the movements cause no difficulties and often pass unnoticed. On the other hand, a comparatively impervious road surface that reduces evaporation, or a structure that affords protection from the sun and heat, permits the moisture to accumulate and the soil to swell. Differential movements then become apparent.

The depth of seasonal variation in moisture content can most readily be judged in some localities by inspection in large-diameter bore holes. In the zone of variation, the soil has a blocky structure and fragments are often slickensided. Below the zone the structure is likely to appear fairly massive and intact, and slickensides are uncommon.

Typical damage to a footing-supported structure is illustrated in Fig. 20.3a. The soil-supported interior floor gradually rises, takes the shape of an irregular dome, and cracks. The swelling clay beneath the floor tends to expand and exert pressure laterally as well as vertically. It therefore tilts the footing walls outward and leads to cracking of the exterior walls of the structure, particularly at the corners. Damage is also often apparent at the connections between the walls and the roof or floors where the movement is restrained. Because the accumulation of sufficient moisture to produce large displacements occurs slowly, the detrimental effects may not be apparent immediately after construction but only after several years.

Deeper foundations, intended to support the structure below the zone of seasonal variation in moisture content, may themselves remain stable. However, if the grade beams supporting the partitions or walls between the piers are left in contact with the underlying soil, they are ultimately forced upward and crack, as shown in Fig. 20.3b. Utilities buried in the soil participate in the vertical and lateral displacements and are particularly vulnerable to differential movements. Water and sewer lines are apt to be broken, whereupon they feed water into the soil and contribute to the swelling.

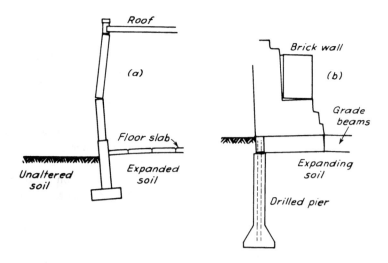

FIGURE 20.3. Damage caused by swelling soil beneath (a) typical footing-supported structure with slab on grade (after Parcher and Means, 1968), and (b) typical pier-supported structure with grade beam in contact with swelling soil.

Design of Foundations in Swelling Soils. Three general methods are available for reducing or avoiding the effects of swelling. These include isolating the structure from the swelling materials, designing a structure that will remain undamaged in spite of the swelling, and elimination of the swelling. All three procedures are in use, either singly or in combination, but the first is by far the most widespread.

Since swelling soils are usually stiff and do not contain free water, they often constitute excellent ground in which to drill holes for the establishment of piers at depths below the zone of seasonal volume change. Where problems with swelling are acute, it has become customary, even for the support of single-family dwellings, to provide drilled-pier foundations terminating in bells that function as anchors in materials that are not subject to significant seasonal movements. The concrete with which the holes are filled is reinforced throughout its length, including the belled section, because the swelling soil above the bells is likely to exert substantial uplift and to create tensile forces in the piers. The piers are attached to grade beams of reinforced concrete that, in turn, support the entire structure, including the ground floor. Inasmuch as the pressure of swelling soil against the bottom of a grade beam or against a floor in contact with the soil ultimately produces large upward forces, provision must be made to prevent the contact or to eliminate the transmission of compressive forces when swelling develops. This requirement is usually satisfied by the use of collapsible forms of cardboard or other frangible material upon which the concrete can be poured, but which crushes at loads only slightly greater than the weight of the wet concrete. Details of such designs are shown in Fig. 20.4.

In some instances the zone of seasonal variations in moisture extends to a depth greater than that below which drilled piers can economically be established and the piers are stopped at a higher level. Inasmuch as the clay in the zone of moisture changes is likely to be highly slickensided, attempts to form the bells for such piers may be unsuccessful because of successive caving of the blocks between slickensides. Delays in forming the bells and placing the concrete, or the presence of even a small amount of moisture seeping along the joints, aggravates the difficulty.

Even if the grade beams and floors of a

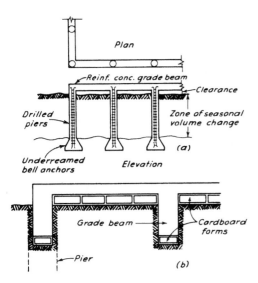

FIGURE 20.4. Details of construction designed to prevent swelling soils from acting directly on grade beams and floors [(a) after Jennings and Henkel, 1949].

structure are not subjected to uplift forces, the swelling soil tends to grip the shafts of the piers and lift them. The force at the junction of shaft and bell may reach a value

$$Q_{\text{uplift}} = \pi d L c_a \qquad 20.2$$

where d and L are respectively the diameter and length of the shaft, and c_a is the adhesion between soil and shaft. If the walls of the drill hole are rough, the adhesion may be considered equal to the shearing strength of the clay. The diameter of the bell required to anchor the pier may be calculated by equating the upward force to the bearing capacity of the soil directly above the bell, or roughly

$$Q_{\text{uplift}} - Q_{\text{dead}} = \frac{q_d}{F} \cdot \frac{\pi}{4} (d_b^2 - d^2) \qquad 20.3$$

where q_d is the net bearing capacity (eq. 18.2), d_b is the diameter of the bell, and F is the desired factor of safety. In calculating the resistance to uplift, D_f is taken conservatively as zero and the dead weight of the pier itself is usually neglected. To keep the uplift to a minimum the shaft is given the smallest practicable diameter, but not less than about one third the diameter of the bell.

The foundation should preferably be laid out to maximize the dead load per pier, and the soil pressures for downward loading at the base of the pier should provide only the minimum acceptable factor of safety; thus the base pressures are utilized as much as possible to counteract the tendency to swell. If there is a likelihood that swelling may occur before the dead load of the superstructure is applied, a factor of safety of at least 1.0 should be provided against heave of the unloaded pier.

The adhesion between soil and pier can be reduced by drilling holes about 4 in. larger in diameter than the shaft, by excavating and concreting the bell, and by casting the shaft in a cardboard form. The space between form and hole may be filled with vermiculite or other materials that possess no appreciable shearing strength.

Drilled piers have little resistance against lateral displacement and should not be expected to restrain downhill movements associated with landslides or creep.

Structures capable of remaining undamaged and undistorted in spite of being supported directly on swelling soils must possess great strength and rigidity. Very small structures can be designed to satisfy these requirements by keeping the stresses within allowable values even if the entire building is assumed to be supported only on a central area equal to about half the plan area of the structure, and also if supported only on the peripheral half of the plan area exclusive of the central area. Such design obviously leads to costly construction. In most large structures there is likely to be little economic advantage in avoiding pier foundations or other deep foundations in view of the cost of providing in the superstructure the additional strength and rigidity required to resist unequal swelling. Hence, it is rarely found worthwhile to attempt to design the structure to withstand the effects of swelling.

The detrimental influence of swelling on any structure can be reduced to some extent, but by no means eliminated, by sur-

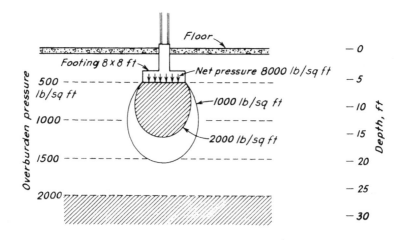

FIGURE 20.5. Diagram illustrating influence on swelling of high contact pressure beneath footing. If net pressure at base of footing is 8000 lb/sq ft and swelling pressure at zero volume change is 2000 lb/sq ft, swelling will be prevented within shaded areas only.

rounding the structure with an impervious apron, usually about 15 ft wide. The apron alters the moisture regime for a limited distance outside the building in a manner similar to the alteration within. Hence, differential behavior at the edges of the structure is minimized. The apron must be expected to rise as the ground swells. It should preferably be made of asphaltic concrete to reduce cracking; any cracks, as well as the joint around the building, should be periodically sealed, especially before the rainy season.

Elimination of the swelling itself can in principle be accomplished in three ways: the ground can be prewetted to a moisture content equal to the equilibrium value; downward loads can be induced large enough to equal or exceed the swelling pressures; or swelling can be inhibited chemically.

Prewetting by flooding the construction area is rarely effective because of the long time required for penetration of the moisture to any great depth, and because of the nonuniformity of the penetration under field conditions. Hence, the procedure is not recommended. On the other hand, if the potentially swelling clay is to be used as a fill over the entire site, compaction by means of comparatively light equipment at a moisture content several per cent above optimum may greatly reduce the swelling. The degree of compaction should not exceed 95 per cent of the Standard Proctor maximum (Art. 1.6). It should be realized that a fill placed according to these requirements will have relatively low supporting capacity with respect to downward loads.

Counteracting the swelling pressures by downward pressures beneath foundation units is not readily accomplished, partly because of the comparative unreliability of the methods of evaluating and predicting the swelling pressures. In lightly loaded structures it may not be possible to develop large enough downward loads to exert the required pressures beneath supports of the smallest practicable size. Furthermore, swelling can be prevented only in a localized zone beneath the footing or pier where the stresses induced by the foundation are concentrated, as illustrated in Fig. 20.5. At a comparatively shallow depth beneath the foundation, the intensity of added stress is small and swelling may occur below this level, even if it is entirely prevented above. In the areas between the footings, swelling is undiminished.

On the other hand, if the entire area can

be covered with a nonswelling material to a depth D_f such that γD_f equals or approaches the swelling pressure, swelling can be effectively prevented; even at smaller values of γD_f the heave may be tolerable. The magnitude of the swelling to be anticipated under different surcharges can be estimated with the aid of the modified swelling-pressure tests described earlier in this article.

Chemical stabilization of swelling soils by the addition of lime may be remarkably effective if the lime can be mixed thoroughly with the soil and compacted at about the optimum moisture content. The appropriate percentage, which usually ranges from about 3 to 8, is estimated on the basis of pH tests and checked by compacting, curing, and testing samples in the laboratory. The lime has the effect of reducing the plasticity of the soil and, hence, its swelling potential. The necessity for intimate mixing restricts the general applicability of lime stabilization to fills. In some localities, however, pressure injection of lime slurry into heavily fissured clays appears to create on the clay fragments a lime-stabilized skin that prevents moisture from entering the fragments and causing swelling. Such a procedure has been found successful in the vicinity of San Antonio, for example, but its applicability under other conditions should not be assumed.

Shrinkage. Differential movements almost the reverse of those encountered in semiarid regions are associated with the presence of soils of high swelling capacity in more humid regions. The natural moisture contents of the soils are likely to be high at the time of construction. After completion, the presence of the building favors drying of the soils, whereupon shrinkage occurs and allows soil-supported floors and sometimes foundations to settle.

The root systems of trees may remove a surprisingly large quantity of water during the growing season and may lead to localized settlements and cracking of nearby structures. The influence of trees may not be significant or apparent in years of normal or near-normal precipitation, but in periods of drought may become of major importance. Indeed, extended dry spells of two or three years, even in regions of only moderately swelling soils, may cause severe shrinkage of the soil and differential settlements of structures that had shown no distress in the preceding several decades. Associated with the settlement and shrinkage may be intense cracking of the ground and appreciable lateral stretching, and corresponding cracking of buildings. The cracks in the ground permit deep penetration of water if rains do occur and lead to further differential movements. Moreover, the cracks in the buildings due to lateral stretching do not tend to close and subsequent dry seasons aggravate the condition. In regions of mild winters, the depth of foundation should be controlled by considerations of shrinkage rather than frost penetration if the subsoil consists of clay having even moderate swelling potential and if normally humid climate may be broken by severe droughts during successive years.

ILLUSTRATIVE DESIGN. DP 20-1. PIER ON SWELLING CLAY

The soil conditions and results of laboratory tests shown on Sh. 1 of DP 20-1 indicate that the swelling potential of the clay is quite high. The indicators are (1) presence of secondary structure (slickensides), (2) low field moisture content relative to the Atterberg limits, (3) high plasticity indices, and (4) moderately large deformations obtained in the swell-pressure tests.

Because of the swelling potential of the soil, a spread footing foundation is considered unsuitable for the support of the building. Large displacements of interior footings would likely occur as the soil attained its equilibrium moisture content beneath the completed structure. Moreover, the performance of exterior footings might be impaired by volume changes caused by seasonal fluctuations in moisture content. The zone of seasonal volume change for the soil in DP 20-1 is presumed to extend to the depth of slickensides, El. 710. However, the

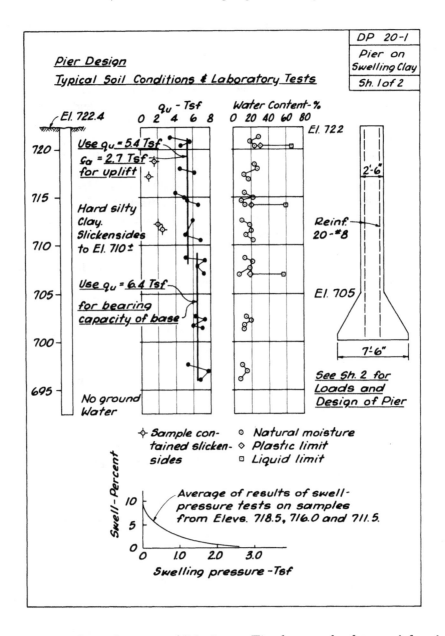

tops of the bells of the pier are established 5 ft lower in order to provide a margin of safety against caving of the slickensided clay into the bell during construction.

The design procedure given on Sh. 2 of the design plate is for the most part a trial and error approach to the determination of the pier dimensions. For example, the diameter of the bell must satisfy certain criteria with respect to the downward loads as well as to the uplift forces caused by the swelling soil. It is not ordinarily possible to determine by inspection which of these criteria governs the diameter of the bell.

The factors of safety used for checking the several conditions, corresponding to different stages of construction and of pier loading, are judgments made by the designer. In the design plate, the weight of the pier has been ignored in the calculation of factors of safety. The error is small in most cases and, with respect to uplift, the omission is on the side of safety. Uplift of the piers themselves prior to the application of the building loads may be avoided by making certain that the factor of safety is at least 1.0. The bell of the pier in the design plate has been proportioned to pro-

```
                                          ┌─────────────────┐
                                          │    DP 20-1      │
                                          ├─────────────────┤
                                          │    Pier on      │
        Pier Design                       │  Swelling Clay  │
        Loads from Superstructure (Bldg. Code) │ Sh. 2 of 2  │
                                          └─────────────────┘
```

<u>Pier Design</u>
<u>Loads from Superstructure (Bldg. Code)</u>
   $DL = 256^K = 128^T$; $LL = 180^K = 90^T$; Total $= 436^K = 218^T$
<u>Uplift caused by Swelling</u>
   Proportion bell to provide $F = 1.2$ against uplift of pier alone, and $F = 2.0$ under the DL of the superstructure. Ignore dead weight of pier.
   <u>Shaft</u>: Use 2'-6" dia. $722 - 705 = $ <u>17'-0" length</u>
                            $(722 - 710 = 12'-0"$ for uplift$)$
   Eq. 20.2: $Q_{up} = 3.14(2.5)(12.0)(2.7) = 254^T$
   Eq. 20.3: $254 = \dfrac{6.2\,(2.7)}{1.2} \times \dfrac{3.14}{4} (d_b^2 - 2.5^2)$
                   $= 11.0\,(d_b^2 - 6.25)$
   <u>Bell</u>: Try $d_b = 5'-6"$ $254 < 11.0\,(30.25 - 6.25)$
                             $254 < 264$
                <u>Use bell dia. = 5'-6" A = 23.8 ft^2</u>
   <u>Reinforcing bars</u>: $A_s = \dfrac{254\,(2)}{40.0}\,(1.2) = 15.2\,in^2$
                  <u>Use 20 #8 $A_S = 15.7\,in^2$</u>
<u>Check F under DL of Superstructure</u>
   Net uplift $= 254 - 128 = 126^T$
        $F = \dfrac{264}{126} = $ <u>$2.1 > 2.0$ ok</u>

<u>Check Bell Area for Total Downward Load</u>
   Assume no uplift on shaft and neglect difference in unit weights of soil and concrete
           $218^T \div 23.8 = $ <u>9.16 Tsf.</u>
           <u>ok, allow $1.5\,(6.4) = 9.6$ Tsf</u> (Eq. 18.6b)

---

vide a factor of safety of 1.2 for this condition. Under the superimposed dead load of the building, the factor of safety against uplift is then found to be greater than 2.0, the desired minimum. The strength of the soil resisting uplift has been taken conservatively as that of the weaker upper zone.

Other judgments involved in the design of the pier include the size of the shaft. The 2.5-ft diameter is about the least that allows an inspector to enter the hole during construction. On the other hand, smaller shafts and special methods of inspection may be used under some soil conditions. In any event, the diameter of the shaft should not be smaller than about one third the diameter of the bell. The use of dimensions adjusted to the next larger 3- or 6-in. size is common practice in the proportioning of piers.

The area of reinforcing bars in the shaft at its junction with the bell has been determined on the basis of a factor of safety of about 1.2 against a yield point of the steel equal to 40,000 psi. This factor of safety is consistent with the previous calculations dealing with the capacity of the soil to resist uplift.

PROBLEMS

1. The foundation for a microwave tower on loess will be subjected to a dead load of 48 tons and a wind load of 154 tons. The site is at Montour, Iowa; the load-settlement curve from a standard load test is given in Fig. 6.21. What dimensions are required for square footings to provide a factor of safety with respect to the critical load of 3 against dead load and 2 against the sum of dead load and live load?

   Ans. The critical load from the load-settlement curve is 2.8 tons/sq ft. The required size is 12 x 12 ft.

2. A drilled pier beneath a building is to carry a dead load of 106 tons and a live load of 142 tons. The upper 20 ft of the surrounding ground are believed to be in the zone of seasonal volume change; the material consists of clay with a high swelling potential and an unconfined compressive strength at its present moisture content of 2.5 tons/sq ft. To resist possible uplift, a bell is to be formed with its base 15 ft below the active zone, in a stiff clay with an unconfined compressive strength of 4 tons/sq ft.

   a. If the diameter of the pier shaft is 3 ft and the weight of the pier is neglected, what diameter of bell is required to prevent uplift of the pier (factor of safety of 1.0) prior to the application of the building load?

   b. After the application of the dead load of the building, what is the factor of safety against uplift?

   c. What factor of safety does the pier have against the full downward load if the uplift does not develop (neglect the skin friction on the shaft)?

   Ans. 7.0 ft; 1.9; 2.8

SUGGESTED READING

Collapsing soils and loess:

J. E. Jennings and K. Knight (1957), "The Additional Settlement of Foundations Due to a Collapse of Structure of Sandy Subsoils on Wetting," *Proc. 4 Int. Conf. Soil Mech.*, London, *1*, 316–319. Proposal of double oedometer test.

J. H. Dudley (1970), "Review of Collapsing Soils," *ASCE J. Soil Mech.*, *96*, SM3, 925–947. Extensive bibliography.

W. A. Clevenger (1956), "Experiences with Loess as Foundation Material," *Trans. ASCE*, *123*, 151–169. Discussion by Peck and Ireland, pp. 171–179.

O. K. Peck and R. B. Peck (1948), "Settlement of Foundation Due to Saturation of Loess Subsoil," *Proc. 2 Int. Conf. Soil Mech.*, Rotterdam, *4*, pp. 4–5. The material was a slightly cemented silt, probably incorrectly described as loess.

A. B. A. Brink and B. A. Kantey (1961), "Collapsible Grain Structure in Residual Granite Soils in Southern Africa," *Proc. 5 Int. Conf. Soil Mech.*, Paris, *1*, 611–614.

Swelling and shrinking:

H. B. Seed, R. J. Woodward, and R. Lundgren (1962), "Prediction of Swelling Potential for Compacted Clays," *ASCE J. Soil Mech.*, *88*, SM3, 53–87.

J. E. Jennings (1953), "The Heaving of Buildings on Desiccated Clay," *Proc. 3 Int. Conf. Soil Mech.*, *1*, Zurich, 390–396.

R. F. Dawson (1959), "Modern Practices Used in the Design of Foundations for Structures on Expansive Soils," *Colo. School of Mines Quarterly*, *54*, 4, 67–87.

J. V. Parcher and R. E. Means (1968), *Soil Mechanics and Foundations*, Columbus, Merrill, 573 pp. Chap. 6, *Foundations on Overconsolidated Desiccated Clay*.

W. H. Ward (1953), "Soil Movement and Weather," *Proc. 3 Int. Conf. Soil Mech.*, Zurich, *1*, 477–482.

M. J. Hammer and O. B. Thompson (1966), "Foundation Clay Shrinkage Caused by Large Trees," *ASCE J. Soil Mech.*, *92*, SM6, 1–17.

M. R. Thompson (1968), "Lime-Treated Soils for Pavement Construction," *ASCE J. Highway Div.*, *94*, HW2, 191–217. Although primarily concerned with lime stabilization of pavement subgrades, the principles and techniques are applicable to compacted stabilized fills for support of structures. Extensive bibliography.

Tower Latino Americana, Mexico City

This 43-story structure, tallest in Mexico City, has a basement with an excavation 42 ft deep. The substructure rests on cast-in-place button-bottom piles founded at a depth of 110 ft on a thin but very dense sand layer located between deposits of exceptionally compressible lacustrine clay. To prevent the excessive heave usually associated with excavation in this city, the hydrostatic pressure in the underlying clay was reduced by pumping from wells draining thin sand layers in the clay; to prevent settlement of the surrounding areas the water was fed into a gravel-filled trough and into injection wells just outside a sheet-pile wall enclosing the site. The successful completion of this foundation without damage to adjacent structures demonstrates the power of soil mechanics in a city noted for spectacular foundation problems.
(Photo courtesy of Professor Leonardo Zeevaert.)

PLATE 21.

CHAPTER 21

Foundations on Nonuniform Soils

21.1. Introduction

In the preceding chapters of Part C, it has generally been assumed that the subsoil is relatively uniform either to a very great depth or else to a limited depth where a firm base is encountered. In reality, such conditions are so unusual as to be considered rare exceptions. Hence, the procedures described in the preceding chapters are not often directly applicable to the solution of practical problems. Nevertheless, they are of value because they can be modified to give reliable indications of the probable behavior of foundations on nonuniform materials.

Most subsoils consist either of definite strata or of more or less lenticular elements. Some of the components of the deposit may consist of fairly resistant and incompressible material, whereas others may be relatively weak and compressible. On the basis of preliminary information, such as that from exploratory borings together with standard penetration tests and simple laboratory tests, one can usually conclude at once whether some parts of the subsoil are sufficiently strong and incompressible to be of no concern. Attention can then be concentrated on the weaker or more compressible members.

The principal task of the designer before he can select the appropriate type of foundation is to determine the influence of the presence of the elements believed to be weak. In general, this may be done by estimating or computing the stresses in the subsoil on the assumption that the subsoil is uniform and elastic. After the physical properties of the doubtful materials have been evaluated on the basis of the exploratory data, the capacity of the doubtful materials to resist the stresses without failure or excessive compression can be determined. The result of this investigation is usually sufficient to permit selection of the appropriate type of foundation. Occasionally, more elaborate exploratory procedures and soil tests may be required to provide the basis for a sound decision.

The computation of stresses may be made by means of Newmark's chart or, under many conditions ordinarily encountered, by some simplified procedure. Although the chart is based on the assumption that the material is homogeneous, the errors in the stresses due to stratification or other irregularities are not likely to be great enough to invalidate the predictions of the probable behavior of the soil.

In the following sections, the more important kinds of nonuniform soil deposits will be discussed.

21.2. Soft or Loose Strata Overlying Firm Strata

When the upper part of the subsoil consists of soft or loose soils, the unsatisfactory character of the materials is likely to be apparent, and the necessity for providing adequate support is rarely overlooked. The principal decision is whether or not a footing foundation can be used. This may be determined by computing the safe load for the upper material on the assumption that it extends to great depth and by estimating the settlement that would arise from the compression of the soft part of the deposit. If the computed safe load is too small or the computed settlement too great, footings must be eliminated from consideration. One alternative is to provide pile or pier support. Another possibility is to reduce the excess load on the subsoil by excavation and to construct a raft foundation.

If piles or piers are adopted, their bearing capacity and behavior may be judged on the basis of the considerations discussed in Chaps. 12 and 18 to 20.

21.3. Dense or Stiff Layer Overlying Soft Deposit

Choice of Foundation. The implications of the presence of a soft deposit at some depth below firm strata are not so obvious as if the soft materials were at shallow depth. If the firm deposit is relatively thin, footings or rafts may exert sufficient pressure to break into the underlying soft soil. A number of failures of this type have occurred. Even if the overlying firm layer is of sufficient thickness to prevent such a failure, the settlement of the structure due to consolidation of the soft deposit may be excessive.

The factor of safety against breaking through the stiff crust may be conservatively estimated by determining the pressures at the upper surface of the soft deposit. The maximum pressure should not exceed the safe load for the soft material as determined by the procedures discussed in the preceding chapters.

If the footings are widely spaced and the firm layer fairly thin with respect to the width of the footings, the stress at the top of the soft layer can be decreased considerably by increasing the size of the footings. On the other hand, if the footings are spaced rather closely and the firm layer is comparatively thick, the distribution of pressure at the top of the soft layer cannot be altered radically by changing the contact pressure. For example, Fig. 21.1 illustrates the distribution of pressure at a depth of 10 ft beneath a large array of square footings spaced 20 ft in both directions. In Fig. 21.1a, each footing exerts a pressure of 2 tons/sq ft; the corresponding maximum pressure at the top of the soft clay layer is 0.84 ton/sq ft. In Fig. 21.1b, the same column loads are transmitted to the stiff clay by footings of twice the area as those in Fig. 21.1a; consequently the contact pressure is reduced to 1 ton/sq ft. The maximum pressure at the top of the soft clay layer, however, is reduced only 19 per cent, to 0.68 ton/sq ft. Even if the same column loads were delivered to a raft covering the entire foundation, the pressure at the top of the soft clay could not be reduced by more than about 24 per cent, to $256/400 = 0.64$ ton/sq ft. If the pressure on an underlying soft layer cannot be reduced to the safe load for the soft material by increasing the size of the footings, either pile or pier support is required, or else material must be excavated to compensate for part of the weight of the building.

Even if the safe load on the soft soil beneath the firm layer is not exceeded, the settlement of a footing or raft foundation may be excessive. The computation of settlement may be made by the procedures already described in connection with confined layers of clay. If the settlement is excessive, one of the other types of foundation mentioned in the preceding paragraph must be adopted.

If the computed settlement is not excessive and if the firm layer is thick enough to prevent a bearing-capacity failure, the footings can be designed as if the soft deposit were not present, by means of the rules given previously for the kind of soil constituting the firm layer.

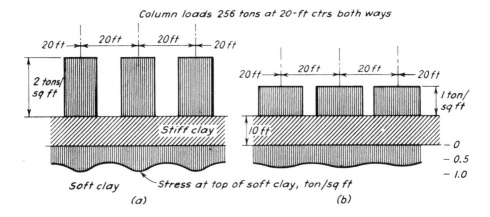

FIGURE 21.1. Effect of reducing contact pressure, beneath closely spaced square footings carrying equal total loads, on stress at top of soft clay beneath stiff clay crust. Footings are spaced at 20-ft centers in both directions; contact pressure in (a) is reduced 50 per cent in (b) by increasing size of footings.

Shortcomings of Load Tests. Load tests have been widely used for estimating the allowable pressure for a given subsoil. Examples of conditions appropriate for the use of the method have been given in several of the preceding chapters. However, the method is likely to be misleading and dangerous if the load tests are made on a firm stratum underlain by softer materials. The reason is illustrated in Fig. 21.2. In this figure, A represents a test plate 1 ft square and B a footing 10 ft square. Both rest on the surface of a stratum of stiff clay 3 ft thick. The stiff clay rests on a deep deposit of soft clay having an unconfined compressive strength of 0.3 ton/sq ft and a compression index equal to 0.27. The stiff clay itself has an unconfined compressive strength of 2 tons/sq ft and may be considered practically incompressible. The failure load for the test plate is approximately 6.2 tons/sq ft, provided the entire surface of sliding is located within the stiff clay layer. Even under this load on the test plate, the maximum stress at the top of the soft layer is only 0.31 ton/sq ft, which is considerably less than the bearing capacity of the soft material. Therefore, the soft material would not influence the failure of the test plate.

The load that may be carried safely by the footing may be estimated in the following manner. The perimeter of the footing is 40 ft and the thickness of the stiff layer is 3 ft. If the footing were to break into the soft soil, a shearing force of $3 \times 40 \times 1.0 = 120$ tons would have to be overcome. This is equivalent to 1.2 tons/sq ft. In addition, the bearing capacity of the soft clay beneath the footing would have to be exceeded. This is equal to about 0.9 ton/sq ft. The ultimate bearing capacity of the large footing cannot exceed the sum of these two components, or 2.1 tons/sq ft, which is much less than that of the test plate. Indeed, the ultimate capacity of the large footing is likely to be less than 2.1 tons/sq ft because the maximum strengths of the stiff and soft clays are not likely to develop simultaneously.

With respect to settlement, the discrepancy is even more striking. At a soil pressure of 1 ton/sq ft, the stress at the top of the soft layer directly beneath the test plate is approximately 0.05 ton/sq ft, and the settlement of the underlying soft clay is practically zero. On the other hand, the stress at the top of the soft clay beneath the center of the large footing is approximately 0.88 ton/sq ft. It decreases with depth, as indicated in the figure. The settlement induced by the stress in the soft layer is about 10 in. Hence, as demonstrated by this example, the settlement of a structure may be excessive even if practically no settlement

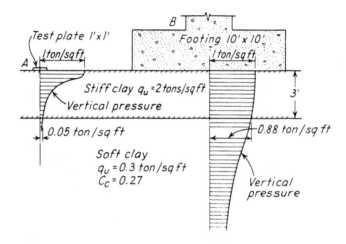

FIGURE 21.2. Diagram illustrating reasons for difference in behavior of test plate 1 ft square and footing 10 ft square if both rest on upper surface of stiff clay crust underlain by soft clay.

occurs during a load test in which the same soil pressure is used that will exist beneath the structure.

Several serious accidents and many instances of excessive settlement have occurred as a result of selecting the allowable soil pressure on the basis of load tests on a stiff crust. If the load-test method is used, it is essential to learn whether the strength of the soil decreases with depth. If it does, load tests must be performed at such levels that the capacity of the softest layers may be investigated. It is generally preferable to determine the safe load on intact clay soils by calculations based on the results of unconfined compression or undrained shear tests.

ILLUSTRATIVE DESIGN. DP 21-1. FOOTINGS ON SAND ABOVE CLAY LAYER

This design plate illustrates the computations necessary to proportion the footings in accordance with the properties of the sand immediately beneath the foundation, and to predict the settlements resulting from the consolidation of the clay that underlies the deposit of sand.

The computations on Sh. 2 are for the most part similar to those in DP 19–1. Likewise, those on Shs. 3 and 4 of this design plate duplicate many of the computations given in DP 18–4. Therefore, a detailed explanation of the computations is unnecessary. However, it should be noted that two sketches of the foundation plan (Sh. 3) are required for computing the increase in pressure at the middle of the clay layer, because the excavated soil will be removed from the entire area of the building, whereas the footings will deliver their loads to the soil beneath six separate areas.

21.4. Alternating Soft and Stiff Layers

If the deposit contains a number of weak layers, bearing-capacity and settlement computations may be made for each. If the structure cannot be supported on footings near the surface of the ground, piles or piers may be used to transmit the loads to one of the firm strata at sufficient depth to provide a satisfactory foundation. This depth can be determined on the basis of the results of the computations. The choice between piles and piers, or of the type of pile to be used, is likely to depend on the difficulty that may be experienced in driving through the firm layers above the bearing stratum. The depth to which piles can be driven in such a deposit can seldom be predicted with accuracy, and any conclusions must be con-

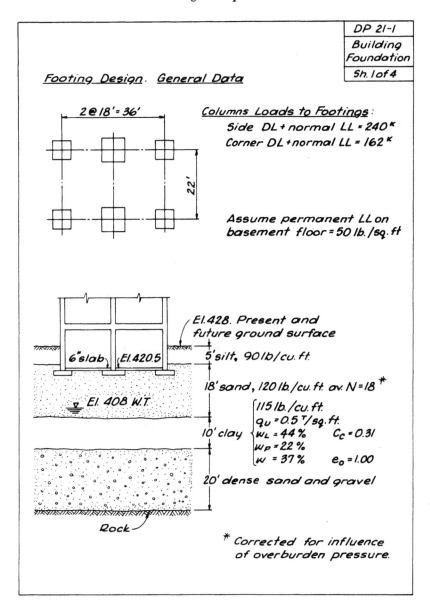

sidered tentative until test piles have been driven.

Excavation to compensate for part or all of the weight of the structure may permit construction of a raft. This alternative should be considered along with the use of piles or piers.

21.5. Irregular Deposits

If the subsoil consists of lenticular or wedge-shaped masses, it is rarely possible to make an accurate estimate of bearing capacity or settlement. In such instances, it is advisable to determine the general character of the deposit by means of numerous subsurface soundings supplemented by a few borings and soil tests. The purpose is to form an opinion regarding the size and distribution of the softer elements of the deposit and to judge the most unfavorable combination of elements that can reasonably be expected. The estimate of settlement should be based on the assumption that the most unfavorable conditions may occur

---

DP 21-1
Building Foundation
Sh. 2 of 4

Footing Design. Required Areas

Side Footings:
  $240^K = 120^T$
   @ say $2 = 60$ sq. ft.
  Try $8'\text{-}0'' \times 8'\text{-}0''$. $A = 64$ sq. ft. Assume depth $= 2'\text{-}0''$
   Column load $= 120^T$ $D_f/B = 0.25$
           $\div 64$ $= 1.88$ T/sq. ft.
   Additional load $2.0\left(\dfrac{0.15 - 0.12}{2}\right) = \dfrac{0.03}{1.91}$ T/sq. ft.
   minus surcharge
   Fig. 19.3, allow 2.0 T/sq. ft.
  Use $8'\text{-}0'' \times 8'\text{-}0''$ for side footings

Corner Footings:
  $162^K = 81^T$
   @ $1.91 = 42.4$ sq. ft.
  Try $6'\text{-}6'' \times 6'\text{-}6''$. $A = 42.3$ sq. ft. Assume depth $= 2'\text{-}0''$
   Column load $= 81^T$ $D_f/B = 0.31$
           $\div 42.3 = 1.92$ T/sq. ft.
   Additional load $= \dfrac{0.03}{1.95}$ T/sq. ft.
   minus surcharge
   Fig. 19.3, allow 2.0 T/sq. ft.
  Use $6'\text{-}6'' \times 6'\text{-}6''$ for corner footings

---

in the most highly stressed portion of the subsoil.

21.6. Excavation and Stability of Slopes in Nonuniform Soils

Conventional Method of Slices. The procedure generally used for estimating the factor of safety of slopes excavated into stratified or nonuniform soils is similar in principle to those described in Art. 18.7 for slopes in homogeneous clays. The surface of sliding is commonly assumed to have a circular shape. However, the sliding mass is divided into a series of vertical slices (Fig. 21.3) in such a manner that the lower boundary of any one slice, such as *de*, is located entirely within a single stratum or lens of soil with values of c and ϕ that may be considered constant. For convenience, vertical boundaries between slices are also established at breaks in the slope, such as points k and m. The sliding mass may then be subdivided by additional vertical boundaries in such a manner as to make the widths of slices as

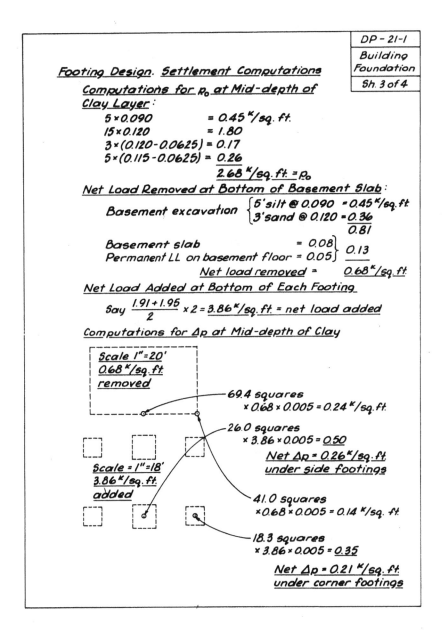

nearly constant as the geometry of the problem will permit. Sufficient accuracy is usually attained with 8 to 15 slices.

Each slice is considered as a free body. The calculations are greatly simplified without serious error if the influence of the forces acting on the vertical sides of the slices is disregarded. Under these conditions the only force considered to be acting above the base de of a slice such as $hdeg$ is its weight ΔW. The driving moment of this slice about O is equal to $\Delta W l_w$. The total driving moment of all slices is $\Sigma \Delta W l_w$ where moments of slices to the left of O are given negative signs.

It is sometimes more convenient to determine the driving moment by resolving the weight ΔW, at the intersection of its line of action with the arc de, into a normal component ΔN and a tangential component ΔT (Fig. 21.3b). These forces may be readily determined from the weight of the slice and functions of the angle α. The line of action of ΔN passes through O; hence,

	DP-21-1
Footing Design. Settlement Computations	Building Foundation Sh 4 of 4

Side footing: $\dfrac{10 \times 12}{1 + 1.00} \times 0.31 \times \log \dfrac{2.68 + 0.26}{2.68} = \underline{0.74''}$

Corner footing: $\dfrac{10 \times 12}{1 + 1.00} \times 0.31 \times \log \dfrac{2.68 + 0.21}{2.68} = \underline{0.60''}$

ΔN has no tendency to produce motion along the arc. The tangential force ΔT, however, has a moment $r\,\Delta T$ tending to cause rotation, and the total driving moment for all the slices is $r\Sigma \Delta T$. In the summation, values of ΔT for slices to the left of the center of rotation have a negative sign, as they tend to oppose rotation.

The forces available to resist motion along de are shown in Fig. 21.3c. If the shearing strength of the soil is expressed by Coulomb's equation

$$s = c + p \tan \phi \qquad 4.2$$

the normal force ΔN creates a frictional force $\Delta F = \Delta N \tan \phi$, which always acts in a direction to oppose the motion. If motion is imminent, the resultant ΔR of the normal and frictional forces is inclined at ϕ to the direction of ΔN. In addition, if the layer possesses cohesion, the sum of the cohesive forces acting along the arc de is $c\hat{l}_{de}$, where $\hat{l}_{de}$ is the curved length of the arc. The cohesive forces also always act in a direction

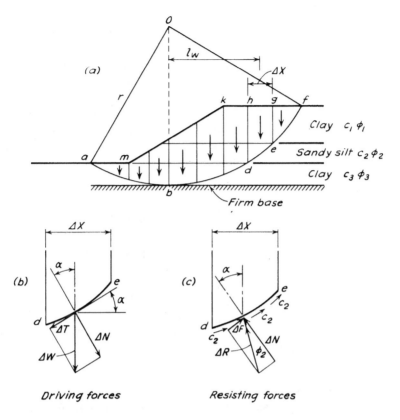

FIGURE 21.3. Conventional method of slices in stratified soil. (*a*) Assumed circular surface of sliding and subdivision of sliding mass into slices. (*b*) Driving forces at base of slice *hdeg*. (*c*) Resisting forces at base of same slice.

to oppose the motion. The moment of the resisting forces available along *de* is then $r(c\hat{l}_{de} + \Delta N \tan \phi)$ and the total ultimate moment is

$$r\Sigma(c\hat{l} + \Delta N \tan \phi)$$

The factor of safety of the entire slope against sliding is then

$$F = \frac{\Sigma(c\hat{l} + \Delta N \tan \phi)}{\Sigma \Delta T} \qquad 21.1$$

The failure may not, of course, take place along the arbitrarily selected arc; it will occur along the arc for which the factor of safety is a minimum. Therefore, various locations for the center *O* and various radii must be selected until the minimum factor of safety is found. This is the calculated factor of safety of the slope.

If one or more layers or lenses intersected by the surface of sliding are in cohesionless sand, the value of *c* along that part of the surface of sliding is taken as zero. Conversely, if plastic clays occur under undrained conditions for which the $\phi = 0$ analysis is applicable (Art. 18.7), *c* may be taken as half the unconfined compressive strength and ϕ equal to zero. Moreover, it should be noted that when ϕ equals zero for the entire arc of failure, eq. 21.1 is equal to eq. 18.12, because $\Sigma \Delta T = Wl_w/r$.

If a section of the surface of sliding, such as *de*, is acted on by a water pressure of average intensity *u*, the analysis is unchanged except that the total normal force ΔN must be reduced by the resultant water pressure before multiplication by $\tan \phi$ to obtain the available frictional resistance ΔF. That is,

$$\Delta F = (\Delta N - u\hat{l}_{de}) \tan \phi \qquad 21.2$$

Modifications of the Method of Slices. Except for moment equilibrium about point O, the conventional method of slices discussed previously does not generally satisfy other criteria of equilibrium. Although, at any factor of safety equal to or greater than unity, each individual slice such as $hdeg$ (Fig. 21.1a) is in equilibrium in a direction normal to the surface de, the tangential forces are not equal and opposite unless $F = 1$. That is, when the slice is not on the verge of failure by sliding along de, the force ΔT tending to cause rotation is less than the resisting force $\tilde{c}l_{de} + \Delta N \tan \phi$. This and other shortcomings, such as disregarding the side forces, have led to the development of numerous modifications of the conventional method of slices (for example, Bishop 1955). However, the properties of the soil are not known with sufficient accuracy in many instances to warrant the refinement of most of the modified procedures.

ILLUSTRATIVE DESIGN DP 21-2. SLOPE STABILITY IN NONUNIFORM SOIL

Computations are given in this design plate to illustrate the conventional method of slices. The factor of safety is determined by means of eq. 21.1. It should be noted that the driving and resisting forces for any given slice are obtained by assuming that the arc on the surface of sliding can be replaced by the chord. Moreover, except for triangular-shaped slices, the centers of gravity are assumed to lie on the vertical centerlines of the slices, and the angles α are determined accordingly. Such assumptions eliminate the tedious procedures of determining centers of gravity and arc lengths and usually furnish sufficiently accurate results. The degree of accuracy may, of course, be improved by increasing the number of slices.

Many electronic computer programs have been developed for the conventional method of slices. Some are capable of accounting for the effects of various conditions of seepage and porewater pressures. Other programs are available for modified methods and for other shapes of the surface of failure (Morgenstern and Price, 1965). Most computer programs are arranged to search numerous circles in order to determine the minimum value of F. The surface selected for analysis in design plate DP 21-2 is very nearly the critical circle for which the factor of safety is a minimum.

PROBLEM

A bed of clay consists of three horizontal strata, each 15 ft thick. The values of c for the upper, middle, and lower strata are, respectively, 600, 400, and 3000 lb/sq ft. The unit weight is 115 lb/cu ft. A cut is excavated with side slopes of 1 (vertical) to 3 (horizontal) to a depth of 20 ft. What is the factor of safety of the slope against failure?

Ans. 1.2.

SUGGESTED READING

Theoretical studies of the bearing capacity of layered clays of different strengths are reported in:

S. J. Button (1953), "The Bearing Capacity of Footings on a Two-Layer Cohesive Subsoil," *Proc. 3 Int. Conf. Soil Mech.*, Zurich, *1*, 332–335.

A. S. Reddy (1967), "Bearing Capacity of Footings on Layered Clays," *ASCE J. Soil Mech. 93*, SM2, 83–99. Shear strength in each layer may vary linearly with depth; strength of soil may differ in horizontal and vertical directions.

Relatively few published accounts describe the design and behavior of foundations on nonuniform soils. Among them are:

E. D. Carlson and S. P. Fricano (1961), "Tank Foundations in Eastern Venezuela," *ASCE J. Soil Mech. 87*, SM5, 69–90.

J. J. Hallenbeck, Jr., and R. E. Johnston (1967), "Pile Foundation for Oakland Coliseum," *Civ. Eng. ASCE, 37*, 1, 57–61.

R. Lennertz (1972), "Settlement of Footings on a Non-Uniform Foundation." *Proc.*

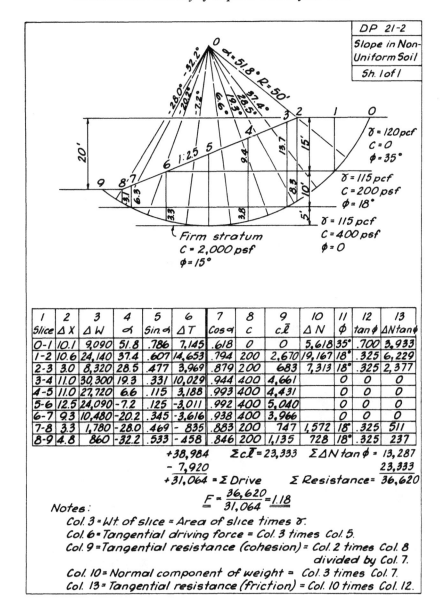

ASCE Conf. on Performance of Earth and Earth-Supported Structures, Purdue, 1, Part 2, 929–938. Development of exploration and design for foundations of apartment building on complex group of clay deposits in Cincinnati.

L. D. Wheeless and G. F. Sowers (1972), "Mat Foundation and Preload Fill, VA Hospital, Tampa." Proc. ASCE Conf. on Performance of Earth and Earth-Supported Structures, Purdue, 1, Part 2, 939–951. Extremely nonuniform subsurface conditions including fairly loose sands, clayey sands, and clays extending deeply into chimney-like holes in limestone containing many solution features.

World Trade Center

The twin steel towers, rising to a record height of 1350 ft or 110 stories, rest on the Manhattan schist at a depth of 70 ft below street level. Estimates of the deflection of the towers under wind forces required evaluation of the modulus of elasticity of the supporting bedrock by means of large scale in-situ testing. Excavation of the entire 16-acre site occupied by the Center involved removal of 16 million cu yd of fill, soil, and rock; continual support of two operating subway tubes; and protection of the surrounding property by means of a slurry wall, tied back with inclined anchors into the rock, and later utilized as the exterior basement wall. (Photo courtesy of Port Authority of New York and New Jersey.)

PLATE 22.

CHAPTER 22

Foundations on Rock

22.1. Basis for Design

Since most unweathered, intact rocks are stronger and less compressible than concrete, the determination of suitable bearing areas or allowable pressures on such materials is a routine matter or may even be unnecessary. However, intact rock masses without weathered zones, joints, or other defects are encountered only rarely. The existence or the location of specific defects often remains unknown until the rock is exposed by excavation or until unexpected behavior of a pile or pier occurs during installation or in a load test. Because uncertainties of this sort are inherent in rock foundations, the principal functions of the designer are to choose a type of foundation that can be adapted or modified to suit the various conditions most likely to be encountered, and to select the appropriate modifications when the actual conditions become known.

22.2. Foundations on Unweathered Rock

In many localities where bedrock can be reached by excavation, the allowable contact pressure is specified by the building code. Current codes, however, differ considerably in their recommendations. Excerpts from a few such documents are condensed in Table 22.1.

The allowable contact pressure on unweathered rock should be based on the inherent strength of the intact rock and on the influence of such defects as joints, shear zones, and solution features. If the rock mass contains virtually no defects, the allowable contact pressure at the surface of the rock may be taken conservatively as the unconfined compressive strength of the intact rock. Most otherwise intact rocks, however, contain one or more sets of joints that may drastically increase the compressibility of the rock mass. Unless the strength of the intact rock is extremely low, roughly equal to or less than that of plain concrete, the allowable contact pressure beneath foundations is governed exclusively by the settlement associated with the defects in the rock, and not by strength. The compressibility is closely related to the spacing and direction of the joints, whether they are tight or open, and whether they are filled or coated with softer materials. If the joints are tight or are not wider than a fraction of an inch, the compressibility is reflected by the RQD (Art. 5.3). Table 22.2 presents values of the allowable contact pressure for jointed rocks on the basis of their RQD. If

Table 22.1 Allowable Pressures on Rock (Tons/Sq Ft) Abstracts from Various Building Codes[a]

Material	Code[b]			
	A	B	C	D
Massive crystalline bedrock including granite, diorite, gneiss, traprock, hard limestone, and dolomite	100	100	$0.2q_u$[c]	10
Foliated rocks such as schist or slate in sound condition	40	40	$0.2q_u$	4
Bedded limestone in sound condition	40	15	$0.2q_u$	4
Sedimentary rock, including hard shales and sandstones	25	15	$0.2q_u$	3
Soft or broken bedrock (excluding shale), and soft limestone	10		$0.2q_u$	
Soft shale	4		$0.2q_u$	

Note: The New York City (1970) code refers specifically to the geological formations in the locality and to their condition. For example, 60 tons/sq ft are allowed on *hard sound rock*, defined as follows: "Includes crystalline rocks such as Fordham gneiss, Ravenswood gneiss, Palisades diabase, Manhattan schist. Characteristics are: the rock rings when struck with a pick or bar; does not disintegrate after exposure to air or water; breaks with sharp fresh fracture; cracks are unweathered and less than $\frac{3}{8}$ in. wide, generally no closer than 3 ft apart; core recovery with a double tube, diamond core barrel is generally 86 per cent or greater for each 5 ft run." Such provisions, based on local experience and conditions, represent excellent practice.

[a] Values do not include increases allowed for embedment.
[b] A = BOCA (1968); B = National Building Code (1967); C = Uniform Building Code (1964); D = Los Angeles (1959).
[c] q_u = unconfined compressive strength.

the design is based on these values, the settlement of the foundation should not exceed 0.5 in., even for large loaded areas.

The RQD for use in Table 22.2 should be the average within a depth below foundation level equal to the width of the foundation, provided the RQD is fairly uniform within that depth. If the upper part of the rock, within a depth of about $B/4$, is of lower quality, the value for this part should be used or the inferior rock should be removed. Since the values in Table 22.2 are based on limiting the settlement, they should not be increased if the foundation is embedded into the rock. Although some building codes arbitrarily allow substantial increases in contact pressure if a pier is drilled into the rock two or three diameters, or allow increases attributed to the development of side friction between the embedded portions of the piers and the rock, such allowances are usually based on the incorrect premise that the capacity of the piers is governed by the bearing capacity rather than the compressibility of the rock.

Table 22.2 Allowable Contact Pressure q_a on Jointed Rock

RQD	q_a[a] (tons/sq ft)	q_a[a] (lb/sq in)
100	300	4170
90	200	2780
75	120	1660
50	65	970
25	30	410
0	10	140

[a] If tabulated value of q_a exceeds unconfined compressive strength q_u of intact samples of the rock, as it might in the case of some clay shales, for instance, take $q_a = q_u$.

Foundations on Unweathered Rock 363

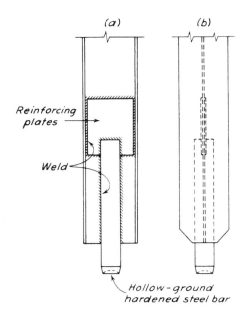

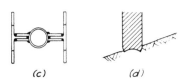

FIGURE 22.1. Oslo points for H-piles to be driven through soft materials to hard rock with sloping surface. (*a*) and (*b*) Details of point. (*c*) Bottom view. (*d*) Sketch showing embedment of point in rock to resist sliding (after Bjerrum, 1957).

A sloping rock surface introduces complications in both design and construction of most types of foundations. Particularly serious difficulties may be encountered in founding a caisson for a bridge pier. Tilting can hardly be avoided if the caisson is advanced by open dredging; the need to use compressed air to complete the caisson may justify the choice of a different type of foundation. In a few localities the bedrock is very hard and its surface extremely irregular; such conditions occur, for example, where steep-walled canyons have been filled with soft sediments, as in Oslo or Gothenberg, or where residual soils have developed over limestone to form a pinnacled rock surface (Fig. 6.32). Piles driven to the rock may slide along the surface, become bent, and fail to develop adequate

capacity. Specially sharpened and hardened steel points for H-piles (Fig. 22.1) have been developed to cope with the conditions in Oslo.

Some poorly indurated shales and siltstones slake and deteriorate rapidly upon exposure. Final excavation in such rocks should be deferred until just before concreting, or the exposed surfaces should be protected; asphalt sprays or pneumatically placed concrete have been used successfully for short-time protection.

Excavation to the rock surface is avoided by the use of piles. If the bedrock is hard, high stresses (Art. 12.5) may be developed in the piles during driving and the tips of the piles may be damaged. Timber piles are especially vulnerable to overdriving; they may be broken or their points broomed.

ILLUSTRATIVE DESIGN. DP 22-1. PIER ON ROCK

Use of the RQD in proportioning piers or footings on rock is illustrated in DP 22-1. The rock is found from NX core samples to be a flat-lying sedimentary sequence of limestones and shales. The overlying soil is a glacial deposit. Since the rock shows little evidence of weathering even at its surface, it is presumed that the weathered upper part of the rock has been removed by glacial action.

The required size of pier decreases with depth because of the increase in rock quality. Choice of a footing 6 ft square at El. 490 or 4 ft square at El. 485 would ordinarily be determined on the basis of economics. The cost of additional rock excavation would probably exceed the savings in cost of a smaller quantity of concrete in the pier. Therefore, the larger pier at the higher elevation is selected as the better design. Moreover, the load will be transferred from the column to the pier through a steel base plate. If the plate has roughly the same plan dimensions as the pier and if the bearing pressure on the concrete is limited to $0.375 f_c'$, the required values of f_c' for the piers at Els. 490 and 485 are $800/0.375 = 2100$ lb/sq in. and $1900/0.375 = 5000$ lb sq in.,

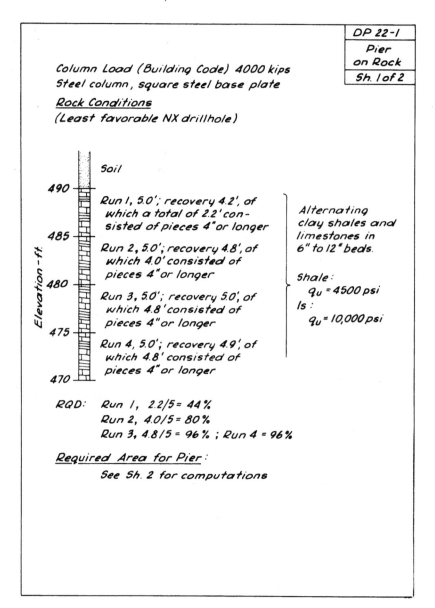

respectively. Therefore, high-strength concrete will be required for the smaller pier; such concrete is usually avoided under the unfavorable conditions likely to prevail for placement of concrete in foundations pits.

It will be observed that the pressures allowed by many building codes, such as those listed in Table 22.1, are less than those determined in DP 22–1. The code values, even after allowances are made for the depth of embedment, are generally overconservative. Some codes permit higher values than those tabulated if an adequate investigation is made, but where no such provision is included the code values may govern the design.

22.3. Treatment of Rock Defects

More or less vertical joints from one to several inches wide are sometimes encountered even in unweathered rocks. They may be either open or clay-filled. Beneath footing foundations such joints are usually of no consequence. They may be cleaned out

```
                                                    DP 22-1
                                                    Pier
      Required Area for Pier                        on Rock
      Pier at surface of Rock:                      Sh 2 of 2
         Assume RQD = 44% for depth ≩ B/4
         Total Load = 4000ᴷ @ say 800 psi *
         4000/(0.8×144) = 34.7 ft² = 5.9² ft², say 6'×6'
      Pier at El. 485:
         Assume RQD = 80% for depth ≩ B/4
                   4000ᴷ @ say 1900 psi *
         4000/(1.9×144) = 14.6 ft² = 3.8² ft², say 4'×4'
         Check qᵤ of shale:
                   1900 < 4500 ok

              Use 6'×6' pier at surface of rock


              * Allowable pressures are
                by interpolation from Table 22.2.
```

to a depth of four or five times their width and filled with *slush grout*, a mixture of about one part of cement and one part of sand by volume with enough water to permit pouring the grout into the joints. Larger spaces, wider at the top, are likely to occur at intersecting joints. These are commonly filled with *dental concrete*, a stiff mixture of lean concrete placed and shaped by shovel (Fig. 22.2a).

The excavation for a heavily loaded pier may in some instances encounter near-vertical joints so wide that they constitute an appreciable fraction of the area of the base. The excavation is then usually deepened until the joints are no longer within the base or until they narrow to an acceptable width. One drilled pier for the John Hancock Building in Chicago, which the designers planned to found on dolomite at a depth of about 135 ft, actually had to be drilled to 191 ft before a pair of intersecting joints passed out of the base.

Many rock masses contain near-hori-

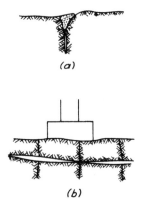

FIGURE 22.2. (a) Dental concrete in joint at surface of rock at foundation level. (b) Horizontal joint responsible for settlement of footing on rock.

zontal joints. Close to the surface, these joints are likely to be open because of the relief of vertical stress due to erosion of formerly overlying rock. If located beneath a footing, as shown in Fig. 22.2b, such joints may lead to uneven and possibly sudden settlement. Where their presence is suspected the rock should be explored to ascertain to what extent and to what depth the joints are open. The investigation is usually carried out by drilling and coring supplemented by the use of pneumatically driven rock drills. The rock above the open joints can sometimes be removed economically and the foundation established at a lower level. Occasionally the spaces can be filled with cement grout. Grouting is usually unsuitable, however, if the joints are filled with clay or other compressible material; attempts to remove the filling by washing are likely to be unsuccessful; use of pressures appreciably greater than the weight of the overburden results either in excessive loss of grout or lifting of the slabs and creation of additional voids. Under these conditions the most suitable foundations may consist of drilled piers established below the open zones.

The significance and appropriate treatment of faults and shear zones can be assessed only upon inspection and possibly further investigation of each specific excavation. The mere presence of such features does not necessarily compromise the suitability of the foundation; although such features are associated with decreased strength and increased compressibility as compared to the surrounding intact rock, their physical properties are often as favorable as those of concrete and far superior to those of strong or stiff soils. On the other hand, in some instances the rock is reduced to fault gouge having the characteristics of soft plastic clay. If the weak materials cannot be avoided or removed, their properties must be determined and the foundations designed or redesigned accordingly. Because of the intense fracturing and crushing, undisturbed samples may be impractical to obtain and, for exceptionally heavy buildings, load tests may be appropriate. At the World Trade Center in New York (Plate 22), for example, the compressibility of the jointed and slightly sheared rock was investigated by the arrangement shown diagrammatically in Fig. 22.3 to permit estimating the deflection of the foundations under dead and wind loads. Expensive tests of this sort are not justifiable for ordinary structures; usually a satisfactory allowance can be made by judgment.

Solution cavities require detailed attention. Since their extent cannot be precisely ascertained (Art. 6.6), a conservative approach is necessary. In bedded limestones, cavities are more likely to occur in some horizons than in others; exploration may be able to establish which horizons are least affected and the foundations carried to a level at which the favorable characteristics are prevalent. In some instances the foundation can be designed to bridge a gap as large as the probable diameter of the largest unsupported area (Fig. 6.31). Beneath heavy and important structures, such as nuclear power plants, the cavities have occasionally been filled with cement grout. Although the effectiveness of the procedure is limited by the presence of clayey materials that cannot usually be washed out successfully, the grouting may, nevertheless, be beneficial because it may provide the slight support necessary to prevent slabs of roof rock above a cavity from becoming detached and starting the formation of a sinkhole;

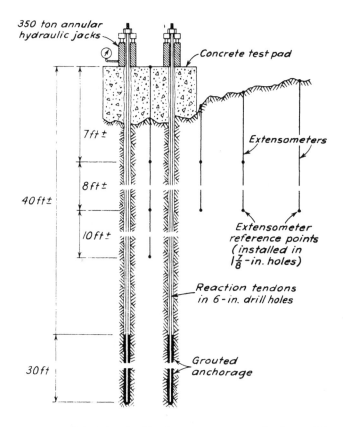

FIGURE 22.3. Simplified sketch of load-test arrangement for determining compressibility of rock, World Trade Center, New York. Concrete test pad is loaded by hydraulic jacks reacting against steel tendons anchored at great depth. Vertical strains are measured between extensometer reference points anchored to rock walls of small drill-holes at suitable locations.

moreover, by filling the voids, the grout eliminates the possibility of a sudden collapse. The grout pressure should not be permitted to exceed that due to the overburden; otherwise the grout merely lifts the layers and creates voids that the costly grout then fills without benefit.

As all rocks with solution features or very open structures are highly pervious, groundwater control is essential where excavation extends below water level. If dewatering is impractical, underwater concrete should be placed only in static water by carefully supervised tremie techniques.

22.4. Foundations on Weathered Rock

The great variety of the physical properties of weathered rock and the nonuniformity of the extent of weathering even at a single site permit few generalizations concerning the design and construction of foundations.

In temperate climates the weathering of indurated sedimentary rocks is most likely to be manifested by the development of soil-like materials along joints and bedding planes. Near the surface the predominant material may be soil in which occasional rock fragments are incorporated; at greater depth the predominant material may be rock surrounded by small amounts of soil. At normal foundation depths, compressible materials may be present along the bedding-plane joints within the zone of influence of the stresses applied to the foundation. The presence, thickness, or physical properties of the soft layers can rarely be determined by sampling and testing, although in some instances these layers may have the characteristics of a normally loaded clay. On the other hand, here and there rock-to-rock contact may exist.

Under these circumstances, the engineer must judge whether the supporting material consists essentially of (1) a matrix of soil in which the rock fragments play a minor role; (2) rock in which the thin seams merely add slightly to the average compressibility; or (3) a material of intermediate character.

If the rock fragments play only a minor role, a footing or raft foundation can be proportioned according to the physical properties of the soil that constitutes the matrix. If support at greater depth is required, piers or piles may be suitable. Heavy-duty equipment (Art. 13.2) such as cable tools may be needed to advance piers to the required depth. Piles must be of types not readily damaged by obstacles (Art. 12.2). They are almost certain to be stopped by blocks of rock underlain by at least some compressible material that may permit settlement of the foundation.

On the other hand, if only thin seams of compressible material are present in a mass consisting primarily of rock, footings on the rock may be entirely adequate. To provide a basis for his judgment, the engineer should estimate the thickness and compressibility of any seams within the depth likely to contribute to the settlement, should calculate the settlement, and should decide whether the settlement is acceptable. Failure by bearing capacity is not likely to occur, but eccentrically loaded individual blocks of rock (Fig. 22.4) may tilt. If this possibility becomes apparent when excavation for a footing exposes the rock, the footing should be enlarged to distribute the load to adjacent blocks. If a shallow foundation on the rock would be unsuitable, piers may be drilled into the rock to such a depth that only acceptable settlements would develop. The increase in cost of such a foundation over one established at higher level may be substantial.

The weathering of less indurated rocks, such as the shales often found in the southwestern part of the United States and in the upper Missouri Valley, often leads to a gradual transition from topsoil, through predominantly clayey or silty soils, to fairly intact shale. The transition continues into unweathered shale generally characterized by a lighter, often gray, color in contrast to the brownish color of the weathered material. The less the degree of weathering, the greater the strength and the less the compressibility. Even in relatively unweathered shale, however, foundations can be designed as if the rock were a heavily overconsolidated clay, according to the applicable procedures of Chaps. 18 and 21. Drilled piers often constitute an expeditious and economical means of establishing foundations at the depth considered necessary to reach adequate support.

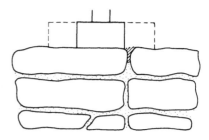

FIGURE 22.4. Block of rock surrounded by compressible material and loaded eccentrically by small footing. Improved support provided by larger footing (dashed lines) extending over several blocks.

Many deeply weathered igneous and metamorphic rocks, such as those often encountered beneath the larger cities of the middle and south Atlantic states, exhibit somewhat erratic variations in strength and compressibility both horizontally and vertically. Drilled piers (Art. 13.3) extending through the overburden and into the rock are widely used. The allowable load on the rock and the depth to suitable material are usually estimated on the basis of cores recovered from borings. Inspection of the bottoms of the shafts has frequently led the engineer to require drilling to a greater depth in the expectation of finding somewhat better rock. In some instances the piers have been carried, at great increase in cost, to depths far greater than anticipated when the estimates of cost were prepared. The difficulty may be minimized by establishing a conservative value of the allowable pressure and by carrying no piers deeper than the estimated depth unless inspection shows the rock conditions to be locally very much less favorable than expected. When these subsurface conditions prevail, the lengths of piles can often be estimated reliably and pile foundations may prove to be suitable and economical.

Chemical weathering under warm, humid conditions may lead to weathered rock with a saprolitic structure (Art. 6.7). Foundations on such materials, which are likely to have the character of collapsible soils, are governed largely by the considerations discussed in Art. 20.2.

Some shales, when exposed, rapidly disintegrate near the fresh surface by desiccation, slaking or swelling. Final excavation in such materials is often deferred to the latest practicable time before concreting. If even this time is too long to prevent the deterioration, the surface of the rock is treated with a bituminous coating or by a few inches of pneumatically placed concrete.

22.5. Excavation in Rock

The removal of rock from basement areas or from space to be occupied by foundations generally requires that the rock first be fragmented by blasting, with the attendant possibility of damage to adjacent properties or previously completed construction. The damage may be caused by dislocation of rock intended to remain in place, by vibrations, and by the impact of pieces of flying rock.

Where space permits, the rock is usually removed to a predetermined depth by drilling, loading, and firing one or more rows of holes to fragment the rock in front of a vertical face (Fig. 22.5a), whereupon the broken rock is excavated. The thickness of rock between the face and the first line of holes is known as the *burden*. The operation is repeated until the entire area has been excavated to the predetermined depth.

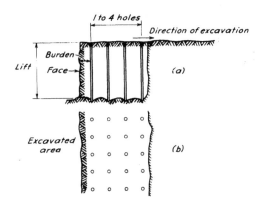

FIGURE 22.5. Typical blasting drill-hole pattern for removing lift of rock in foundation excavation. (a) Section through lift. (b) Plan.

If the excavation is shallow, the operation is carried out in a single lift; deep excavations may require several lifts.

Good practice in foundation blasting usually consists of drilling small holes of 2- to 4-in. diameter at a spacing of 4 to 8 ft (Fig. 22.5b); in contrast, holes of 6 to 8 in. at a spacing of 20 ft or more are customarily used for production blasting in quarries or large road cuts. The holes extend about 2 ft below the proposed bottom of the lift except for the final lift, in which they should terminate 1 or 2 ft above grade to prevent loosening of the foundation rock. Lift thicknesses are usually 12 to 15 ft, and seldom exceed 20 ft. Relatively small holes and close spacing are used in foundation blasting because they result in better fragmentation of the rock and facilitate its removal by excavating machinery. They also reduce the breakage beyond and below the excavation limits.

The holes are usually charged with dynamite at a powder factor ranging from 0.5 to 1.0 lb/cu yd for relatively weak and relatively strong rock, respectively, and the charge is fired with millisecond delays. Alternatively, the charge may be fired with primacord with millisecond-delay connectors. The delays minimize ground motions near the excavation and reduce damage to adjacent property. The shot-holes around the perimeter of the excavation are drilled at a spacing less than that of the interior holes, usually about 2 ft.

Controlled blasting techniques such as line drilling, cushion blasting, and presplitting are available if it is necessary to protect the integrity of the rock just outside the excavation. In *line drilling*, the main excavation is carried up to two or three rows from the boundary and the remaining rock is slabbed off, by the use of delays, a row at a time. The spacing of the shot holes is decreased toward the boundary and the holes are less heavily loaded. The perimeter holes are spaced at two to three diameters and are not loaded; the shock waves break the thin ribs of rock between the holes. In *cushion blasting* the main excavation is carried up as close as possible to the boundary, usually to the last adjacent row of blast holes. The perimeter holes are drilled at a reduced spacing equal to about 0.8 times the burden and are loaded lightly with charges well distributed vertically in each hole; between the charges the holes are filled with sand. The charges in all the holes are fired simultaneously. In *presplitting* or *preshearing* the perimeter holes are drilled at about 1½- to 3-ft centers, and are loaded and fired before any of the adjacent main excavation areas are blasted. By this procedure, a tensile fracture through the perimeter holes is induced, beyond which the effect of the subsequent main excavation is minimized.

The degree of success of the foregoing methods for reducing overbreakage beyond the boundary of the excavation depends to a great extent on the quality of the rock, its uniformity, and the nature and orientation of its jointing. Specification of the most suitable procedures requires detailed knowledge of the geological features of the site, as well as mature experience.

In some localities or under some circumstances, blasting may not be permitted. More expensive methods involving wedging may be necessary. Holes are drilled at about 2-ft centers, and cracks are produced between the holes by means of expanding hydraulic jacks inside adjacent holes or by feather-wedges driven into the holes with a jackhammer. Pneumatic hole punchers or rock breakers mounted on backhoes or tractors may sometimes be suitable.

The necessity for supporting an excavated face in rock depends on the height of the face and on the nature and orientation of such features as bedding planes, joints, shear zones, faults, or schistosity. Planar defects of these types may be so oriented as to form the boundaries of wedges that can descend as units into the excavation. Under these circumstances, rock bolts or cable tendons may be required. The lengths, spacing, and directions of the bolts can be determined rationally only after the details of the rock structure are determined. In some instances, the thickness of the excavation lifts may have to be limited to permit installation of bolts before the depth of excavation becomes too great for stability.

PROBLEMS

1. The unweathered quartzite below excavation level for a multistory building has an RQD of 30 for the upper 5 ft and 70 for the next 20 ft. A load of 1500 tons is delivered to the rock through a square reinforced concrete pedestal. What would be the size of the pedestal to restrict the settlement to about 0.5 in.? *Ans.* 6.4 x 6.4 ft.

2. If the upper 5 ft of rock were removed in the excavation described in Prob. 1, what would be the size of the pedestal? *Ans.* 3.7 x 3.7 ft.

3. A shale with an RQD of 90 and an unconfined compressive strength of 50 tons/sq ft is to support a column load of 1000 tons transmitted to the rock through a circular pier drilled a few feet into the shale. What diameter of shaft is required if the settlement is not to exceed 0.5 in.? *Ans.* 5.0 ft.

SUGGESTED READING

A discussion of the geological conditions influencing the properties of rock as a foundation material, including a review of the RQD as a measure of rock quality and data concerning the compressibility of rock, is found in *Rock Mechanics in Engineering Practice*, edited by Stagg and Zienkiewicz, London, Wiley, 1968, Chap. 1 by D. U. Deere, "Geologic Considerations," pp. 1-20. The original paper on RQD is D. U. Deere (1963), "Technical Description of Rock Cores for Engineering Purposes," *Felsmechanik und Ingenieur Geologie*, *1*, 1, 16-22.

Details of blasting for rock excavation are extensively treated in *Blasters' Handbook*, E. I. du Pont de Nemours and Co., Inc., Wilmington, Delaware (15th ed., 1969).

Examples of foundation design and construction in rock include:

W. F. Swiger and H. M. Estes (1959), "Major Power Station Foundation in Broken Limestone," *ASCE J. Soil Mech.*, *85*, SM5, 77-86. Discussion by R. B. Peck (1960), Vol. 86, SM1, 95-98. Different foundation types for adjacent units of power plant.

R. E. White (1943), "Heavy Foundations Drilled into Rock," *Civ. Eng. ASCE*, *13*, 1, 19-22.

L. R. Squier (1970), "Mat Foundation Spans Rubble Channel," *Civ. Eng. ASCE*, *40*, 8, 61-62.

PART D

Design of Foundations and Earth-Retaining Structures

The final step in the design of a foundation is the structural analysis and design of its various parts.

Most modern substructures consist of reinforced-concrete elements designed in accordance with the same fundamental principles that apply to reinforced-concrete members in general. Hence, in Part D it is presumed that the reader already possesses a thorough knowledge of statics, strength of materials, and principles of reinforced-concrete design. Only those aspects of design peculiar to foundation engineering will be discussed.

In many instances, at least preliminary structural designs of a foundation must be prepared to serve as a basis for estimates of cost before the most suitable type of foundation can be selected. Therefore, the information in Part D may often be required before all the steps outlined in Part C can be carried out.

The designer is likely to have the impression that those aspects of foundation engineering dealing with soils are based on highly empirical procedures, whereas he may regard the design of reinforced-concrete members as having a more satisfactory theoretical basis. Yet, more mature reflection on the origin of current procedures for analyzing and designing concrete members leads to the conclusion that modern design codes have comparatively little basis in theory and have been derived to a considerable degree from the results of tests in the laboratory and in the field. Without proper appreciation of this fact, the designer is likely to lack a sense of proportion and to devote relatively too little consideration to the characteristics of the soil deposits at a site.

Arthur N. Talbot (1857–1942)

Teacher and research worker associated for over 50 years with the University of Illinois and influential in the formation and development of its Engineering Experiment Station. Beginning in 1903 he directed a comprehensive investigation of reinforced concrete in all its structural forms. His investigation of footings, from 1908 to 1912, furnished the basis for procedures of design and represented the only extensive experimental work on this subject prior to 1944. (Photo courtesy of Mrs. Warren G. Goodell.)

PLATE 23.

CHAPTER 23

Individual Column and Wall Footings

23.1. Basis for Design Procedure

The design of ordinary concentrically loaded footings is based on the assumption that the soil pressure against the bottom of the footing is uniformly distributed. There is much indirect evidence that this is a satisfactory and generally conservative assumption, although there have been few if any reliable field observations of the actual distribution of contact pressures.

The rules for design of footings, like those for all types of concrete members, have a primarily empirical basis. Current methods originated with an extensive series of tests made by A. N. Talbot and reported in 1913. Many of Talbot's findings are still reflected in the present code of the American Concrete Institute. The principal changes since 1913 take account of the increased strength of concrete and reinforcing steel. For example, comprehensive investigations of footing behavior carried out by F. E. Richart at the University of Illinois in 1944 demonstrated that the allowable bond stress could be increased for reinforcing bars with more effective types of deformations. This determination and other observations from the Richart tests led to several revisions in the ACI Building Code of 1951; major revisions were also made in 1963 and 1971.

Since that time, a systematic procedure of updating has been in effect.

Procedures for the analysis and design of concrete and steel structures evolved during the 1960s from those based on elastic behavior (working-stress method) to those based on plastic behavior (strength method). In the working-stress method, the stresses are computed for the loads that can reasonably be expected to act on the structure, and these stresses are compared with allowable stresses determined by applying a factor of safety to the ultimate strength. On the other hand, in strength design, the margin of safety is introduced by multiplying the loads by *load factors*, and the forces, moments, and shears induced in the members by the factored loads are compared with the ultimate strength of the members. The ultimate strengths may additionally be modified by *capacity factors* that depend on various considerations, such as workmanship.

The selection of loads and factors of safety applicable to foundations has been discussed in Art. 17.3. Further references to this subject occur in other chapters of Part C. The foundation engineer must deal with stress–strain characteristics and ultimate strengths of soil and rock irrespective of the method of design. In some instances he may be concerned with safe pressures or allowable pile loads under working loads specified by a building code; at other times, or perhaps even on the same project, he may need to determine factors of safety against ultimate strengths for exceptional overloads.

In most instances when strength procedures are used for the structural design of footings and pile caps, it is necessary first to proportion the foundation or to determine the number or arrangement of piles, in accordance with procedures described in Part C, for working loads without application of load factors. Thereafter, the load factors must be applied and the soil pressures and pile reactions produced by the factored loads must be calculated and used for analysis.

Design plates throughout Part D illustrate basic principles of the structural design of certain types of foundations. The strength method of design is used in most of the illustrations.

23.2. Critical Sections

The procedure for designing footings, like that for other reinforced-concrete members, is related to the manner in which the member may fail. Experiments have indicated that several different types of structural behavior may be exhibited before the ultimate strength of a footing is reached. At some strength less than the ultimate, the deformations and cracks in the footing are likely to be so great that the footing is no longer useful. This is referred to as the stage of primary failure. The nature of the primary failure is sometimes difficult to determine because of the interdependence of various factors.

If inadequate tensile reinforcement has been provided at the bottom of the footing, the primary failure is evidenced by excessive yielding of the bars. On the other hand, if the development length of the reinforcement is inadequate, the primary failure may take the form of slipping of the bars. In both these instances, the ultimate failure is usually one of shear around the loaded area.

If the primary failure of a square column footing occurs as the result of shear, a section of concrete, flaring out downward from the column, is punched from the rest of the footing. Cracks form initially on the under side of the footing while the stresses in the steel are still low. As the shear increases, the cracks propagate upward toward the edges of the column, and the footing fails in diagonal tension before the average stress in the tensile steel reaches the yield point. If the shape of the footing is rectangular, and the ratio of length to width is large, the crack may extend across the full width of the footing.

The preceding discussion of the modes of failure of a footing indicates that the design or stress analysis may appropriately be based on the moment and shear at certain critical sections where failure due to excessive strains in flexure or diagonal tension may be initiated. The critical section for shear is commonly located at a distance $d/2$ from the face of the column, pedestal, or wall, as shown in Fig. 23.1. Hence the length of the section de is equal to the depth of the footing plus the width of the column. The total shear used in computing the unit stress for a square footing is the sum of the forces on area $cdef$ in Fig. 23.1a. If the footing is rectangular, the total shear on $cdefgh$ is used for critical section de (Fig. 23.1b); the shear stress on $d'e'$ must also be checked for the total shear on $d'e'gh$.

If the part of the structure resting on the footing proper consists of reinforced concrete, the critical section for flexure and for development length is ordinarily assumed to extend across the footing at the face of the column, pedestal, or wall, as shown by ab and jk in Fig. 23.1. The stresses in the concrete and steel are computed from the forces and moments acting on this section. For footings under masonry walls, it is common practice to investigate flexural stresses at a section under the wall; usually the quarter point is chosen. If the column load is transferred to the footing through a base plate, it is reasonable, because of the flexibility of

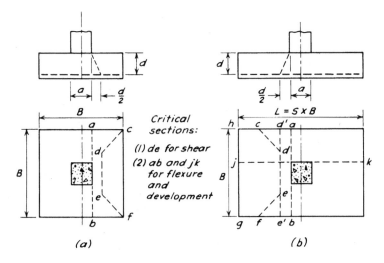

FIGURE 23.1. Critical sections for shear, flexure, and development of reinforcement in (a) square and (b) rectangular footings.

the plate, to use a section midway between the edge of the plate and the face of the column.

The foregoing comments concerning critical sections have made it apparent that the stresses computed in an analysis of a spread footing are only average values found from total moments or shears acting on sections many feet in extent. It is impossible to predict, under field conditions, the actual variation of these stresses along any section of the footing. It is fortunate, therefore, that footings, like other indeterminate structures, have considerable ability to redistribute the moments and shears before failure occurs at the sections that may be overstressed.

23.3. Placing of Reinforcement

In a wall footing the main reinforcement is placed at right angles to the wall and should be spaced uniformly. Furthermore, some longitudinal reinforcement is desirable to assist in bridging soft spots associated with the almost certain variation in soil conditions along the length of the footing. A steel percentage of 0.2 to 0.3 is adequate for this purpose in all except unusual cases.

The reinforcement in square footings is usually placed in directions parallel to the edges. In each direction the bars are commonly spaced uniformly. Likewise, the reinforcement in the long direction of rectangular footings is usually distributed uniformly across the full width. On the other hand, tests on rectangular footings have demonstrated that the bars in the short direction should be spaced more closely near the center than at the outside. The following equation is often used for determining the distribution of the steel:

$$n_b = \left(\frac{2}{S+1}\right) n_t \qquad 23.1$$

in which n_b = number of bars placed in a central width equal to the short dimension of the footing

n_t = total number of bars required by moment at the critical section

S = ratio of long side to short side of the footing

Special care is required to insure good concrete at the bottom of footings, particularly if the bottom of the excavation is below the water table. The excavation should be dry when the concrete is placed. Moreover, the reinforcement should be protected by at least 3 in. of concrete.

23.4. Depth of Spread Footings

It is evident that the structural design of footings is largely a matter of trial. Revisions may be necessary as the design progresses. Usually the depth determined by shear is adequate to satisfy the requirements of flexure. If shear on the critical section de (Fig. 23.1) is assumed to control the depth of a square footing, the following equation is useful in selecting the depth d:

$$\frac{d}{B} = \frac{\sqrt{4C^2+4C+k^2(1+4C+4C^2)}-k(1+2C)}{2(1+C)} \quad 23.2$$

The constant C is evaluated from the net soil pressure q_n (lb/sq ft) and the shear stress v_c (lb/sq in.), for strength design, by the expression $C = q_n/576v_c$. In this expression v_c is adjusted by the appropriate capacity factor. The net soil pressure q_n, effective in producing shear and moment in the footing, is the factored column load divided by the footing area. The value of k is the ratio of the least column or pedestal width a to the footing width B; thus $k = a/B$.

23.5. Procedure for Design and Use of Minimum-Depth Curves

The curves of Fig. 23.2 have been developed from eq. 23.2 to give the designer a

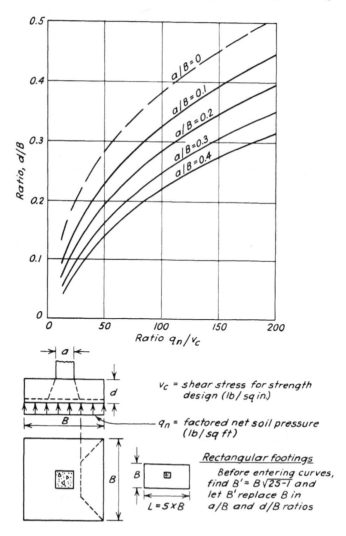

FIGURE 23.2. Curves for selecting depth of footing as determined by shear, provided two-way behavior prevails.

rapid indication of the minimum depth of footing. If a depth equal to or greater than that indicated by Fig. 23.2 is used, the shear stress corresponding to the factored column load will not exceed the ultimate value. Although the curves have been constructed for square footings, they can be utilized equally well for rectangular footings, as long as two-way behavior controls, by using an adjusted value of B. The equation for making this adjustment is

$$B' = B\sqrt{2S - 1} \qquad 23.3$$

in which B' is the adjusted width of the footing, which replaces B in both the a/B and d/B ratios, and S is the ratio L/B.

The complete design procedure, exclusive of the juncture of the column and footing, can be summarized as follows:

1. Proportion the area of the footing in accordance with the methods given in Part C.
2. Apply load factors to the column loads and compute the necessary ratios for entering Fig. 23.2.
3. Select the trial depth from Fig. 23.2, and check the unit shear stress. For rectangular footings, the unit shear stress corresponding to one-way action across the footing should be checked also.
4. Select a trial moment arm for the tension–compression couple and calculate trial values of the total tensile force T and the total compressive force C due to moment.
5. Select the reinforcement based on the above trial value of T. The amount of reinforcement is often governed by minimum values specified in codes for flexure and for shrinkage stresses.
6. Check the compressive stress block and repeat steps 4 and 5 if greater accuracy is believed necessary.
7. Check the development length of the reinforcement.

ILLUSTRATIVE DESIGNS. DP 23-1. COLUMN AND WALL FOOTINGS

The two footings designed in DP 23-1 illustrate the application of previous discussions of critical sections, placing of the reinforcement, and design procedure. Both footings were proportioned in DP 18-1. The computations in this design plate indicate that the previously assumed depths of the footings were several inches too large. However, any revisions of the depths in DP 18-1 would not be justified in view of the negligible effect of depth on the required areas.

The design computations and allowable stresses used in these examples are for the most part in agreement with the strength methods recommended in the ACI Building Code adopted in 1971. One deviation, however, is that the minimum reinforcement for flexure has been set conservatively at 0.5 per cent even though the yield strength of the steel is 60 kips/sq in.

23.6. Isolated Column Footings on Piles

Various types of piles were described in Chap. 12. The selection of the proper type of pile for given soil conditions was given considerable attention in Parts B and C.

Reinforced-concrete footings are used with all types of piles and serve as pile caps as well as supports for the columns.

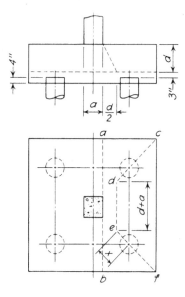

FIGURE 23.3. Concrete footing on piles, showing critical section for design.

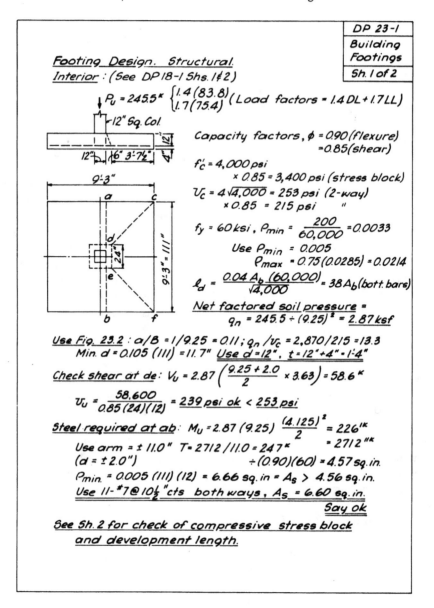

The piles commonly project 3 or 4 in. into the footing as shown in Fig. 23.3, and about 3 in. of concrete should separate the bottom reinforcement and the tops of the piles.

In general, the procedure for designing footings supported by piles closely parallels that used for footings on soil. Any differences are due to the concentrated reactions from the piles instead of the relatively uniform pressure from the soil. Although the locations of the piles in the field are likely to be at least several inches from their theoretical positions, it is common practice to take the critical section for shear at the same location as for footings on soil. As shown in Fig. 23.3, a section de, located at a distance equal to one half the depth of the footing from the face of the column, is ordinarily used for investigating diagonal tension. The critical section ab for flexure and development length may be assumed at the face of the column as in the case of footings on soil.

If the center of a pile is one half-diameter or more outside the critical section, the

```
                                                    ┌──────────────┐
                                                    │   DP 23-1    │
                                                    │  Building    │
  Footing Design. Structural                        │  Footings    │
  Interior: Cont'd from Sh.1                        │  Sh. 2 of 2  │
                                                    └──────────────┘
     Check compressive stress block at ab:
     C = T = 247ᵏ
         ÷ 3.4 = 72.6 sq.in.
             ÷ 111" = 0.65" < 2.0" ok
     Arm = 12 − 0.65/2 = 11.67" > 11.0" but since ρ_min.
        controls A_s, no reduction in A_s is possible
     Check development length:
        ℓ_d = 38 A_b = 38 (0.60) = 22.8" ok < 4'-0"
  Wall: (See DP 18-1 Shs. 1 & 2)
        P_u = 8.7ᵏ/ft. = 1.4 (3.8) + 1.7 (2.0)
           12" Masonry Wall
                Net factored Soil pressure =
                   q_n = 8.7/4.25 = 2.05 ksf
                Depth required for shear
                   V_u = 2.05 (1.38) = 2.83ᵏ/ft. of wall

                   v_u = 2√4,000 (0.85) = 108 psi  d = 2,830/(108(12)) = 2.18"
                      Use Min. d = 6", t = 6" + 4" = 10"
     Steel required at ¼ Pt. of wall:
        M_u = 2.05 (1.88)²/2 = 3.62 ᶦᵏ = 43.5 "ᵏ
        Say arm = 5.0"  T = 43.5/5.0 = 8.70ᵏ
                   ÷ (60 × 0.9) = 0.16 sq.in./ft.
        ρ_min. = 0.005 × 12 × 5 = 0.30 sq.in./ft. = A_s
                   Use #5 @ 12" ctrs. A_s = 0.31
     Longitudinal steel
        Say 0.0025 × 51 × 6 = 0.76 sq.in.
                   Use 4 - #4 @ abt. 16" ctrs. A_s = 0.80
     Compressive stress and development ok by inspection
```

entire reaction of the pile should be assumed effective in producing moment or shear on the section. The reaction from any pile located one half-diameter or more inside the section probably contributes very little to the moment or shear; hence, it may be considered as zero. For intermediate positions, a straight-line interpolation is commonly used to estimate the appropriate portion of pile reaction for analysis and design. By reference to Fig. 23.3, it becomes apparent that the shear and moment on ab will be produced by two full pile reactions, whereas the shear on de will be one pile reaction (two halves), provided the centers of the piles are more than half the pile diameter from the points d and e. For example, if the distance x is only one fourth of the pile diameter, three fourths of a pile reaction are assumed to contribute to the shear on section de.

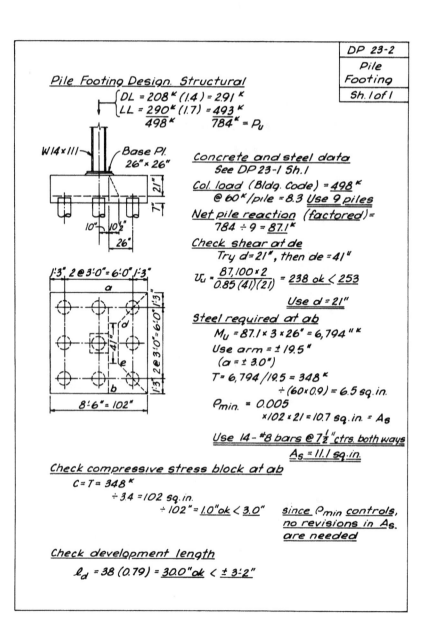

DP 23-2
Pile Footing
Sh. 1 of 1

<u>Pile Footing Design. Structural</u>

DL = 208^k (1.4) = 291^k
LL = 290^k (1.7) = 493^k
498^k 784^k = P_u

<u>Concrete and steel data</u>
See DP 23-1 Sh.1

<u>Col. load</u> (Bldg. Code) = <u>498^k</u>
@ 60^k/pile = 8.3 <u>Use 9 piles</u>

<u>Net pile reaction</u> (<u>factored</u>) =
784 ÷ 9 = <u>87.1^k</u>

<u>Check shear at de</u>
Try d = 21", then de = 41"
$v_u = \dfrac{87,100 \times 2}{0.85(41)(21)}$ = <u>238 ok</u> < <u>253</u>

<u>Use d = 21"</u>

<u>Steel required at ab</u>
M_u = 87.1 × 3 × 26" = 6,794 "^{k}
Use arm = ± 19.5"
(a = ± 3.0")
T = 6,794/19.5 = 348^k
÷ (60 × 0.9) = 6.5 sq. in.
ρ_{min} = 0.005
× 102 × 21 = 10.7 sq. in. = A_s

<u>Use 14 - #8 bars @ 7½" ctrs. both ways</u>
<u>A_s = 11.1 sq. in.</u>

<u>Check compressive stress block at ab</u>
C = T = 348^k
÷ 3.4 = 102 sq. in.
÷ 102" = <u>1.0" ok</u> < <u>3.0"</u>

since ρ_{min} <u>controls, no revisions in A_s are needed</u>

<u>Check development length</u>
ℓ_d = 38 (0.79) = <u>30.0" ok</u> < ± 3'-2"

ILLUSTRATIVE DESIGN. DP 23-2. PILE-SUPPORTED COLUMN FOOTING

The selection of the allowable pile load of 30 tons, arbitrarily used in DP 23-2, was discussed in Chap. 12 and Part C. A section halfway between the face of the steel column and the edge of the base plate has been selected as critical for tension and development of the reinforcement. The section determined in this manner is approximately 10 in. from the center line of the column. The critical section for shear is located a distance from the center line of 10 in. plus half the footing depth.

SUGGESTED READING

Principles of reinforced-concrete design, not discussed in this chapter, are described in standard texts on that subject. Among these are

P. M. Ferguson (1973), "Reinforced Concrete Fundamentals." New York, Wiley, 3rd ed., 750 pp.

G. Winter and A. H. Nilson (1973), "Design of Concrete Structures." New York, McGraw-Hill, 8th ed., 615 pp.

The Building Code of the American Concrete Institute is adopted by incorporation into most building ordinances throughout the United States and is kept current by frequent revisions.

Foundations Subject to Moment

Elevated water tank, Hennepin, Illinois. This structure with a capacity of 300,000 gal, is representative of many similar structures subjected to a large overturning moment due to wind. The transmission towers shown in the photograph are also subject to wind loading and, if a line should break, to large unbalanced longitudinal forces. Both sets of forces produce moment on the foundation; sufficient uplift resistance must be provided to prevent rise of those footings that tend to be lifted because of the moment. (Photo courtesy of Chicago Bridge & Iron Co.)

PLATE 24.

CHAPTER 24

Footings Subjected to Moment

24.1. Introduction

Chapter 23 dealt with footings supporting concentrically located columns that produced either uniform soil pressures or uniform pile reactions. However, many foundations must resist not only vertical loads but also moment about one or both axes. The moment M may exist at the bottom of a concentrically placed column (Fig. 24.1a), whence it is transferred to the footing, or it may be produced by a vertical load P located eccentrically at a distance e from the centroid of the base of the footing (Fig. 24.1b). If the footings in a and b have the same dimensions in plan, the soil reactions are identical provided $M = P \times e$. Footing b is then said to be equivalent to a. The substitution of an eccentric load for the real moment and column load sometimes simplifies the computations.

Other common foundations that must resist moment are those for retaining walls, abutments, and bridge piers. The moment on a retaining wall is due primarily to the active earth pressure, whereas the moments on the foundations for bridge piers are produced primarily by wind and traction on the superstructure. Foundations for these structures are discussed in more detail in Art. 24.6 and in Chap. 26.

In the discussion of concentrically loaded footings (Chap. 23), particular attention was given to the choice of critical sections for investigation of moments and shears; the computations of the moments and shears themselves involved no difficulty because the soil reaction or the load per pile was considered uniform. In contrast, the principal problem in the design of eccentrically loaded footings is the determination of the soil pressure or the load per pile, inasmuch as these quantities are no longer uniform. Once they have been determined, the selection of critical sections and the computation of stresses due to moment and shear are carried out as for concentrically loaded

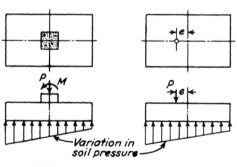

(a) Actual footing (b) Equivalent footing

FIGURE 24.1. Spread footing subjected to moment.

footings. Hence, Chap. 24 is concerned primarily with the computation of soil pressure or load per pile.

Fundamental to all computations in this chapter are the laws of statics. Regardless of the methods of analysis, the distribution of vertical soil pressure on the base of a footing must satisfy the requirements of statics that (1) the total upward soil reaction must be equal to the sum of the downward loads that act on the base, and (2) the moment of the resultant vertical load about any point must be equal to the moment of the total soil reaction about the same point. In addition, an adequate horizontal soil reaction, usually provided by shear along the base, must be available to oppose the resultant horizontal load.

Ordinary footings are commonly assumed to act as rigid structures. This premise leads to the conclusion that the vertical settlement of the soil beneath the base must have a planar distribution because a rigid foundation remains plane when it settles. A planar distribution of soil pressure follows from a second assumption that the ratio of pressure to settlement is constant. Neither of these assumptions is strictly valid, but each is generally considered sufficiently accurate for ordinary problems of design.

24.2. Resultant Within Middle Third

A footing subjected to moment is shown in Fig. 24.2. The forces acting on the footing, including the weight of the footing itself, have been resolved into ΣV and ΣH. It is apparent that the moment ΣM acting about the center of gravity of the base may be expressed as $\Sigma M = \Sigma H \times h = \Sigma V \times e$. Solving for e, we obtain

$$e = \Sigma M / \Sigma V \qquad 24.1$$

Equation 24.1 is of fundamental importance because it determines the position of the resultant of all forces acting above (or below) the base of a footing, regardless of how complicated the conditions of loading may be. In some problems, such as the design of retaining walls, it may be more convenient to choose a center of moments elsewhere than

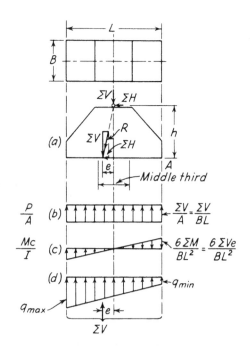

FIGURE 24.2. Forces at base of footing. Resultant inside middle third.

at the center of the base; nevertheless, eq. 24.1 remains applicable. For example, if the point A in Fig. 24.2 is taken as the center of moments, the value of ΣM must include $\Sigma V \times L/2$. When this total moment is divided by ΣV, the position of the resultant is found unchanged at a distance of $L/2 + e$ from A.

If the vertical load ΣV acts alone, the soil pressure is uniform as shown in Fig. 24.2b. The horizontal force ΣH, if acting alone, produces a shear that must be resisted by the soil at the base and a moment that corresponds to the distribution of pressure shown in c. The maximum pressure caused by the moment is obtained by dividing the moment by the section modulus of the area of the base. The combined effect of ΣV and ΣM corresponds to the distribution of soil pressure shown in d. Adding, for each edge of the pedestal, the pressures in b and c, we have

$$q_{max} = \frac{\Sigma V}{BL} + \frac{6\Sigma Ve}{BL^2} = \frac{\Sigma V}{BL}\left(1 + \frac{6e}{L}\right) \qquad 24.2$$

and

$$q_{\min} = \frac{\Sigma V}{BL} - \frac{6\Sigma Ve}{BL^2} = \frac{\Sigma V}{BL}\left(1 - \frac{6e}{L}\right) \quad 24.3$$

Equations 24.2 and 24.3 are often expressed as

$$q = \frac{\Sigma V}{BL}\left(1 \pm \frac{6e}{L}\right) \quad 24.4$$

Equation 24.4 is no more than a special form of the basic formula for stress on a section subjected to a direct load P and a moment M. The formula is commonly expressed in strength of materials in the form,

$$f = \frac{P}{A} \pm \frac{Mc}{I}$$

24.3. Resultant Outside Middle Third

If ΣH (Fig. 24.2) is increased, the moment at the base of the footing becomes larger and increases the soil pressures indicated in c. When the extreme pressures in c become equal numerically to those in b, $q_{\min}$ becomes zero. For this condition

$$\frac{\Sigma V}{BL} = \frac{6\Sigma Ve}{BL^2}$$

Solving for e, we find $e = L/6$, which indicates that the resultant must fall within the middle third of the base in order for compression to exist beneath the entire base. This condition is generally considered desirable.

Figure 24.3 shows the distribution of the soil pressure beneath the footing when the resultant is outside the middle third of the base. According to the laws of statics, the total upward force must be equal to and colinear with ΣV. These two conditions may be expressed by

$$\Sigma V = \frac{q_{\max} Bx}{2} \quad 24.5$$

and

$$x/3 = (L/2) - e \quad 24.6$$

From these expressions the maximum soil pressure $q_{\max}$ can be computed.

Inspection of eq. 24.5 reveals that the maximum soil pressure is merely twice the average pressure produced by ΣV acting

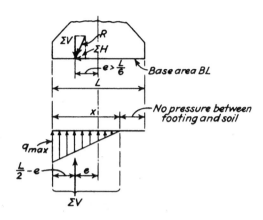

FIGURE 24.3. Forces at base of footing. Resultant outside middle third.

on the area Bx. The same result may be obtained from eq. 24.2 if the dimension L is replaced by the distance x and, since e must be measured from the center of gravity of the area under compression, if e is replaced by $x/6$. With these substitutions, eq. 24.2 becomes identical with eq. 24.5.

The preceding sections of this chapter concerning the position of the resultant vertical load ΣV with respect to the middle third of the base pertain only to rectangular footings subjected to moment about one axis. More generally, the locus of the most eccentric points of application of ΣV corresponding to compression over the entire base is known as the *kern*. For a circular base, the kern is at the quarter-point rather than the third-point of the diameter.

Most footings are proportioned so that the resultant of the loads determining their plan dimensions falls within the kern and the soil reaction is everywhere compressive. The point of application of the factored loads on which the structural design of the footing is based, however, may lie outside the kern and the footing must be designed structurally for a distribution of pressure corresponding to that shown in Fig. 24.3. This situation is particularly likely to occur in the design of the base slab of a cantilever retaining wall; It is discussed, together with a conservative simplification in the calculations, in Chap. 26 in connection with DP 26–1.

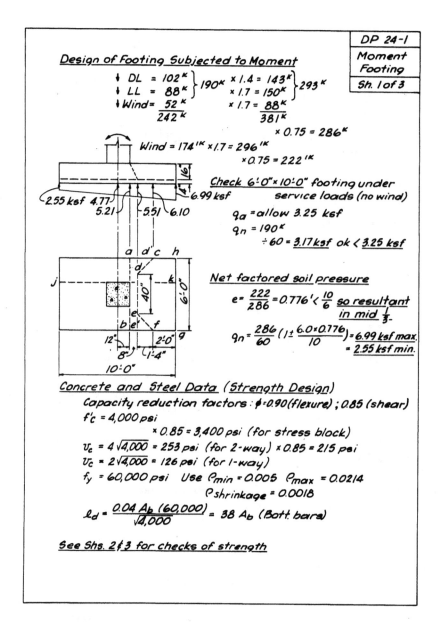

ILLUSTRATIVE DESIGN. DP 24-1. FOOTING SUBJECTED TO MOMENT

The computations in DP 24-1 represent common procedure for the design of rectangular footings subjected to moment about one axis. The footing is first proportioned so that the allowable soil pressure is not exceeded under service loads. Load factors are then applied to the loads and overturning moment, and the corresponding factored soil pressures are calculated. Since the footing loads also include those caused by wind, and since the probability of simultaneous application of all loads is small, only 75 per cent of the factored loads and external moment is used for the structural design of the footing. The critical sections are chosen in accordance with the rules in Chap. 23. The shear and moment at the critical sections are computed in accordance with the nonuniform distribution of pressure against the base. This is the principal difference between the com-

```
┌─────────────────────────────────────────────────────────┐
│                                        ┌──────────────┐ │
│                                        │   DP 24-1    │ │
│    Design of Footing Subjected to Moment│   Moment     │ │
│    Try d = 16"  t = 16" + 4" = 20"     │   Footing    │ │
│    Check 2-way shear at de:            │   Sh. 2 of 3 │ │
│                                        └──────────────┘ │
│      Average qₙ on cdef ≈ 5.86 ksf                      │
│                         × 6.22 ft² = 36.5ᵏ              │
│      Average qₙ on cfgh = 6.55 ksf                      │
│                         × 12.0 ft² = 78.5ᵏ              │
│                                      115.0ᵏ = Vᵤ        │
│                                                         │
│       vᵤ = 115,000 / [0.85(40)(16)] = 211 psi ok < 253 psi │
│                                                         │
│    Check 1-way shear at d'e':                           │
│      Average qₙ on d'e'gh = (5.51 + 6.99)/2 = 6.25 ksf  │
│                              × 20.0 ft² = 125.0ᵏ = Vᵤ   │
│                                                         │
│       vᵤ = 125,000 / [0.85(72)(16)] = 127.7 psi ≈ 127 psi, say ok │
│                                              Use d = 16"│
│    Steel required at ab:                                │
│      Use triangular variations of qₙ for convenience.   │
│      (6.99/2)(4×6) = 83.9ᵏ                              │
│                    × (2/3)(4) = 223.7 ¹ᵏ                │
│      (5.21/2)(4×6) = 62.5ᵏ                              │
│                    × (1/3)(4) = 83.4 ¹ᵏ                 │
│                                 307.1 ¹ᵏ = 3,685 "ᵏ = Mᵤ│
│      Use arm = ± 14.5"   T = 3,685 ÷ 14.5 = 254ᵏ        │
│              (d = ± 3.0")    ÷ (60 × 0.9) = 4.70 sq. in.│
│                                                         │
│              ρ = 4.70/(72×16) = 0.0041 < 0.005          │
│                                                         │
│         Min. Aₛ = 0.005 (72)(16) = 5.76 sq. in.         │
│                                                         │
│      Use 13 - #6 @ 5½" ctrs., long way, Aₛ = 5.72 sq. in.│
│                                                Say ok   │
└─────────────────────────────────────────────────────────┘
```

putations illustrated in DP 24–1 and those in DP 23–1.

If the designer assumes that two-way behavior controls and wishes to use Fig. 23.2 for determining the depth of footing, q_n should be taken as the average pressure on the area contributing to the shear at section de. The computations in DP 24–1 (Sh. 1) indicate that the average pressure on area $cdefgh$ after load factors are applied is about 6.4 kips/sq ft, whence $q_n/v_c = 29.8$. Since the footing is rectangular, $S = 10/6 = 1.67$, $B' = 6\sqrt{2(1.67) - 1} = 9.15$ ft, and $a/B' = 0.219$. By means of these values and Fig. 23.2, the appropriate value of d/B' is found to be 0.135. Hence, for two-way action, the depth of the footing should not be less than about 15 in. On the other hand, when the footing in DP 24–1 is checked against one-way action it is found that a depth of 16 in. is required.

The amount of reinforcement in the long direction of the footing is governed by flexure at the face of the column. On the

```
                                                            ┌──────────┐
                                                            │ DP 24-1  │
        Design of Footing Subject to Moment                 │ Moment   │
                                                            │ Footing  │
        Steel required at jk:                               │ Sh. 3 of 3│
                                                            └──────────┘
```

Design of Footing Subject to Moment

Steel required at jk:

   Use average q_n from $\mathcal{C}$ to k over entire area.

$$\frac{4.77 + 6.99}{2}(2 \times 10) = 117.6^k$$

$$\times 1.0 = 117.6^{\prime k} = 1411^{\prime\prime k} = M_u$$

$$T = 1411 \div 14.5 = 97.3^k$$

$$\div (60 \times 0.9) = 1.80 \text{ sq.in.}$$

   Min. $A_s = \rho_{shrinkage} = 0.0018\,(120)(16) = 3.46$ sq.in.

      Note: 3.46 sq.in. $>$ 1.33 (1.80) therefore $\rho_{shrinkage}$ ok for flexure also

   <u>Use 8 - #6 @ abt. 16½ ctrs., short way, $A_s = 3.52$ sq.in.</u>

Check compressive stress block at ab:

$$C = T = 254^k \quad (\text{see Sh. 2})$$
$$\div 3.4 = 74.7 \text{ sq.in.}$$
$$\div 72 = 1.04^{\prime\prime} \text{ ok} < 3.0^{\prime\prime}$$

$$\text{Arm} = 16.0 - \frac{1.04}{2} = 15.48^{\prime\prime} > 14.5^{\prime\prime} \text{ but } A_s \text{ is min. so say ok}$$

Check development length

$$\ell_d = 38\,A_b = 38\,(0.44) = \underline{16.7^{\prime\prime} \text{ ok}} < \pm 1'\text{-}10^{\prime\prime} \text{ short way}$$
$$< \pm 3'\text{-}10^{\prime\prime} \text{ long way}$$

---

other hand, the steel required in the short direction is determined by shrinkage stresses for which the minimum percentage is 0.18 for bars having a yield point of 60,000 lb/sq in. The reinforcement calculated in this manner is found in this example to be more than one third greater than that required for flexure. Hence, the amount of shrinkage steel is considered to satisfy the requirement for flexure in the short direction, even though the area of steel is less than $\rho = 0.5$ per cent. These minimum limitations for shrinkage and flexure correspond to the 1971 ACI Code.

The need for a concentration of steel in the central portion of the footing, as reflected in eq. 23.1 for concentrically loaded footings, is partly offset by determining the required reinforcement in the short direction on the basis of the higher than average soil pressure near the edges. Consequently, the reinforcement in the short direction is distributed uniformly.

24.4. Moment About Both Axes

Article 24.2 has shown that the soil pressures beneath footings may be estimated by means of the common formula for stress

$$\frac{\Sigma V}{A} \pm \frac{\Sigma M c}{I}$$

provided the moment acts about only one axis and the entire area of the base is in compression. When moments act simultaneously about both axes, the pressures may be computed by means of eq. 24.7 provided the entire area of the base is in compression.

$$q = \frac{\Sigma V}{A} \pm \frac{\Sigma M_1 c_1}{I_1} \pm \frac{\Sigma M_2 c_2}{I_2} \quad 24.7$$

The location of the maximum and minimum soil pressures may be determined readily by observing the directions of the moments. Likewise, the proper signs in eq. 24.7 may be determined by inspection for any other point on the base of the footing.

When moments act simultaneously about both axes, only part of the base may be in compression. This condition exists when the minimum soil pressure determined by eq. 24.7 has a negative value indicating tension that cannot be developed between the soil and concrete. Under this circumstance, the computation of the maximum soil pressure must be based, as in Art 24.3, on the area actually in compression. Equation 24.7 is no longer applicable.

Although several analytical and graphical methods are available for the solution of this problem, the following procedure of trial and error has the advantage of simplicity. It is illustrated in Fig. 24.4.

1. Find e_1 and e_2 by means of eq. 24.1, and let ΣV act through point A. The moments M_1 and M_2 may then be disregarded because the same soil pressure will act against the base of the equivalent eccentrically loaded footing.
2. Use eq. 24.7 to compute fictitious pressures at the corners as if tension could exist at D.
3. Locate the line of zero pressure on the basis of the pressures computed in step 2.
4. Select a trial axis of zero pressure, ZZ, approximately parallel to the line determined in step 3 but somewhat closer to point A.
5. Compute the moment of inertia about ZZ of the area assumed to be in compression.
6. Find $q_{max} = \Sigma V z b / I_{zz}$. The pressures at the other corners may be computed in like manner by substituting the appropriate dimensions for b.
7. Compare the magnitude of the total soil reaction (volume of the pressure diagram) with the vertical load ΣV, and compare the point of application of the total soil reaction (center of gravity of the volume of the pressure diagram) with the position of point A. These comparisons constitute a check on the requirements of statics.
8. If the discrepancies found in step 7 are too large, another axis ZZ may be assumed, and the computations in steps 5 to 7 made again. However, a great degree of refinement cannot be justified in view of the uncertainties associated with the various assumptions that must be made in solving any problem of this type.

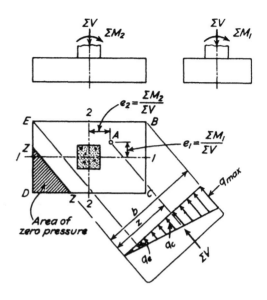

FIGURE 24.4. Footing subjected to moments about both axes with zero pressure under part of base.

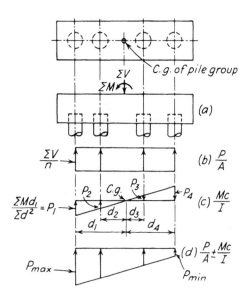

FIGURE 24.5. Computation of pile reaction.

24.5. Footings Having Unsymmetrical Shapes

The principles discussed in this chapter, except for the remarks pertaining to Fig. 24.4, apply only to footings having in plan at least one axis of symmetry. Occasionally an unsymmetrical plan may be necessary. If the resultant load does not coincide with the center of gravity of the area of the footing, the computation of the soil pressures becomes a problem involving bending on an unsymmetrical section. Theoretically, eq. 24.7 is not applicable even though the entire base may be in compression. However, unless the footing is greatly unsymmetrical, the errors involved in using eq. 24.7 are tolerable for design.

24.6. Moment on Pile Footings

The reactions exerted by piles beneath a footing subject to moment are calculated in a manner similar to that described in the preceding articles concerning the pressure under soil-supported footings. Pile caps, such as those shown in Figs. 24.5 and 24.6, are commonly assumed to act as rigid structures. A planar distribution of the vertical settlement of the piles follows as a result of this assumption. Finally, if the ratio of reaction to settlement is assumed to be constant, the loads in the piles vary in the same planar fashion. Neither the assumption of the rigid cap nor the supposition that reaction is directly proportional to settlement is strictly valid, but each is generally considered sufficiently accurate for the purposes of design.

The analysis of a moment-resistant group of piles is illustrated in Fig. 24.5. If there were no moment and ΣV acted through the center of gravity of the four piles, the loads in the piles would be as shown in b. On the other hand, if there were no resultant vertical force and only ΣM acted on the group of piles, the loads would be as shown in c.

The loads shown in d represent the total reaction and are the sums of those in b and

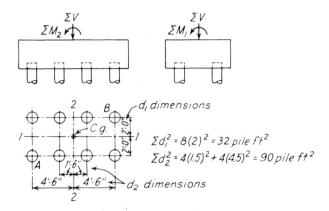

FIGURE 24.6. Group of piles subjected to direct load and to moments about both axes.

c. The same loads would have been produced by ΣV acting eccentrically at a distance e to the left of the center of gravity.

From statics, it is evident that the resisting moment of the reactions furnished by the piles (Fig. 24.5c) must equal the applied moment, ΣM. The following equation expresses this relationship if resisting moments at the junctions of the piles and the cap either do not exist or are disregarded.

$$\Sigma M = P_1 d_1 + P_2 d_2 + P_3 d_3 + P_4 d_4 \quad 24.8a$$

If the variation in pile reactions shown in c is assumed to be linear, then

$$P_1/d_1 = P_2/d_2 = P_3/d_3 = P_4/d_4$$

or

$$P_2 = P_1 d_2/d_1$$
$$P_3 = P_1 d_3/d_1$$
$$P_4 = P_1 d_4/d_1$$

Substituting these values of P_2, P_3, and P_4 in eq. 24.8a, we have

$$\Sigma M = P_1 d_1^2/d_1 + P_1 d_2^2/d_1 + P_1 d_3^2/d_1 + P_1 d_4^2/d_1 \quad 24.8b$$

Solving for P_1,

$$P_1 = \frac{\Sigma M d_1}{d_1^2 + d_2^2 + d_3^2 + d_4^2} = \frac{\Sigma M d_1}{\Sigma d^2} \quad 24.9$$

Similarly, the part of the load on any other pile due to moment may be computed by means of eq. 24.9 if d_1 is replaced by the distance from the pile to the center of gravity of the group.

The total reaction on any pile, found by adding the load shown in Fig. 24.5c to that in Fig. 24.5b, may be expressed in the form

$$P = \frac{\Sigma V}{n} \pm \frac{\Sigma M d}{\Sigma d^2} \quad 24.10$$

where P = total pile reaction resulting from moment and direct load
ΣV = sum of vertical loads acting on the foundation
ΣM = sum of moments about the center of gravity of the group. ΣM is sometimes expressed as $\Sigma V e$

n = number of piles in the group
d = distance from the center of gravity of the group to pile in question
Σd^2 = sum of the squares of the distances to each pile from the center of gravity of the group

Inspection of eq. 24.10 reveals that it is no more than a special form of the basic formula for stress on a section or for pressure beneath a soil-supported footing when either is subjected simultaneously to direct load and moment. The number of piles n is substituted for the area, and the term Σd^2 replaces the moment of inertia of the area. For this reason Σd^2 is sometimes called the moment of inertia of the group of piles. The analogy between the terms of the two equations is shown in Fig. 24.5.

Most groups of piles contain several rows. Furthermore, moment about both axes is not uncommon. Equation 24.11 applies to these conditions.

$$P = \frac{\Sigma V}{n} \pm \frac{\Sigma M_1 d_1}{\Sigma d_1^2} \pm \frac{\Sigma M_2 d_2}{\Sigma d_2^2} \quad 24.11$$

The subscript of the moment denotes the centroidal axis about which the moment acts. The subscript of the distance indicates the centroidal axis to which the distance from the pile is measured. These symbols are shown in Fig. 24.6.

If the moments have the directions shown in Fig. 24.6, it is apparent that pile A carries the greatest load whereas pile B carries the least. Both ΣM_1 and ΣM_2 increase the reaction at A and decrease that at B. Thus, it is possible to select by inspection the proper signs in the application of eq. 24.11 to any pile.

The determination of Σd^2 for large groups of piles may be considerably simplified by the use of eq. 24.12, which applies to a single row of piles with equal spacing.

$$\Sigma d^2 \text{ (one row)} = \frac{s^2}{12} n_1(n_1^2 - 1) \quad 24.12$$

where s = spacing of piles in the row
n_1 = number of piles in the row

24.7. Piles Subjected to Tension

Ordinarily the piles beneath a footing are expected to act in compression, and only nominal provision is made to anchor them to the footing. This condition exists whenever all the pile reactions computed in accordance with eqs. 24.10 and 24.11 are positive. If some of the reactions are negative but the piles are not anchored to the cap, the situation is analogous to that described in Arts. 24.3 and 24.4, which deal with footings having only part of their bases in compression. If the piles corresponding to the negative reactions cannot resist the tensile forces, the compression in the other piles is increased.

On the other hand, piles are often used specifically to resist tensile forces beneath several common types of structures such as towers, gas storage tanks, and tall stacks. Beneath such structures the tensile forces are usually temporary and are almost always caused by the moment due to wind. Under these conditions, if the piles are

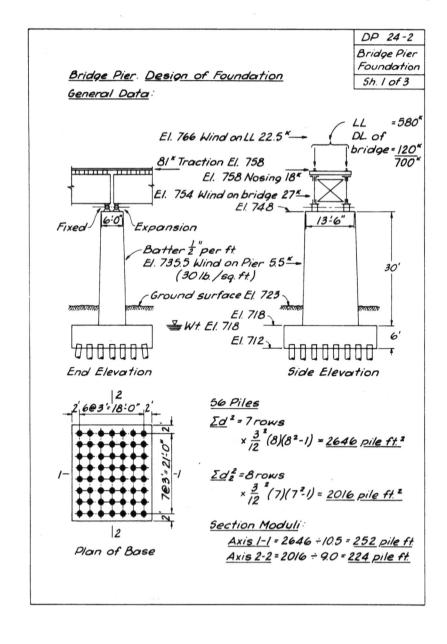

capable of withstanding tension and are adequately anchored to the cap, the loads in each pile may be computed by means of eqs. 24.10 and 24.11.

ILLUSTRATIVE DESIGN. DP 24–2. BRIDGE PIER

The base of a bridge pier is a common example of a footing subjected to vertical loads together with moment about both axes. The vertical loads are due to the dead weight and live load of the superstructure and to the weight of the pier itself. Moments and shears on the foundation are produced by horizontal forces such as centrifugal force and those due to traction, nosing, wind, current, and ice. For the most unfavorable combination of these loadings, the allowable soil pressure or pile reaction beneath the base is ordinarily increased from 25 to 50 per cent above the value permitted under dead plus live load.

A typical pier for a single track E–72 rail-

DP 24-2 — Bridge Pier Foundation — Sh. 2 of 3

Bridge Pier. Design of Foundation
Pile Reactions Due to Vertical Loads:

Shaft: Top area $= 13.5 \times 6$ $= 81.00$ sq. ft.
      Mid " $\times 4 = 14.75 \times 7.25 \times 4 = 427.75$
      Bott. " $= 16.0 \times 8.5$ $= 136.00$
                          $\overline{644.75}$ sq. ft.
                              $\times \frac{30}{6} = 3224$ cu. ft.
          3224 cu. ft. $\times 0.15 = 484^K$

Footing: $22 \times 25 \times 6 = 3300$ cu. ft.
                          $\times 0.15 = 495^K$

Earth: $22 \times 25 = 550$ sq. ft.
      $15.8 \times 8.3 = 131$
            $\overline{419}$ sq. ft.
             $\times 5 = 2095$ cu. ft.
                 $\times 0.12 = 251^K$

DL (Superstructure): $\underline{120^K}$
                              1350^K

Buoyancy: $22 \times 25 \times 6 = 3300$ cu. ft.
                  $\times 0.0625 = \underline{206^K}$
                             1144^K = total DL

LL: $\underline{580^K}$
                               1724^K = total DL + LL
                               $\div 56 = 30.8^K$/pile

Pile Reactions Due to Moment:
Traction: $81^K \times 46' = 3726^{IK}$
           $\div 224 =$ 16.6^K/pile

Transverse
wind: On bridge $27^K \times 42' = 1134^{IK}$
                     $\div 252 = 4.5^K$/pile
      On LL $22.5^K \times 54' = 1212^{IK}$
                     $\div 252 = 4.8$ 9.8^K/pile 29.7^K/pile
      On end of shaft
         $5.5^K \times 23.5' = 129^{IK}$
                     $\div 252 = 0.5$

Nosing: $18^K \times 46' = 828^{IK}$
            $\div 252 =$ 3.3^K/pile

Maximum pile reaction: 60.5^K/pile
Minimum pile reaction: $30.8^K - 29.7^K =$ 1.1^K/pile

Bridge Pier. Design of Foundation   DP 24-2
                                              Bridge Pier
Soil Pressure if Piles are Omitted: Foundation
  Base 22' × 25'   Area = 550 sq. ft   Sh. 3 of 3
    Section Moduli: Axis 1-1 = $\frac{1}{6}(22)(25)^2$ = 2290 ft.^{3}
                     Axis 2-2 = $\frac{1}{6}(25)(22)^2$ = 2020 ft.^{3}
  Vertical load on base = 1724^K
  Moments on Base:
        1134^{1K}
        1212
         129
         828
  M_{1-1} = 3303^{1K} M_{2-2} = 3726^{1K}

Maximum soil pressure: $\frac{1724}{550} + \frac{3303}{2290} + \frac{3726}{2020}$ = 6.4^K/sq. ft. *

Minimum soil pressure: 3.13 − 1.44 − 1.84 = −0.15^K/sq. ft. *

  * Since the minimum soil pressure is negative
    indicating a small tension, the maximum
    soil pressure should be computed by other
    methods described in Art. 24.4. However,
    in this instance, greater accuracy is not
    warranted.

road bridge is shown in Sh. 1, DP 24–2. The transverse spacing of the girders and the dimensions of the base plates for the bridge shoes determine the size of the top of the pier shaft. The forces that must be considered in the analysis and design of the structure are given; a discussion of the basis for determining the magnitude of each force is beyond the scope of this text.

The object of the design plate is to demonstrate the application of the principles discussed in Arts. 24.4 and 24.6. Hence the computations are limited mainly to those for determining the critical pile reactions and for evaluating the pressures against the base if piles are omitted. The number of piles and dimensions of the base are usually determined by trial and revision. The computations on Sh. 2 show that the pile loads resulting from traction, nosing, and wind amount to almost 50 per cent of the total reaction on the most highly stressed pile. Therefore, the total number of piles required is considerably greater than that determined by dividing the total vertical load by the allowable capacity per pile. The vertical loads in the piles are found by means of eq. 24.11; it should be noted that the computation of the moment of inertia of the group of piles is simplified considerably by the use of eq. 24.12. In Sh. 3, the distribution of pressure beneath the base if no piles are used is found by means of eq. 24.7.

As indicated on Sh. 1, most pile foundations for bridge piers contain some batter piles to provide greater stability against horizontal forces. However, computations concerned with this phase of the design have been omitted because the analysis of foundations containing batter piles is explained in Art. 26.8 as part of the discussion of pile-supported retaining walls.

A complete structural design of the base is not included because the choice of critical sections and the subsequent computations would involve the duplication of much that is shown in previous design plates. The computations for the shaft are also omitted because this phase of the design is exclusively a problem in reinforced-concrete analysis and design. Therefore, the pile reactions and soil pressures resulting from loads multiplied by load factors have not been calculated.

PROBLEMS

1. If the vertical load on the footing shown in DP 24–1 remains 242 kips, compute the maximum moment that can be resisted on the assumption that the area between the soil and the footing is entirely in compression. What is the magnitude of the maximum soil pressure for this condition of loading?
Ans. 404 ft-kips; 8.1 kips/sq ft.

2. Compute the maximum soil pressure acting against the base of the footing shown in DP 24–1 if the vertical column reaction is 180 kips and the moment is 360 ft-kips.
Ans. 6.6 kips/sq ft.

3. A vertical load of 135 kips is applied at the center of a footing 5 × 9 ft in plan.
   a. How much moment in the direction of the long axis can the footing resist before the soil pressure exceeds 5 kips/sq ft;
   b. How much moment in the same direction can the footing resist before the soil pressure exceeds 8 kips/sq ft?
Ans. 135 ft-kips; 337 ft-kips.

4. Compute the maximum and minimum soil pressures at the base of the footing shown. The weight of the footing itself is 3.4 kips. Compute the magnitude of the pressures at the other two corners.

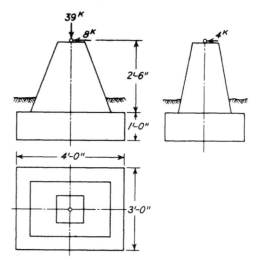

Ans. 12.6; 1.1; 8.1; 5.6 kips/sq ft.

5. Compute each of the pile reactions beneath the footing shown in Fig. 24.6 if $\Sigma V = 162$ kips, $M_1 = 90$ ft-kips, and $M_2 = 236$ ft-kips. Neglect the weight of the footing itself. Check sum of pile reactions against ΣV.
Ans. Top row: 26.5; 18.6; 10.8; 2.9 kips. Bottom row: 37.7; 29.8; 22.0; 14.1 kips.

6. Substitute the pile foundation in the figure below for the foundation of the bridge pier as shown in DP 24–2, and compute the maximum and minimum pile reactions. Compute the maximum and minimum soil pressure beneath the base if the piles are omitted. The elevations of top and bottom of the base are the same as in DP 24–2.

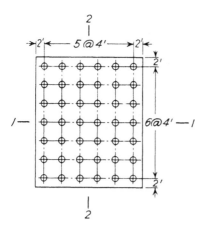

Ans. Pile loads: Maximum = 76.0 kips/pile; minimum = 8.8 kips/pile. Soil pressure: Maximum = 5.1 kips/sq ft; minimum = 0.2 kip/sq ft.

SUGGESTED READING

The design of bridge piers and their foundations, in which horizontal forces and moments assume particular importance, is usually carried out in accordance with the current specifications of the American Association of State Highway Officials (AASHO) or the American Railway Engineering Association (AREA).

The specific references are:

American Association of State Highway Officials, "*Standard Specifications for Highway Bridges,*" Washington, D. C. New editions are brought out from time to time.

American Railway Engineering Association, "*AREA Manual of Recommended Practice,*" A.A.R., Chicago. Loose-leaf document continually brought up to date.

Raft Foundation

Base slab for Grant Park Underground Garage, Chicago. The completed structure, with two floors of parking, is 360 by 1200 ft in plan. It is underlain by more than 30 ft of soft to medium clays, and is covered by fill that supports the restored park and by Michigan Boulevard, a major thoroughfare. Since the structure weighs less than the displaced soil, settlement has been negligible. (Photo courtesy of Chicago Park District.)

PLATE 25.

CHAPTER 25

Combined Footings and Rafts

25.1. Purpose of Combined Footings

The need for combined footings arises most commonly either when two columns are spaced so closely that individual footings are not practicable or when a wall column is so close to the property line that it is impossible to center an individual footing under the column.

A combined footing is so proportioned that the centroid of the area in contact with the soil lies on the line of action of the resultant of the loads applied to the footing; thus, the distribution of soil pressure is fairly uniform. In addition, the dimensions of the footing are chosen such that the allowable soil pressure is not exceeded. When these two criteria are satisfied, the footing should neither settle nor rotate excessively.

In the design of a combined footing that rests on piles, the total number of piles is found by dividing the sum of the vertical loads on the footing, including the weight of the footing, by the safe load per pile. The piles are then arranged in a pattern such that the centroid of the group is close to the line of action of the resultant vertical load. Consequently, all piles under the footing are subjected to approximately equal loads.

25.2. Combined Footings of Rectangular and Trapezoidal Shapes

A combined footing is usually given a rectangular shape (Fig. 25.1a) if the rectangle can extend beyond each column whatever distance is necessary to make the centroid of the rectangle coincide with the point at which the resultant R of the column loads intersects the base. If the footing is to support an exterior column at the property line where the projection is limited, a rectangular shape ordinarily can still be used, provided the interior column carries the greater load. The length L of the combined footing is established by adjusting the projection d of the footing beyond the interior column. The width is then calculated by dividing the sum of the vertical loads by the product of the length and the allowable soil pressure.

If, for some reason, the footing cannot project the required distance beyond one or both columns, a trapezoidal combined footing (Fig. 25.1b) is commonly used. The location of the resultant of the column loads establishes the position of the centroid of the trapezoid. The length L is usually limited by the property line at one end and

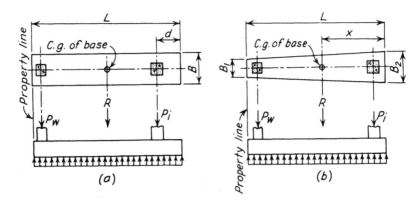

FIGURE 25.1. Diagram illustrating forces acting on combined footings of (a) rectangular and (b) trapezoidal shape.

adjacent construction at the other. The dimensions B_1 and B_2 can be determined from the solution of two simultaneous equations: one equation expresses the location of the centroid; the other equates the sum of the loads to the product of the allowable soil pressure and the area of the footing. In this manner the two criteria discussed in Art. 25.1 are satisfied.

The solution of the two equations mentioned in the preceding paragraph leads to the following expressions for the widths of a trapezoidal footing.

$$B_1 = \frac{2A}{L}\left(\frac{3x}{L} - 1\right) \qquad 25.1$$

and

$$B_2 = (2A/L) - B_1 \qquad 25.2$$

where A is the area determined by dividing the sum of the loads by the allowable soil pressure; other symbols are shown in Fig. 25.1b. Inspection of eq. 25.1 shows that the width B_1 is zero when x is equal to one third the length of the footing. If this condition exists, a triangular footing is needed to satisfy the requirement of uniform soil pressure. In tentative designs, whenever the distance x approaches or is less than $L/3$, the length L should be increased by increasing the projection at the wide end.

25.3. Cantilever Footings

A third relatively common type of combined construction is the cantilever footing (Fig. 25.2). It is designed to support a wall column near its edge without causing nonuniform soil pressure. The cantilever princi-

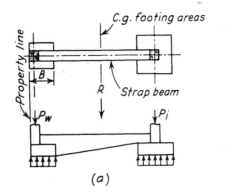
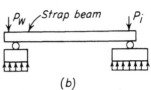

FIGURE 25.2. Diagram illustrating (a) forces acting on cantilever footing. (b) Principle of cantilever footing.

ple, largely concealed in actual footings of this type, is illustrated in Fig. 25.2b. The combined footing may be regarded as two individual footings connected by a strap beam. It may be inferred from Fig. 25.2b that the proportioning of the areas of the two individual footings is a problem in statics if the allowable soil pressure is known and if the dimension B of the wall footing is either fixed or assumed. Furthermore, the centroid of the two areas must lie on the line of action of the resultant load. This requirement may not be obvious because the two areas are usually found more or less independently from reactions determined by statics.

25.4. Choice of Column Loads

Except for the choice of the allowable soil pressure, the most difficult step in proportioning a combined footing is the selection of the appropriate column loads. Generally, it is considered good practice to proportion the footing for a uniform soil pressure under the dead load plus only the amount of live load that is likely to govern the settlement. If the footing is to rest on clay, the appropriate live load is the average likely to act on the footing over a period of years; if on sand, it is the maximum probable value. In either instance, the centroid of the footing is made to lie on the line of action of the resultant of column loads that consist of dead load plus only a fraction of the live load specified in the building code.

Even though a combined footing may be so proportioned that uniform soil pressure is produced by dead load plus a reduced live load (Fig. 25.3a), the size of the footing should be large enough to provide a factor of safety of between 2 and 3 against a bearing capacity failure of the soil under the larger service loads specified by the building code (Art. 17.3). Moreover, to design the footing structurally by the strength method, soil pressures must be calculated that correspond to column loads multiplied by appropriate load factors (Art. 23.1). The resultant of neither of these two greater loadings is likely to coincide with the center of gravity of the footing proportioned to the reduced loading. Consequently, the soil pressure for structural design (Fig. 25.3b) is likely to be nonuniform.

25.5. Structural Design of Combined Footings

The rigidity of ordinary combined footings is in general somewhat less than that of most individual spread footings. Nevertheless, the design of a combined footing is commonly based on the assumption that the soil pressure beneath the footing has a planar distribution. This assumption is usually satisfactory and conservative in view of the other uncertainties associated with the problem.

The principal reinforcement in a combined footing is placed in the longitudinal direction. The amount of steel is usually

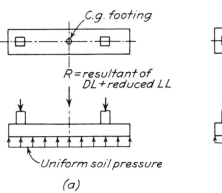
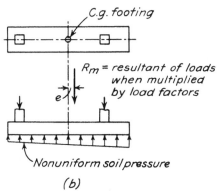

FIGURE 25.3. Forces to be considered (a) in determining dimensions of combined rectangular footing and (b) in making structural design of the footing.

determined by assuming that the footing acts as a one-way slab. However, transverse reinforcement is also provided at the bottom of the footing near the columns. The critical section for transverse bending is taken at the faces of the columns or pedestals. The transverse reinforcement is divided into groups proportionate in sectional area to the column loads, and the reinforcement at each column is placed uniformly within a band having an arbitrarily chosen width often taken as the width of the footing at the column. Finally, shear reinforcement is almost always necessary to resist the high diagonal tension adjacent to the columns.

The procedure for the design of a combined footing such as that shown in Fig. 25.3 may be summarized as follows:

1. Determine the column loads appropriate for considerations of settlement. These consist of dead load plus only a portion of the live load specified for design of the columns.

2. Using the resultant of the loads in step 1, select the plan dimensions of the footing to obtain a uniform soil pressure that does not exceed the pressure appropriate for this condition of loading.

3. Using the column loads specified in the building code (without load factors) and the plan dimensions determined in step 2, calculate the corresponding soil pressure. If the maximum soil pressure under this loading exceeds the value considered appropriate for this condition of loading, the width of the footing must be increased whereas the position of the centroid must remain unchanged.

4. Compute the soil pressure beneath the footing corresponding to the column loads multiplied by appropriate load factors.

5. Draw the shear and moment diagrams for the footing when it is subjected to the maximum conditions of step 4.

6. Using step 5 as the basis for design, determine the depth of the footing and the necessary amount of reinforcing steel at appropriate sections.

ILLUSTRATIVE DESIGN. DP 25-1a, 25-1b, and 25-1c. COMBINED FOOTINGS

The preceding articles of this chapter have discussed in some detail the various types of combined footings. The purpose of these design plates is to present typical computations that demonstrate the application of some of the principles associated with the proportioning and structural design of three common types of combined footings.

The computations given in DP 25-1a deal with a rectangular combined footing. The footing is first proportioned for uniform soil pressure under dead load plus a reduced live load. Second, the variation in soil pressure is determined for the conditions of loading specified by the building code by methods discussed in Art. 24.2. The maximum soil pressure under this loading is found to be less than the safe soil pressure. Hence, no revisions in the plan dimensions of the footing are needed. Next, the soil pressures caused by the column loads multiplied by load factors are calculated, and the corresponding shear and moment diagrams are drawn. These represent the basis for the design computations carried out in accordance with strength design.

The proportions of a trapezoidal combined footing are determined in DP 25-1b. Equations 25.1 and 25.2 are used to obtain appropriate end widths for uniform soil pressure under dead load plus reduced live load. In order to determine the soil pressures under the other conditions of loading, the location of the center of gravity of the footing is obtained by means of the equation

$$\bar{x} = \frac{L}{3}\left(\frac{2B_1 + B_2}{B_1 + B_2}\right) \qquad 25.3$$

The moment of inertia of the footing is conveniently determined by the parallel-axis relationship

$$I_{cg} = I_{B_2} - A\bar{x}^2 \qquad 25.4$$

Finally the soil pressures are obtained by means of the basic equation

$$q = \frac{\Sigma V}{A} \pm \frac{Mc}{I} \qquad 25.5$$

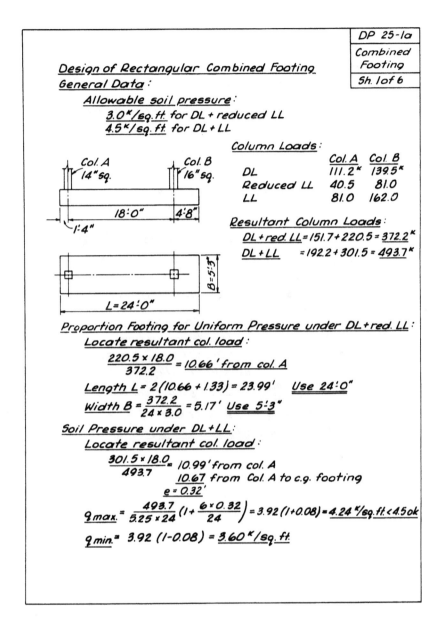

The computations given in DP 25-1c demonstrate the method of proportioning cantilever footings. Reference is made to Fig. 25.2b for illustration of the cantilever principle. Otherwise, the computations are believed to be self-explanatory.

25.6. Basis for Design of Raft Foundations

The conditions under which a raft foundation is suitable were discussed in Part B, and the methods for selecting the allowable soil pressure and for estimating the settlement were discussed in Part C. This article deals with the structural design of the raft itself.

In its simplest common form a raft consists of a reinforced-concrete slab that supports the columns and walls of a structure and that distributes the load therefrom to the underlying soils. Such a slab is usually regarded and designed as an inverted con-

```
                                                    DP 25-1a
                                                    Combined
    Structural Design of Footing                    Footing
    Factored column loads                           Sh. 2 of 6
        Col. A : 1.4 (111.2) = 155.7ᵏ    Col. B: 1.4 (139.5) = 195.3
                1.7 (81.0) = 137.7              1.7 (162.0) = 275.4
                           293.4ᵏ                          470.7ᵏ
                           470.7
              Resultant = 764.1ᵏ
    Factored soil pressures:
```

Locate resultant: $\dfrac{470.7 \times 18}{764.1} = 11.09'$ from col. A

10.67 from col. A to c.g.

$\overline{0.42'} = e$

$q_{max} = \dfrac{764.1}{5.25 \times 24}\left(1 + \dfrac{6 \times 0.42}{24}\right) = 6.06(1 + 0.105) = \underline{6.70 \text{ ksf}}$

$q_{min} = \phantom{\dfrac{764.1}{5.25 \times 24}\left(1 + \dfrac{6 \times 0.42}{24}\right) =} 6.06(1 - 0.105) = \underline{5.42 \text{ ksf}}$

Concrete and steel data:

Capacity reduction factors: $\phi = 0.90$ (flexure); 0.85 (shear)

$f'_c = 3,000$ psi

$\phantom{f'_c = 3,000 \text{ psi}} \times 0.85 = 2,550$ psi (for stress block)

$v_c = 2\sqrt{3000} = 110$ psi (for 2-way action)

$f_y = 40,000$ psi $\rho_{min} = 0.005$; $\rho_{shrinkage} =$ say 0.002

$\phantom{f_y = 40,000 \text{ psi }}\rho_{max} = 0.0278$ (0.75 × balanced section)

$\ell_d = \dfrac{0.04 A_b (40,000)}{\sqrt{3,000}} = 29.2 A_b$ (bottom bars)

$\ell_d = 1.4 \times 29.2 A_b = 40.9 A_b$ (top bars)

See Sh. 3 for shear and moment diagrams

tinuous flat-slab floor supported without upward deflection at the columns and walls. The soil pressure acting against the slab is commonly assumed to be uniformly distributed and equal to the total of all column loads, multiplied by appropriate load factors, and divided by the area of the raft. The moments and shears in the slab are determined by the use of appropriate coefficients listed in specifications for the design of flat-slab floors.

On account of the erratic variations in compressibility of almost every soil deposit, there are likely to be correspondingly erratic deviations of the soil pressure from the average value. Since the moments and shears are determined on the basis of the average pressure, it is considered good practice to provide the slab with more than the theoretical amount of reinforcement and to use the same percentage of steel at top and bottom.

The flat-slab analogy has been widely used, often with complete success. On the

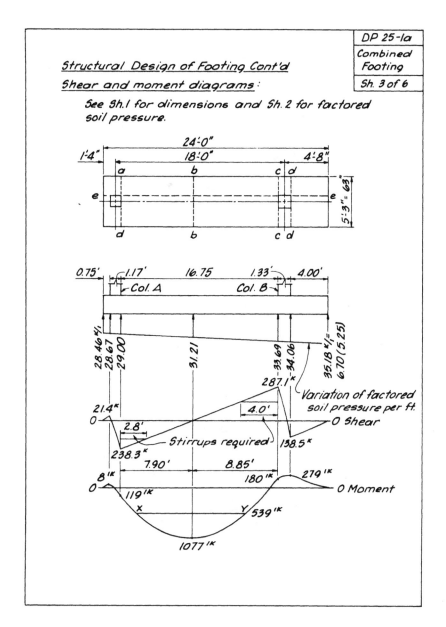

other hand, it has frequently led to structural failure not only of the slab but also of the superstructure. Therefore, its limitations must be clearly understood.

The analogy is valid only if the differential settlement between columns will be small and, furthermore, if the pattern of the differential settlement will be erratic rather than systematic. These limitations are necessary because the design of a flat-slab floor is based on the tacit assumption that there will be insignificant differential settlements among the points of support at columns or walls.

Furthermore, even if deep-seated or systematic settlements are negligible, the flat-slab analogy is likely to lead to uneconomical or unconservative design unless the columns are more or less equally loaded and equally spaced. If the downward loads on some areas are on the average much heavier than on others, differential settlements may lead to substantial redistribution of moments in the slab. Under these circum-

```
                                                              ┌─────────────┐
                                                              │  DP 25-1a   │
    Structural Design of Footing Cont'd                       │  Combined   │
    Longitudinal reinforcement:                               │   Footing   │
      Top bars at b-b:                                        ├─────────────┤
        Try d = 26" and arm d - a/2 = 24"                     │  Sh. 4 of 6 │
        T = 1077 × 12/24 = 538.5                   Use 20-#8 bars
                    ÷ (40 × 0.9) = 15.0 sq.in.    A_s = 15.8 sq.in.
        Check ρ: 15.8/(63×26) = 0.0096 ok < 0.0278 max.
                                  ok > 0.005 min.   Use d = 26"
      Check compressive stress block at b-b:
        C = T = 538.5 ᴷ           a = 211/63 = 3.35" ok < 4.0"
               ÷ 2.55 = 211 sq.in. say accuracy of A_s ok
      Bott. bars at d-d: T = 279 × 12/24 = 139.5ᴷ
                              ÷ (40 × 0.9) = 3.9 sq.in.
        Check ρ: 3.9/(63×26) = 0.00238 > 0.002, shrinkage ok
                                < 0.005, so increase A_s
                                        by 1/3 then A_s =
        Use 7-#8 bars  A_s = 5.53 sq.in.            5.2 sq.in.
    Transverse reinforcement:
      Bott. bars at e-e:  M_u = 6.06 (24) (1.96)²/2 = 279 ᴵᴷ
        T = 279 × 12/24 = 140 ᴷ
                ÷ (40 × 0.9) = 3.9 sq.in.
        Check ρ: 3.9/(say 2(63)(26)) = 0.0012 < 0.002, so use
                                            A_s for shrinkage
        A_s = 0.002 (2)(63)(26) = 6.6 sq.in.
                        × 293.4/764.1 = 2.5 sq.in. at col. A
                        × 470.7/764.1 = 4.1 sq.in. at col. B
        At col. A, use 8-#5 @ 10" ctrs. A_s = 2.48 sq.in.
        At col. B, use 14-#5 @ 6" ctrs. A_s = 4.34 sq.in.
```

stances, rafts are sometimes designed as if they rested on a bed of closely and equally spaced elastic springs of equal stiffness. The contact pressure q beneath any small area is then proportional to the deflection of the springs in that area and thus to the settlement S. The constant of proportionality

$$k = q/S \qquad 25.6$$

is called the *modulus of subgrade reaction*. It has the units of force per unit volume. Although the theory has been well developed for calculating moments and shears in the raft for a subgrade with properties represented by a constant value of modulus k, the value of k for real soils depends not only on the stress-deformation characteristics of the soil but also in a complex manner on the shape and size of the loaded area and the magnitude and position of nearby loaded areas. In some instances, values of k that might appear reasonable lead to computed soil pressures greater than the bearing capacity of the subsoil. Hence, evaluation

```
┌─────────────────────────────────────────────────────────┬──────────────┐
│                                                         │  DP 25-1a    │
│  Structural Design of Footing Cont'd.                   │  Combined    │
│  Stirrups:                                              │  Footing     │
│   V_c = 110(0.85)(63)(26) = 153ᵏ < 238.3ᵏ = V_u         │  Sh. 5 of 6  │
│                           < 287.1ᵏ therefore            │              │
│                           stirrups req'd. both cols.    │              │
│                                                                        │
│   Use #5 ⊓   A_v = 4 × 0.31 = 1.24 sq.in.                              │
│                        × 40 = 49.6ᵏ                                    │
│   At Col. A: 238.3ᵏ                                                    │
│              153.0     S = 49.6(26)/85.3 = 15.1" > d/2 = 13" Max.      │
│              ─────                                                     │
│               85.3     Use 3 spaces @ 12"                              │
│   At Col. B: 287.1                                                     │
│              153.0     S = 49.6(26)/134.1 = 9.6"                       │
│              ─────                                                     │
│              134.1                                                     │
│                        Use 3 spaces @ 9" &                             │
│                        2 spaces @ 12"                                  │
│  Development lengths of reinforcement:                                 │
│   Longitudinal reinforcement                                           │
│      Top bars: Top layer @ Col. A: Use hook                            │
│                           @ Col. B: Extend bars                        │
│                           40.9(0.79) = 32.3" beyond c-c,               │
│                           say 2'-0" outside ℄ Col. B.                  │
│                Bott. layer: Extend Min. = 26" = d beyond               │
│                           pts. X & Y (see Mom. Diag., Sh. 3)           │
│                           Use 16'-0" bars from inside                  │
│                           Col. A to 9" from inside Col. B.             │
│      Bott. bars @ Col. B:                                              │
│         ℓ_d = 29.2(0.79) = 23.1" ok < ± 3'-10" outside d-d             │
│                                  & < ± 3'-3" inside ℄ Col. B           │
│   Transverse Reinforcement:                                            │
│         ℓ_d = 29.2(0.31) = 9.1" ok < ± 1'-10" outside e-e              │
│                                                                        │
│  See Sh. 6 for Sketch of Footing.                                      │
└────────────────────────────────────────────────────────────────────────┘
```

of k for design, and even the judgment regarding the applicability of the concept of modulus of subgrade reaction to a particular project, require mature consideration and are fraught with uncertainty.

Adequate structural design of the slab of a raft foundation by the flat-slab analogy or by the use of a modulus of subgrade reaction is, unfortunately, no guarantee that the deflections of the raft will actually be unimportant. Indeed, if the structure covers a fairly large area and significantly increases the stresses in an underlying deposit of compressible clay or silt, it is likely to experience large systematic differential settlements. These cannot be avoided merely by providing great strength in the slab; it is also necessary to provide stiffness. However, a stiff foundation is likely to be subjected to bending moments far in excess of those corresponding to the flat-slab or subgrade-modulus analyses. These moments may be so great as to require deep beams, trusses, or even utilization of the superstructure to

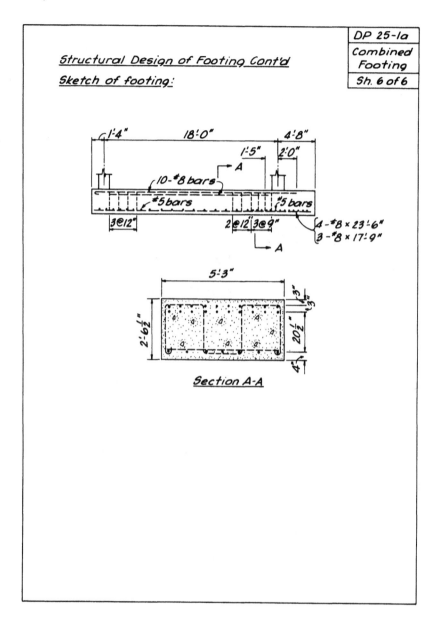

provide the necessary strength. The raft foundation must now consist of two almost independent elements: the base slab, which may still be designed by the flat-slab analogy; and the stiffening members, which have the function of preventing most of the differential settlement of the points of support for the base slab.

The design of the stiffening members is a difficult structural problem for which no straightforward procedure as yet exists. Experience and mature judgment are essential.

Increasing the stiffness of a raft foundation above a compressible soil leads to a redistribution of the soil pressure against the base slab; the raft should not deflect excessively when acted upon by the redistributed, nonuniform pressure. Unfortunately, no reliable rational estimate of the real distribution of pressure can be made because of the extremely complex relationships among stress, strain, and time, not only for soils but also for building frames.

According to theory, if a rigid structure

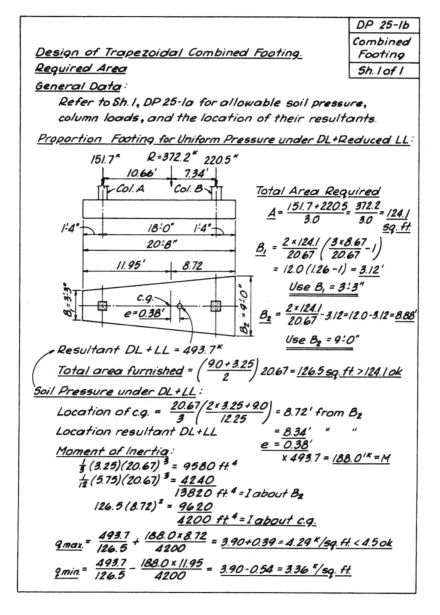

rested on a subgrade with a constant modulus of elasticity, the pressure on the base of the structure would vary from a minimum near the middle of the base to a maximum at the edges. This knowledge has occasionally been used as a basis for estimating the soil pressure for design of a stiff raft foundation above a subsoil of compressible clay or silt. However, since it is unlikely that the edge pressure will exceed twice the average pressure, it is considered conservative to design the stiffening elements and the slab of such a raft foundation for two conditions: a soil pressure uniform over the entire slab; and a pressure varying in some arbitrary fashion from a minimum at the middle to twice the average at the edges. The average pressure is, of course, the same for both conditions. In any part of the foundation the strength is made adequate for whichever distribution leads to the more severe conditions, and for the nonuniform distribution the distortion of the foundations must not be excessive.

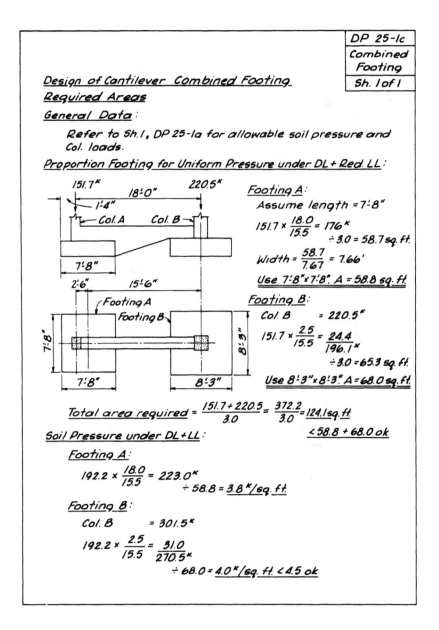

This is a logical basis for design, but it may often be too conservative and hence uneconomical. The choice of the most suitable edge pressure often taxes the ability of the most experienced foundation engineer.

These paragraphs suggest that the design of stiff rafts above sand, stiff clay, or other relatively incompressible materials is a somewhat complex but otherwise routine structural problem. On the other hand, if the subsoil contains highly compressible layers, the problem is by no means routine.

As an alternative to the relatively great expense of a stiff raft of large size above a compressible deposit, substantial economy can be realized by designing a flexible raft and superstructure that can deform without structural damage into the shape corresponding to the compression of the subsoil. Obviously this alternative cannot be chosen if architectural or functional considerations demand a relatively unyielding structure. On the other hand, many types of structures, such as large steel tanks and one- or two-story steel-frame industrial buildings with corrugated metal or asbestos siding, can experience large deformations without any detrimental consequences. It may often prove preferable to accept the deformations if the cost of a stiff foundation can be avoided.

The design of a flexible raft foundation cannot readily be based on the calculation of stresses in the slab. Instead, it is necessary to estimate, on the basis of a settlement forecast, the maximum curvature to which the raft may be subjected, and to select the thickness of the slab and the amount of reinforcement such that the slab will not develop cracks large enough to be unsightly or to permit serious leakage of groundwater even if it is deformed in accordance with the estimated curvature. As a rough guide, the quantity of steel may be taken as 1 per cent in each of two directions at right angles to each other, equally divided between the top and bottom of the slab. The thickness of the slab should not generally be greater than about 0.01 times the radius of curvature, but local increases in thickness near columns and walls may be required to prevent shear failures. In any particular project, many factors enter into the final choice of the criteria for design, and expert judgment and experience are required. Hence, the design of a flexible raft above compressible deposits, like that of a stiffened raft, is not a routine problem.

PROBLEM

1. Given the following columns loads, for which uniform soil pressure is desired under reduced live load.

	Col. A (kips)	Col. B (kips)
DL	94.4	116.6
Reduced LL	46.2	74.8
LL	92.4	149.6

Use the allowable soil pressure given in DP 25-1a and:
   a. Proportion a rectangular combined footing to support these column loads.
   b. Proportion a trapezoidal combined footing to support these column loads.
   c. Proportion a cantilever combined footing to support these column loads.

SUGGESTED READING

The many considerations in evaluating the coefficient of subgrade reaction are discussed in a definitive paper by K. Terzaghi (1955b), "Evaluation of Coefficients of Subgrade Reaction," *Géotechnique*, 5, 4, 297–326. Mathematical solutions of various problems of practical importance, based on the premise that the subsoil possesses a constant k, are contained in M. Hetenyi (1946), *Beams on Elastic Foundation*, Ann Arbor, University of Michigan Press, 255 pp. A more general but approximate method of calculating the distribution of contact pressure between a rigid raft and an elastic subgrade is given by L. Barden (1962), "Distribution of Contact Pressure Under Foundations," *Géotechnique*, 12, 3, 181–198.

The difference between the usual assumptions and reality is exemplified, however, in C. Y. Teng (1949), "Determination of the Contact Pressure Against a Large Raft Foundation," *Géotechnique*, 1, 4, 222–228. This paper leaves no doubt that the design of a large raft foundation is by no means a routine matter.

Use of the concept of the modulus of subgrade reaction, tempered by experience and judgment, is illustrated in S. V. DeSimone, and J. P. Gould (1972), "Performance of Two Mat Foundations on Boston Blue Clay." *Proc. ASCE Conf. on Performance of Earth and Earth-Supported Structures*, Purdue, 1, Part 2, 953–980.

Charles Augustin Coulomb (1736–1806)

French army engineer. After impairing his health by nine years of service in the tropics he devoted himself to purely scientific research and became immortal by his fundamental contributions to our knowledge of friction, electricity, and magnetism. However, at the beginning of his career as a scientist, in 1773, he published a paper on the earth pressure against retaining walls. The theory contained in this paper is distinguished by its clarity and simplicity, and by its realistic assumptions derived from field observations regarding the shape of the surface of rupture. Today it is still the basis for the design of retaining walls. (Cliché des musées nationaux français.)

PLATE 26.

CHAPTER 26

Retaining Walls and Abutments

26.1. Introduction

A properly designed retaining wall or abutment must satisfy two almost independent requirements. First, to make the structure safe against failure by overturning and excessive settlement, the pressure beneath the base must not exceed the allowable soil pressure; furthermore, the structure as a whole must have an adequate factor of safety with respect to sliding along its base or along some weak stratum below its base. The structure is proportioned, and its overall stability is checked, for working loads and for earth pressures unmodified by load factors. Second, the entire structure as well as each of its parts must possess adequate strength. In this phase of the design, load factors are ordinarily applied. The corresponding pressures and forces provide the basis for checking the ultimate structural strength at various critical sections. Thus, the entire procedure is similar in principle to that used in the previous chapters of Parts C and D for the design of footings.

Retaining walls and abutments of the gravity and semigravity types are sometimes used, especially where a high degree of permanence under unfavorable climatic conditions is desired. The design of such structures, however, is relatively simple in comparison with that of cantilever walls and abutments. Therefore, this chapter is concerned primarily with the latter type. Backfilling and drainage, both matters of outstanding importance with respect to the successful performance of retaining walls, have been discussed in Art. 14.2.

In general, the procedure for the design of retaining walls and abutments, like that for many other structures, is essentially one of trial and correction. Tentative dimensions must be assumed before either the stability or the structural strength can be investigated. After analysis, certain dimensions may have to be revised before a satisfactory design is obtained.

26.2. Proportions of a Cantilever Retaining Wall

Base. The base of the ordinary cantilever retaining wall should be as narrow as possible for economy, but at the same time it must be wide enough to provide adequate stability against overturning and sliding, and to reduce the soil pressure to a tolerable value. The ratio of the width of the base to the overall height of the wall commonly varies from 0.40 to 0.65. The smaller ratio is appropriate if the base is supported by firm soil and if the backfill has a horizontal surface and consists of clean sand or gravel. On the other hand, as the strength of the subsoil or backfill decreases, and as the slope of the backfill increases, the ratio may approach or even exceed 0.65. Furthermore, the width of the base is influenced by additional loads on the fill behind the wall, such as those due to a railroad, highway, or structure.

The thickness of the base is a function of the shears and moments at sections located at the front and back faces of the stem. Therefore, the thickness is significantly influenced by the position of the stem on the base. If the stem is located so that the projection of the toe from the front face of the wall is approximately $\frac{1}{3}$ the width of the base, the thickness of the base commonly lies in the range of $\frac{1}{12}$ to $\frac{1}{8}$ the height of the wall.

The depth of the base below the ground surface in front of the wall should be sufficient to avoid the movements associated with freezing and thawing of the soil. The necessary depth varies from a few inches in the southernmost states to as much as 8 ft in some of the northern regions of the United States. Even if the depth of frost is small, the base should preferably be placed below the zone of seasonal volume change caused by the variation in moisture content of the subsoil. This is particularly important if swelling clays are encountered. In many instances it becomes necessary to establish the base at a depth much greater than these minimum values to reach a stratum of soil adequate to withstand the pressures imposed upon it.

Stem. The thickness of the stem must be sufficient to resist safely the shears and moments due to the earth pressure against the back of the wall. For this reason, the strength and slope of the backfill have considerable influence on the thickness of the stem. The thickness at the top of the wall should be great enough to permit easy placement of the concrete. The critical section for shear and moment is at the junction of the stem with the base. In order to provide adequate strength at this section it is customary to increase the thickness of the stem with depth by $\frac{1}{4}$ to $\frac{3}{4}$ in./ft.

26.3. Summary of Forces Acting on Retaining Walls

The forces for the stability analysis of a cantilever retaining wall are shown in Fig. 26.1a. The principal unfactored forces are the earth pressure P_A against the vertical section ab through the heel, the earth pressure P_P against the vertical section cd through the toe, the soil pressure ΣV, which acts vertically on the base db, the shear along

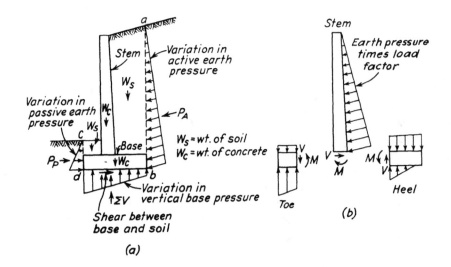

FIGURE 26.1. Cantilever retaining wall. (a) Forces considered in stability analysis. (b) Forces acting upon the principal structural elements of the wall.

the base db, and the weights of the various parts of the wall and of the masses of earth above the base.

The three principal structural elements of the wall for investigation of strength are shown in Fig. 26.1b. The forces acting on each of these parts correspond to factored loads and are, therefore, different from those used in the initial stability analysis. For strength design it is convenient and conservative to assume that the vertical pressure against the base is distributed uniformly over the front third of the base. This procedure is demonstrated in DP 26-1.

26.4. Earth Pressure

Rankine's Theory. The pressure exerted by an earth backfill against a retaining wall can be computed with reasonable accuracy on the basis of theory only for conditions rarely encountered in practice. In the first place, the designer must know what materials are to be used for the backfill and in what state they will be placed. This implies careful supervision of the backfilling operations. Moreover, the backfill, once it is placed, must be protected in such a way that its physical properties remain constant. These conditions cannot be satisfied economically on any but the largest and most important projects; ordinarily the designer can hope to learn in advance of construction little more than the general type of backfill material. Hence, theoretical earth-pressure calculations can rarely be justified for a particular retaining wall because the physical characteristics of the backfill are not usually known. It is, therefore, preferable to estimate the earth pressure on the basis of charts or rules having a partly theoretical, partly empirical basis. Nevertheless, a knowledge of earth-pressure theory permits recognition of the more important variables and their influence on the earth pressure and serves to sharpen the judgment of the engineer. Consequently, presentation of the charts and rules will be preceded by a brief account of the principles of the theories of earth pressure. For the most part, the discussion will be limited to the pressure of dry cohesionless sand having a shearing resistance $s = p \tan \phi$. This value of s corresponds to Coulomb's equation (eq. 4.2), in which c is set equal to zero.

An idealized deposit of dry cohesionless sand with a horizontal ground surface is shown in Fig. 26.2a. The sand extends infinitely in horizontal directions and to infinite depth. At point A in the interior of the deposit the vertical pressure on a horizontal plane is

$$p_v = \gamma z \qquad 26.1$$

where γ is the unit weight of the deposit and z is the depth. The horizontal pressure on vertical planes at point A is considered to be

$$p_h = k_0 p_v = k_0 \gamma z \qquad 26.2$$

where k_0 is known as the *coefficient of earth pressure at rest*. The value of k_0 depends on the relative density and method of deposition of the sand. It can be determined experimentally; for a sand deposited in horizontal layers without tamping it ranges between 0.4 and 0.5 (Terzaghi, 1920).

The rupture lines (Art. 4.5) in the Mohr diagram for the sand are shown in Fig. 26.2b; they are inclined at ϕ to the horizontal axis. The pressures p_v and p_h are represented by points on the horizontal axis because there are no shearing stresses on horizontal planes in the deposit; hence, there are no shearing stresses on vertical planes and p_v and p_h are both principal stresses.

The circle of stress, circle 1, corresponding to the principal stresses p_v and $k_0 p_v$, does not touch the rupture lines because the deposit is nowhere in a state of failure. On no plane or planes in the deposit is the shearing strength of the sand fully mobilized.

Point A (Fig. 26.2c) is located in a portion of the deposit bounded by two vertical planes ab and $a'b'$ separated by a distance L. If by some imaginary process the deposit is now stretched horizontally, so that planes ab and $a'b'$ move farther apart, the horizontal pressure p_h will tend to decrease; p_v on the other hand must remain equal to the weight γz of the overburden. The corresponding circle of stress, circle 2 in

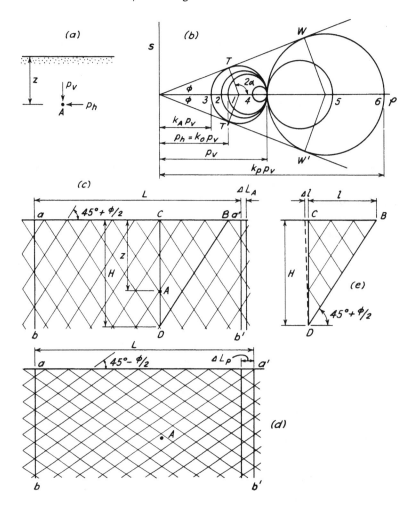

FIGURE 26.2. Deformation conditions associated with Rankine's earth-pressure theory. (a) Stresses at point A beneath horizontal surface of semi-infinite body of sand. (b) Circles of stress and rupture diagram illustrating active and passive states of stress. (c) Planes of slip in active Rankine state associated with horizontal extension of sand. (d) Planes of slip in passive Rankine state associated with horizontal compression of sand. (e) Planes of slip in local active Rankine zone behind vertical surface CD.

Fig. 26.2b, is larger than circle 1 for earth pressure at rest. As the stretching is continued, the circle of stress becomes larger until it touches the rupture lines and becomes the *rupture circle*, circle 3 in Fig. 26.2b. At this stage, failure must occur within the deposit. The failure takes the form of slip between the particles along two sets of planes corresponding to the points of tangency T and T' of the rupture circle. According to the geometry of the rupture diagram (Art. 4.5), the planes of slip are inclined at $45° + \phi/2$ to the plane on which the major principal stress acts; that is, they rise (Fig. 26.2c) at $45° + \phi/2$ to the horizontal. Further stretching of the deposit can cause only further slip along the failure planes because no circles of stress larger than circle 3 can exist if the major principal stress is p_v. Hence, no smaller horizontal pressure can exist at A than that corresponding to the left-hand extremity of the rupture circle 3. This minimum lateral pressure is known as the *active earth pressure* p_A.

Its value can be determined, from the geometry of Fig. 26.2b, as

$$p_A = k_A p_v = \frac{1 - \sin \phi}{1 + \sin \phi} p_v \qquad 26.3$$

The coefficient

$$k_A = \frac{1 - \sin \phi}{1 + \sin \phi} \qquad 26.4a$$

is known as the *coefficient of active earth pressure*. It may also be expressed, by trigonometric transformation, as

$$k_A = \tan^2 (45° - \phi/2)$$
$$= \frac{1}{\tan^2 (45° + \phi/2)} \qquad 26.4b$$

For a given sand at a given relative density, a definite strain $\Delta L_A/L$ is required to produce the active state (Terzaghi, 1934). For a dense sand it is on the order of 0.1 per cent; for a loose sand it is several times greater.

If the semiinfinite deposit of sand is compressed instead of being stretched, so that planes ab and $a'b'$ (Fig. 26.2d) are moved closer together, the horizontal pressure p_h increases while p_v remains constant. Consequently, the circle of stress becomes smaller (circle 4) and, when $p_h = p_v$, reduces to a point. As horizontal compression continues, the horizontal pressure exceeds the vertical pressure and becomes the major principal stress (circle 5). Eventually, the circle of stress touches the rupture lines and becomes the *rupture circle*, circle 6 in Fig. 26.2b. Failure then occurs within the deposit along two sets of planes corresponding to the points of tangency W and W'. According to the geometry of the rupture diagram, these planes are inclined at $45° + \phi/2$ to the plane on which the major principal stress acts. Since the major principal stress now is the horizontal stress, the failure planes (Fig. 26.2d) are inclined at $45° - \phi/2$ to the horizontal. Further compression of the deposit can cause only further slip along the failure planes because no circles of stress larger than circle 6 can exist if the minor principal stress is p_v. Hence, no larger horizontal pressure can occur at A than that corresponding to the righthand extremity of the rupture circle 6. This maximum lateral pressure is known as the *passive earth pressure* p_P. Its value can be determined from the geometry of Fig. 26.2b, as

$$p_P = k_P p_v = \frac{1 + \sin \phi}{1 - \sin \phi} p_v \qquad 26.5$$

The coefficient k_P is known as the *coefficient of passive earth pressure*. It may also be expressed as

$$k_P = \tan^2 (45° + \phi/2) = 1/k_A \qquad 26.6$$

For a given sand at a given relative density, a definite strain $\Delta L_P/L$ is required to produce the passive state. The magnitude of the necessary strain is several times larger than the tensile strain required to produce the active state.

According to eqs. 26.3 and 26.5, both the active and passive earth pressure increase in direct proportion to the depth below the surface. The total pressure on a unit width of a vertical plane extending from the surface to a depth H is, therefore,

$$P_A = \tfrac{1}{2} k_A \gamma H^2 \qquad 26.7$$

or

$$P_P = \tfrac{1}{2} k_P \gamma H^2 \qquad 26.8$$

The theory discussed in the preceding paragraphs was originally developed by Rankine (1857). Equation 26.7 has frequently been used to calculate the active earth pressure of a sand backfill with horizontal surface against a vertical retaining wall with height H. However, examination of Fig. 26.2c demonstrates that the state of stress associated with Rankine's theory for these conditions requires that there be no shearing stresses on vertical planes. Since the backs of real walls are rough and shearing stresses may develop, the Rankine theory can for most conditions provide only an approximation.

In reality no semiinfinite masses of sand exist. If, however, the active state of stress can be induced in a wedge-shaped zone such as CBD (Fig. 26.2c), the earth pressure against the vertical plane CD with height H is correctly given by eq. 26.7. The

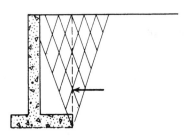

FIGURE 26.3. Active Rankine state of stress behind cantilever retaining wall.

required strain $\Delta l/l$ (Fig. 26.2e) is the same as that needed for the active state, $\Delta L_A/L$ (Fig. 26.2c). It is apparent that the strain can be produced by rotation of the plane CD about its lower end. The zone CBD is often designated as the *failure wedge*.

Coulomb's Earth-Pressure Theory. Rankine's theory can be modified to take account of cohesion (Resal, 1910), and to take account of a sloping ground surface. It can also allow for a uniform surcharge over the ground surface and a free water surface parallel to the surface of the backfill. The theory can be applied directly to the common cantilever retaining wall illustrated in Fig. 26.3 because the vertical section through the heel of the wall corresponds to section CD in Fig. 26.2c. Nevertheless, many other problems commonly encountered cannot be solved within the framework of Rankine's theory or of more general theories based on the state of stress within the failure wedge (Brinch Hansen, 1953; Sokolovski, 1960). Satisfactory solutions can often be obtained, however, by means of the so-called *wedge theories* in which the equilibrium of the failure wedge, such as BCD (Fig. 26.2e), is investigated without regard to the state of stress within the wedge. As a matter of fact, the best known of the wedge theories was developed by Coulomb in 1776, almost a century before Rankine published his state-of-stress solution.

Coulomb recognized that the resultant earth pressure P_A acting on the face CD of a wall retaining the earth is determined by the equilibrium of a wedge such as BCD (Fig. 26.4a). He concluded on the basis of observations of slips behind retaining structures that the surface of sliding BD is likely to have a slight curvature but, as an approximation, be replaced the curved surface by a plane. At the instant of slip of the wedge BCD along the failure surface, the resultant F of the normal and frictional forces on BD is inclined at ϕ to the normal to the failure surface. Therefore, if the direction of the failure surface is known, the direction of F is also known. Coulomb further assumed that the resultant active earth pressure P_A can act at any arbitrarily assigned angle to the normal to the face CD of the structure, providing the angle does not exceed δ, the angle of friction between the backfill material and the material comprising the wall. Thus, for any specific problem the direction of P_A can be considered known. Since the weight W of the wedge can be determined in direction and magnitude, the force triangle (Fig. 26.4b) can be drawn and the magnitude of P_A determined.

However, it is not known in advance whether BD is actually the plane on which the slip will occur. If BD is not the critical plane, the shearing resistance is not developed along it to the fullest extent; that is, the inclination of F to the normal to BD is less than ϕ and the value of P_A is less than that corresponding to some other plane on which the full shearing strength of the material is required for equilibrium. Hence, to find the correct active earth pressure it is necessary to assume several surfaces of sliding and to construct a force triangle for each of the corresponding wedges, each time on the assumption that the shearing

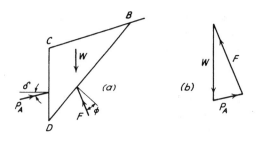

FIGURE 26.4. (a) Forces on failure wedge in Coulomb's earth-pressure theory. (b) Force triangle for evaluation of P_A.

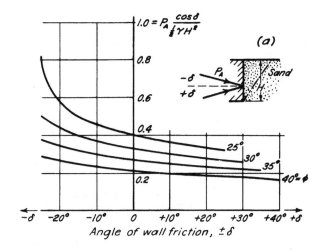

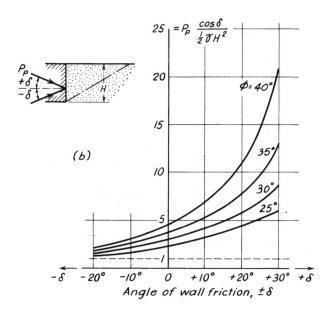

FIGURE 26.5. Values of (a) active and (b) passive earth pressure according to Coulomb's theory, for different values of ϕ and δ.

resistance is fully developed on the assumed surface. The maximum value of P_A so determined is the Coulomb active earth pressure, and the corresponding wedge is the failure wedge.

Coulomb actually approached the problem analytically as a problem in maxima and minima. He found, for example, that for a horizontal ground surface and for $\delta = 0$,

$$P_A = \tfrac{1}{2} k_A \gamma H^2 \qquad 26.7$$

where

$$k_A = \frac{1 - \sin \phi}{1 + \sin \phi} \qquad 26.4a$$

These equations are identical with those derived by Rankine. Moreover, the boundary of the failure wedge was found to rise at $45° + \phi/2$ with the horizontal.

For a horizontal ground surface and a vertical wall, but with the resultant earth pressure acting not horizontally but at an angle δ with respect to the normal to the wall, Coulomb's theory leads to values of active and passive pressure given by the diagrams in Figs. 26.5a and 26.5b, respectively.

Trial Wedge Method. For many practical problems, it is convenient to assume a series of plane surfaces of failure, to construct the corresponding force triangles, and to determine graphically the value of the active earth pressure. A simple procedure for assembling all the force triangles in a single diagram is known as the *trial wedge method*. By means of this procedure the active earth pressure can be determined for a variety of conditions, including irregularly shaped ground surfaces. Partial submergence of the backfill can also be taken into account. For these reasons, the trial wedge method is widely used in practice.

The trial wedge method is illustrated in Fig. 26.6. A wall with an inclined back, supporting a backfill with irregular surface, is shown in Fig. 26.6a. The active earth pressure is assumed to act at an arbitrarily chosen angle β to the normal to the back of the wall. The lines *D1*, *D2*, etc., represent various assumed positions of the plane surface of sliding; that is, they represent the boundaries of various trial wedges of failure. The force triangles are assembled in Fig. 26.6b. The weight of the wedge *OD1* is represented by vector *01*, of *OD2* by vector *02*, etc. Each of the rays *01′*, *02′*, etc., corresponding to the vector *F* in Fig. 26.4b, is parallel to the resultant force on the assumed surface of sliding. For convenience, the directions of the rays may be determined in the following manner. In the cross section (Fig. 26.6a), construct a vertical through point *D*. Determine the angle (θ_1, θ_2, etc.) between this vertical and each of the assumed surfaces of rupture. In the force diagram (Fig. 26.6b), lay off a horizontal line through *O* and establish a reference line *OR* making an angle ϕ below the reference line. With *O* as a center, lay off below the reference line the successive angles θ_1, θ_2, etc.

Each force triangle is then completed by drawing from the lower end of the appropriate weight vector a line parallel to P_A. In this manner the points 1′, 2′, etc., are determined. A curve, known as the *earth-pressure locus*, is then drawn to connect these points.

To determine the active earth pressure, a tangent is constructed to the earth-pressure locus, parallel to the weight vectors. From the point of tangency *T* a vector is drawn to the line of weight vectors, parallel to P_A. The length of this vector represents the active earth pressure, and the value of θ

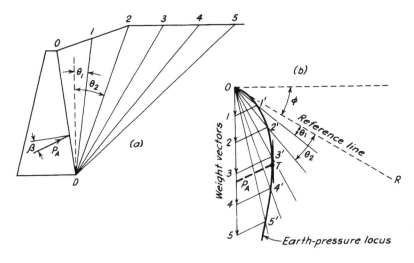

FIGURE 26.6. Trial wedge method for determining active earth pressure. (a) Trial surfaces of sliding *D1*, *D2*, etc. (b) Combined force diagram.

corresponding to OT determines the position of the plane of rupture.

If a free horizontal water surface exists within the backfill, the weights of those parts of the wedges below the water surface are taken to be the submerged weights. The earth pressure is then calculated by the same procedure as for dry sand, but to the earth pressure must be added the hydrostatic pressure of the water.

Distribution of Earth Pressure. The state-of-stress theories, such as Rankine's earth-pressure theory, furnish the intensity of the active earth pressure at any depth. Hence, the distribution of earth pressure or the point of application of the resultant active earth pressure is known. On the other hand, the wedge theories deal only with the equilibrium of the weight of the failure wedge and the resultant external forces acting upon it. The point of application of the earth pressure is not known and must be determined independently.

If the surface of the backfill within the failure wedge is plane and carries no surcharge, the point of application of the resultant active earth pressure may be taken at the lower third-point of the back of the wall. If the surface of the backfill is plane and carries a uniformly distributed surcharge q (Fig. 26.7a), the surcharge may be converted into an equivalent height of fill $H_s = q/\gamma$, where γ is the unit weight of the backfill material, and the height of the wall is considered to be $H' = H + H_s$. The center of pressure may then be taken at $H'/3$ above the base of the wall. If the surface of the backfill has an irregular shape, the point of application of P_A may be located approximately by assuming it to coincide with the intersection O' of the back of the wall and a line drawn through the centroid O of the failure wedge parallel to the surface of sliding (Fig. 26.7c).

The direction of P_A must also be established before a solution can be carried out by a wedge theory. Where slip is likely to occur along the face of the wall in contact with the backfill, as, for example, in Fig. 26.6a, the angle β should ordinarily be taken as the angle of friction δ between the wall material and the backfill; for concrete retaining walls, the value $\delta = \frac{2}{3}\phi$ is usually a reasonable approximation. For cantilever walls (Fig. 26.3), where slip is more likely to occur within the backfill than along the contact between concrete and backfill, the direction of P_A is approximately the same as the slope of the surface of the backfill.

Pressures Due to Line Loads and Concentrated Loads. The pressure against the vertical back of a wall due to a line load q' per unit of length (Fig. 26.8a) parallel to the crest of the wall has been found by large-scale tests (Gerber, 1929; Spangler, 1938) to vary with depth below the top of the wall in the manner shown in Fig. 26.8b. Similarly, the magnitude and distribution of earth pressure acting horizontally against a vertical line on the back of a wall directly opposite a concentrated load Q have been found to vary as shown in Fig. 26.8c.

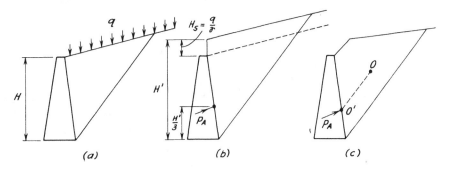

FIGURE 26.7. (a) and (b) Method of determinging point of application of earth pressure against wall with sloping backfill carrying uniform surcharge q per unit of area. (c) Method for wall supporting irregularly shaped backfill.

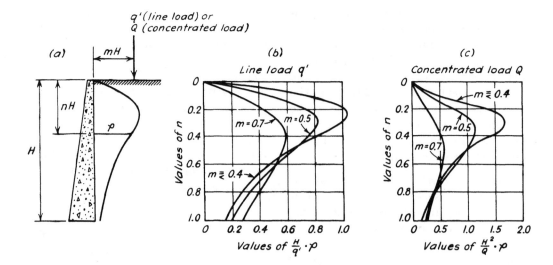

FIGURE 26.8. Earth pressure against vertical wall due to line load or concentrated load on surface of horizontal backfill. (a) Dimensions and variables. (b) Pressures due to live load q per unit of length parallel to back of wall. (c) Pressures along vertical line on back of wall directly opposite point of application of concentrated load Q.

Charts for Estimating Backfill Pressure. The wedge theories, including the trial wedge method, can be modified with little difficulty to take account of cohesion on the assumption that the shearing strength of the soil is expressed as

$$s = c + p \tan \phi \qquad 4.2$$

Procedures are available for dealing with most of the conditions encountered in practice (Huntington, 1957). Nevertheless, since the designer rarely knows in advance what materials are to be used for the backfill and in what state they will be placed, it is usually preferable to determine the earth pressure for design of ordinary retaining walls, with heights not exceeding about 20 ft, by means of charts such as those shown in Fig. 26.9. These charts are based partly on theory and partly on studies of the performance of satisfactory and unsatisfactory walls. In order to use Fig. 26.9, the backfill material must be classified into one of the four categories listed in the diagram.

The magnitude of the passive pressure can also be calculated by the wedge theories. Values may be estimated by means of Fig. 26.5b, if ϕ can be approximated and δ taken as $\tfrac{2}{3}\phi$.

For the materials included in the categories described in Fig. 26.9 the intensity of both the active and passive pressures can be assumed to increase directly with the depth below the ground surface.

The passive pressure P_P is often disregarded in the stability analyses of retaining walls because of the possibility that the soil in front of the wall may not yet be in position when the backfill is being placed or may at some time be removed by excavation or scour. Furthermore, the strength of the soil may be lowered as a result of frost action, infiltration of water, or cracking due to shrinkage.

When earth pressures have been measured by means of pressure cells installed in the back faces of retaining walls, they have been found to exceed the active values and the points of application have been at higher elevations than the foregoing procedures would predict. These apparent discrepancies merely indicate that a well-designed retaining wall with an adequate factor of safety does not deflect or rotate far enough under

Vertical Pressure Against Base

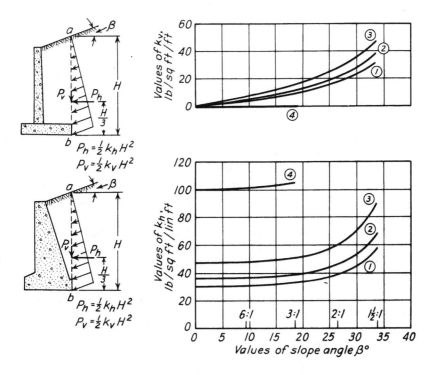

FIGURE 26.9. Chart for estimating pressure of backfill against retaining walls supporting backfills with plane surface. Use of chart limited to walls not over about 20 ft high. (1) Backfill of coarse-grained soil without admixture of fine particles, very permeable, as clean sand or gravel. (2) Backfill of coarse-grained soil of low permeability due to admixture of particles of silt size. (3) Backfill of fine silty sand, granular materials with conspicuous clay content, and residual soil with stones. (4) Backfill of very soft or soft clay, organic silt, or silty clay.

working stresses to permit the development of the active state. Before such a wall would fail, however, unless it were unusually rigid and on an unusually rigid foundation, it would experience the required movement and the pressures would decrease to those corresponding to the active state. Hence, the stability analysis of retaining walls is properly based on the active earth pressure.

26.5. Vertical Pressure Against Base

Experience has shown that most failures of retaining walls have been the result of misjudgment of the foundation conditions. Therefore, a careful evaluation of the strength and compressibility of the soil beneath both the base and the backfill is the most important single step in the design of a retaining wall. All the factors discussed in Part C in connection with the selection of suitable types of foundations and the choice of appropriate soil pressures or pile loads are as worthy of consideration with respect to retaining walls as to any other type of structure.

The pressure against the base of a retaining wall is commonly assumed to vary in planar fashion, as shown in Fig. 26.1a. The position of the resultant of all the forces acting on the supporting material at the level of the base can be found by eq. 24.1. For ordinary walls supported by soil the resultant corresponding to unfactored loads is commonly required to intersect the base within the middle third; hence, the entire area beneath the base is theoretically subjected to compression. If this condition exists, the magnitude of the soil pressure against the base can be determined by eq. 24.4.

Two useful criteria for proportioning soil-supported walls may be established from the preceding discussion: (1) the eccentricity of the resultant force, measured from the center line of the base, should not exceed one sixth the width of the base, and (2) the maximum pressure should not exceed the allowable soil pressure. One or the other of these two criteria commonly controls the width of the base.

On the other hand, retaining walls may be supported by rock, in which case the first criterion given above is commonly changed to permit larger eccentricities. However, in order to provide adequate safety against overturning, most designers prefer to limit the eccentricity to one fourth the width of the base. That is, the resultant force must intersect the plane of the bottom of the base within the middle half, even though the pressure at the toe may be considerably less than the allowable pressure for the rock. When the resultant lies outside the middle third, the maximum pressure at the toe must be computed by eqs. 24.5 and 24.6, because compression does not exist over the entire area of the base.

26.6. Forces Resisting Sliding

According to Fig. 26.1a, the horizontal component of the unfactored earth pressure P_A must be resisted by the shear between the soil and the base and by the passive earth pressure of the soil in contact with the front of the structure. The ratio between the resisting forces and the horizontal component of P_A is known as the *factor of safety against sliding*. This ratio should be not less than 1.5. Moreover, the passive earth pressure should be disregarded in computing the factor of safety unless local conditions permit reliable evaluation of its lower limiting value and unless the existence of the pressure is assured during the placing of the fill behind the wall.

The shearing resistance between the base and the soil is greatly influenced by the character of the soil. If the surface of contact between the concrete and soil is rough, the maximum shearing strength of the soil can be counted on. Procedures for determining the shearing strength of soils of different types have been discussed in Chap. 4. However, in the absence of tests, the total shearing resistance between the base and a soil that derives most of its strength from internal friction may be taken as the normal force ΣV times a coefficient of friction selected from the following values. For coarse-grained soil without silt, the coefficient of friction may be taken as 0.55; for coarse-grained soil with silt, 0.45; and for silt, 0.35. If the base of the retaining wall rests on clay, the shearing resistance against sliding should be based on the cohesion of the clay, which can be conservatively estimated as one half the unconfined compressive strength. If the clay is stiff or hard, its surface should be roughened before the concrete base is placed.

If the factor of safety against sliding is less than 1.5, the design should be revised. The resistance to sliding may be increased by the use of a key that projects into the soil below the base, as shown in Fig. 26.10, or the base may be widened to increase the surface of sliding. For the same volume of concrete, a key is ordinarily considered to be somewhat more effective than an increase in base width, but, on the other hand, the width of the base can usually be extended at less cost.

The effectiveness of short keys is often overestimated. Consideration of the equilibrium of the block of soil $bcde$ (Fig. 26.10) leads to the conclusion that the total horizontal force acting on the key can be no larger than the sum of the force P_K and the shearing force S developed on the surface

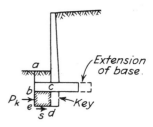

FIGURE 26.10. Horizontal forces resisting movement of key beneath retaining wall.

de. However, S is not likely to be much greater than the shearing resistance available along *bc* if no key were employed. For this reason, any additional horizontal force gained by the use of a key can be no larger than P_K. Hence the additional resistance to sliding offered by the projection of a key below the base can be determined only by an evaluation of the lower limiting passive resistance of the soil against the surface *be* during the time the backfill is being placed and throughout the subsequent life of the retaining wall. Any estimate of the passive resistance against the plane *abe*, or on any portion of this plane is, at best, only a crude approximation. Moreover, the excavation for a key is likely to disturb the subsoil during construction and in some instances may conceivably do more harm than good.

On the other hand, if the base of a retaining wall is supported by rock or by a very stiff cohesive soil, a key may provide an effective means of creating additional resistance to sliding. If one or the other of these subsoil conditions exists, a larger, more reliable passive resistance P_K (Fig. 26.10) acts against the plane *be*. Moreover, the shearing resistance S developed on the surface *de* is likely to exceed the shearing resistance available along *bc* if no key were employed.

26.7. Summary of Procedure for Design of Cantilever Retaining Wall

The design of a retaining wall can be summarized by the following steps, of which the first five relate to proportioning and stability, and the final two constitute the strength investigation.

1. Choose tentative proportions for the structure, including dimensions for the stem and the base as well as the position of the stem on the base.
2. Estimate the magnitude of all the forces acting above the bottom of the base, as indicated in Fig. 26.1*a*.
3. Determine the point of intersection of the resultant of the forces found in step 2 with the plane of the bottom of the base. The location of this point constitutes a check on the stability of the wall with respect to overturning.
4. Determine the magnitude of the foundation pressure against the base.
5. Check the factor of safety against sliding.
6. Apply load factors to the earth pressure and other loads and compute the corresponding pressures, reactions, shears, and moments.
7. Calculate the ultimate strengths at critical sections of the elements shown in Fig. 26.1*b*.

The computations involved in steps 3 to 7 almost always indicate necessary revisions in the tentative dimensions of step 1.

ILLUSTRATIVE DESIGN. DP 26-1. CANTILEVER RETAINING WALL

The application of the principles discussed thus far in Chap. 26 is demonstrated in this design plate. It is noted that the computations may be divided into two separate steps: (1) the stability analysis in which the soil pressures beneath the base are determined and the factor of safety against sliding along the base is estimated, and (2) the structural design, in which the strengths of the cantilever elements (Fig. 26.1*b*) are made adequate to resist the applied forces.

It may be noted that the maximum soil pressure at the toe computed by methods discussed in Art. 24.2 is somewhat less than the allowable value. This would suggest that the width of the base might be reduced. However, further investigation shows that very little, if any, reduction in width can be made without resulting in an inadequate factor of safety against sliding.

The computations relating to the ultimate strength of the various critical sections are typical for reinforced-concrete design and are similar to those given in previous design plates in Part D. However, the procedure for determining the factored soil pressures against the base for the ultimate strength design of the toe and heel does not correspond to eq. 24.5 or to the distribution

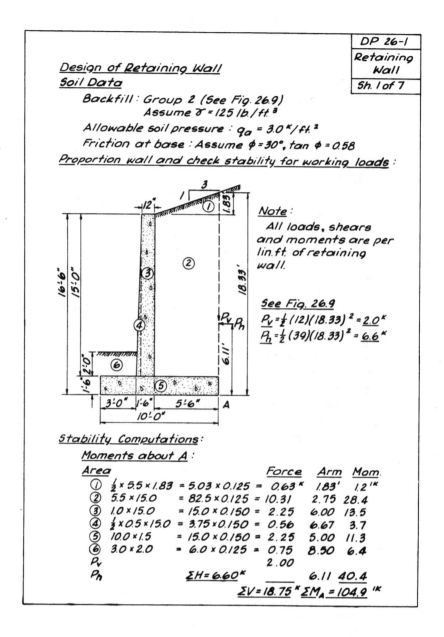

shown in Fig. 26.1b. Instead, the conservative assumption is made that the sum of the vertical loads is uniformly distributed over the front third of the base. The soil reaction against the base is therefore analogous to the compressive stress block utilized in the flexural analysis of reinforced-concrete beams and beam-columns.

26.8. Pile-Supported Retaining Walls

Introduction. A retaining wall is commonly established on a pile foundation when the soil for a considerable depth is too weak or compressible to provide adequate support for the structure. Furthermore, if excessive dimensions of the base are required for an adequate factor of safety with respect to sliding or to keep the vertical pressure within the allowable limit, it may be more economical to provide the retaining wall with a pile foundation, as shown in Fig. 26.11.

If the horizontal force against the retaining wall cannot otherwise be resisted, some

```
                                                    DP 26-1
   Stability computations cont'd.:                 Retaining
      Location of resultant:                         Wall
           From point A, 104.9/18.75 = 5.6'         Sh. 2 of 7
                       then e = 5.6 − 10/2 = 0.6' ok < 10/6
      Soil pressure at toe of base:
           q_max = 18.75/10 (1 + 6×0.6/10) = 1.875 (1+0.36) = 2.55 k/ft.²
                                                    ok < 3.0 k/ft.²
      Check F against sliding
           Shear available along base = 18.75 k × 0.58 = 10.9 k
           Passive force at toe: (See Fig. 26.5b)
               Use δ = 2/3 (30°) = 20°, P_p (cos δ / ½ σ H²) = 5.8
           P_p = 5.8 (.125)(3.5)² / 2(.940) = 4.7 k
           Min. F = 10.9/6.6 = 1.7,   Max. F = (10.9+4.7)/6.6 = 15.6/6.6 = 2.4
                                                    ok without key
   Structural Design:
      Load factors: Stem — use 1.7 P_h (Fig. 26.9)
                    Base (toe and heel) — distribute
                    ΣV uniformly over front B/3
      Concrete and steel data:
           Capacity reduction factors: 0.90 (flexure); 0.85 (shear)
           f'_c = 3,000 psi × 0.85 = 2,550 psi (for stress block)
           v_c = 2 √3,000 = 110 psi
           f_y = 40,000 psi  ρ_min = 0.005 ;  ρ_max = 0.0278
                             ρ_shrinkage = 0.002
           ℓ_d = 0.04 A_b (40,000) / √3,000 = 29.2 A_b (bott. bars)
                                      × 1.4 = 40.9 A_b (top bars)
```

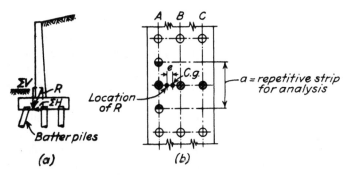

FIGURE 26.11. Pile foundation beneath retaining wall. (*a*) Section. (*b*) Foundation plan.

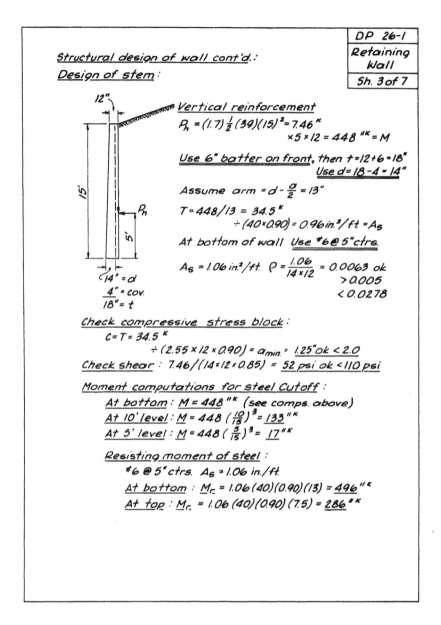

of the piles are ordinarily driven on a batter. The arrangement of the piles indicated in Fig. 26.11 is typical for retaining walls approximately 15 to 20 ft in height, subjected to moderate horizontal loads. For higher walls and more extreme conditions of loading, it is sometimes necessary to use the same number of piles along all rows and to provide batter piles in more than one row.

The design of a batter-pile foundation, as described in the following paragraphs, is based on the assumption that the only horizontal forces to be resisted are those that act above the base level of the retaining wall. In many instances, possibly in most, the lateral forces against that part of the subsoil in which the piles are buried are very much greater than those acting on the wall itself. The batter piles may then be ineffective. The circumstances under which this may occur have been discussed in Art. 18.8.

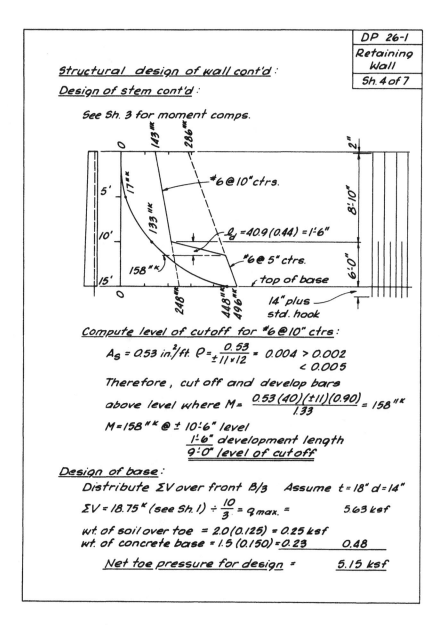

Vertical Forces on Piles. If the piles are arranged in plan as shown in Fig. 26.11b, the centroid of the piles usually lies very near the line of action of the resultant R, and the loads on the piles are approximately equal. The loads on the vertical piles in rows B and C, as well as the vertical component of the load on the piles in row A, are commonly determined by eq. 24.10. It is evident from Fig. 26.11 that the strip a containing four piles of the foundation is repetitive. For this reason, the section properties of these four piles, along with the loads and moment acting upon strip a, can be used in eq. 24.10. Figure 26.11 indicates that the resultant force R lies to the left of the centroid of the piles. Hence, the vertical load per pile in row A is larger than that in rows B and C. On the other hand, if the eccentricity were to lie to the right of the centroid, the piles in rows B and C would be subjected to a greater load than those in row A. The

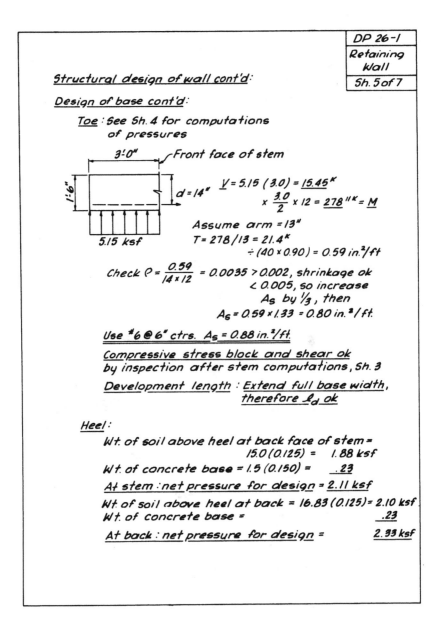

magnitude of the moment that exists at the junction of the piles and the base is the product of the summation of the vertical loads ΣV and the eccentricity e. The eccentricity is measured from the centroid of the piles, not from the centerline of the base, unless the two coincide.

Horizontal Forces on Piles. The horizontal load ΣH at the base of a pile-supported retaining wall is transferred to the piles and thence to the soil in which the piles are embedded. Thus the vertical piles may not appreciably alter the factor of safety of the wall against sliding on its base or within some weak deposit beneath its base. If the strength of the soil in which the piles are located is inadequate to provide the necessary factor of safety, the available resistance may be supplemented by the use of batter piles. Usually the resistance of the batter piles and of the soil beneath the wall are considered to act simultaneously; this presupposes that before failure sufficient lateral

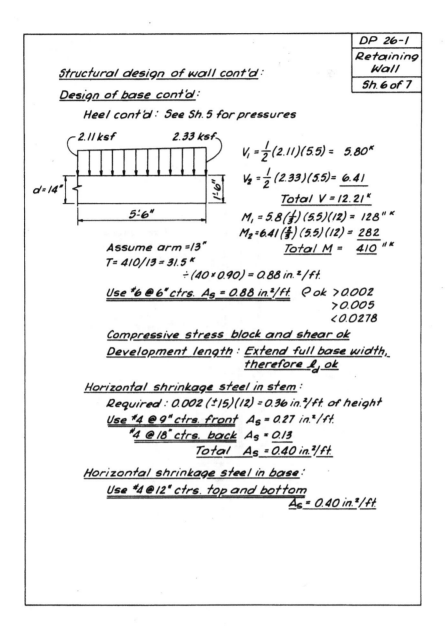

movement of the wall can take place to develop the ultimate strength of both elements.

It is commonly assumed that the vertical load carried by a batter pile is the same as it would be if the pile were vertical. Thus, the vertical component of load in a batter pile may be computed by means of eq. 24.10. If the line of action of the resultant load P (Fig. 26.12a) is made to coincide with the axis of the pile, the horizontal component of this resultant can be determined readily if the vertical component and the slope of the pile are known. The force diagram for a single batter pile is shown in Fig. 26.12b.

Steps in Design. In the design of a batter-pile foundation, two criteria must be satisfied: (1) the axial load in any pile must not exceed the safe load per pile, and (2) the sum of the horizontal components of the forces in the batter piles must equal the applied force ΣH. Since in a batter pile the relationship between the axial load and its horizontal component depends on the slope or

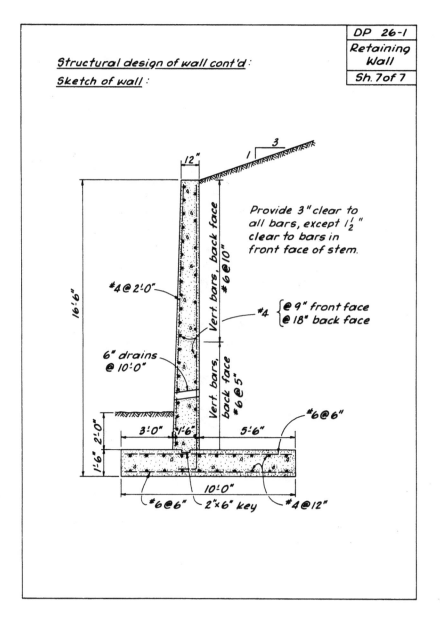

batter, which is not initially known, the design is most conveniently made by a method of trial and correction. The steps in this method for a typical problem are as follows.

1. Make a tentative layout (Fig. 26.13a) of the foundation piles for the section at the base of the wall.

2. Compute the vertical forces V_A, P_B, and P_C in each pile by means of eq. 24.10.

3. Compare the vertical forces in the batter piles with the safe axial load for the piles. The maximum vertical force in a batter pile should be limited tentatively to a value about 8 per cent less than the safe axial load to allow for the influence of the batter.

4. Construct the force polygon (Fig. 26.13b). According to this diagram, if the piles in row A are to resist the entire horizontal force ΣH, they must have the in-

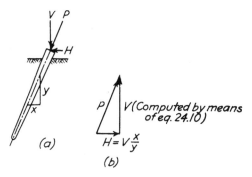

FIGURE 26.12. (a) Batter pile. (b) Force diagram for single batter pile.

clination of line FG. Thus the required batter is

$$m = 12 \frac{\Sigma H}{\Sigma V'} \qquad 26.9$$

where m = batter of the piles expressed in inches (horizontal) per foot (vertical)

ΣH = horizontal force to be resisted by batter piles

$\Sigma V'$ = summation of the vertical forces acting upon the batter piles. For the conditions shown in Fig. 26.13, $\Sigma V' = 2V_A$

5. Check whether the required batter is reasonable (see Art. 12.1). If it is not, a new layout is required or else batter piles must be used in more than one row as explained in the following paragraph.

6. Compute the axial load in the batter piles, and compare with the safe axial load per pile. If the safe load is exceeded, a new layout is required.

If the horizontal force ΣH is large, it may be necessary to batter more than one row of piles. If batter piles are used in row B as well as row A, and if the same slope is desired in both rows, a line connecting E and G (Fig. 26.13b) gives the appropriate slope. Hence, it is apparent that eq. 26.9 remains applicable, provided the vertical summation $\Sigma V'$ is adjusted to include P_B. In some designs the slope of the batter piles differs from row to row. In many instances, either engineering judgment or tests of the soil into which the piles are driven may lead the designer to the decision that only a fraction of the total horizontal force need be resisted by batter piles. In any event, a force diagram similar to that shown in Fig. 26.13b is helpful in establishing the proper slopes for the different rows.

The value m as given by eq. 26.9 is commonly expressed to the nearest inch. More refinement cannot be justified in view of the many assumptions and uncertain field conditions involved in this type of problem.

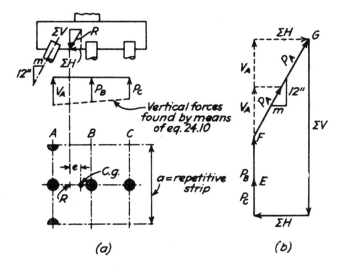

FIGURE 26.13. Diagrams illustrating method of computation of forces acting in piles beneath retaining wall. (a) Arrangement of piles. (b) Force polygon.

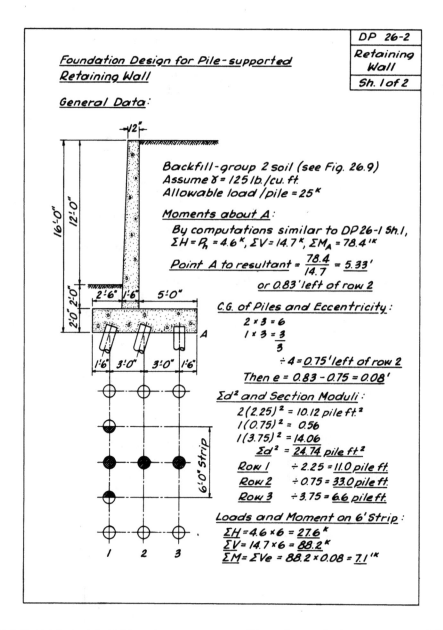

ILLUSTRATIVE DESIGN. DP 26-2. BATTER-PILE FOUNDATION

This design plate illustrates a common method of design for pile foundations in which some of the piles are driven on a batter to increase the stability of the structure against horizontal loads. A cantilever retaining wall has been chosen for the example because a similar soil-supported structure has been designed in DP 26-1.

First, the vertical components of the pile reactions are found in accordance with the discussion of pile foundations subjected to moment (Art. 24.6). Second, a value of 2 kips is estimated as the horizontal load that may be assigned to each pile as a consequence of the strength of the soil into which the piles are driven, combined with the flexural stiffness of the piles. Next, the values of the batter for rows 1 and 2 are determined conveniently by a force polygon sim-

```
┌─────────────────────────────────────────────────────┬──────────────┐
│                                                     │  DP 26-2     │
│  Foundation Design for Pile-supported               │  Retaining   │
│  Retaining Wall                                     │  Wall        │
│                                                     │  Sh. 2 of 2  │
└─────────────────────────────────────────────────────┴──────────────┘
```

Vertical Components of Pile Reactions:

Row 1: $\dfrac{88.2}{4} + \dfrac{7.1}{11.0} = 22.1 + 0.6 = \underline{22.7^K}$

Row 2: $\dfrac{88.2}{4} - \dfrac{7.1}{33.0} = 22.1 - 0.2 = \underline{21.9^K}$

Row 3: $\dfrac{88.2}{4} - \dfrac{7.1}{6.6} = 22.1 - 1.1 = \underline{21.0^K}$

Horizontal Loads on Piles:

Assume 2^K/pile resisted by moment in piles combined with passive resistance of soil.

$\times 4 = 8.0^K$ So batter piles for $27.6 - 8.0 = \underline{19.6^K}$

$20.7^K > 19.6^K$ ok

$\overline{5.5\ 7.6\ 7.6}$

- 6.9^K
- $\div 4 = 1.7^K$/pile $< 2.0^K$ ok
- Row 1: batter 4"/ft.
- $\Sigma V = 88.2^K$
- Row 2: batter 3"/ft.

22.7^K, 22.7^K, 21.9^K, 21.0^K

$\Sigma H = 27.6^K$

Maximum Pile Reaction:

$\sqrt{22.7^2 + 7.6^2} = \sqrt{575} = \underline{24.0^K/\text{pile} < 25^K\ \text{ok}}$

---

ilar to Fig. 26.13*b*. Finally, the resultant load per pile in row 1 is compared with the allowable value.

As in DP 26–1, a load factor would be applied to the earth pressure for the structural design of the various elements of the retaining wall. Since the application of the factored earth pressure usually results in the relief of most, if not all, of the compression in the piles beneath the heel, the pile-supported base slab may be designed in a manner similar to that demonstrated for the footing in DP 26–1. That is, it may be assumed that the sum of the vertical forces is resisted by the piles beneath the toe, whereas the heel resists the weight of the backfill without the benefit of pile support.

26.9. Abutments

General. The principles governing the analysis of the stability and strength of cantilever retaining walls are for the most part appli-

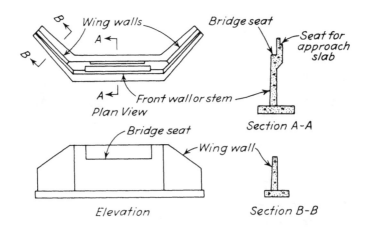

FIGURE 26.14. Typical cantilever abutment for highway bridge.

cable to cantilever abutments and wing walls. In this article the discussion of cantilever retaining walls is extended to include some of the problems associated with the design of cantilever abutments. A typical cantilever abutment with wing walls is shown in Fig. 26.14. Other types of abutments have been described in Art. 14.3.

The forces that are commonly considered in the design of the individual parts of an abutment are dependent to a considerable degree on the tendency of these parts to act together as one unit. For example, the wing walls are generally assumed to behave as ordinary retaining walls and they are designed accordingly. Yet, it is apparent that the force P_A (Fig. 26.15a) must cause appreciable shear, moment, and direct tension on vertical sections near the corner A. Therefore, if continuity is maintained at this junction, the action of the walls in this region differs considerably from that of pure vertical cantilevers. Moreover, as explained in Art. 26.4, some tilt is necessary to establish the condition of active earth pressure in the soil behind a retaining wall; otherwise, the total earth pressure may be appreciably larger than the assumed active value.

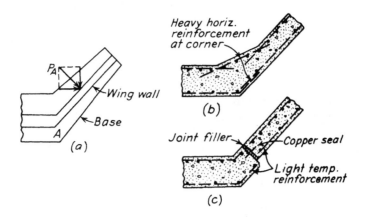

FIGURE 26.15. (a) Detail at junction of front wall and wing wall of abutment. (b) Reinforcement required for monolithic construction. (c) Joint required for independent action.

Abutments Built Monolithically. Where no joint has been provided between the wing walls and the front of abutments, field evidence has indicated a tendency to crack near the corner A (Fig. 26.15a). However, most of the failures can be attributed to inadequate structural strength in this region. A detail of the corner for monolithic construction is shown in Fig. 26.15b. The proper amount of horizontal reinforcement and the appropriate thickening of the concrete at the corner can be determined only by judgment. Both should increase as the angle between the wing walls and the front of the abutment increases. If a detail similar to that shown in Fig. 26.15b is used, and if the structure is otherwise properly designed, there is no reason to expect unsatisfactory performance of the abutment at the corner.

Abutments with Joints at Wing Walls. Some engineers consider it better practice as well as more economical design to provide a joint at each corner as shown in Fig. 26.15c. Such a joint has no structural strength and permits the wing walls as well as the front wall of the abutment to act as true vertical cantilevers. It also accommodates the movements associated with temperature changes. The offset in the concrete at the front of the joint permits considerable relative movement without detracting from the appearance of the structure.

Action of Base Slab. The base of an abutment such as that shown in Fig. 26.14 is commonly cast as a continuous unit extending beneath the wing walls and the front wall. Nevertheless, in the structural analysis, regardless of the presence or absence of joints at the corners, a fictitious joint is commonly assumed to extend across the base at the junction of each wing wall with the front of the abutment. Thus, the abutment is assumed to consist of three entirely independent units.

The assumptions for design, as outlined in the preceding paragraph, are generally considered conservative from the standpoint of the stability of the structure as a whole because the resisting moment of the entire area of the base acting as a unit is greater than that of the assumed three components acting separately, and the resistance increases rapidly with increasing angle between the wing walls and the front of the abutment. On the other hand, continuity of the base beneath the entire abutment and continuity between the wings and the front face undoubtedly reduce the tilt of the walls. The influence of this reduction in tilt upon the earth pressure acting against the walls is usually ignored. This involves a relatively small error, but one that is on the unsafe side.

Loads in Addition to Earth Pressure. The preceding discussion has indicated that wing walls are commonly designed as simple cantilever retaining walls; therefore, the design procedure is identical to that given in design plates DP 26-1 and 26-2. On the other hand, the front section of the abutment must resist certain forces in addition to the normal active earth pressure against the back of the wall. Two important additional forces are the reaction of the superstructure of the bridge and the increase in earth pressure due to wheel loads on the backfill adjacent to the wall. The remainder of this article is devoted to a discussion of the influence of these additional loads.

The reactions from the superstructure may be transmitted to the bridge seat of an abutment in several ways. Roller and rocker bearings provide for expansion and contraction of the bridge and are assumed to transmit only vertical forces to the abutment. On the other hand, fixed bearings at the end of the bridge subject the abutment to horizontal as well as vertical reactions.

The number and spacing of the beams, girders, or trusses that comprise the superstructure determine the number and location of the concentrated reactions that must be resisted by the abutment.

Although the vertical and horizontal reactions from the superstructure represent more or less concentrated loads, they are commonly assumed to be distributed over the entire length of the front wall of the abutment. That is, the sum of the reactions, either horizontal or vertical, is divided by the length of the wall to obtain a load per foot to be used in both the stability analysis and the structural design. This procedure is probably sufficiently exact for most design purposes. However, in the design of low abutments where the reactions from the superstructure are widely spaced, considerable judgment must be exercised in the establishment of a reasonable width over which each reaction is distributed.

The earth pressure against the back of an abutment is increased whenever wheel loads are transmitted to the fill immediately behind the wall. The magnitude of this increase depends on the type of soil and the position of the wheels relative to the abutment, as well as on the wheel loads themselves. An allowance for this increase in earth pressure must be included in the analysis and design of the abutment. Usually the wheel loads are assumed to be equivalent to a uniformly distributed load often taken as 240 lb/sq ft for H–10 highway loading and as 140 lb/sq ft for Cooper's E–10 railway loading. The uniform surcharge is commonly considered as an additional layer of backfill with a height H_s (Fig. 26.16). The corresponding additional horizontal pressure is assumed to be uniformly distributed and equal to $k_h H_s$, where k_h is a coefficient depending on the type of backfill. Values of k_h may be determined by means of Fig. 26.9. Since the additional pressure $k_h H_s$ is uniformly distributed, the resultant force P_{hs} is assumed to act at midheight of the vertical section ab.

In the preceding discussion it has been assumed that the wheel loads can bear directly upon the fill behind the abutment. However, on highways with concrete pavement, the use of approach slabs adjacent to abutments is common practice. These are heavily reinforced concrete slabs, seated on the back of the abutment wall, and extending some 15 or 20 ft back of the abutment. The principal purpose of such construction is to provide a slab with enough structural strength to bridge the depression caused by any differential settlement between the fill behind the abutment and the abutment itself. A typical approach slab is shown in Fig. 26.17. For this type of construction, it should be noted that most of the reaction due to the wheel loads on the approach slab is transmitted to the abutment at the seat which provides the support for the slab. Therefore, for design purposes, the force P_{hs} (Fig. 26.16) may usually be re-

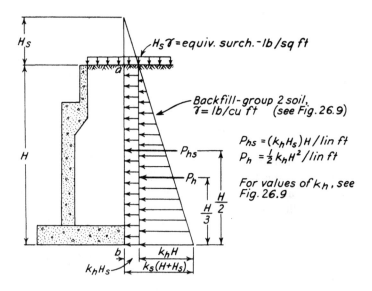

FIGURE 26.16. Diagram illustrating horizontal forces acting against cantilever abutment.

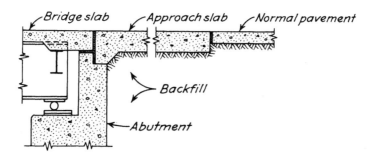

FIGURE 26.17. Approach slab for highway bridge abutment.

placed by the vertical reaction of the approach slab upon the abutment.

Regardless of whether or not an approach slab is utilized, the design of abutments for highway bridges should consider the possibility that wheel loads may be located directly over the parapet wall, as shown in Fig. 26.18. It is apparent that wheels in this position will cause moment at section a–a. Therefore, in some cases, considerable reinforcement may be required at the front of the wall to provide adequate strength against a backward movement of the portion of the abutment above this section.

ILLUSTRATIVE DESIGN. DP 26-3. CANTILEVER BRIDGE ABUTMENT

This design plate is concerned only with the stability of the front section of a cantilever bridge abutment. It is noted that many of the computations correspond to those given in DP 26–1. However, the reactions at the bridge seat as well as the surcharge resulting from loads on the fill behind the abutment are additional factors peculiar to this type of structure. Therefore, the principal purpose of this design plate is to demonstrate the effects of these additional loads as discussed in the preceding section.

PROBLEMS

1. Compute the active earth-pressure P_A against a vertical wall 15 ft high if $\delta = 0°$. The backfill, with a horizontal ground

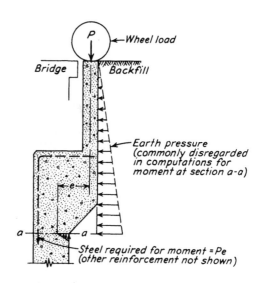

FIGURE 26.18. Diagram illustrating cause of moment near top of cantilever abutment due to wheel load directly over parapet wall.

surface, consists of sand having a friction angle $\phi = 35°$ and a unit weight of 114 lb/cu ft.
Ans. 3470 lb/ft.

2. Determine P_A for the wall of Prob. 1 if $\delta = +20°$. Use Fig. 26.5a.
Ans. 3280 lb/ft.

3. For a backfill of clean sand and gravel, what value of P_A would be estimated from Fig. 26.9 for the wall of Prob. 1?
Ans. 3380 lb/ft.

4. If the ground surface behind the wall in Prob. 1 rises at an angle of 20° to the horizontal, approximately what is the value of P_A?

```
┌─────────────────────────────────────────────────────────────┬──────────────┐
│                                                             │   DP 26-3    │
│   Design of Bridge Abutment. Stability                      │    Bridge    │
│   Analysis of Front Wall                                    │   Abutment   │
│   General Data:                                             │   Sh. 1 of 2 │
└─────────────────────────────────────────────────────────────┴──────────────┘
```

General Data:
Backfill - group I soil (see Fig. 26.9). Assume $\gamma = 110$ lb/cu. ft.
Equivalent surcharge = 3 ft.
Allowable soil pressure = 3.5 ^{k}/sq. ft.
Allowable shear between soil and base = 0.45ΣV

Note:
All loads and moments are per lin. ft. of abutment.

$P_{hs} = 30(3.0)(12) = 1.08^k$

$P_h = \frac{1}{2}(30)(12^2) = 2.16^k$

Stability Computations:
Moments about A

Area		Force	Arm	Mom.
①	25 × 10.5 × 0.110	= 2.89^k	1.25'	3.61$^{\prime k}$
②	1.0 × 5.5 × 0.110	= 0.61	3.00	1.83
③	½ × 1.0 × 1.0 × 0.110	= 0.06	2.83	0.15
④	0.75 × 3.0 × 0.150	= 0.34	2.88	0.98
⑤	2.0 × 1.0 × 0.150	= 0.30	3.50	1.05
⑥	½ × 1.0 × 1.0 × 0.150	= 0.08	3.17	0.25
⑦	10 × 6.5 × 0.150	= 0.97	4.00	3.88
⑧	7.0 × 1.5 × 0.150	= 1.58	3.50	5.53
L_v		5.26	3.88	20.40
L_h		0.24	9.00	2.16
P_{hs}		1.08	6.00	6.48
P_h		2.16	4.00	8.64
		$\Sigma H = 3.48^k$	$\Sigma V = 12.09^k$	$\Sigma M_A = 54.96^{\prime k}$

Ans. From curves 1 (Fig. 26.9), 3880 lb/ft.

5. A vertical anchor wall 9 ft high is pulled horizontally against a mass of sand with a horizontal ground surface. The sand has a unit weight of 120 lb/cu ft and a value of $\phi = 33°$. As it is pulled, the wall tends to rise with respect to the sand. The angle of wall friction is approximately $\frac{2}{3}\phi$. What is the horizontal component of the passive earth pressure?

Ans. From Fig. 26.5 and $\delta = 22°$, 7770 lb/ft.

6. If the anchor wall in Prob. 5 can be prevented from rising, what would be the horizontal component of the passive earth pressure?
Ans. 36,400 lb/ft.

7. A vertical retaining wall 15 ft high supports a cohesionless fill that weighs 110 lb/cu ft. The backfill rises from the crest of the wall at an angle of 20°

```
┌─────────────────────────────────────────────────────────┬──────────────┐
│                                                         │   DP 26-3    │
│   Design of Bridge Abutment. Stability                  │    Bridge    │
│   Analysis of Front Wall                                │   Abutment   │
│                                                         ├──────────────┤
│                                                         │   Sh. 2 of 2 │
└─────────────────────────────────────────────────────────┴──────────────┘
```

Stability Computations:

Location of Resultant

From point A, $\dfrac{54.96}{12.09} = 4.55'$ then $e = 4.55 - \dfrac{7.0}{2} = \underline{1.05'} < \dfrac{7}{6}$ ok

Soil Pressure at Base:

At toe $q_{max.} = \dfrac{12.09}{7}\left(1 + \dfrac{6 \times 1.05}{7}\right)$

$\underline{q_{max.}} = 1.73(1 + 0.9) = \underline{3.3^k/sq.\,ft.} < 3.5$ ok

At heel $\underline{q_{min.}} = 1.73(1 - 0.9) = \underline{0.2^k/sq.\,ft.}$

Sliding:

Shear available along base $= 12.09^k \times 0.45 = \underline{5.43^k}$

Factor of safety $= \dfrac{5.43}{3.48} = \underline{1.6} > 1.5$ ok

---

with the horizontal. If $\phi = 28°$ and $\delta = +20°$, what is the total active earth pressure against the wall? Use the trial-wedge graphical construction.
Ans. 5700 lb/ft.

SUGGESTED READING

B. Baker (1881), "The Actual Lateral Pressure of Earthwork." *Min. Proc. Inst. Civ. Eng.*, London, **65**, 140–186. Experiences of one of the greatest of civil engineers, a contemporary of Rankine and Boussinesq, leading him to the conclusion "that the laws governing the lateral pressure of earthwork are not at present satisfactorily formulated."

K. Terzaghi (1934), "Large Retaining-Wall Tests. I," *Eng. News-Record*, **112**, 5, 136–140. First of a series of five articles presenting the fundamental relations between the displacements of a rigid wall and the pressure against the wall, and explaining after half a century the basis for Sir Benjamin Baker's disenchantment with earth-pressure theory. Terzaghi's large-scale tests verified remarkably his small-scale but fundamentally correct tests conducted with homemade equipment at Robert College in Istanbul in 1919, and reported first in "Old Earth-Pressure Theories and New Test Results," *Eng. News-Record (1920)*, 85, 14, 632–637. This article is one of a small group that marks the beginning of modern soil mechanics.

Details of classical earth-pressure theory and its application to retaining walls under a variety of conditions are thoroughly treated in W. C. Huntington (1957), *Earth Pressures and Retaining Walls*, New York, Wiley, 534 pp.

Stanley D. Wilson (1912–   )

Eminent consultant on foundations, dams, landslides, and problems of soil dynamics; leader in the application of full-scale field observations to the improvement of design and to the diagnosis of difficulties requiring remedial action. His development of a variety of equipment for measuring strains and deformations in earth masses has been a significant factor in obtaining present knowledge of the deflections and movements of such structures as flexible bulkheads and the support systems of braced and tied-back cuts.

Plate 27.

CHAPTER 27

Flexible Earth-Retaining Structures

27.1. Behavior of Flexible Earth-Retaining Structures

According to Art. 26.4, most gravity and cantilever retaining walls are capable of rotating about their bases far enough to satisfy the strain requirements for the active state of stress in the failure wedge (Fig. 26.2e). The total earth pressure against the wall is then equal to the active earth pressure, which, for a backfill with a plane horizontal surface, may be calculated by eq. 26.7. The distribution of the earth pressure with depth is linear and may be calculated by eq. 26.3.

In contrast, the walls of anchored bulkheads (Fig. 27.1a), braced cuts (Fig. 27.1b), and tie-back cuts (Fig. 27.1c) usually consist of members with comparatively small flexural rigidity, but they are supported at various elevations by anchors or struts and are also usually supported by embedment into the ground below the lowest excavation level. The supports impose restraints on the movement of the walls. Hence, as excavation progresses in front of the walls, or as filling takes place behind anchored bulkheads, the walls deform and translate in characteristic fashions indicated by the dashed lines in Fig. 27.1. The pattern of the deformations has no resemblance to that shown in Fig. 26.2e. Usually the deformations near the tops of the walls are considerably less than those corresponding to the active Rankine state, whereas those near the bottoms are greater. Consequently, the magnitude of the earth pressure against the walls differs somewhat from the active earth pressure, and the distribution of pressure with depth may differ substantially

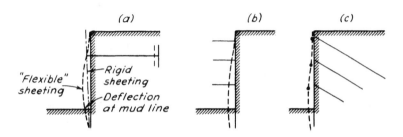

FIGURE 27.1. Typical patterns of deformation of vertical walls of (*a*) anchored bulkhead, (*b*) braced cut, and (*c*) tie-back cut.

from the linear distribution indicated by eq. 26.3.

The actual earth pressure against the back of a flexible vertical support and the loads in the supporting members depend to a considerable extent not only on the properties of the soil being supported, but also on the sequence of construction operations. They are particularly influenced by the relation between the depth at which supports are installed and the depth of excavation at the same time. Therefore, the pressures used for design cannot be determined exclusively by theory but, since they are influenced by the manner in which the contractor carries out his work, must be modified by experience and the results of observations and measurements on full-sized structures. The following sections in this chapter describe procedures that have been found generally satisfactory for the three principal types of flexible earth-retaining structures.

27.2. Anchored Bulkheads

Modes of Failure. Bulkheads usually consist of steel sheet piles, although reinforced concrete piles and timber Wakefield piles (Fig. 13.3) are sometimes used. The piles are commonly supported near their tops by horizontal beams or wales attached to steel tie rods extending to an anchorage. Three common types of anchorages are shown in Fig. 27.2. The lower ends of the sheet piles are supported by their embedment in the underlying soil.

Most bulkheads are located on the waterfront. In some instances, the piles may be driven, the anchorage constructed, and backfill placed on the land side. The structure is then known as a *fill* bulkhead. Under other circumstances, the sheet piles may be driven, the anchorage installed, and the soil in front of the bulkhead removed by excavation. The structure is then known as a *dredge* bulkhead. In any event, only granular material is usually permitted immediately behind the bulkhead; otherwise the lateral pressures are extremely large. In the remainder of this article, it is assumed that the backfill consists of cohesionless sand down to the level corresponding to the ground surface on the water side of the bulkhead. Below that level, known as the *mud line* or *dredge level*, the ground in which the piles are embedded is assumed to consist either of sand or of clay.

An anchored bulkhead may fail in one of at least three ways:

1. If the bulkhead is founded above weak

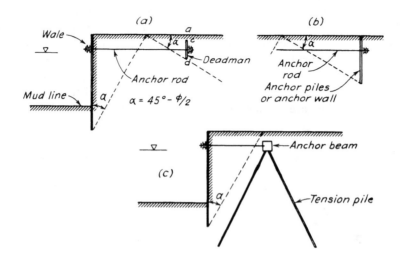

FIGURE 27.2. Types of anchorages for anchored bulkheads. (*a*) Deadman or anchor plates. (*b*) Anchor wall or row of anchor piles. (*c*) Anchor beam supported by batter piles.

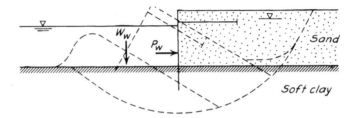

FIGURE 27.3. Deep-seated failure of bulkhead due to inadequate bearing-capacity of underlying weak soil.

cohesive material, the underlying soil may experience a bearing-capacity failure under the unbalanced weight of the material behind the sheeting. The movements may take the form of a general rotational failure involving the sheeting and even the anchorage (Fig. 27.3). This possibility may be investigated by the procedures described in Arts. 18.7 and 21.6. The force tending to cause the bearing-capacity failure consists of the full weight of the soil and water behind the bulkhead. The weight W_w and lateral pressure P_w of the water in front of the bulkhead must be included as resisting forces.

2. The anchorage may fail either because the force in the anchor rods has been underestimated, the resistance of the anchorage has been overestimated, or the anchorage has been located too close to the bulkhead and has moved together with the bulkhead toward the water. In some instances, settlement of the ground beneath the anchor rods, due to the compression of deep underlying layers of compressible material, requires the rods to support much of the weight of the overlying material and surcharge loads. The rods are thus subjected to additional tension that may cause them to break.

3. The toe embedment may be inadequate, whereupon the soil in front of the embedded portion may rupture or experience excessive lateral movement. If the groundwater level behind the bulkhead is higher than that in front, a condition not uncommon after a heavy rain or in areas where tidal fluctuations occur, water tends to flow downward behind the bulkhead and upward in front of the embedded portion, as shown by the flow net in Fig. 2.9a. The upward seepage forces may appreciably reduce the strength of the soil in front of the embedded portion of the sheet piles, particularly in fine or silty sands. Many bulkheads have stood for years until they experienced an exceptionally great difference in the hydraulic head, whereupon they failed.

The foregoing three modes encompass the great majority of bulkhead failures. In contrast, failure of the sheet piles by bending under the influence of the earth pressure is extremely rare. The several reasons for this favorable situation are discussed subsequently.

The details of structural design require careful consideration to avoid other types of failure. For example, wales are sometimes placed on the land side of the sheet piles. The bolts between the piles and wales are then required to resist in tension all the force in the anchor rods. Therefore, such bolts may fail if, for any reason, the anchorage force has been underestimated.

Loads on Sheeting. The loads acting on the land side of the sheeting of an anchored bulkhead are shown in Fig. 27.4a if the sheeting is embedded entirely in sand below the mud line, and in Fig. 27.4b if the material below the mudline is a clay with an undrained shear strength c or an unconfined compressive strength q_u. The sand above the mud line has a friction angle ϕ_1; that below the mud line in Fig. 27.4a has a friction angle ϕ_2.

For the bulkhead in contact only with sand (Fig. 27.4a), the pressure diagram

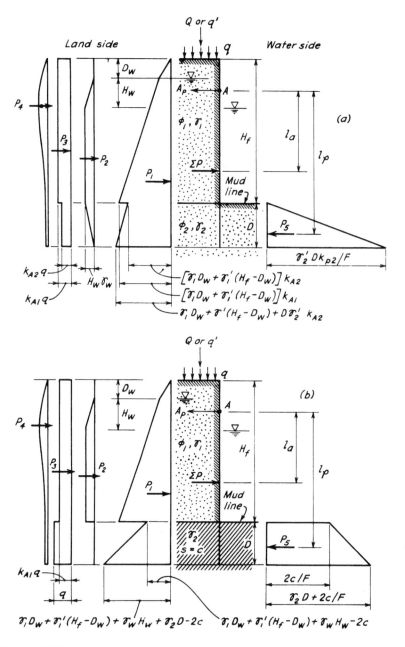

FIGURE 27.4. Loads against anchored bulkhead. (a) Bulkhead retaining sand above mud line and embedded in sand below. (b) Bulkhead retaining sand above mud line and embedded in plastic clay below. P_1 represents resultant horizontal pressure due to weight of soil behind bulkhead; P_2 represents unbalanced horizontal water pressure; P_3 is resultant lateral pressure due to uniformly distributed surcharge q per unit of area; P_4 is resultant force due to concentrated load Q or line load q' acting on the surface of soil behind the bulkhead. Resisting force P_5 represents mobilized portion of passive pressure of soil in front of embedded portion of sheet piles.

450

represented by P_1 corresponds to the active earth pressure computed in accordance with eq. 26.3 in which the vertical intensity of pressure p_v (eq. 26.1) includes the full weight of the sand above the water table and the submerged weight of the sand below the water table. Use of eq. 26.3 implies a linear increase of earth pressure with depth, contrary to the strain requirements mentioned in Art. 27.1. The error is taken into account by making suitable adjustments as described subsequently.

The pressure represented by P_2 is the unbalanced water pressure associated with the maximum difference in water level, compatible with conditions at the site, behind and in front of the sheeting. If the soil below the mud line is permeable (Fig. 27.4a), the flow net resembles Fig. 2.9a and the unbalanced water pressure decreases from $\gamma_w H_w$ at the mud line to zero at the bottom of the sheet piles.

The portion of the pressure diagram represented by P_3 indicates the additional active earth pressure accompanying any uniformly distributed surcharge q per unit of area that may be acting on the ground surface. It has a constant intensity with depth equal to

$$p_q = k_A q \qquad 27.1$$

The irregularly shaped pressure diagram P_4 represents the horizontal pressure due to vertical line loads q'/lin ft parallel to the top of the sheeting, or to concentrated loads Q acting on the ground surface. The magnitudes of these pressures may be determined from the charts, Fig. 26.8.

The sheet piles must be in equilibrium under working conditions. The horizontal forces on the land side are assumed to correspond to the active state of stress because comparatively little movement of the piles away from that side is required to mobilize the shearing strength of the soil and develop active conditions (Art. 26.4). In contrast, since much more movement is required to develop the passive state of stress, the pressure P_5 on the water side of the wall in Fig. 27.4a is restricted to a fraction $1/F$ of the passive pressure P_P (eq. 26.8). Thus, F represents the factor of safety against failure by exceeding the passive resistance, and $P_5 = P_P/F$. The unit weight to be used in eq. 26.8 is the submerged unit weight γ', unless the value of H_w is so great that the effective unit weight of the sand is reduced appreciably by the upward seepage pressures caused by the rising flow of water in the sand in front of the sheet piles (Fig. 2.9a). The reduction $\Delta\gamma'$ may be calculated on the basis of a flow net. Such calculations lead to the approximation that $\Delta\gamma'$ in pounds per cubic foot may be taken as $20H_w/D$.

For the bulkhead in Fig. 27.4b, the pressures on the land side in the clay beneath the sand are determined on the premise that undrained conditions prevail. Therefore, according to eq. 4.7a, if p_v and p_h are assumed to be principal stresses,

$$p_h = p_v - 2c \qquad 27.2$$

Again, on the water side, only the fraction $1/F$ of the shearing resistance is considered to be mobilized. Hence,

$$p_h = p_v + 2c/F \qquad 27.3$$

In these expressions, p_h and p_v are total stresses, in keeping with the undrained analysis (Art. 4.8). Thus, at the top of the clay at the mud line, on the land side

$$p_h = \gamma_1 H_f - 2c$$

and on the water side

$$p_h = \gamma_w(H_f - D_w - H_w) + 2c/F$$

The net pressure, or the difference between the values of p_h on the land and water sides, is unaltered if the same quantity is subtracted from both values. For convenience, the water pressure $\gamma_w(H_f - D_w - H_w)$ is subtracted from both, whereupon the value of p_h on the land side becomes $\gamma_1 D_w + \gamma_1'(H_f - D_w) + \gamma_w H_w - 2c$, and that on the water side merely $2c/F$. These values are shown on Fig. 27.4b. Since the term $2c$ has been subtracted from the vertical pressure to obtain the horizontal pressure against the land side of the embedded portion in diagram P_1, it is not subtracted again from the surcharge q in diagram P_3.

The diagram representing pressure P_2 does not extend below the mud line in a plastic clay in which undrained conditions prevail, because the unbalanced water pressure is fully taken into account by the effect of the surcharge $\gamma_w H_w$ acting vertically on the surface of the clay.

In addition to the loads shown in Fig. 27.4, the sheet piles are subjected to the anchor pull A_p.

Design of the bulkhead requires determination of the depth of embedment D necessary to prevent a toe failure, and of the magnitude A_p of the anchor pull. These quantities are determined by consideration of the equilibrium of the sheet piles.

Equilibrium of Sheet Piles. The required depth of embedment is calculated by taking moments of all the horizontal forces acting on the sheeting about the point of application of the anchor pull. Experience has shown the depth of embedment found in this fashion to be satisfactory in thoroughly explored, relatively homogeneous soils.

Although in principle the calculation of the moments about the point of application of the anchor pull is simple, determination of the depth of embedment in sand involves a cubic equation best solved by trial. After the depth of embedment D has been determined, the anchor pull may be calculated by equating to zero the sum of the moments of all horizontal forces on the sheeting about the point of application of the passive resistance. Alternatively, A_p may be calculated by equating to zero the sum of all horizontal forces on the sheeting.

To take account of the inevitable variations in the strength and compressibility of the materials in front of the embedded portion of the sheet piles, it is considered good practice to drive the piles to a depth of embedment 20 per cent greater than the value of D calculated on the basis of a selected factor of safety. Furthermore, to take account of the difference in the real distribution of pressure against the sheeting from that computed by ignoring the influence of the flexibility of the sheeting, the calculated value of the anchor pull is increased by 20 per cent before the anchorage is designed.

Anchorage. A typical dead-man anchor is shown in Fig. 27.2a, and an anchor wall is shown in Fig. 27.2b. Anchorages of this type depend on the passive pressure of the soil for their stability. The soil providing the resistance should not be within the zone in which the strain conditions correspond to the active state behind the bulkhead; otherwise, all the material between the anchorage and the bulkhead may move as a body toward the water. It is generally satisfactory to locate the anchorage at least as far back as indicated by the geometry in Figs. 27.2a and 27.2b. The net resistance of the anchor wall (Fig. 27.2b) is the difference between the active pressure on the back side and the passive pressure on the front side. The wall need not extend to the ground surface; if the distance ac (Fig. 27.2a) from the ground surface to the top of the wall is not greater than one third the distance ad from the ground surface to the bottom of the wall, the resistance may be calculated as if the anchorage extended over the entire height ad. No wall friction should be depended on. Hence the resistance $P_P - P_A$ may be computed by means of eqs. 26.7 and 26.8. Because of the necessity for a sufficiently strong anchorage, the factor of safety should not be less than 2.5 unless the loads and soil conditions are very well known.

Anchorages may also consist of batter piles in the form of A-frames. Such anchorages may be located closer to the face of the bulkhead than an anchor wall. The forces in the piles may be determined by means of a simple force polygon. It may be difficult, however, to obtain the required tensile resistance in the back pile. Furthermore, if the anchorage is too stiff, because of the rigidity of the pile system and the relative shortness of the anchor rods, the outward movement of the bulkhead at the anchor level may be so restricted that the earth pressures against the back of the wall are considerably greater than the active pressures assumed in the design. If the movement at the end of the anchor rod at the wale is not at least 0.1 per cent of the distance H_f from the top of the sheet pile down to the mud line, the earth-pressure distribution shown in Fig. 27.4 should not be used. A more conservative distribution of pressures, such as those discussed in connection with braced cuts (Art. 27.3), should be substituted.

Maximum Moment in Sheet Piles. Numerous field observations as well as comprehensive laboratory tests (Tschebotarioff, 1949; Rowe, 1952) have demonstrated that the bending moments in the sheet piles of anchored bulkheads are much smaller than those that would be calculated if the forces shown in Fig. 27.4 were applied to the sheet piles. The error leads to excessive cost and in some instances would indicate moments so large that conventional sheet piles could not be successfully used.

The smaller bending moments are a consequence of at least three factors. Because the sheet piles used in practice are comparatively flexible, their deflection at the mud line (Fig. 27.1a), as compared to that at the tip, is relatively greater than it would be if the sheet piles were rigid. Hence, the resultant resisting pressure P_5 acts at a higher level, closer to the mud line, than indicated in Figs. 27.4a or 27.4b. This has the effect of shortening the span l_p; since the bending moments in the sheeting are a function of l_p^3, the shortening has a significant influence on the moments between the anchorage and the dredge line. Furthermore, the earth pressure against the sheet piles is reduced at those locations where the piles deflect the most; the pressure is redistributed to the more unyielding locations such as the points of attachment of the anchorage system or the embedded portions of the piles. The redistribution is greater for dredge than for fill bulkheads. Finally, the tendency for the portions of the sheet piles above the anchor point A (Fig. 27.4) to push against the soil and thus to increase the earth pressure above A on the land side reduces the moment in the sheet piles below A.

In practice, field observations have demonstrated that bulkheads for which the depth of embedment has been determined in accordance with the procedures outlined in the preceding sections are likely to develop a point of reversal of curvature or contraflexure very close to the dredge line. By assuming a point of contraflexure at this level and by calculating moments in the sheet piles due to the forces shown in Fig. 27.4 above this level, satisfactory agreement has been found between the calculated and observed moments. Hence, this procedure is recommended whenever the material at the dredge line is reasonably firm or dense. If the material at and just below the dredge line is loose or weak, the point of contraflexure should be taken conservatively one or two feet below the dredge line. Moreover, if the piles derive their toe resistance mainly from penetration into hard material at a shallow depth underlying soft compressible deposits, the point of contraflexure should be taken at the top of the hard material.

When the moments in the sheet piles are estimated on the basis of an assumed point of contraflexure, the corresponding anchor pull may differ from that determined previously for design of the anchorage. The design of the anchorage should, nevertheless, not be altered.

Practical Considerations. Anchored bulkheads are commonly built along the waterfront where recently deposited soft organic soils overlie firmer materials. The sheet piles are driven, supported by anchor rods, and backfilled with sand. As the fill approaches from the land side, the soft material may accumulate in the form of mud waves in front of the advancing sand and become trapped against the lower portion of the sheet piles. It may then be covered with sand. Since the pressure of the nearly liquid organic material is much greater than that of sand, the bulkhead may deflect excessively or may even fail. Hence, care should be taken to remove all organic material from behind the bulkhead for a distance at least equal to the height H_f. Preferably, in order to avoid unequal settlement, the organic material

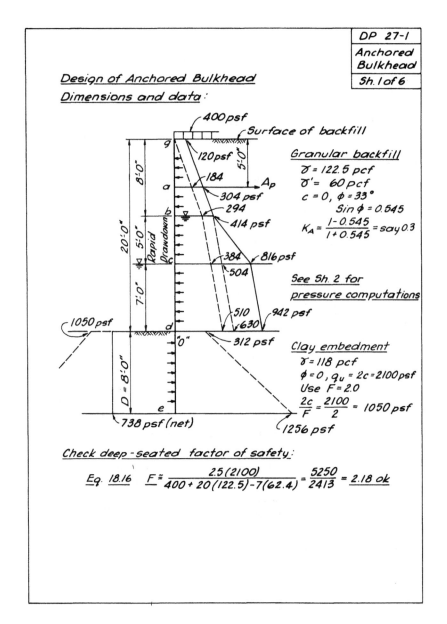

should be removed from beneath the entire backfill area.

Where anchored bulkheads are used to retain backfills of clay or organic soils, or where such materials constitute the natural ground behind a dredged bulkhead, the lateral pressures may be excessive. Rather than to design for such pressures, it may be advisable to replace the cohesive materials for a horizontal distance behind the bulkhead equal at least to $H_f/2$ and preferably equal to H_f. The pressures against the bulkhead can then be calculated as if the entire backfill were sand. If the cohesive materials cannot be removed, the bulkhead may be designed for lateral pressures corresponding to the apparent pressure envelope for braced cuts (Art. 27.3 and Fig. 27.7d), but such a design is by no means routine and careful attention must be given to the consequences of the relatively large movement associated with the soft soil (Terzaghi, 1954).

If compressible material remains beneath the anchor rods, the ensuing settle-

```
                                                    ┌─────────────┐
                                                    │  DP 27-1    │
         Compute horizontal pressures               │  Anchored   │
            Active pressure at:                     │  Bulkhead   │
                                                    │  Sh. 2 of 6 │
               g.  400 psf × 0.3 = 120 psf          └─────────────┘
               a.  122.5 (5.0)(0.3) = 184 + 120 = 304 psf
               b.  122.5 (8.0)(0.3) = 294 + 120 = 414 psf
               c.  122.5 (8.0) = 980
                    60.0 (5.0) = 300
                                 1280
                                       × 0.3 = 384 + 120 = 504 psf
                    62.4 (5.0) =                   312
                                                   816 psf

               d.  122.5 (8.0) = 980
                    60.0 (12.0) = 720
                                  1700
                                       × 0.3 = 510 + 120 = 630 psf
                    62.4 (5.0) =                   312
                                                   942 psf (fill)

                              400 surcharge
                                       1700
                    62.4 (5.0) = 312
                                 2412
                                 2100
                                  312 psf (clay)
               e.              (312 + 118 D) psf

          Passive pressures at:

               d.  2c/F =       1050 psf

               e.              (1050 + 118 D) psf

          Net pressure d to e:
                                1050 + 118 D
                              -( 312 + 118 D)
                                 738 psf net
```

ment, as indicated previously, causes a downdrag on the rods and subjects them to additional tension. The tension may be avoided by placing the anchor rods in corrugated culvert pipes so arranged that the pipes may settle several inches before they come into contact with the rods.

The stress in the anchor rods should not exceed the yield stress of the steel divided by the selected factor of safety. The rods should be upset to a diameter such that the threads for bolts and turnbuckles do not reduce the area below that on which the capacity of the rods is based.

ILLUSTRATIVE DESIGN. DP 27-1. ANCHORED BULKHEAD

Most of the procedures described in Art. 27.2 for the design of anchored bulkheads are demonstrated in DP 27-1.

The pressures against the back and against the embedded portion of the wall are computed in accordance with Fig.

```
                                                              DP 27-1
                                                              Anchored
  Equilibrium of wall                                         Bulkhead
    Forces and moments about a:                               Sh. 3 of 6

                        Force        Arm          Moment
    gab:  ½(294)(8)     1176 lb.    0.33'      +  390ᶦᵏ
          (120)(8)       960        1.00                      -960ᶦᵏ
    bc:   ½(414)(5)     1035        4.67          4830
          ½(816)(5)     2040        6.33         12910
    cd:   ½(816)(7)     2856       10.33         29500
          ½(942)(7)     3297       12.67         41770
                                                +89,400
                                                -   960
                        11,364 lb.               88,440ᶦᵏ

                        -738 D ; 15.0 + D/2 ; - (11,070 D + 369 D²)
```

Compute depth of embedment, D:

ΣM at point $a = 0 = 88,440 - 11,070 D - 369 D^2$)

$D^2 + 30 D = 240$ or $(D+15)^2 = 465$

$D + 15 = \sqrt{465} = 21.6 \quad D = 6.6$

Use $D = 6.6 \times 1.2 =$ say $8'\text{-}0''$

Compute anchor pull, A_p:

$\Sigma H = 0 = 11,364 - 738(6.6) - A_p$

$A_p = 11,364 - 4,871 = 6,493$ lb. per ft.

For size of rod, wales, and anchorage, use

$A_p = 6,493 \times 1.2 = 7.79^K$ per ft.

Determine size of anchor rod:

Assume PDA27; Rods @ $6 \times 1'\text{-}4'' = 8'\text{-}0''$ ctrs.
(ok, see sh. 4)

$$A_s = \frac{7.79 \times 8.0}{22.0} = 2.83 \text{ sq. in.}$$

Use $2''$ dia. upset to $2½''$ $A_s = 3.14$

27.4b. The 5-ft differential in water level assumed for design increases the required embedment by 20 to 25 per cent.

For the computation of moment in the piles of the bulkhead wall, the point of contraflexure is assumed 1 ft below the mud line. The anchor pull corresponding to this assumption is about 11 to 12 per cent less than that computed for overall equilibrium.

Several of the unknowns evaluated in the design plate, such as the point of maximum moment in the bulkhead wall and the depth of piles in the anchor wall, are found by trial and error. Such a procedure is preferred for accuracy when the solution of quadratic or cubic equations involves small differences of large numbers.

27.3. Braced Cuts

Modes of Failure in Sand. Typical bracing systems for deep cuts are shown in Fig. 8.3. In sand above the water table, failures have occurred almost exclusively by the buckling of struts, one after the other, in a progres-

```
                                                    ┌─────────────┐
                                                    │   DP 27-1   │
   Determine size of sheet piles                    │  Anchored   │
                                                    │  Bulkhead   │
      Assume point of contraflexure 1'-0"           ├─────────────┤
      below d at point "0".                         │  Sh. 4 of 6 │
                                                    └─────────────┘
   Compute revised reactions, shears, and moments:
   Forces and moments about a: (See Sh. 3)
      ΣP_A  gabcd = 11.36^k        ΣM = 88.39^{1k}
             d"0" = -0.74 × 15.5 = -11.47
                    ─────────       ──────
                      10.62^k        76.92^{1k}

      Reaction at "o" = 76.92 ÷ 16.0 = 4.81^k
      Revised A_p = A_p' = 10.62 - 4.81 = 5.81^k

   Find point of V=0:
      V at point c:                      +5.81 = A_p'
         120 + 414
      - ─────────── (8.0) = 2.14^k
              2

         414 + 816
      - ─────────── (5.0) = 3.08        -5.22
              2                         ──────
                                        +0.59^k = V_c

   Distance below point c to V=0:
      0.59 ÷ say 0.822 = 0.72' below point c

   Compute maximum moment (at V=0)
      +4.81 (8.00 - 0.72) =         +35.0^k
      +0.74 (7.50 - 0.72) =         + 5.0      +40.0^k

      -½ (.829) ⅓ (7.00 - 0.72)² = - 5.4^k
      -½ (.942) ⅔ (7.00 - 0.72)² = -12.4       -17.8
                                              ──────
                          Max. Mom =          +22.2^{1k}

   Section modulus required:
              22.2 × 12
      S = ─────────────── = 10.7 in.³/ft
                 25

                              Use PDA27  S = 10.7 in.³
                              (Alt. PZ27  S = 30.2 in.³)
```

sive manner. The buckling is often preceded by local crippling of the wales if they have been inadequately stiffened to accept the concentrated loads from the struts. Failures by bending of sheet or soldier piles, or of wales, are rare. Moreover, in sand above the water table there is no danger of a general heave of the bottom of the excavation. In a few instances, the sheet piles or soldier piles have settled excessively in loose sand as a result of loss of ground during excavation (Art. 16.1), whereupon the system of bracing has distorted enough to cause local crippling of the connections followed by the failure of struts and general collapse. With the exception of the latter type of failure, which can be avoided by driving the sheet or soldier piles deeply enough to develop adequate vertical resistance, failures in the bracing systems of cuts in sand above the water table can be prevented by adequate structural design of the various members for the earth pressures to which the system will be exposed.

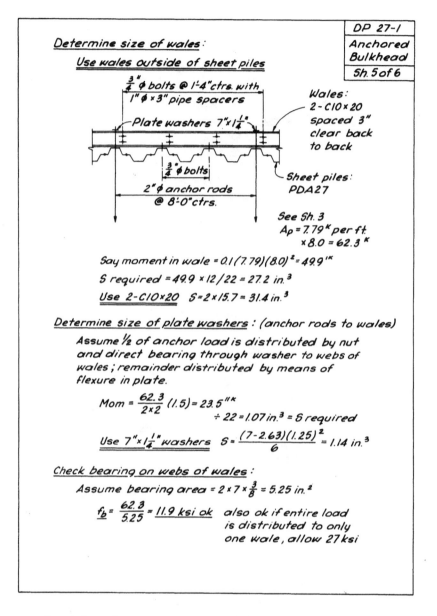

Cuts extending below the groundwater table in sand are preferably dewatered in advance of excavation and bracing (Chap. 9). Sheet-pile walls may be relatively impervious as compared to the sand; if the water level remains at a higher level outside the sheeting than within the excavation, a flow pattern similar to that illustrated in Fig. 2.9a is established. The seepage forces inside the cut at the bottom are directed upward and may cause instability of the soil expected to provide lateral support for the embedded portion of the sheet piles. The sand may even become quick. Under these circumstances, vigorous inward movement of the lower portions of the sheet piles may occur and the system of bracing may collapse.

If a pervious stratum, overlain by less pervious material, is located beneath excavation level and is not drained in advance, it may remain under excess hydro-

```
                                                    ┌─────────────┐
                                                    │   DP 27-1   │
        Design anchorage                            │  Anchored   │
            Use continuous sheet pile wall          │  Bulkhead   │
            Determine depth of penetration, d_a:    │   Sh.6 of 6 │
                                                    └─────────────┘
```

Design anchorage

Use continuous sheet pile wall

Determine depth of penetration, d_a:

$k_a = 0.3$ (Sh.1) $k_p = \dfrac{1+0.545}{1-0.545} =$ say 3.4

$k_{net} = k_p - k_a = 3.4 - 0.3 = 3.1$

For $F = 2.5$, $P_{pnet} = 2.5 A_p = 2.5(1.2)(6,193) = 19,479$ lb/ft.
(see Sh.3)

$P_{pnet} = \tfrac{1}{2}(3.1)(122.5) d_a^2 = 19,479$

$d_a = \sqrt{\dfrac{2(19,479)}{3.1(122.5)}} = \sqrt{102.6} = 10.13$

Since depth extends below water table, P_{pnet} will be reduced somewhat

Try $d_a = 10'-6''$ (2'-6'' below W.T.)

$P_{pnet} = \tfrac{1}{2}(3.1)(122.5)(10.5)^2 - \tfrac{1}{2}(3.1)(62.5)(2.5)^2$
$= 20,934 - 605 = 20,330$ lb/ft.
$\div 6,493 = 3.1 = F$ ok

Use $d_a = 10'-6''$

Determine size of anchorage piles:

Connect rods at 7'-0" depth

Say moment $= \tfrac{1}{6}(3.1)(122.5)(7)^3 = 21.7$ ᴵᴷ/ft.

S required $= \dfrac{21.7 \times 12}{25} = 10.4$ in.^{3}/ft.

Use PDA27 $S = 10.7$ in.^{3}

Wales and connection details:

Use same as for main wall (see Sh.5)

Locate anchorage: $45° - \tfrac{\phi}{2} = 28.5°$ tan $= 0.54$ cot $= 1.84$

$28.0(0.54) + 10.5(1.84) = 15.1 + 19.3 =$ say $35'-0''$ Min.

---

static pressure, as indicated in Fig. 27.5, and may cause a blowup of the bottom of the excavation. Such a condition may lead to a catastrophic failure.

Loads on Struts in Sand. Since most open cuts are excavated in stages within the confines of sheet-pile walls or walls consisting of soldier piles and lagging, and since struts are inserted progressively as the excavation deepens, the walls are likely to deform as shown in Fig. 27.1b. Little inward movement can occur at the top of the cut after the first strut is inserted. The pattern of deformation differs so greatly from that required for the active Rankine state (Fig. 26.2e) that the distribution of earth pressure associated with retaining walls is not a satisfactory basis for design. The pressures against the upper portions of the walls are substantially greater than those indicated by eq. 26.3. Moreover, because the

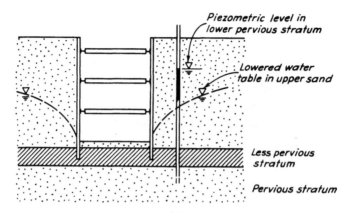

FIGURE 27.5. Uplift pressure in undrained pervious stratum beneath less pervious soil in bottom of braced cut.

pressures depend to a considerable extent on the manner in which the work is executed, knowledge of realistic earth pressures must be based on measurements on actual projects.

A considerable body of information has accumulated concerning the loads in struts of braced excavations. For comparison of the results from different cuts and for design of struts in new cuts, it has been found convenient to convert the strut loads to equivalent pressures. This can be done by the simple but approximate procedure of dividing the strut load by the area of the portion of the sheet-pile or soldier-pile wall extending halfway to the neighboring struts both vertically and horizontally. It has been found that, even in a single open cut in which the work has been executed in expert fashion, the loads in equally spaced struts at a given level vary over a wide range and, correspondingly, the pressure diagrams for struts in various vertical profiles differ from each other. Since it is not possible to predict which of apparently identically situated struts will experience the greatest loads, conservative use of the empirically derived pressure diagrams for design requires that each strut be proportioned as if it would be subjected to the maximum load indicated by any of the pressure diagrams. Hence, for design of struts, it is appropriate to use a pressure envelope that encloses all the pressure diagrams derived from observations. Such an envelope is designated an *apparent pressure envelope*. Thus, an apparent pressure envelope represents a fictitious pressure distribution for estimating the maximum strut loads in a system of bracing. It does not, however, indicate the magnitude or distribution of loading on the sheeting or wales. A general form of such a diagram is shown in Fig. 27.6.

The apparent pressure envelope is used to calculate strut loads for design by the simple procedure indicated in Fig. 27.6. The elevations of the various struts in a cut are chosen, at least tentatively, for convenience to avoid interference with the structure to be built within the enclosure, and to prevent excessive deformations during excavation. Once the elevations have been selected, the load for which any given strut should be designed is determined from

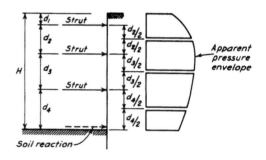

FIGURE 27.6. Diagram illustrating method of calculating strut loads from apparent pressure diagram.

the apparent pressure envelope. For the arrangement of struts shown in Fig. 27.6, the areas tributary to each strut are indicated. If no strut is inserted at the very bottom of the excavation, the tributary portion of the apparent pressure envelope is considered to be taken as a soil reaction inside the embedded portion of the sheet or soldier piles. No consideration is given in the procedure to the actual continuity of the sheet piles or soldier piles, because no continuity was assumed when the apparent pressure envelopes were developed from the measured strut loads.

For cuts in dry or moist sand, the apparent pressure envelope may be considered a simple rectangle (Fig. 27.7b) in which the magnitude of the pressure is $0.65\gamma H \tan^2(45° - \phi/2)$. The diagram may also be used for determining the strut loads in a drained sand if the free water surface has been lowered at least to the level of the bottom of the cut.

Modes of Failure in Clay. Excavations inside braced cuts in clay are generally made rapidly with respect to the rate at which the water content of the clay can adjust to the new stress conditions. Hence, undrained or $\phi = 0$ conditions prevail (Art. 4.8). As the depth of the cut increases, the soil outside the walls behaves like a surcharge with respect to the clay inside the enclosure and causes the soil beneath the excavation to rise. The movement occurs even if the sheeting is comparatively stiff and extends a considerable distance below the bottom of the cut, unless a firm base exists within a short distance beneath excavation level. If the cut becomes too deep with respect to the strength of the clay, the heave of the bottom may be uncontrollable, settlements of the surrounding ground surface may become excessive, and the bracing system may collapse.

In clays as well as in sands, failures by bending of wales or of sheet piles or soldier piles are unusual. If the possibility of bottom heave does not exist, the principal type of failure to be guarded against is the buckling of struts or the crippling of wales where the strut reactions occur.

Base Failure of Cuts in Clay. The strength of the clay beneath the bottom of the cut at any given level of excavation has a decisive influence on the behavior of the bracing system and the surrounding soil. If the un-

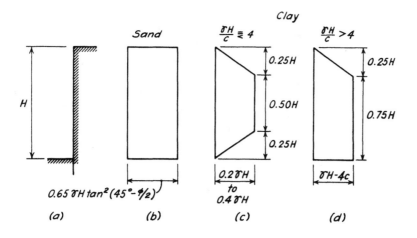

FIGURE 27.7. Apparent pressure diagrams for calculating loads in struts of braced cuts. (a) Sketch of wall of cut. (b) Diagram for cuts in dry or moist sand. (c) Diagram for clays if $\gamma H/c$ is less than 4. (d) Diagram for clays if $\gamma H/c$ is greater than 4 provided $\gamma H/c_b$ does not exceed about 4, where c is the average undrained shearing strength of the soil beside the cut, and c_b the undrained strength of the clay below excavation level.

drained shearing resistance below base level is denoted by c_b, the ability of the soil to withstand the surcharge γH of the clay outside the excavation is given approximately by the bearing-capacity equation (eq. 18.2), and

$$q_d = c_b N_c \qquad 27.4$$

where N_c ranges from 5 to 6, depending on the plan dimensions of the cut, as shown in Fig. 18.2. If the depth of excavation is great enough to induce a bearing-capacity failure, $q_d = \gamma H$ and $N_c = \gamma H/c_b$. Experience has demonstrated that if $\gamma H/c_b$ is less than about 6, movements of the bracing sytem and heave of the clay below base level are small. If $\gamma H/c_b$ reaches about 8, the movements of even a well-designed bracing system become intolerably large. At values exceeding about 8, the bracing is likely to collapse because of the large inward movements of the clay outside the embedded portion of the sheet piles or soldier piles and the uncontrollable upward heave of the clay beneath excavation level. Simple braced open excavations should not be attempted under these conditions.

Loads on Struts in Clay. If the base stability of an open cut in clay is adequate for all depths of excavation, as indicated by a value of $\gamma H/c_b$ less than about 6, the strut loads can be determined from apparent pressure envelopes (Fig. 27.6). The apparent pressure envelope to be used depends on the value of $\gamma H/c$, where c is the average undrained shearing strength of the clay alongside the cut. If this ratio is equal to or less than about 4, the behavior of the clay beside the cut is essentially elastic and the load in the struts depends primarily on the deflection of the sheeting permitted during excavation and bracing. The corresponding apparent pressure envelope is shown in Fig. 27.7c. In most instances, the width of the apparent pressure envelope can be considered equal to $0.3\gamma H$. The load in the lowermost braces is relatively small and the lower portion of the apparent pressure envelope decreases in width as shown in the figure.

If the ratio $\gamma H/c$ exceeds about 4, the apparent pressure envelope may be taken as the diagram shown in Fig. 27.7d, provided the width of the envelope is greater than that in Fig. 27.7c. Otherwise, the value from Fig. 27.7c governs, regardless of the value of $\gamma H/c$. The increased value of $\gamma H/c$ is associated with inelastic behavior of the clay near the bottom of the cut; consequently, the apparent pressure envelope is not truncated as is that shown in Fig. 27.7c.

The diagram (Fig. 27.7d) may be used for values of $\gamma H/c$ as great as 10 or 12. On the other hand, if $\gamma H/c_b$ exceeds about 7 and base failure is imminent, the strut loads may be much larger than indicated by the diagram. Hence, the stability of the base should always be investigated before an estimate is made of the strut loads.

Depth of Embedment of Sheet Piles or Soldier Piles. The portion of the sheet or soldier piles embedded below the bottom of the excavation reduces the inward movement associated with the last stages of excavation and, hence, the settlement of the adjacent ground surface. The extent of the reduction depends primarily on the properties of the soil beneath excavation level and, to a much smaller extent, on the stiffness of the piles. If the subsoil consists of clay to a considerable depth and if the base stability factor $\gamma H/c_b$ is large, even a great depth of penetration of heavy piling has a negligible influence on the movements. On the other hand, if the soft soil is underlain within a few feet by a firm material in which the piles can be embedded, the beneficial effect may be substantial.

As the depth of excavation of a braced cut increases, the soil outside the cut tends to settle and to drag the piles down by negative skin friction. The settlement can be reduced if the piles can be driven far enough into a bearing stratum to develop point resistance at least equal to the dragdown forces. Moreover, the piles are subjected to downward loads caused by the weight of the bracing system and often, in urban areas, to the weight of decking and street traffic that must be maintained during excavation. For these reasons the embedment of sheet or soldier piles may be increased beyond that otherwise required. It may be advantageous and economical to extend the piles to reach a firm stratum even if the depth of embedment would appear to be unnecessarily great.

Other Considerations. According to the preceding paragraphs, each strut in a braced cut should be proportioned as if it would be subjected to the maximum load indicated by the apparent pressure diagram. However, the deflection of the wales and sheet piles between struts induces shearing stresses in the soil, which transfer part of the load directly to the areas of more rigid support. Therefore, the wales and the sheeting or soldier piles need not be designed for the full bending moments corresponding to the pressure given by the diagram. Experience has suggested that the design of such members is adequate if the members are considered continuous over their supports and are proportioned for $\frac{2}{3}$ of the intensity of load determined from the apparent pressure diagrams.

Wales often consist of H sections with horizontal webs, as shown in Fig. 8.7. Wherever the struts deliver their reactions to such wales, stiffeners are necessary to prevent crippling of the webs under the full design loads in the struts. Because the bracing system is assembled and often fabricated in the field, allowance must be made in design for eccentricities and uneven bearing. The dimensions of the bracing system should also allow ample tolerances so that the bracing system will not encroach on the permanent structure to be constructed within the cut. Since sheet piles or soldier piles cannot be driven to perfect alignment, the wales should be set out several inches from the vertical members to permit wedges or other blocking to be placed to make up for the irregularities. Furthermore, since the sheet piles or soldier piles move inward during excavating and bracing, an allowance should be made for such movements.

To reduce the general movements of the bracing system as much as possible, the struts should be prestressed as soon as they are installed (Art. 8.5). Prestress loads between about 40 and 70 per cent of the anticipated maximum strut load are customary.

The inward lateral movements of the walls of the excavation, and consequently the settlement of the adjacent ground surface, increase with increasing vertical distance between struts. Hence, if the movements are likely to be excessive, the distance between struts should be restricted and excavation should be permitted to extend no deeper than necessary to permit installation of each of the struts. In plastic clays, to keep the movements to a minimum, the vertical distance between struts should not exceed $2c/\gamma$, where c is the average undrained shearing strength of the clay for a depth of about $B/2$ below the level of the preceding strut, and B is the width of the cut.

The wales and struts of the bracing system for a large cut constitute heavy vertical loads that require support. Usually each line of bracing is hung from a beam spanning the cut. It is customary to provide not only hangers but also vertical spacers, capable of carrying compression, at each point of support so that buckling either upward or downward is resisted. Horizontal bracing is also required in wide cuts to prevent lateral buckling of the struts. If the cut is too wide, the bracing may be supported by piles driven to suitable embedment or bearing below final excavation grade.

Inclined Bracing. In the foregoing paragraphs, the vertical walls of the cut are considered to be supported by horizontal struts. In wide excavations, cross-lot bracing may be impractical, and inclined braces or rakers (Art. 8.4) are often used. Preferably, the load in the rakers is transferred to a completed portion of the foundation or basement slab of the structure (Fig. 8.4). Otherwise, an inclined footing or *kicker block*, supported by the soil, usually provides the reaction. The capacity of soil-supported kicker blocks is often seriously overestimated.

The ultimate capacity of an inclined kicker block (Fig. 27.8a) can be estimated, for saturated clays or for sands, respectively, by equations analogous to eqs. 18.2 and 19.2. The bearing-capacity factors depend, however, on the inclination α between the face of the kicker block and the horizontal. They also depend on the ratio D_f/B. Values of the bearing-capacity factor N_{cq}, appropriate for clays, may be obtained from Fig. 27.8b, and the ultimate bearing capacity determined from

$$q_d = cN_{cq} \qquad 27.5$$

Even if a high factor of safety, such as 3 or more, is applied to this value, progressive movement of kicker blocks on soft or medium clay is likely to occur. If such materials cannot be avoided for support, provision should be made to maintain the load in the inclined bracing. Similarly, for sands, values of the bearing-capacity factor $N_{\gamma q}$ may be obtained from Fig. 27.8c and inserted in

$$q_d = \tfrac{1}{2}B\gamma N_{\gamma q} \qquad 27.6$$

For sands, D_f/B should be taken as zero unless the wedge of soil abc, excavated for installation of the kicker block, is backfilled to the level of the surrounding ground surface.

To restrict movements of the bracing system, the factor of safety of kicker blocks should be no less than 2.5 even though their use is in temporary construction.

The force diagram (Fig. 27.8d) indicates that under the influence of the raker load produced by the earth pressure, the sheet-

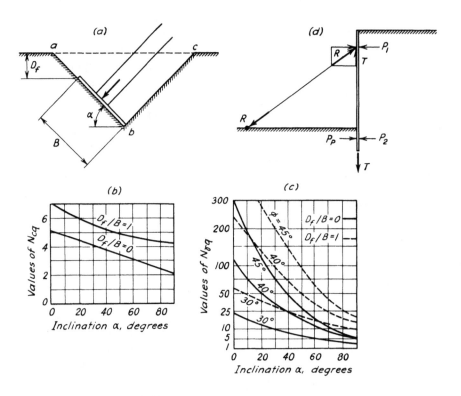

FIGURE 27.8. (a) Geometry of kicker block for inclined strut. (b) Bearing-capacity factors for kicker block in saturated clay. (c) Bearing-capacity factors for kicker block in sand for various angles of internal friction, ϕ. (d) Simplified diagram of forces acting on sheeting in raker system; P_1 and P_2, components of active earth pressure; P_P, passive pressure; T, tension in pile to be resisted by uplift resistance of embedded portion; R, raker load [(b) and (c) after Meyerhof, 1953].

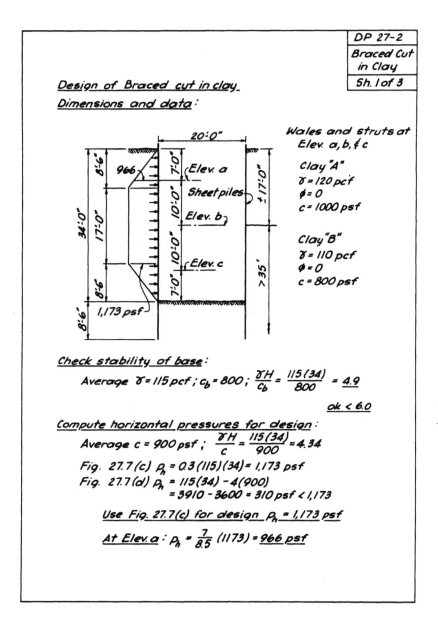

ing tends to move upward with respect to the soil beneath excavation level. Adequate embedment must be provided to resist the uplift. The apparent pressure diagrams (Fig. 27.7) may be used to estimate the horizontal components of the raker loads.

ILLUSTRATIVE DESIGN. DP 27-2. BRACED CUT

The section chosen for illustration is assumed to be located in the central portion of a long cut; hence, the wales are designed as beams. Near the ends of the cut the wales may also carry axial loads derived from the earth pressure acting on the end walls; under these circumstances they must be designed as beam-columns. Since the blocking between the sheet piles and the wales provides considerable restraint against vertical deflection of the wales, buckling of the compression flange does not usually influence the design.

Although the webs of the wales may be

```
                                                        ┌─────────────┐
                                                        │   DP 27-2   │
   Determine size of sheet piles                        │ Braced Cut  │
     Cantilever moment at Elev. a:                      │  in Clay    │
       M = say ⅔(⅙)(966)(7.0)² = 5.2 ᶦᵏ/ft.             │  Sh. 2 of 3 │
     Check moment at Elev. b:                           └─────────────┘
       M = say ⅔(1/10)(1173)(10)² = 7.8 ᶦᵏ/ft. > 5.2 ᶦᵏ/ft.
       S req'd = (7.80 × 12)/25 = 3.7 in³/ft.    Use PMA 22
   Design wales and struts:                      S = 5.4 in³/ft.
```

$$M = \text{say } \tfrac{2}{3}\left(\tfrac{1}{6}\right)(966)(7.0)^2 = 5.2\ ^{\text{ik}}/\text{ft}$$

$$M = \text{say } \tfrac{2}{3}\left(\tfrac{1}{10}\right)(1173)(10)^2 = 7.8\ ^{\text{ik}}/\text{ft} > 5.2\ ^{\text{ik}}/\text{ft}$$

$$S_{\text{req'd}} = \frac{7.80 \times 12}{25} = 3.7\ \text{in}^3/\text{ft}$$

Use PMA 22, $S = 5.4\ \text{in}^3/\text{ft}$

Load per ft. at

Elevs. a & c: $\tfrac{1}{2}(966)(7.0) = 3{,}381$ lb/ft.
$\tfrac{1}{2}(966 + 1173)(1.5) = 1{,}604$
$(1173)(3.5) = 4{,}106$
Total $= 9{,}091$ lb/ft.

Elev. b: $(1173)(10) = 11{,}730$ lb/ft.

Assume struts spaced at 20'-0" max.

Moment in wale at Elev. b = say $\tfrac{2}{3}\left(\tfrac{1}{10}\right)(11.73)(20)^2 = 313\ ^{\text{ik}}$

$S_{\text{req'd}} = \dfrac{313 \times 12}{22} = 171\ \text{in}^3$

Use W24 × 76 wales at Elev. b $S = 176\ \text{in}^3$

See Sh. 3 for design of struts and bearing stiffeners

---

theoretically adequate to carry the strut reactions without crippling, field connections in braced cuts are rarely fitted to the close tolerances expected in steel bridges or buildings. Consequently, the connections are likely to introduce eccentric loadings into the wales. Stiffeners are provided to assure that the wales will not be deformed in spite of the inevitable field inaccuracies.

27.4 Tieback Bracing Systems

Modes of Failure. Almost all failures of tieback bracing systems have been the result of inadequate anchorage. In some instances, the anchors have been located too close to the wall, whereupon the entire mass of soil behind the wall, including the anchors, has descended into the cut. In others, individual anchors have pulled out of the stable soil in which they were embedded.

As the excavation progresses and the walls move in (Fig. 27.1c), the soil outside the cut settles and exerts a downward drag on sheet piles or soldier piles. A similar tendency has been pointed out in connection with braced cuts. If the tieback anchors

```
                                                    DP 27-2
                                                    Braced Cut
    Design wales and struts cont'd.                 in Clay
        Strut at Elev. b: P = 1.173 (10)(20) = 235ᵏ Sh. 3 of 3
        Say Fy = 36 ksi  KL = 16.0'

    Use W10×54 struts  Allow P = 253ᵏ
    (Alternate 10"⌀ extra strong pipe Allow P = 291ᵏ)

    Design bearing stiffeners:
        Provide area of stiffeners + web of wale equal
        to area of strut.
            Strut area =              15.9 sq. in.
            Wale web = say 10 × 7/16 = 4.4
                                      11.5 sq. in. req'd.
    Use 2-4"×3/4" pls. each side    A_b = 12.0 sq. in.
```

are inclined as in Fig. 27.1c, the force in the anchors adds to the downward load on the walls. Hence, there is a tendency for the walls to settle and, when the cut nears its maximum depth, for the point capacity of the sheet piles or soldier piles to be exceeded. The downward movements of the piles may appreciably reduce the stresses in the relatively short lowermost anchors and destroy their effectiveness. The load in the lower anchors is then transferred to some of the upper anchors. If the upper anchors are thus overloaded, a general failure may be initiated. Hence, adequate point capacity or embedment of the vertical piles is vital to the stability of such a tieback system.

Anchorage. The lateral movements associated with the construction of a tieback anchorage (Fig. 27.1c) resemble those for a braced cut (Fig. 27.1b). The limited observational data presently available indicate that the loads in the anchors may be determined by assuming that their horizontal components are given by the apparent pressure diagrams for braced cuts (Fig. 27.7). If wales are used, they may be proportioned as suggested for braced cuts; sheet piles or soldier piles may also be so proportioned.

The choice between horizontal or inclined tiebacks depends on several circumstances. If the strength of the soil increases with depth or if there are particularly resistant layers at a reasonable depth below weaker materials, inclined anchorages deriving their support from the strong material may be preferable. On the other hand, if the sheet piles or soldier piles do not encounter adequate resistance to restrict their downward movement, the lower anchors in a series of inclined tiebacks may become destressed and ineffective.

Anchors consisting of steel rods or tendons inserted in small-diameter holes and grouted to the surrounding supporting material can usually be installed at any inclination. It is advantageous to locate them in cohesive or slightly cohesive members of the deposit to reduce difficulties associated with caving of the holes. Moreover, larger-diameter anchors, terminating in enlargements or bells that provide the pull-out resistance, must be so located that the bells can be formed in cohesive materials. Equipment for drilling such shafts and forming the bells cannot usually be operated horizontally; an angle of 30° or more with the horizontal is generally required. To satisfy these two conditions, belled anchors are almost always inclined. The choice among the alternative types of anchors depends primarily on economy, on the presence or absence of cohesive zones, and on the equipment available in a given locality.

In any event, the tiebacks must develop their anchorage within the stable soil behind the failure wedge that contributes to the active earth pressure on the wall. The wedge may be considered to have the dimensions shown in Fig. 27.9. Within this wedge, as indicated by the dashed line, no support for the anchors should be assumed. Indeed, it is preferable within this distance to backfill the anchor holes with material that will not develop a positive bond between the anchorage and the surrounding soil. Beyond the dashed line, grouted an-

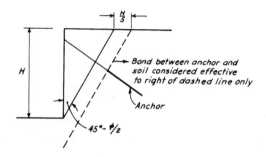

FIGURE 27.9. Limits beyond which the anchorages of tiebacks must be located to avoid general failure.

chors must extend far enough to develop the required resistance by bond, and belled anchors must develop the necessary bearing capacity.

Properly grouted anchor rods or tendons may be expected to develop an ultimate bond resistance equal approximately to the shearing resistance of the surrounding soils. The pull-out resistance for estimating the uplift resistance of belled anchors may be computed by eq. 20.3 and the procedures suggested for estimating the uplift resistance of belled piers in swelling soils. Such calculations for either type of anchor are justifiable only for making preliminary estimates of cost. The final conclusions concerning the capacity of individual anchors must be based on pull-out tests in the field.

It is good practice to prestress anchors for the same reasons that struts are prestressed in braced cuts. The equipment for prestressing can also be used for performing pull-out tests to determine the ultimate capacity of enough anchors to assure adequacy of the design. On many jobs, test anchors are required to develop a pull-out capacity of 150 per cent or more of the design load, and all anchors are prestressed to the design load. Indeed, one of the principal advantages of a tieback bracing system is the ability at low cost to test each anchor at least to its expected working load. Defective anchors can be lengthened, repaired, or replaced by others if necessary. In this fashion, the possibility of failure by inadequacy of individual anchors can be eliminated.

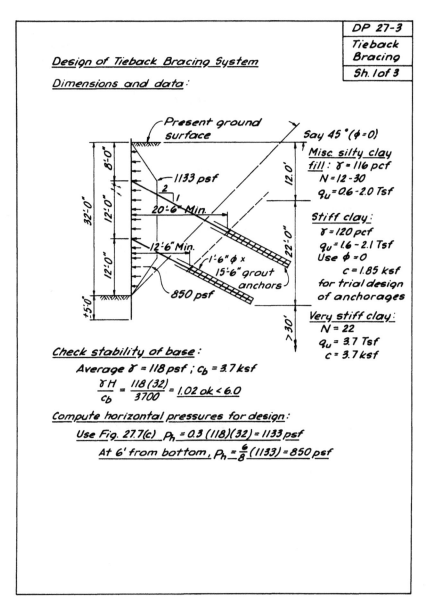

Practical Considerations. To avoid the noise associated with pile driving, tieback systems are often constructed by predrilling vertical holes into which soldier piles are lowered. If necessary to keep the holes from collapsing, drilling mud is used, the piles are inserted, and low-strength concrete is placed by tremie methods so as to displace the drilling mud from the bottom. The low-strength concrete can be chipped away as excavation proceeds, in order to insert the lagging. The procedure has the additional advantage that at their bottoms the soldier piles are embedded in a concrete cylinder with a bearing area much greater than that of the piles themselves. The increased bearing area substantially reduces the likelihood of bearing capacity failure and excessive settlement.

Details of the connection between soldier piles and lagging, with or without wales, have been developed to suit a wide variety of conditions. Similarly, many different designs are available for the tiebacks them-

> **DP 27-3**
> **Tieback Bracing**
> **Sh. 2 of 3**
>
> <u>Determine size of soldier piles</u>:
>
> <u>Space piles on 6'-0" ctrs.</u>
>
> <u>Cantilever moment</u>: $M = 6(\frac{1}{6})(1133)(8)^2 = \underline{72.6}^{\prime k}$
>
> <u>Section modulus required</u>:
>
> $S = \frac{72.6 \times 12}{22} = 39.6 \, in.^3$ <u>Use W14×30 S=41.9 in.^{3}</u>
>
> <u>Check for interior moment</u>:
>
> $M = \text{Say } 6(\frac{2}{3})(\frac{1}{10})(1133)(12)^2 = \underline{65.3}^{\prime k} \, ok < 72.6^{\prime k}$
>
> <u>Determine size of wood lagging</u>:
>
> Say $M = \frac{2}{3}(\frac{1}{8})(1.133)(5.5)^2 = 2.86^{\prime k}$
> $\qquad\qquad\qquad\qquad\qquad \times 12 = 34.3^{\prime\prime k}$
>
> S req'd = 34.3
> $\qquad \div$ say 1.8 ksi = 19.1 in.^{3}
>
> $S = \frac{1}{6}(12)t^2 = 19.1 \quad t = \sqrt{\frac{19.1 \times 6}{12}} = 3.1''$
>
> $\qquad\qquad\qquad$ <u>Use 3" lagging</u>
>
> <u>Compute loads in tiebacks</u>:
>
> <u>At 8 ft. level</u>: $H = 6\left[\frac{1}{2}(1133)(8) + (1133)(6)\right]$
> $\qquad\qquad = 6[4532 + 6798] = 68.0^k$
> $\qquad\qquad\qquad\qquad\qquad \times \frac{\sqrt{5}}{2} = \underline{76.0^k = R}$
>
> <u>At 20 ft. level</u>: $H = 6\left[1133(10) + \frac{1133+850}{2}(2)\right]$
> $\qquad\qquad = 6[11330 + 1983] = 79.9^k$
> $\qquad\qquad\qquad\qquad\qquad \times \frac{\sqrt{5}}{2} = \underline{89.3^k = R}$
>
> <u>Determine size of tiebacks</u>
>
> Use $4 - \frac{1}{2}''$ (7-wire) strands (ASTM A-416)
>
> $\quad$ Working load = 0.6 ultimate
> $\qquad\qquad\qquad = 4(0.6)(41.3) = \underline{99.1^k \, ok > 89.3^k}$

selves. Some procedures are patented or are specialties of various construction organizations. If the general requirements discussed above are satisfied with respect to the stability of the installation, the most economical or convenient of the special systems may be safely used.

Since tiebacks extend beyond the walls to be supported, they may pass beyond the property lines. Occasionally, the responsibilities or legal issues involved preclude the otherwise advantageous use of a tieback system.

ILLUSTRATIVE DESIGN. DP 27-3. TIEBACK BRACING SYSTEM

The soldier piles and lagging illustrated in this example could as well have been used for the braced cut of DP 27-2; sheet piles could have been used in DP 27-3. The design procedures for the walls are

> **DP 27-3**
> **Tieback Bracing**
> **Sh. 3 of 3**
>
> <u>Estimate development length for anchorage:</u>
>
> Say $\frac{C}{F} = \frac{1.85}{1.5} = 1.23$ ksf
>
> Max load = 89.3^k (see Sh. 2)
>
> $\div 1.23 = 72.6$ sq. ft. = Shear area required
>
> Use 1'-6" $\phi \times$ 15'-6"   A = 73.0 sq. ft.
>
> <u>Check capacity by applying 134^k to test anchors</u>
>
> <u>Determine size of wales:</u>
>
> Assume horizontal component of tieback load (79.9^k) is applied at center of 6'-0" between soldier piles.
>
> Say $M = \frac{2}{3}\left(\frac{1}{8}\right)(79.9)(6) = 40.0$ '^{k}
>
> S req'd. $= \frac{40.0 \times 12}{22} = 21.8$ in.^{3}
>
> Use 2 - C9 $\times$ 15   S = 22.6 in.^{3}
>
> <u>Bearing stiffeners:</u>
>
> $H_{test\ load} = 134 \times \frac{2}{\sqrt{5}} = 120^k$
>
> A_b req'd. $= \frac{120}{27} = 4.44$ sq. in.
>
> Webs of wales say $2(6)\left(\frac{5}{16}\right) = 3.7$ sq. in.
>
> Use 2 - 2" $\times \frac{3}{8}$" Pls. each channel = 3.0
>
> $A_b = 6.7$ sq. in. ok > 4.44
>
> <u>Locate stiffeners to provide efficient transmission of test load, as well as permanent load, between anchors and wales.</u>

essentially the same whether the support consists of struts or tiebacks.

Although lagging is designed by many engineers by the method shown on Sh. 2, the results are likely to be overly conservative because the soil pressures are concentrated on the outside flanges of the stiffer soldier piles. Hence, the dimensions of the lagging are customarily selected by experienced engineers or contractors on the basis of judgment.

Calculation of the length of embedment of the grouted portion serves only for design of the first test anchors. The actual dimensions or allowable loads and spacing of the anchors for the final design are established on the basis of the test results.

SUGGESTED READING

General

G. B. Sowers and G. F. Sowers (1967), "Failures of Bulkhead and Excavation Bracing," *Civ. Eng. ASCE*, 37, 1, 72–77.

L. Bjerrum, C. J. Frimann Clausen, and J. M. Duncan (1972), "Earth Pressures on Flexible Structures—a State-of-the-Art Report," *Proc. 5 European Conf. Soil Mech.*, Madrid, 2, 169–196.

Anchored Bulkheads

K. Terzaghi (1954), "Anchored Bulkheads," *Trans. ASCE*, 119, 1243–1324. Comprehensive treatment.

P. W. Rowe (1952), "Anchored Sheet Pile Walls." *Proc. Inst. Civ. Eng.*, London, 1, Part 1, 27–70. Influence of flexibility of piles and other factors on design.

G. Tschebotarioff and E. R. Ward (1957), "Measurements with Wiegmann Inclinometer on Five Sheet Pile Bulkheads," *Proc. 4th Int. Conf. Soil Mech.*, London, 2, 248–255. Field verification of proximity of point of contraflexure to dredge line.

Braced Cuts

K. Terzaghi (1941), "General Wedge Theory of Earth Pressure," *Trans. ASCE*, 106, 68–97. One of the classics of soil mechanics; the first paper to explain and express quantitatively the influence of the deformations associated with braced cuts on the magnitude and distribution of the earth pressure.

R. B. Peck (1969b), "Deep Excavations and Tunneling in Soft Ground," *Proc. 7th Int. Conf. Soil Mech.*, Mexico, 225–290. State-of-the-art volume. Summary of field measurements and procedures for design of bracing for cuts in various soils.

J. G. Thon and R. C. Harlan (1971), "Slurry Walls for BART Civic Center Subway Station," *ASCE J. Soil Mech.*, 97, SM9, 1317–1334. Recent example of braced excavation for subway station.

Tieback systems

W. L. Shannon and R. J. Strazer (1970), "Tied-Back Excavation Wall for Seattle First National Bank," *Civ. Eng. ASCE*, 40, 3, 62–64.

T. D. Wosser and R. Darragh (1970), "Tiebacks for Bank of America Building Excavation Wall," *Civ. Eng. ASCE*, 40, 3, 65–67.

T. H. Hanna and G. A. Matallana (1970), "The Behavior of Tied-Back Retaining Walls," *Canadian Geot. Jour.*, 7, 4, 372–396. Results of model tests to investigate interdependence of earth-pressure distribution, wall movements, and anchor loads.

Bibliography

Aas, G. (1965), "A Study of the Effect of Vane Shape and Rate of Strain on the Measured Values of *in-situ* Shear Strength of Clays," *Proc. 6 Int. Conf. Soil Mech.*, Montreal, *1*, 141–145.

Adams, J. I. (1965), "The Engineering Behaviour of a Canadian Muskeg." *Proc. 6 Int. Conf. Soil Mech..* Montreal, *1*, 3–7.

AASHO (1970), *Standard Specifications for Highway Materials and Methods of Sampling and Testing.* 10th ed., Wash., D. C.. Am. Assn. State Hwy. Officials, Part I, 523 pp., Part II, 817 pp.

AREA, *Manual of Recommended Practice*, Chicago, Assoc. Amer. Railroads, loose-leaf document continuously brought up to date.

ASCE (1966), "Bibliography on Chemical Grouting," Committee on Grouting, *ASCE J. Soil Mech.*, *92*, SM6, 39–66.

ASCE (1957), "Chemical Grouting," *ASCE J. Soil Mech.*, *83*, SM4, Paper No. 1426, 106 pp.

ASCE (1958), "Cement and Clay Grouting of Foundations," *ASCE J. Soil Mech.*, *84*, SM1, Papers No. 1544–1552.

ASCE (1972), "Subsurface Investigation for Design and Construction of Foundations of Buildings," Task Committee for Foundation Design Manual. Part I, *ASCE J. Soil Mech.*, *98*, SM5, 481–490; Part II, No. SM6, pp. 557–578; Parts III and IV, No. SM7, pp. 749–764.

ASP (1960), *Manual of Photographic Interpretation*, Am. Soc. Photogrammetry, Wash., D. C., 868 pp.

ASTM (1952), "Symposium on Surface and Subsurface Reconnaissance," *ASTM Spec. Tech. Publ. 122*, 228 pp.

ASTM (1957), "Vane Shear Testing of Soils," *ASTM Spec. Tech. Publ. 193*, 70 pp.

ASTM (1966), "Testing Techniques for Rock Mechanics," *ASTM Spec. Tech. Publ. 402*, 297 pp.

ASTM (1970), "Special Procedures for Testing Soil and Rock for Engineering Purposes," *ASTM Spec. Tech. Publ. 479*, 630 pp.

ASTM (1971), "Sampling of Soil and Rock," *ASTM Spec. Tech. Publ. 483*, 193 pp.

Atterberg, A. (1908), *Studien auf dem Gebiet der Bodenkunde* (Studies in the Field of Soil Science), *Landw. Versuchsanstalt*, Vol. 69.

Atterberg, A. (1911), *Über die Physikalishe Bodenuntersuchung und Über die Plastizität der Tone* (On the Investigation of the Physical Properties of Soils and on the Plasticity of Clays), *Int. Mitt. für Bodenkunde*, *1*, 10–43.

Bagnold, R. A. (1941), *The Physics of Blown Sand and Desert Dunes*, New York, Wm. Morrow, 265 pp.

Baker, B. (1881), "The Actual Lateral Pressure of Earthwork." *Min. Proc. Inst. Civ. Eng.*, London, 65, 140-186.

Baker, C. N., Jr., and F. Khan (1971), "Caisson Construction Problems and Correction in Chicago," *ASCE J. Soil Mech.*, 97, SM2, 417-440.

Barden, L. (1962), "Distribution of Contact Pressure Under Foundations," *Géotechnique*, 12, 3, 181-198.

Barkan, D. D. (1962), *Dynamics of Bases and Foundations* (translated from the Russian by L. Drashevska; translation edited by G. P. Tschebotarioff), New York, McGraw-Hill, 434 pp.

Basore, C. E., and J. D. Boitano (1969), "Sand Densification by Piles and Vibroflotation," *ASCE J. Soil Mech.*, 95, SM6, 1303-1323.

Belcher, D. J., L. E. Gregg, D. S. Jenkins, and K. B. Woods (1946), "Origin and Distribution of United States Soils," Map included in *Preliminary Technical Development Rep. 52*, Civil Aero. Admin., Wash., D. C.

Bishop, A. W. (1948), "A New Sampling Tool for Use in Cohesionless Sands Below Ground Water Level," *Géotechnique*, 1, 2, 125-131.

Bishop, A. W. (1955), "The Use of the Slip Circle in the Stability Analysis of Slopes," *Géotechnique*, 5, 1, 7-17.

Bishop, A. W., and L. Bjerrum (1960), "The Relevance of the Triaxial Test to the Solution of Stability Problems," *Proc. ASCE Research Conf. on Shear Strength of Cohesive Soils*, pp. 437-501.

Bishop, A. W., and D. J. Henkel (1962), *The Measurement of Soil Properties in the Triaxial Test*, 2nd ed., London, Edward Arnold, 228 pp.

Bjerrum, L. (1957), "Norwegian Experiences with Steel Piles to Rock," *Géotechnique*, 7, 2, 73-96.

Bjerrum, L. (1967), "Engineering Geology of Norwegian Normally-Consolidated Marine Clays as Related to Settlements of Buildings," *Géotechnique*, 17, 2, 83-117.

Bjerrum, L., and A. Eggestad (1963), "Interpretation of Loading Test on Sand," *Proc. European Conf. Soil Mech.*, Wiesbaden, 1, 199-203. Reprinted in *NGI Publ. 58*, 23-27.

Bjerrum, L., and N. E. Simons (1960), "Comparison of Shear Strength Characteristics of Normally Consolidated Clays," *Proc. ASCE Research Conf. on Shear Strength of Cohesive Soils*, pp. 711-726.

Bjerrum, L., C. J. Frimann Clausen, and J. M. Duncan (1972), "Earth Pressures on Flexible Structures—a State-of-the-Art Report," *Proc. 5 European Conf. Soil Mech.*, Madrid, 2, 169-196.

Bjerrum, L., I. J. Johannessen, and O. Eide (1969), "Reduction of Negative Skin Friction on Steel Piles to Rock," *Proc. 7 Int. Conf. Soil Mech.*, Mexico, 2, 27-34.

Bjerrum, L., A. Casagrande, R. B. Peck, and A. W. Skempton, eds. (1960), *From Theory to Practice in Soil Mechanics; Selections from the Writings of Karl Terzaghi*, New York, Wiley, 425 pp.

Brinch Hansen, J. (1953), *Earth Pressure Calculation*, Copenhagen, Danish Technical Press, 271 pp.

Brink, A. B. A. and B. A. Kantey (1961), "Collapsible Grain Structure in Residual Granite Soils in Southern Africa," *Proc. 5 Int. Conf. Soil Mech.*, Paris, 1, 611-614.

Brinkhorst, W. H. (1936), "Settlement of Soil Surface Around Foundation Pit," *Proc. 1 Int. Conf. Soil Mech.*, Cambridge, Mass., 1, 115-119.

Brzezinski, L. S. (1969), "Behavior of an Overpass Carried on Footings and Friction Piles," *Canadian Geot. Jour.*, 6, 4, 369-382.

Burmister, D. M. (1951), "Identification and Classification of Soils," *ASTM Spec. Tech. Publ. 113*, pp. 3-24.

Button, S. J. (1953), "The Bearing Capacity of Footings on a Two-Layer Cohesive Subsoil," *Proc. 3 Int. Conf. Soil Mech.*, Zurich, *1*, 332–335.

Cadling, L., and S. Odenstad (1950), "The Vane Borer," *Proc. Swedish Geot. Inst.*, No. 2, 88 pp.

Carlson, E. D., and S. P. Fricano (1961), "Tank Foundations in Eastern Venezuela," *ASCE J. Soil Mech.* 87, SM5, 69–90.

Carson, A. B. (1961), *General Excavation Methods*, New York, F. W. Dodge Corp., 392 pp.

Casagrande, A. (1932a), "Research on the Atterberg Limits of Soils," *Public Roads*, *13*, 121–136

Casagrande, A. (1932b), "The Structure of Clay and Its Importance in Foundation Engineering," *J. Boston Soc. Civil Engrs.*, *19*, 4, 168–221.

Casagrande, A. (1935), "Seepage Through Dams," *J. New England Water Works Assoc.*, *51*, 2, 131–172. Reprinted in *Contributions to Soil Mechanics, 1925–1940*, Boston Soc. Civ. Eng., 1940, and as Harvard Univ. Soil Mech. Series No. 5.

Casagrande, A. (1936), "The Determination of the Pre-Consolidation Load and Its Practical Significance," *Proc. 1 Int. Conf. Soil Mech.*, Cambridge, Mass., *3*, 60–64.

Casagrande, A. (1947), "The Pile Foundation for the New John Hancock Building in Boston," *J. Boston Soc. Civ. Eng.*, *34*, 4, 297–315. Reprinted in *Contributions to Soil Mechanics, 1941–1953*, Boston Soc. Civ. Eng. 1953, 147–165; also as Harvard Soil Mech. Series No. 30.

Casagrande, A. (1948), "Classification and Identification of Soils," *Trans. ASCE*, *113*, 901–991.

Casagrande, A. (1949), "Soil Mechanics in the Design and Construction of The Logan Airport," *J. Boston Soc. Civil Engrs.*, *36*, 2, 192–221.

Casagrande, A. (1971), "On Liquefaction Phenomena," Report of lecture as prepared by P. A. Green and P. A. S. Ferguson, *Géotechnique*, *21*, 3, 197–202.

Casagrande, A., and R. E. Fadum (1940), "Notes on Soil Testing for Engineering Purposes, "*Harvard Univ. Grad. School of Engineering Publ. No. 8*, 74 pp.

Casagrande, A., and R. E. Fadum (1944), "Application of Soil Mechanics in Designing Building Foundations," *Trans. ASCE*, *109*, 383–416.

Casagrande, A., and R. C. Hirschfeld (1960), "Stress-Deformation and Strength Characteristics of a Clay Compacted to a Constant Dry Unit Weight," *Proc. ASCE Research Conf. on Shear Strength of Cohesive Soils*, pp. 359–417.

Casagrande, A., and S. D. Wilson (1951), "Effect of Rate of Loading on Strength of Clays and Shales at Constant Water Content," *Géotechnique*, *2*, 3, 251–263.

Casagrande, L. (1952), "Electro-Osmotic Stabilization of Soils," *J. Boston Soc. Civ. Eng.*, *39*, 1, 51–83.

Casagrande, L. (1966), "Subsoils and Foundation Design in Richmond, Va.," *ASCE J. Soil Mech.*, *92*, SM5, 109–126.

Casagrande, L., and S. Poulos (1969), "On the Effectiveness of Sand Drains," *Canadian Geot. Jour.*, *6*, 3, 287–326.

Cedergren, H. R. (1967), "Seepage, Drainage, and Flow Nets," New York, Wiley, 489 pp.

Chellis, R. D. (1961), *Pile Foundations*, 2nd ed., New York, McGraw-Hill, 704 pp.

Clevenger, W. A. (1956), "Experiences with Loess as Foundation Material," *Trans. ASCE*, *123*, 151–169. Discussion by Peck and Ireland, pp. 171–179.

Condron, T. L., and E. R. Math (1932), "Investigating a Foundation in Soft Soil," *Civ. Eng. ASCE*, 2, 4, 237–241.

Crawford, C. B. (1964), "Some Characteristics of Winnipeg Clay," *Canadian Geot. Jour.*, 1, 4, 227–235.

Crawford, C. B., and J. G. Sutherland (1971), "The Empress Hotel, Victoria, British Columbia. Sixty-five Years of Foundation Settlements," *Canadian Geot. Jour.*, 8, 1, 77–93.

Cummings, A. E. (1940), "Dynamic Pile Driving Formulas," *J. Boston Soc. Civ. Eng.*, 27, 6–27; reprinted in *Contributions to Soil Mechanics 1925–1940*, Boston Soc. Civ. Eng., 1940, pp. 392–413.

Cummings, A. E. (1949), "Lectures on Foundation Engineering," *U. of Ill. Eng. Exp. Sta. Circ. No. 60*, 142 pp.

Cummings, A. E., G. O. Kerkhoff, and R. B. Peck (1950), "Effect of Driving Piles into Soft Clay," *Trans. ASCE*, 115, 275–285. Discussions by Avery and Wilson, pp. 322–331; Rutledge, pp. 301–304; Zeevaert, pp. 286–292.

D'Appolonia, D. J. (1971), "Effects of Foundation Construction on Nearby Structures," *Proc. 4 Panamerican Conf. Soil Mech.*, Puerto Rico, 1, 189–236.

D'Appolonia, D. J., E. D'Appolonia, and R. F. Brissette (1968), "Settlement of Spread Footings on Sand," *ASCE J. Soil Mech.*, 94, SM3, 735–760. Discussions in Vol. 95, No. SM3, pp. 900–916, and Vol. 96, No. SM2, pp. 754–762.

D'Appolonia, D. J., and T. W. Lambe (1971), "Floating Foundations for Control of Settlement," *ASCE J. Soil Mech.*, 97, SM6, 899–915.

D'Appolonia, E., C. E. Miller, Jr., and T. M. Ware (1953), "Sand Compaction by Vibroflotation," *Proc. ASCE*, 79, Separate No. 200, 23 pp.

D'Appolonia, D. J., R. V. Whitman, and E. D'Appolonia (1969), "Sand Compaction with Vibratory Rollers," *ASCE J. Soil Mech.*, 95, SM1, 263–284.

Davisson, M. T. (1970a), "Static Measurements of Pile Behavior," *Proc. Conf. on Design and Installation of Pile Foundations and Cellular Structures*, Lehigh Univ., Envo Publ. Co., pp. 159–164.

Davisson, M. T. (1970b), "Design Pile Capacity," *Proc. Conf. on Design and Installation of Pile Foundations and Cellular Structures*, Lehigh Univ., Envo Publ. Co., pp. 75–85.

Davisson, M. T. (1970c), "Lateral Load Capacity of Piles," *Hwy. Res. Rec.*, 333, 104–112.

Davisson, M. T. (1973), "High Capacity Piles," in *Innovations in Foundation Construction*, Soil Mech. Div., Ill. Sect., ASCE, Chicago, pp. 81–112.

Dawson, R. F. (1959), "Modern Practices Used in the Design of Foundations for Structures on Expansive Soils," *Colo. School of Mines Quarterly*, 54, 4, 67–87.

Deere, D. U. (1957), "Seepage and Stability Problems in Deep Cuts in Residual Soils, Charlotte, N. C.," *Proc. AREA*, 58, 738–745.

Deere, D. U. (1963), "Technical Description of Rock Cores for Engineering Purposes," *Felsmechanik und Ingenieurgeologie*, 1, 1, 16–22.

Deere, D. U. (1968), "Geological Considerations," *Chap. 1* in K. G. Stagg and O. C. Zienkiewicz, *Rock Mechanics in Engineering Practice*, New York, Wiley, pp. 1–20.

DeSimone, S. V., and J. P. Gould (1972), "Performance of Two Mat Foundations on Boston Blue Clay," *Proc. ASCE Conf. on Performance of Earth and Earth-Supported Structures*, Purdue, 1, Part 2, 953–980.

Dudley, J. H. (1970), "Review of Collapsing Soils," *ASCE J. Soil Mech.*, 96, SM3, 925–947.

Eades, J. L., and R. E. Grim (1966), "A Quick Test to Determine Lime Requirements for Lime Stabilization," *Hwy. Res. Rec.*, 139, 61–72.

Eggestad, A. (1963), "Deformation Measurements Below a Model Footing on the Surface of Dry Sand," *Proc. European Conf. Soil Mech.*, Wiesbaden, *1*, 233-239. Reprinted in *NGI Publ. 58*, pp. 29-35.

Endo, M., A. Minou, T. Kawasaki, and T. Shibata (1969), "Negative Skin Friction Acting on Steel Pipe Pile in Clay," *Proc. 7 Int. Conf. Soil Mech.*, Mexico, *2*, 85-92.

Feda, J. (1961), "Research on the Bearing Capacity of Loose Soil," *Proc. 5 Int. Conf. Soil Mech.*, Paris, *1*, 635-642.

Feld, J. (1965), "Tolerance of Structures to Settlement," *ASCE J. Soil Mech. 91*, SM3, 63-77.

Feld, J. (1968), *Construction Failure*, New York, Wiley, 399 pp.

Fellenius, B. H., and B. B. Broms (1969), "Negative Skin Friction for Long Piles Driven in Clay," *Proc. 7 Int. Conf. Soil Mech.*, Mexico, *2*, 93-98.

Fenneman, N. M. (1931), *Physiography of Western United States*, New York, McGraw-Hill, 534 pp.

Fenneman, N. M. (1938), *Physiography of Eastern United States*, New York, McGraw-Hill, 714 pp.

Ferguson, P. M. (1973), *Reinforced Concrete Fundamentals*, New York, Wiley, 3nd ed., 750 pp.

Fisk, H. N. (1947), "Fine-Grained Alluvial Deposits and Their Effects on Mississippi River Activity," *2*, Waterways Exp. Sta., 74 plates.

Fletcher, G. F. A. (1965), "Standard Penetration Test: Its Uses and Abuses," *ASCE J. Soil Mech.*, *91*, SM4, 67-75.

Flint, R. F. chm. (1945), "Glacial Map of North America," Geol. Soc. Am., New York.

Flint, R. F. (1971), *Glacial and Quaternary Geology*, New York, Wiley, 893 pp.

Gammon, K. M. and G. F. Pedgrift (1962), "The Selection and Investigation of Potential Nuclear Power Station Sites in Suffolk," *Proc. Inst. Civil Engrs.*, London, *21*, 139-160.

Gauntt, G. C. (1962), "Marina City—Foundations," *Civ. Eng. ASCE*, *32*, December, 61-63.

Gerber, E. (1929), *Untersuchen über die Druckverteilung im örtlich Belasteten Sand*, Dissertation, Technische Hochschule, Zurich.

Gerwick, B. C. Jr., (1970), "Current Construction Practices in the Installation of High-Capacity Piling," *Hwy. Res. Rec.*, *333*, 113-122.

Gibbs, H. J., and W. G. Holtz (1957), "Research on Determining the Density of Sands by Spoon Penetration Testing," *Proc. 4 Int. Conf. Soil Mech.*, London, *1*, 35-39.

Gill, G. W. (1959), "Waterproofing Buildings Below Grade," *Civ. Eng., ASCE*, *29*, 1, 3-5.

Glossop, R. (1960), "The Invention and Development of Injection Processes, Part I: 1802-1850," *Géotechnique*, *10*, 3, 91-100; and "Part II: 1850-1960," *Géotechnique*, *11*, 4 (1961), 255-279.

Gray, R. E., and J. F. Meyers (1970), "Mine Subsidence and Support Methods in Pittsburgh Area." *ASCE J. Soil Mech.*, *96*, SM4, 1267-1287.

Grim, R. E. (1959), "Physico-Chemical Properties of Soils: Clay Minerals," *ASCE J. Soil Mech.*, *85*, SM2, 1-17.

Grim, R. E. (1962), *Applied Clay Mineralogy*. New York, McGraw-Hill, 422 pp.

Hagerty, D. J., and R. B. Peck (1971), "Heave and Lateral Movements Due to Pile Driving," *ASCE J. Soil Mech.*, *97*, SM 11, 1513-1532.

Hallenbeck, J. J., Jr., and R. E. Johnston (1967), "Pile Foundation for Oakland Coliseum," *Civ. Eng. ASCE*, *37*, 1, 57-61.

Hammer, M. J., and O. B. Thompson (1966), "Foundation Clay Shrinkage Caused by Large Trees," *ASCE J. Soil Mech.*, *92*, SM6, 1-17.

Hanna, T. H., and G. A. Matallana (1970), "The Behavior of Tied-Back Retaining Walls," *Canadian Geot. Jour.*, 7, 4, 372–396.

Hanrahan, E. T. (1954), "An Investigation of Some Physical Properties of Peat," *Géotechnique*, 4, 3, 108–123.

Hedefine, A., and L. G. Silano (1968), "Newport Bridge Foundations," *Civ. Eng. ASCE*, 38, 10, 37–43.

Heiland, C. A. (1940), *Geophysical Exploration*, Englewood Cliffs, N. J., Prentice-Hall, 1013 pp.

Hendron, A. J. (1963), *The Behavior of Sand in One-Dimensional Compression*. Ph.D. thesis, Univ. of Illinois, Urbana, 283 pp.

Hetenyi, M. (1946), *Beams on Elastic Foundation*, Ann Arbor, University of Michigan Press, 255 pp.

HRB (1965), "Geophysical Methods and Statistical Soil Surveys in Highway Engineering—6 Reports," *Hwy. Res. Rec.*, 81, 60 pp.

HRB (1968), "Conference on Loess: Design and Construction," *Hwy. Res. Rec.*, 212, 38 pp.

Hirsch, T. J., L. L. Lowery, H. M. Coyle, and C. H. Samson, Jr. (1970), "Pile-Driving Analysis by One-Dimensional Wave Theory: State of the Art," *Hwy. Res. Rec.*, 333, 33–54.

Holtz, W. G., and H. J. Gibbs (1951), "Consolidation and Related Properties of Loessial Soils," *ASTM Spec. Tech. Publ. 126*, 9–26.

Holtz, W. G., and C. A. Lowitz (1965), "Effects of Driving Displacement Piles in Lean Clay," *ASCE J. Soil Mech.*, 91, SM5, 1–13.

Hubbard, P. G. (1955), "Field Measurement of Bridge-Pier Scour," *Proc. Hwy. Res. Bd.*, 34, 184–188.

Hunter, J. W. (1948), "Site Exploration for Foundations at Portsmouth," *Proc. 2 Int. Conf. Soil Mech.*, Rotterdam, 2, 159–162.

Hunter, A. H., and M. T. Davisson (1969), "Measurement of Pile Load Transfer," *ASTM Spec. Tech. Publ. 444*, pp. 106–117.

Huntington, W. C. (1957), *Earth Pressures and Retaining Walls*, New York, Wiley, 534 pp.

Huntington, W. C. (1963), *Building Construction*, 3rd ed., New York, Wiley, 734 pp.

Hvorslev, M. J. (1937), *Über die Festigkeitseigenschaften Gestörter Bindiger Böden* (On the Strength Properties of Remolded Cohesive Soils), *Danmarks Naturvidenskabelige Samfund, Ingeniørvidenskabelige Skrifter*, Series A, No. 45, Copenhagen, 159 pp.

Hvorslev, M. J. (1948), *Subsurface Exploration and Sampling of Soils for Civil Engineering Purposes*. Waterways Exp. Sta., Vicksburg, Miss., 465 pp.

Hvorslev, M. J. (1960), "Physical Components of the Shear Strength of Saturated Clays," *Proc. ASCE Research Conf. on Shear Strength of Cohesive Soils*, pp. 169–273.

Ireland, H. O. (1957), "Pulling Tests on Piles in Sand," *Proc. 4 Int. Conf. Soil Mech.*, London, 2, 43–45.

Ireland, H. O., O. Moretto, and M. Vargas (1970), "The Dynamic Penetration Test: A Standard That is not Standardized," *Géotechnique*, 20, 2, 185–192.

ISSMFE (1969), "Engineering Properties of Lateritic Soils," *Proc. Specialty Session 1, 7 Int. Conf. Soil Mech.*, Asian Inst. Tech., Bangkok, 1, 207 pp; Vol. 2 (1970), 203 pp.

Jennings, J. E. (1953), "The Heaving of Buildings on Desiccated Clay," *Proc. 3 Int. Conf. Soil Mech.*, 1, Zurich, 390–396.

Jennings, J. E., and Henkel, D. J. (1949), "The Use of Underreamed Pile Foundations on Expansive Clay Soils in South Africa." *Nat'l Bldg. Research Institute, South African CSIR, Bull. No. 3*, 9–15.

Jennings, J. E., and K. Knight (1957), "The Additional Settlement of

Foundations Due to a Collapse of Structure of Sandy Subsoils on Wetting," *Proc. 4 Int. Conf. Soil Mech.*, London, *1*, 316–319.

Johannessen, I. J., and L. Bjerrum (1965), "Measurement of the Compression of a Steel Pile to Rock Due to Settlement of the Surrounding Clay," *Proc. 6 Int. Conf. Soil Mech.*, Montreal, *2*, 261–264.

Johnson, S. J. (1970a), "Precompression for Improving Foundation Soils," *ASCE J. Soil Mech.*, *96*, SM1, 111–144.

Johnson, S. J. (1970b), "Foundation Precompression with Vertical Sand Drains," *ASCE J. Soil Mech.*, *96*, SM1, 145–175.

Kallstenius, T. (1963), "Studies on Clay Samples Taken with Standard Piston Sampler," *Proc. Swedish Geot. Inst.*, 21, Stockholm, 210 pp.

Kaplar, C. W. (1970), "Phenomenon and Mechanism of Frost Heaving," *Hwy. Res. Rec.*, *304*, 1–13.

Kézdi, Á. (1965), "Deep Foundations," *Proc. 6 Int. Conf. Soil Mech.*, Montreal, *3*, 256–264.

Klohn, E. J. (1963), "Pile Heave and Redriving," *Trans. ASCE*, *128*, Part I, 557–577. Discussion by Olko, pp. 578–587.

Kolb, C. R., and W. G. Shockley (1959), "Engineering Geology of the Mississippi Valley," *Trans. ASCE*, *124*, 633–656.

Kotzias, P. C., and A. C. Stamatopoulos (1969), "Preloading for Heavy Industrial Installations," *ASCE J. Soil Mech.*, *95*, SM6, 1335–1355.

Krinitzsky, E. L., and W. J. Turnbull (1967), "Loess Deposits of Mississippi," *Geol. Soc. Amer. Spec. Paper 94*, 64 pp.

Lambe, T. W. (1951), *Soil Testing for Engineers*. New York, Wiley, 165 pp.

Lambe, T. W. (1959), "Physico-Chemical Properties of Soils: Role of Soil Technology," *ASCE J. Soil Mech.*, *85*, SM2, 55–70.

Lambe, T. W., and H. M. Horn (1965), "The Influence on an Adjacent Building of Pile Driving for the M.I.T. Materials Center," *Proc. 6 Int. Conf. Soil Mech.*, Montreal, *2*, 280–284.

Lambe, T. W., and R. V. Whitman (1969), *Soil Mechanics*. New York, Wiley, 553 pp.

Larsen, E. S., and H. Berman (1934), "The Microscopic Determination of the Nonopaque Minerals," 2nd ed. *U.S. Dept. of Interior Bull. 848*, 266 pp.

Laursen, E. M. (1955), "Model-Prototype Comparison of Bridge Pier Scour," *Proc. Hwy. Res. Bd.*, *34*, 188–193.

Leet, L. D., and S. Judson (1971), "*Physical Geology*," 4th ed., Englewood Cliffs, N. J., Prentice-Hall, 687 pp.

Legget, R. F., ed. (1961), *Soils in Canada*, Royal Soc. Canada, *Spec. Publ. 3*, Univ. of Toronto Press, 229 pp.

Legget, R. F. (1962), *Geology and Engineering*, 2nd ed., New York, McGraw-Hill, 884 pp.

Lennertz, R. (1972), "Settlement of Footings on a Non-Uniform Foundation." *Proc. ASCE Conf. on Performance of Earth and Earth-Supported Structures*, Purdue, *1*, Part 2, 929–938.

Le Roy, L. W., and H. M. Crain, eds. (1949), "Subsurface Geologic Methods, a Symposium," *Colorado School of Mines Quarterly*, *44*, 3, Golden, 826 pp.

Linell, K. A., F. B. Hennion, and E. F. Lobacz (1963), "Corps of Engineers' Pavement Design in Areas of Seasonal Frost," *Hwy. Res. Rec.*, *33*, 76–136.

Lockwood, M. G. (1954), "Ground Subsides in Houston Area," *Civ. Eng. ASCE*, *24*, 6, 48–50.

Lumb, P. (1962), "The Properties of Decomposed Granite," *Géotechnique*, *12*, 3, 226–243.

Lynch, T. J. (1960), "Pile Driving Experiences at Port Everglades," *ASCE J. Soil Mech.*, *86*, SM2, 41–62.

MacDonald, D. H., J. deRuiter, and T. C. Kenney (1961), "The Geotechnical Properties of Impervious Fill Materials in Some Canadian Dams," *Proc. 5 Int. Conf. Soil Mech.*, Paris, 2, 657–662.

Mansur, C. I., and R. I. Kaufman (1960), "Dewatering the Port Allen Lock Excavation." *ASCE J. Soil Mech.*, 86, SM6, 35–55.

Mansur, C. I., and R. I. Kaufman (1962), "Dewatering," Chap. 3 in *Foundation Engineering*, G. A. Leonards, ed., New York, McGraw-Hill, pp. 241–350.

Mesri, G. (1973), "Coefficient of Secondary Compression," *ASCE J. Soil Mech.*, 99, SMI, 123–137.

Meyerhof, G. G. (1951), "The Ultimate Bearing Capacity of Foundations," *Géotechnique*, 2, 4, 301–332.

Meyerhof, G. G. (1953), "The Bearing Capacity of Foundations Under Eccentric and Inclined Loads," *Proc. 3 Int. Conf. Soil Mech.*, Zurich, 1, 440–445.

Meyerhof, G. G. (1955), "Influence of Roughness of Base and Ground-Water Conditions on the Ultimate Bearing Capacity of Foundations," *Géotechnique*, 5, 3, 227–242.

Meyerhof, G. G. (1963). "Some Recent Research on the Bearing Capacity of Foundations," *Canadian Geot. Jour.*, 1, 1, 16–26.

Meyerhof, G. G. (1970), "Safety Factors in Soil Mechanics," *Canadian Geot. Jour.*, 7, 4, 349–355.

Meyerhof, G. G., and G. Y. Sebastyan (1970), "Settlement Studies on Air Terminal Building and Apron, Vancouver International Airport, British Columbia." *Canadian Geot. Jour.* 7, 4, 433–456.

Miller, R. P. (1965), "Engineering Classification and Index Properties for Intact Rock," Ph.D. thesis, Univ. of Illinois, Urbana, 333 pp.

Mitchell, J. K. (1970), "In-Place Treatment of Foundation Soils."*ASCE J. Soil Mech.*, 96, SM1, 73–110.

Mohr, H. A. (1943), *Exploration of Soil Conditions and Sampling Operations*, Soil Mechanics Series No. 21, 3rd ed., Graduate School of Engineering, Harvard University, 63 pp.

Mohr, E. C. J., and F. A. van Baren (1954), *Tropical Soils*, New York, Interscience Publ., 498 pp.

Moore, R. W. (1961), "Observations on Subsurface Exploration Using Direct Procedures and Geophysical Techniques," *Proc. 12th Annual Symp. on Geology as Applied to Highway Engineering, U. of Tenn. Eng. Exp. Sta. Bull. 24*, pp. 63–87.

Moorhouse, D. C., and G. L. Baker (1969), "Sand Densification by Heavy Vibratory Compactor," *ASCE J. Soil Mech.*, 95, SM4, 985–994.

Moran, Proctor, Mueser, and Rutledge (1958), *Study of Deep Soil Stabilization by Vertical Sand Drains*, U. S. Dept. of Commerce, Office Tech. Serv., Wash., D. C., 192 pp.

Morgenstern, N. R., and V. E. Price (1965), "The Analysis of the Stability of General Slip Surfaces," *Géotechnique*, 15, 1, 79–93.

Newmark, N. M. (1942), "Influence Charts for Computation of Stresses in Elastic Foundations," *Univ. of Illinois Eng. Exp. Sta. Bull. 338*, 28 pp.

Nordlund, R. L. (1963), "Bearing Capacity of Piles in Cohesionless Soils,' *ASCE J. Soil Mech.*, 89, SM3, 1–35.

Nordlund, R. L., and D. U. Deere (1970),"Collapse of Fargo Grain Elevator," *ASCE J. Soil Mech.*, 96, SM2, 585–607.

Olson, R. E., and G. Mesri (1970), "Mechanisms Controlling the Compressibility of Clays," *ASCE J. Soil Mech.*, 96, SM6, 1863–1878.

O'Neill, M. W., and L. C. Reese (1970), "Behavior of Axially Loaded Drilled Shafts in Beaumont Clay, Part One—State of the Art," Research Rep. 89-8, Center for Hwy. Res., U. of Texas, Austin, 147 pp.

Orrje, O., and B. Broms (1967), "Effects of Pile Driving on Soil Properties," *ASCE J. Soil Mech.*, 93, SM5, 59–73.

Osterberg, J. O. (1952), "New Piston Type Soil Sampler," *Eng. News-Rec.*, 148, 77–78.

Paige, S., chm. (1950), *Application of Geology to Engineering Practice*, Berkey Volume, Geol. Soc. Amer., 327 pp.

Parcher, J. V., and R. E. Means (1968), *Soil Mechanics and Foundations*, Columbus, Merrill, 573 pp.

Parola, J. F. (1970), "Mechanics of Impact Pile Driving," Ph.D. Thesis, Univ. of Ill. Urbana, 236 pp.

Parsons, J. D. (1959), "Foundation Installation Requiring Recharging of Ground Water," *ASCE J. Constr. Div.*, 85, CO2, 1–21.

Parsons, J. D. (1966), "Piling Difficulties in the New York Area," *ASCE J. Soil Mech.*, 92, SM1, 43–64.

Peck, O. K., and R. B. Peck (1948), "Settlement of Foundation Due to Saturation of Loess Subsoil," *Proc. 2 Int. Conf. Soil Mech.*, Rotterdam, 4, 4–5.

Peck, R. B. (1948), "History of Building Foundations in Chicago," *Univ. of Ill. Eng. Exp. Sta. Bull.* 373, 64 pp.

Peck, R. B. (1953), "Foundation Exploration—Denver Coliseum," *Proc. ASCE*, 79, Separate No. 326, 14 pp.

Peck, R. B. (1958), "A Study of the Comparative Behavior of Friction Piles," *Hwy. Res. Bd. Special Report 36*, 72 pp.

Peck, R. B. (1960), "Major Power Station Foundation in Broken Limestone: Discussion," *ASCE J. Soil Mech.*, 86, SM1, 95–98.

Peck, R. B. (1969a), "Advantages and Limitations of the Observational Method in Applied Soil Mechanics," *Géotechnique*, 19, 2, 171–187.

Peck, R. B. (1969b), "Deep Excavations and Tunneling in Soft Ground," *Proc. 7 Int. Conf. Soil Mech.*, Mexico. State-of-the-art volume, pp. 225–290.

Peck, R. B., and F. G. Bryant (1953), "The Bearing-Capacity Failure of the Transcona Elevator," *Géotechnique*, 3, 5, 201–208.

Peck, R. B., and H. O. Ireland (1957), "Backfill Guide," *ASCE J. Struct. Div.*, 83, ST4, 10 pp.

Peck, R. B., H. O. Ireland, and C. Y. Teng (1948), "A Study of Retaining Wall Failures," *Proc. 2 Int. Conf. Soil Mech.*, Rotterdam, 3, 296–299.

Peck, R. B., and W. C. Reed (1954), "Engineering Properties of Chicago Subsoils," *Univ. of Ill. Eng. Exp. Sta. Bull.* 423, 62 pp.

Pettijohn, F. J. (1957), *Sedimentary Rocks*, 2nd ed., New York, Harper, 718 pp.

Poulos, H. G. (1968), "Analysis of the Settlement of Pile Groups," *Géotechnique*, 18, 4, 449–471.

Proctor, R. R. (1933), "Four Articles on the Design and Construction of Rolled-Earth Dams," *Eng. News-Record*, 111, 245–248, 286–289, 348–351, 372–376.

Rankine, W. J. M. (1857), "On the Stability of Loose Earth," *Phil. Trans. Roy. Soc.*, London, 147, Part 1, 9–27.

Reddy, A. S. (1967), "Bearing Capacity of Footings on Layered Clays," *ASCE J. Soil Mech.* 93, SM2, 83–99.

Reiche, P. (1950), *A Survey of Weathering Processes and Products*, Univ. of New Mexico Press, 95 pp.

Riggs, L. W. (1966), "Tagus River Bridge — Tower Piers," *Civ. Eng. ASCE*, 36, 2, 41–45.

Rosenqvist, I. Th. (1959), "Physico-Chemical Properties of Soils: Soil-Water Systems," *ASCE J. Soil Mech.*, 85, SM2, 31–53.

Rowe, P. W. (1952), "Anchored Sheet Pile Walls." *Proc. Inst. Civ. Eng.*, London, 1, Part 1, 27–70.

Rutledge, P. C. (1944), "Relation of Undisturbed Sampling to Laboratory Testing," *Trans. ASCE*, *109*, 1155–1183.

Salley, J. R., and R. B. Peck (1969), "Tolerable Settlements of Steam Turbine-Generators," *ASCE J. Power Div.*, *95*, PO2, 227–252.

Sanger, F. J. (1968), "Ground Freezing in Construction," *ASCE J. Soil Mech.*, *94*, SM1, 131–158.

Sanglerat, G. (1972), *The Penetrometer and Soil Exploration*, Amsterdam, Elsevier, 464 pp.

Scheidig, A. (1931), *Versuche über die Formänderung von Sand und ihre Anwendung auf die Setzungsanalyse von Bauwerken* (Tests on the deformation of sand and their application to the settlement analysis of buildings), M.S. thesis, Vienna.

Schmertmann, J. H. (1955), "The Undisturbed Consolidation Behavior of Clay," *Trans. ASCE*, *120*, 1201–1227.

Schmertmann, J. H. (1970), "Static Cone to Compute Static Settlement Over Sand," *ASCE J. Soil Mech.*, *96*, SM3, 1011–1043.

Schousboe, I. (1972), "Suggested Design and Construction Procedures for Pier Foundations," *ACI Jour.*, *69*, 8, 461–480.

Scott, E. W., Jr. (1948), "Philadelphia Conducts Extensive Subsurface Exploration Prior to Airport Expansion." *Civ. Eng. ASCE*, *18*, 2, 44–46.

Seed, H. B., J. K. Mitchell, and C. K. Chan (1960), "The Strength of Compacted Cohesive Soils," *Proc. ASCE Research Conf. on Shear Strength of Cohesive Soils*, pp. 877–964.

Seed, H. B., R. J. Woodward, and R. Lundgren (1962), "Prediction of Swelling Potential for Compacted Clays," *ASCE J. Soil Mech.*, *88*, SM3, 53–87.

Shannon, W. L., and R. J. Strazer (1970), "Tied-Back Excavation Wall for Seattle First National Bank," *Civ. Eng. ASCE*, *40*, 3, 62–64.

Shannon, W. L., S. D. Wilson, and R. H. Meese (1962), "Field Problems: Field Measurements," in *Foundation Engineering*, G. A. Leonards, ed., New York, McGraw-Hill, pp. 1025–1080.

Skempton, A. W. (1942), "An Investigation of the Bearing Capacity of a Soft Clay Soil," *J. Inst. Civil Engrs.*, London, *18*, 307–321; discussions, pp. 567–576.

Skempton, A. W. (1944), "Notes on the Compressibility of Clays," *Quart. J. Geol. Soc.*, London, *C*, pp. 119–135.

Skempton, A. W. (1948), "Vane Tests in the Alluvial Plain of the River Forth Near Grangemouth," *Géotechnique*, *1*, 2, 111–124.

Skempton, A. W. (1951), "The Bearing Capacity of Clays," *Proc. British Bldg. Research Congress*, *1*, 180–189.

Skempton, A. W. (1953), "Discussion on Piles and Pile Foundations," *Proc. 3rd Int. Conf. Soil Mech.*, Zurich, *3*, 172.

Skempton, A. W. (1959), "Cast-*in-situ* Bored Piles in London Clay," *Géotechnique*, *9*, 4, 153–173.

Skempton, A. W. (1960), "Terzaghi's Discovery of Effective Stress," in *From Theory to Practice in Soil Mechanics*, New York, Wiley, 42–53.

Skempton, A. W. (1961), "Effective Stress in Soils, Concrete and Rocks," *Pore Pressure and Suction in Soils*, London, Butterworths, pp. 4–16.

Skempton, A. W. (1971), "The Albion Mill Foundations," *Géotechnique*, *21*, 3, 203–210.

Skempton, A. W., and D. H. MacDonald (1956), "The Allowable Settlement of Buildings," *Proc. Inst. Civ. Eng.*, London, *5*, 3, Part 3, 727–784.

Skempton, A. W., R. B. Peck, and D. H. MacDonald (1955), "Settlement Analyses of Six Structures in Chicago and London," *Proc. Inst. C. E.*, London, *4*, Part I, July, 525–544.

Smith, G. D. (1942), "Illinois Loess," *Univ. of Ill. Agr. Exp. Sta. Bull. 490*, 45 pp.

Sokolovski, V. V. (1960), *Statics of Soil Media*. Translated from Russian by D. H. Jones and A. N. Schofield. London, Butterworths, 237 pp.

Sowers, G. F. (1954), "Soil Problems in the Southern Piedmont Region," *Proc. ASCE*, 80, Separate 416, 18 pp.

Sowers, G. F. (1969), "The Safety Factor in Excavations and Foundations," *Hwy. Res. Rec.*, 269, 23–34.

Sowers, G. F., C. B. Martin, L. L. Wilson, and M. Fausold, Jr. (1961), "The Bearing Capacity of Friction Pile Groups in Homogeneous Clay from Model Studies," *Proc. 5 Int. Conf. Soil Mech.*, Paris 2, 155–159.

Sowers, G. B., and G. F. Sowers (1967), "Failures of Bulkhead and Excavation Bracing," *ASCE Civ. Eng.*, 37, 1, 72–77.

Spangler, M. G. (1938), "Horizontal Pressures on Retaining Walls Due to Concentrated Surface Loads," *Iowa Eng. Exp. Sta. Bull. 140*, 79 pp.

Spock, L. E. (1962), *Guide to the Study of Rocks*, 2nd ed., New York, Harper, 298 pp.

Squier, L. R. (1970), "Mat Foundation Spans Rubble Channel," *Civ. Eng. ASCE*, 40, 8, 61–62.

Stagg, K. G., and O. C. Zienkiewicz, eds. (1968), *Rock Mechanics in Engineering Practice*, New York, Wiley, 442 pp.

Steinman, D. B. (1945), *The Builders of the Bridge*, New York, Harcourt, 457 pp.

Stermac, A. G., K. G. Selby, and M. Devata (1969), "Behavior of Various Types of Piles in a Stiff Clay," *Proc. 7 Int. Conf. Soil Mech.*, Mexico, 2, 239–245.

Swiger, W. F. (1960), "Control of Ground Water in Excavations," *ASCE J. Constr. Div.*, 86, CO1, 41–53.

Swiger, W. F., and H. M. Estes (1959), "Major Power Station Foundation in Broken Limestone," *ASCE J. Soil Mech.*, 85, SM5, 77–86.

Taylor, A. W. (1959), "Physico-Chemical Properties of Soils: Ion Exchange Phenomena," *ASCE J. Soil Mech.*, 85, SM2, 19–30.

Taylor, D. W. (1937), "Stability of Earth Slopes," *J. Boston Soc. Civil Engrs.*, 24, 197–246.

Taylor, D. W. (1948), *Fundamentals of Soil Mechanics*. New York, Wiley, 700 pp.

Teng, C. Y. (1949), "Determination of the Contact Pressure Against a Large Raft Foundation," *Géotechnique*, 1, 4, 222–228.

Terzaghi, K. (1920), "Old Earth-Pressure Theories and New Test Results," *Eng. News-Record*, 85, 14, 632–637.

Terzaghi, K. (1922), *Der Grundbruch an Stauwerken und Seine Verhütung"* (The failure of dams by piping and its prevention), *Die Wasserkraft*, 17, 445–449. Reprinted in *From Theory to Practice in Soil Mechanics*, New York, Wiley, 1960, pp 114–118.

Terzaghi, K. (1923), *Die Berechnung der Durchlässigkeitsziffer des Tones aus dem Verlauf der Hydrodynamischen Spannungserscheinungen*, Akademie der Wissenschaften in Wien; Sitzungsberichte. Mathematisch-naturwissenschaftliche Klasse. Part IIa, 132, 3/4 125–138. Reprinted in *From Theory to Practice in Soil Mechanics*, New York, Wiley, 1960, pp. 133–146.

Terzaghi, K. (1925), *Erdbaumechanik auf Bodenphysikalischer Grundlage*, Vienna, Deuticke, 399 pp.

Terzaghi, K. (1929), "Soil Studies for the Granville Dam at Westfield, Mass." *J. New Engl. Water Works Assoc.*, 43, 2, 191–223.

Terzaghi, K. (1934), "Large Retaining-Wall Tests. I," *Eng. News-Record*, 112, 5, 136–140.

Terzaghi, K. (1936), "Relation Between Soil Mechanics and Foundation Engineering; Presidential Address," *Proc. 1st Int. Conf. Soil Mech.*, Cambridge, Mass., 3, 13–18. Reprinted in *From Theory to Practice in Soil Mechanics*, New York, Wiley, 1960, pp. 62–67.

Terzaghi, K. (1941), "General Wedge Theory of Earth Pressure." *Trans. ASCE*, 106, 68–97.

Terzaghi, K. (1943), *Theoretical Soil Mechanics*, New York, Wiley, 510 pp.

Terzaghi, K. (1951), "The Influence of Modern Soil Studies on the Design and Construction of Foundations," Bldg. Research Congr., London, 1951, Div. 1, Part III, pp. 139–145. Reprinted in *From Theory to Practice in Soil Mechanics*, New York, Wiley, 1960, pp. 68–74.

Terzaghi, K. (1953), "Origin and Functions of Soil Mechanics," *Trans. ASCE, CT*, 666–696.

Terzaghi, K. (1954), "Anchored Bulkheads," *Trans. ASCE*, 119, 1243–1324.

Terzaghi, K. (1955a), "Influence of Geological Factors on the Engineering Properties of Sediments," in *Economic Geology*. Fiftieth Anniversary Volume, pp. 557–618.

Terzaghi, K. (1955b), "Evaluation of Coefficients of Subgrade Reaction," *Géotechnique*, 5, 4, 297–326.

Terzaghi, K. (1957), "Address at Opening Session of Fourth International Conference on Soil Mechanics and Foundation Engineering," *Proc. 4th Int. Conf. Soil Mech.*, London, 3, 55–58. Reprinted in *From Theory to Practice in Soil Mechanics*, New York, Wiley, pp. 75–78.

Terzaghi, K. (1960), "Landforms and Subsurface Drainage in the Gačka Region in Yugoslavia," in *From Theory to Practice in Soil Mechanics*, New York, Wiley, pp. 81–105.

Thompson, M. R. (1968), "Lime-Treated Soils for Pavement Construction," *ASCE J. Highway Div.*, 94, HW2, 191–217.

Thon, J. G., and R. C. Harlan (1971), "Slurry Walls for BART Civic Center Subway Station," *ASCE J. Soil Mech.*, 97, SM9, 1317–1334.

Thornburn, T. H. (1969), "Geology and Pedology in Highway Soil Engineering," *Reviews in Engineering Geology*, II, D. J. Varnes and G. Kiersch, eds., Geological Society of America, Boulder, Colo., pp. 17–57.

Thornburn, T. H., D. J. Hagerty, and T. K. Liu (1970), "Engineering Soil Report, Will County, Illinois," *Univ. of Ill. Eng. Exp. Sta. Bull. 501*, 195 pp.

Thornbury, W. D. (1954), *Principles of Geomorphology*, New York, Wiley, 618 pp.

Thornbury, W. D. (1965), *Regional Geomorphology of the United States*, New York, Wiley, 609 pp.

Thornley, J. H. (1959), *Foundation Design and Practice*, New York, Columbia Univ. Press, 298 pp.

Tomlinson, M. J. (1957), "The Adhesion of Piles Driven in Clay Soils," *Proc. 4 Int. Conf. Soil Mech.*, London, 2, 66–71.

Trask, P. D., ed. (1950), *Applied Sedimentation*, New York, Wiley, 707 pp.

Tschebotarioff, G. F. (1949), "Large Scale Earth Pressure Tests with Model Flexible Bulkheads," Final report to Bureau of Yards and Docks, U. S. Navy, Princeton Univ., 272 pp.

Tschebotarioff, G., and E. R. Ward (1957), "Measurements with Wiegmann Inclinometer on Five Sheet Pile Bulkheads," *Proc. 4 Int. Conf. Soil Mech.*, London, 2, 248–255.

Turnbull, W. J., and C. R. Foster (1958), "Stabilization of Materials by Compaction," *Trans. ASCE*, 123, 1–15.

U. S. Bureau of Reclamation (1947), "Laboratory Tests on Protective

Filters for Hydraulic and Static Structures," *Earth Materials Laboratory Rept. EM-132*, Denver, 28 pp.

U. S. Bureau of Reclamation (1963), *Earth Manual*, 1st ed. revised, Wash., D. C., 783 pp.

USDA (1938), *Soils and Men*, U. S. Dept. Agriculture Yearbook, Wash., D. C., G.P.O., 1232 pp.

USDA (1951), *Soil Survey Manual*, Handbook No. 18, U. S. Dept. Agriculture, Wash., D. C., G.P.O., 503 pp.

USDA (1960), *Soil Classification, a Comprehensive System*, U. S. Dept. Agriculture, Wash., D. C., G.P.O., 265 pp.

Van der Veen, C. (1953), "The Bearing Capacity of a Pile," *Proc. 3 Int. Conf. Soil Mech.*, Zurich, 2, 84–90.

Van der Veen, C., and L. Boersma (1957), "The Bearing Capacity of a Pile Pre-Determined by a Cone Penetration Test," *Proc. 4 Int. Conf. Soil Mech.*, London, 2, 72–75.

Vesič, A. (1963), "Bearing Capacity of Deep Foundations in Sand," *Hwy. Res. Rec.*, 39, 112–153.

Vesič, A. (1969), "Experience with Instrumented Pile Groups in Sand, *ASTM Spec. Tech. Publ. No. 444*, pp. 177–222.

Ward, W. H. (1953), "Soil Movement and Weather," *Proc. 3 Int. Conf. Soil Mech.*, Zurich, 1, 477–482.

WES (1950), "Undisturbed Sand Sampling Below the Water Table," Corps of Engrs., *Waterways Exp. Sta. Bull. 35*, 19 pp.

WES (1953), "Unified Soil Classification System," Corps of Engrs., *Waterways Exp. Sta. Tech. Mem. 3-357*, 30 pp.

Werblin, D. A. (1960), "Installation and Operation of Dewatering Systems," *ASCE J. Soil Mech.*, 86, SM1, 47–66.

Wheeless, L. D., and G. F. Sowers (1972), "Mat Foundation and Preload Fill, VA Hospital, Tampa." *Proc. ASCE Conf. on Performance of Earth and Earth-Supported Structures*, Purdue, 1, Part 2, 939–951.

Whitaker, T. (1957), "Experiments with Model Piles in Groups," *Géotechnique*, 7, 4, 147–167.

Whitaker, T., and R. W. Cooke (1965), "Bored Piles with Enlarged Bases in London Clay," *Proc. 6 Int. Conf. Soil Mech.*, Montreal, 2, 342–346.

White, E. E. (1962), "Underpinning," Chap. 9 in *Foundation Engineering*, G. A. Léonards, ed., New York, McGraw-Hill, pp. 826–893.

White, L. S. (1953), "Transcona Elevator Failure: Eye-Witness Account," *Géotechnique*, 3, 5, 209–214.

White, R. E. (1943), "Heavy Foundations Drilled into Rock," *Civ. Eng. ASCE*, 13, 1, 19–22.

White, R. E. (1962), "Caissons and Cofferdams," Chap. 10 in *Foundation Engineering*, G. A. Leonards, ed., New York, McGraw-Hill, pp. 894–964.

Willis, E. A. (1946), "Discussion: A Study of Lateritic Soils," *Proc. Hwy. Res. Board*, 26, 589–591.

Winter, G., and A. H. Nilson (1973), *Design of Concrete Structures*, 8th ed., New York, McGraw-Hill, 615 pp.

Woods, K. B., R. D. Miles, and C. W. Lovell, Jr. (1962), "Origin, Formation, and Distribution of Soils in North America," Chap. 1 in *Foundation Engineering*, G. A. Leonards, ed., New York, McGraw-Hill, pp. 1–65.

Woodward, R. J., R. Lundgren, and J. D. Boitano (1961), "Pile Loading Tests in Stiff Clays," *Proc. 5 Int. Conf. Soil Mech.*, Paris, 2, 177–184.

Woodward, R. J., W. S. Gardner, and D. M. Greer (1970), *Drilled Pier Foundations*, New York, McGraw-Hill, 288 pp.

Woolf, D. O. (1950), *The Identification of Rock Types*, U. S. Dept. Commerce, Bur. Public Roads, Wash., D. C., 11 pp.

Wosser, T. D., and R. Darragh (1970), "Tiebacks for Bank of America Building Excavation Wall," *Civ. Eng. ASCE*, 40, 3, 65–67.

Yong, R. N., and B. P. Warkentin (1966), *Introduction to Soil Behavior*, New York, MacMillan, 451 pp.

Zeevaert, L. (1949), "An Investigation of the Engineering Characteristics of the Volcanic Lacustrine Clay Deposit Beneath Mexico City." Ph.D. thesis, U. of Ill., Urbana, 234 pp.

Zeevaert, L. (1957a), "Consolidation of Mexico City Volcanic Clay," *ASTM Spec. Tech. Publ. 232*, pp. 28–32.

Zeevaert, L. (1957b), "Foundation Design and Behaviour of Tower Latino Americana in Mexico City," *Géotechnique*, 7, 3, 115–133.

Zeevaert, L. (1957c), "Compensated Friction-Pile Foundation to Reduce the Settlement of Buildings on the Highly Compressible Volcanic Clay of Mexico City," *Proc. 4 Int. Conf. Soil Mech.*, London, 2, 81–86.

Name Index

Aas, G., 122, 473
Adams, J. I., 78, 473
AASHO, 14, 16, 26, 27, 29, 36, 398, 473
AREA, 398, 473
ASCE, 473
ASP, 473
ASTM, 5, 8, 9, 14, 15, 16, 24, 27, 28, 36, 42, 112, 113, 122, 215, 473
Atterberg, A., 21, 48, 473
Avery, S. B., Jr., 476

Bagnold, R. A., 160, 474
Baker, B., 443, 444, 474
Baker, C. N., Jr., 243, 304, 474
Baker, G. L., 330, 480
Barden, L., 413, 474
Barkan, D. D., 259, 474
Basore, C. E., 330, 474
Baumann, F., 184
Belcher, D. J., 124, 474
Berkey, C. P., 159
Berman, H., 12, 479
Bishop, A. W., 101, 109, 121, 305, 358, 474
Bjerrum, L., 78, 93, 96, 161, 268, 305, 329, 363, 472, 474, 478
Boersma, L., 323, 485
Boitano, J. D., 330, 474, 485
Boussinesq, J., 288
Brinch Hansen, J., 420, 474
Brink, A. B. A., 346, 474
Brinkhorst, W. H., 258, 474
Brissette, R. F., 329, 476
Broms, B., 304, 305, 477, 481
Bryant, F. G., 304, 481
Brzezinski, L. S., 305, 474
Burmister, D. M., 36, 474
Button, S. J., 358, 475

Cadling, L., 122, 475
Carlson, E. D., 358, 475
Carson, A. B., 174, 475
Casagrande, A., 2, 22, 23, 27, 36, 43, 55, 60, 61, 63, 72, 77, 78, 94, 95, 96, 119, 161, 304, 305, 474, 475
Casagrande, L., 182, 305, 475

Cedergren, H. R., 56, 475
Chan, C. K., 482
Chellis, R. D., 227, 475
Clevenger, W. A., 346, 475
Condron, T. L., 166, 476
Cooke, R. W., 304, 485
Corps of Engineers, 27
Coulomb, C. A., 413, 420
Coyle, H. M., 227, 478
Crain, H. M., 122, 479
Crawford, C. B., 79, 305, 476
Cummings, A. E., 202, 227, 257, 304, 476
Curtis, L., 230

D'Appolonia, E., 200, 329, 330, 476
D'Appolonia, D. J., 200, 259, 305, 329, 330, 476
Darcy, H., 39, 40, 53
Darragh, R., 472, 486
Davisson, M. T., 215, 222, 224, 227, 230, 476, 478
Dawson, R. F., 79, 346, 476
Deere, D. U., 30, 31, 36, 112, 158, 304, 476, 480
deRuiter, J., 133, 479
DeSimone, S. V., 413, 476
Devata, M., 305, 483
Dudley, J. H., 346, 476
Duncan, J. M., 472, 474
duPont, E. I. de Nemours, 476

Eades, J. L., 199, 476
Eggestad, A., 329, 474, 477
Eide, O., 305, 474
Endo, M., 305, 477
Estes, H. M., 370, 483

Fadum, R. E., 43, 56, 72, 304, 475
Fausold, M., Jr., 305, 483
Feda, J., 329, 477
Feld, J., 259, 266, 477
Fellenius, B. H., 162, 305
Fellenius, W., 162
Fenneman, N. M., 159, 383, 477
Ferguson, P. M., 383, 477
Fisk, H. N., 148, 477
Fletcher, G. F. A., 121, 477
Flint, R. F., 124, 160, 477

Foster, C. R., 200, 484
Fricano, S. P., 358, 475
Frimann Clausen, C. J., 472, 474

Gammon, K. M., 166, 477
Gardner, W. S., 243, 485
Gauntt, G. C., 304, 477
Gerber, E., 423, 477
Gerwick, B. C., Jr., 227, 477
Gibbs, H. J., 66, 121, 477, 478
Gill, G. W., 190, 477
Glossop, R., 182, 477
Goodell, W. G., 374
Gould, J. P., 413, 476
Gray, R. E., 259, 477
Greer, D. M., 243, 485
Gregg, L. E., 124, 485
Grim, R. E., 35, 199, 476, 477

Hagerty, D. J., 160, 304, 477, 484
Hallenbeck, J. J., Jr., 358, 477
Hammer, M. J., 346, 477
Hanna, T. H., 472, 478
Hanrahan, E. T., 78, 478
Harlan, R. C., 482, 484
Hazen, A., 38, 40, 48
Hedefine, A., 243, 478
Heiland, C. A., 122, 478
Hendron, A. J., 95, 478
Henkel, D. J., 95, 101, 341, 474, 478
Hennion, F. B., 56, 479
Hetenyi, M., 413, 478
Highway Research Board, 26, 122, 478
Hirsch, T. J., 227, 478
Hirschfeld, R. C., 94, 475
Holtz, W. G., 66, 121, 304, 477, 478
Horn, H. M., 304, 479
Hubbard, P. G., 330, 478
Hunter, A. H., 330, 478
Hunter, J. W., 166, 478
Huntington, W. C., 189, 424, 444, 478
Hvorslev, M. J., 95, 101, 102, 107, 121, 138, 478

Ireland, H. O., 121, 248, 301, 306, 325, 346, 378, 481
ISSMFE, 478

Jenkins, D. S., 124, 474
Jennings, J. E., 334, 341, 346, 478
Johannessen, I. J., 305, 474, 479
Johnson, S. J., 182, 200, 305, 358, 479
Johnston, R. E., 358, 477
Judson, S., 159, 479

Kallstenius, T., 121, 479
Kantey, B. A., 346, 474
Kaplar, C. W., 56, 479
Kaufman, R. I., 182, 330, 480
Kawasaki, T., 305, 477
Kenney, T. C., 133, 479

Kerkhoff, G. O., 304, 476
Kézdi, Á., 329, 479
Khan, F., 243, 304, 474
Kiersch, G., 122
Klohn, E. J., 304, 479
Knight, K., 346, 478
Kolb, C. R., 160, 479
Kotzias, P. C., 200, 479
Krinitzsky, E. L., 160, 479

Lambe, T. W., 35, 56, 304, 305, 476, 479
Larsen, E. S., 12, 479
Laursen, E. M., 330, 479
Leet, L. D., 159, 479
Legget, R. F., 160, 479
Lennertz, R., 358, 479
Leonards, G. A., 122, 159, 182
LeRoy, L. W., 122, 479
Linell, K. A., 56, 479
Liu, T. K., 160, 484
Lobacz, E. F., 56, 479
Lockwood, M. G., 259, 479
Lovell, C. W., Jr., 159, 485
Lowery, L. L., 227, 478
Lowitz, C. A., 304, 478
Lumb, P., 161, 479
Lundgren, R., 346, 482, 485
Lynch, T. J., 330, 479

MacDonald, D. H., 133, 266, 305, 479, 482
Mansur, C. I., 182, 330, 480
Martin, C. B., 305, 483
Matallana, G. A., 472, 478
Math, E. R., 166, 476
Means, R. E., 340, 346, 481
Meese, R. H., 122, 482
Mesri, G., 12, 73, 74, 480
Meyerhof, G. G., 78, 266, 305, 310, 314, 329, 464, 480
Meyers, J. F., 259, 477
Miles, R. D., 159, 485
Miller, R. P., 31, 100, 330, 476, 480
Minou, A., 305, 477
Mitchell, J. K., 200, 480, 482
Mohr, H. A., 104, 121, 480
Mohr, E. C. J., 161, 480
Moore, R. W., 119, 480
Moorhouse, D. C., 331, 480
Moran, D. E., 182, 228, 480
Moretto, O., 121, 478
Morgenstern, N. R., 358, 480
Mueser, W. H., 182, 228, 480

Newmark, N. M., 288, 289, 349, 480
Nilson, A. H., 383, 485
Nordlund, R. L., 304, 330, 480

Odenstad, S., 122, 475
Olson, R. E., 12, 480
O'Neill, M. W., 238, 243, 480

Orrje, O., 304, 481
Osterberg, J. O., 121, 481

Paige, S., 159, 481
Parcher, J. V., 340, 346, 481
Parola, J. F., 221, 481
Parsons, J. D., 330, 481
Peck, O. K., 346, 481
Peck, R. B., 132, 160, 161, 166, 190, 248, 259, 266, 283, 301, 304, 305, 308, 346, 370, 472, 474, 476, 477, 481, 482
Pedgrift, G. F., 166, 477
Pettijohn, F. J., 161, 481
Poulos, H. G., 305, 481
Poulos, S., 182, 475
Prandtl, L., 271
Price, V. E., 358, 480
Proctor, C. S., 182, 480
Proctor, R. R., 14, 36, 481

Rankine, W. J. M., 419, 481
Reddy, A. S., 358, 481
Reed, W. C., 132, 160, 481
Reese, L. C., 238, 243, 480, 481
Reiche, P., 160, 481
Résal, J., 420
Richart, F. E., 375
Riggs, L. W., 243, 481
Root, J. W., 185
Rosenqvist, I. Th., 35, 481
Rowe, P. W., 453, 472, 481
Rutledge, P. C., 79, 182, 476, 480, 482

Salley, J. R., 266, 482
Samson, C. H., Jr., 227, 478
Sanger, F. J., 182, 482
Sanglerat, G., 115, 121, 482
Scheidig, A., 86, 482
Schmertmann, J. H., 61, 78, 121, 482
Schousboe, I., 243, 482
Scott, E. W., Jr., 166, 482
Sebastyan, G. Y., 78, 305, 480
Seed, H. B., 95, 346, 482
Selby, K. G., 305, 483
Shannon, W. L., 122, 472, 482
Shibata, T., 305, 477
Shockley, W. G., 160, 479
Silaño, L. G., 243, 478
Simons, N. E., 93, 474
Skempton, A. W., 44, 56, 62, 80, 92, 93, 97, 161, 190, 266, 271, 279, 304, 305, 329, 474, 482
Smith, G. D., 160, 483
Sokolovski, V. V., 420, 483
SooySmith, W., 230, 262
Sowers, G. B., 471, 482
Sowers, G. F., 161, 266, 305, 359, 471, 483, 485
Spangler, M. G., 423, 483
Spock, L. E., 159, 483
Squier, L. R., 370, 483
Stagg, K. G., 31, 36, 370, 483
Stamatopoulos, A. C., 200, 479

Steinman, D. B., 243, 483
Stermac, A. G., 305, 483
Stevenson, R., 176
Strazen, R. J., 472, 482
Sutherland, J. G., 305, 476
Swiger, W. F., 182, 370, 483

Talbot, A. N., 374, 375
Taylor, A. W., 35, 483
Taylor, D. W., 71, 298, 299, 483
Teng, C. Y., 248, 301, 413, 481, 483
Terzaghi, K., 44, 49, 56, 59, 70, 78, 81, 135, 149, 159, 161, 266, 298, 308, 310, 413, 417, 419, 443, 444, 454, 472, 483, 484
Thompson, M. R., 200, 346, 484
Thompson, O. B., 346, 477
Thon, J. G., 472, 484
Thornburn, T. H., 122, 160, 484
Thornbury, W. D., 159, 484
Thornley, J. H., 227, 484
Tomlinson, M. J., 283, 305, 484
Trask, P. D., 160, 484
Tschebotarioff, G., 453, 472, 484
Turnbull, W. J., 160, 200, 479, 484

USBR, 27, 36, 49, 484, 485
USDA, 25, 29, 36, 129, 137, 144, 160, 485

Van Baren, F. A., 161, 480
Van der Veen, C., 323, 485
Vargas, M., 121, 478
Varnes, D. J., 122
Vesić, A., 329, 485
Vitrivius, 204

Ward, E. R., 472, 484
Ward, W. H., 346, 485
Ware, T. M., 330, 476
Warkentin, B. P., 35, 56, 486
WES, 36, 485
Werblin, D. A., 182, 485
Wheeless, L. D., 359, 485
Whitaker, T., 304, 305, 485
White, E. E., 252, 485
White, Lazarus, 250
White, L. S., 304, 484
White, R. E., 243, 250, 370, 485
White, W. A., 22
Whitman, R. V., 35, 200, 330, 476, 479
Willis, E. A., 158, 485
Wilson, L. L., 305, 483
Wilson, S. D., 96, 122, 446, 475, 476, 482
Winter, G., 383, 485
Woods, K. B., 124, 159, 474, 485
Woodward, R. J., 243, 279, 346, 482, 485
Woolf, D. O., 36, 485
Wosser, T. O., 472, 486

Yong, R. N., 35, 56, 486

Zeevaert, L., 140, 160, 254, 305, 348, 476, 486
Zienkiewicz, O. C., 31, 36, 370, 483

Subject Index

Page numbers in boldface indicate definitions of terms or major discussions.

A-frame anchorage, 452
AASHO classification, 25, 27, 128
　compaction test, 15, 198
Abrasion, 126
Abutment, 229, 247ff, 385
　base slab, 439
　cantilever, 441
　design, 437
　gravity, 248
　joints, 439
　loads, 439
　pile-bent, 248
　spill-through, 248
　types, 248
　U, 248
ACI Building Code, 375, 379, 390
Acker drill, 121
Active earth pressure, 385, 418ff
Active Rankine state, 418ff, 459
Adhesion between pile and soil, 214
Adsorbed ion, 21ff
Aeolian soil, 126
Aerial photography, 128, 130, 137, 139, 144
Aggregate properties, 8
　relations among, 17
Air bubble, 42
Airfield classification system, 27
Airphoto, 135, 137
　interpretation, 128
　pattern, 130ff, 137, 139, 144
Albion Mill, 190
Allowable contact pressure, 361
　on jointed rock, 362
Allowable load on rock, 362, 368
Allowable pile load, 383
Allowable settlement, 266
Allowable soil pressure, 186ff
　on clay, 273, 276
　on rafts on sand, 319
Alluvial fan, 147
Alluvial soil, 126, 130
Alpine glacier, 129
Alteration, hydrothermal, 30

Alumina octahedron, 10
Aluminum-micarta pile cushion, 220, 222
Aluminum oxide, 155
Ambassador Bridge, 228
Analysis, effective stress, 101
　total stress, 101
Anchor, 172
　beam, 448
　grouted, 468
　pile, 214, 448
　plate, 448
　prestressed, 468
　rod, 455
　test, 468
　wall, 448
Anchorage, A-frame, 452
　bulkhead, 448, 452
　of tieback, 467
Anchored bulkhead, see Bulkhead
Andesite, 31
Angle, contact, 48
　of internal friction, 90
　of wall friction, 423
Angular particles, 66
Anion, 21
Anthracite coal, 31
Aplite, 31
Apparent cohesion, 48, 99, 118, 170
Apparent pressure envelope for strut loads, 460
Approach slab, 440
Area ratio, 108
Arkose, 31
Arlington, Ore., 100
Armco pipe pile, 205
Asbestos pile cushion, 220
Asphalt spray, 363
Atterberg limits, 20ff; see also Limit, Atterberg
Auditorium Building, 187
Auger, 231, 237, 238
　boring, 103
　bucket, 104, 117
　helical, 104
　hollow-stem, 104, 115

491

Iwan, 104
pile, 206
post-hole, 104
power, 117, 230
samples, 106

Backfill, 197, 247, 301
Backswamp deposit, 145
Bailer, 106, 237
Bank of America, 472
Baraboo, Wis., 100
Barre, Vt., 100
BART, 472
Basalt, 31, 100
Base failure of cut in clay, 461
Base width of retaining wall, 415
Basement wall, 173
Baton Rouge, La., 147
Batter pile, 203, 225, 302, 430, 432, 435, 448
   slump of concrete, 213
Beach deposit, 147
Beach ridge deposit, 136
Beam, anchor, 448
   T-, 188
Bearing capacity, 166, 264, 307, 311, 313
   of clay, 270ff, 304
   factor, 271, 278, 310, 464
   failure, 256, 265, 269, 271, 308, 318, 350
   of pier on sand, 322
   of pile cluster, 283
   of sand, 307ff, 329
   ultimate net, 271
Bearing pile in dense sand, 322ff
Bearpaw shale, 332
Bedding plane, 30, 44, 99, 149, 370
   joints, 367
Bedford, Ind., 100
Bedrock, 164
   unweathered, 149ff
Belgium, 296
Belled pier, 172, 230, 236, 239ff, 281
Belling bucket, 239
Bending strength of pile, 302
Bent, pile, 225ff
Bentonite, 106, 237
Berm, 233
Bit, chopping, 104
   drilling, 106
Bituminous coal, 31
Bituminous coating, 369
Black cotton soil, 159
Blasting, 257, 369
Block sampling, 113
Blocky structure, 339
Blow-up of excavation, 459
Boiling, 46, 326
Bond strength, 65
Bore-hole camera, 117
Boring, 102ff
   auger, 103

depth, 164
exploratory, 103ff
hollow-stem auger, 105
number of, 164
rotary, 106
in rock, 109ff
wash, 104
Boston, 78, 305, 413
Boulder, 4, 235
   penetration test in, 115
   residual, 153
Boulder, Colo., 101
Bounce chamber, 211
Boundary condition, flow net, 51
Braced cut, 259, 446, 456ff
   in clay, 461ff
   crippling of wale, 456, 461
   design, 463, 465
   failure of base in clay, 461
   lateral movement adjacent to, 463
   in sand, 459ff
   settlement adjacent to, 463
   tieback system, 466ff
Braced trench, 173
Bracing, 169ff
   cross-lot, 463
   inclined, 172, 463
   prestressing, 463
Braided stream, 144ff
Bridge pier, 325, 385, 395
   progressive movement in clay, 301
   scour, 330
Bridge trestle, 226
Breccia, 31
British Columbia, 78, 131, 133, 135, 305
Brunspile, 206
Bucket auger, 104, 117
Buckling of pile, 213
   of strut, 456, 461
Building code, 187, 215, 265, 361, 362
Bulkhead, 446
   anchorage, 452
   anchored, 447ff, 472
   design, 452, 455
   dredge, 448
   failure, 449
   fill, 448
   loads, 449ff
   settlement of anchor rods, 455
   sheet-pile moment, 453
Bulking, 49, 198
Buoyancy of raft, 320
Buried channel, 130, 140
Buried valley, 139ff

c/p ratio, 93, 98
Cable tendon, 370
Cable-tool drilling, 106
Caisson, 229, 232ff, 243, 256, 281, 326, 363
   floating, 234

open dredged, 243
pneumatic, 235, 243, 262, 327
Calcium carbonate, 147, 153
California, 147, 358
Caliper log, 153
Camden, N. J., 228
Camera, bore-hole, 117
Canada, 9, 124, 126, 131, 133, 135, 160, 176, 337
Cantilever abutment, 441
Cantilever footing, 185ff, 402, 405
Cantilever retaining wall, 246, 415, 427
Capacity factor (concrete design), 375, 378
Capacity of pile, 215, 218, 225
  to resist lateral load, 227
  in sand, 323
Capillarity, 47, 179
Capillary moisture, 99, 118, 190
Capillary phenomena, 47ff
Capillary rise, 47, 48
Capillary zone, 118
Carbonation, 126
Carbon dioxide, 126
Carbonic acid, 126
Cased hole, 229
Cased pile, 205
  Franki, 205
Casing, 104, 235ff
  dimensions, 112
  drilled pier, 240ff
Cast-in-place concrete pile, 205ff
  effect of slump, 213
Castle Rock. Colo., 9
Catalyst, 181
Cation exchange, 21
  capacity of clay minerals, 21
Cave, 32, 152
Caving, 235
Cavity, 99
  in limestone, 366
  solution, 44
Cedar Rapids, Iowa, 146
Cedar River, Iowa, 145, 146
Cellular cofferdam, 232
Cementation, 159, 170
Cemented rock, 30
Cemented sand, 236, 333
Cement grout, 181
Central Valley, Cal., 147
Channel, buried, 130
Channel markings, 130
Chart, for correction of N-values, 312
  plasticity, 22
  for pressure on retaining wall, 424, 425
  triangular, 25, 127
Chemical alteration, 65
Chemical grouting, 182
Chemical injection, 181
Chemical stabilization, 22, 343
Chemical weathering, 125, 369
Chicago, 79, 129, 132, 133, 160, 184, 187, 190, 230, 243, 262, 304, 305, 365, 400
Chicago Bridge and Iron Co., 384
Chicago clay, 184
Chicago method, 239
Chicopee, Mass., 146, 149
Chlorite, 151
Chopping bit, 104
Circle, Mohr, *see* Mohr's circle
Classification, AASHO, 25, 27, 128
  airfield, 27
  engineering, 128
  MIT, 25, 30
  of rock, 30, 32
  of soil, 24
  textural, 25, 28
  triangular chart, 25
  Unified system, 25, 27ff, 36, 128
  visual, 107
Classification tests, 4, 7
Clay, 5, 164, 196, 247
  allowable soil pressure, 273, 276
  bearing capacity, 270ff, 304
  braced cut, 461ff
  Chicago, 184
  cone resistance, 115
  consistency, 20
  consolidation of sensitive, 64
  contact pressure, 277
  creep strength, 95
  critical slope, 298
  effect of lowering water table, 257
  erratic, 290
  excavation, 256, 297ff
  expansive, 79
  extrasensitive, 20, 65, 66
  failure of base of cut, 461
  fat, 22
  fissured, 98, 273, 297
  footings on, 270ff
  foundations on, 269ff
  gross soil pressure, 277
  highly plastic, 96
  inorganic, 22, 27
  joints, 297
  lacustrine, 79, 348
  lateral displacement due to vertical load, 301ff
  lean, 22
  Leda, 138
  montmorillonitic, 334
  net allowable soil pressure, 272
  net ultimate soil pressure for piers, 278
  normally loaded, 61, 66, 73, 290, 292
  open cut, 461ff
  organic, 7, 22, 27
  overconsolidated, 61, 63, 67, 73, 93, 98, 290, 292
  Paulding, 136
  penetration resistance, 114
  penetration tests, 115
  permeability, 43

pier design, 277ff
preloaded, 61, 63, 67, 73, 93, 98, 290, 292
progressive lateral movement, 301
quick, 20, 137
raft on, 276ff
remolded, 59, 162
residual, 155
resistivity of, 120
sensitive, 20, 290
settlement, 265, 286ff
shear strength, 97, 102
skin friction on piers, 281
slickensides, 297
soft, 91, 242
softening, 170
stability of slopes, 98, 170, 297ff, 300, 305
static capacity of friction piles, 283
stiff, 236, 299
strength, 94, 97, 102
stress-deformation characteristics, 90
swelling, 67, 195, 199
ultimate bearing capacity, 270ff
undisturbed, 162
undrained strength reduction factor, 96
varved, 138
Clay-filled joints, 151
Clay grouting, 181
Clay minerals, 10, 35, 127, 155
  cation exchange capacity, 21
Clay slurry, 181, 326
Clay-slope stability number, 298
Clay structure, disturbance, 291
Clay-tile drain, 189
Clay till, 133
Cleavage, vertical, 142
Climate, 127
Closed-ended steel pipe pile, 208
Coal, anthracite, 31
  bituminous, 31
Coarse-grained soil, 4, 8, 13, 27, 99, 133
Cobbles, 153, 235
Cobi pile, 205
Coefficient, of active earth pressure, 419
  of compressibility, 73
  of consolidation, 70, 294
  of curvature, 9, 28
  of earth pressure at rest, 417
  of passive earth pressure, 419
  of permeability, 40, 43, 70, 177, 326
  of secondary consolidation, 73
  of subgrade reaction, 413
  uniformity, 9, 28, 30
  of volume compressibility, 70, 73
Cofferdam, 54, 229ff, 243
  cellular, 232
  double wall sheet-pile, 232
  leakage, 232
  single wall, 244
  single wall sheet pile, 232
Cohesion, 5, 13, 177, 235

apparent, 48, 99, 118
of loess, 141
of sand, 257
Cohesionless sand, 176, 196
Cohesionless silt, 307
Cohesionless soil, 198
Cohesive granular material, 196
Cohesive soil, 198, 237
Collapsible soil, 65, 164, 333ff
Colloidal particle, 10
Color, 7
Colorado, 9, 101, 116, 134, 136, 142, 143, 335
Colluvial soil, 126
Column footing, 185
Columnar jointing, 32
Combined footing, 185ff, 401ff
Compacted fill, 94, 193ff, 198
  foundations on, 193ff
Compaction, by flooding, 197
  control, 198ff
  of fill, 195ff
  modified AASHO, 16
  per cent, 15, 16, 18
  piles, 200, 323ff
  standard AASHO, 16
  test, 14ff, 36
  by vibration, 327, 330
  100%, 15
Compensated foundation, 188, 277, 291, 353
Composite pile, 209
Composite shore deposit, 150
Composition, mineralogical, 10, 21
Compressed air, 235, 256
  caisson, 243, 327
  pier construction, 230
Compressibility, 193, 270
  coefficient, 73
  evaluation in practice, 66
  of rock, 151
Compression index, 62, 290
Compression test, confined, 59ff
  triaxial, 83ff
  unconfined, 19ff, 92
Compressive strength, intact rock, 100
  triaxial, of sand, 85
  unconfined, of clay, 19
Compressive wave, 218
Concentrated load, earth pressure due to, 423
Concrete, dental, 365
  design, 383
  deterioration, 147
  encasement of steel piles, 208
  guide tube, 240
  pneumatically placed, 363, 369
  segregation, 240
  slump, 213, 240, 242
  tremie, 240
  vibration, 240
Concrete pier, 240
Concrete pile, 205ff, 219

Armco pipe, 205
auger, 206
Bruns precast sectional, 206
cased Franki, 205
cast-in-place, 205
Cobi, 205
deterioration, 207
Franki, 205, 206
Fuentes, 206
pedestal, 206
precast, 205ff
precast sectional, 206, 207
prestressed, 207
Raymond step-taper, 205
sectional precast, 206, 207
uncased Franki, 205
Union Metal monotube, 205
working load, 207
working stress, 213
Concretion, 19, 159
Cone penetration test, 113, 115, 133, 136
Cone resistance, 115
Cone test, Dutch, 113, 115, 198, 323
  static, 121
Confined compression test, 59; *see also* Consolidation, test
Confining pressure, 86
Conglomerate, 31
Consistency, 19, 121
  measurement, 112ff
Consolidation, 58, 68ff, 92, 96, 98, 179, 180, 182, 192, 226, 282, 285
  coefficient, 70, 294
    of secondary, 73
  characteristics of, collapsible soil, 65ff
    residual soil, 65
    sand, 66
    sensitive clay, 64ff
  degree, 71ff
  differential equation, 70
  drainage by, 50
  paired, 333ff
  primary, 72, 73
  rate, 73
  secondary, 73, 195
  by surcharging, 192
  test, 42, 59, 270, 290
  theory, 68ff, 81
Consolidation pressure, initial, 68
Consolidated undrained test, 88
Constant head permeameter, 41
Construction, damage due to, 254ff
  methods, 167, 255
  of piers, 244, 326
  procedures, influence on design, 259
  slopes, 169
Contact angle, 48
Contact moisture, 48
Contact pressure, 351
  on clay, 277

Continental glacier, 129
Continental deposit, 144, **146**
Continuous back drain, retaining wall, 247
Continuous footing, **185**
Continuous sample, 164
Contraction during shear, 126
Contraflexure, point in sheetpile bulkhaed, 453, 456
Control observations, 196
Control of compaction, 198ff
Controlled blasting, 370
Controlled fill, 193
Core, 165, 212; *see also* Rock core
Core barrel, 109, 111, 112, 164
  double tube, 111
  shot, 117
  single tube, 111
  triple tube, 111
Core catcher, 108, 109
Core drill, 109
Core lifter, 111
Correction factor for water table in sand, 313
Corrosion of steel pile, 208
Coulomb, earth pressure theory, **420**
  equation, 87, 417
Counterfort retaining wall, 246
Cracks in soil, 19, 20, 30, 93, 159, 177
Cracking, precast concrete pile, 218
Creep, 95, 341
Creosoted timber pile, 204
Crib wall, 247
Critical circle, **298**, 358
Critical frequency of vibrations, 327
Critical height of slope, 299
Critical hydraulic gradient, 45
Critical pressure, 335
Critical section (concrete design), 376
  of pile, 213
  of pile footing, 379
Critical void ratio, 86
Cross-lot bracing, 463
Crushing of grains, 66, 83
Culpeper, Va., 100
Curvature, coefficient of, 9, 28
Cushion blasting, 370
Cushion blocks, pile driving, 220, 221
Cuttings from drilling, 106, 110
Cut, braced, *see* Braced cut
Cut, open, *see* Open cut
  tieback, 446, 447
Cutting shoe, 107
Cyclic mobility, 95
Cylindrical pipe pile, 224

Dacite, 31
Damage due to adjacent construction, 254ff
  due to future construction, 259
  due to swelling and shrinking, 332, 339, 340
  to point of pile, 208
Damp proofing, 189ff

Darcy's law, 39, 51
Dead load on pier, 341
Deadman, 448, 452
Decay of piles, 204
Deep-seated settlement, 194ff, 199
Deep well pump, 180, 326
Defects in piers, 241
Deformation condition for, Rankine theory, 418
  open cut in sand, 459
Deformation of rock, 99
Degree of, compaction, 16ff, 337, 342
  consolidation, 68, 71, 72
  disturbance, 108
  saturation, 12, 17, 18, 47
Delft, 115
Delmag pile hammer, 210
Delta deposit, 146, 149
Dense sand, 66, 88, 325
Dense silt, 325
Dense or stiff layer over soft deposit, 350ff
Densest state, 13, 14
Densification, by blasting, 200
  by pile driving, 195
  by vibroflotation, 195
Density, dry, 12, 18
  relative, 11, 13, 14, 112ff, 117, 118, 198, 307, 308
  soil aggregate, 13
Density index, 13ff, 66, 198
Dental concrete, 365
Denver, Colo., 9, 116, 134, 136, 142, 160
Depletion, zone of, 127
Deposit, backswamp, 145
  beach, 147
  beach ridge, 136
  braided stream, 147
  composite shore, 150
  compressible, 188
  continental, 144, 146
  delta, 146, 149
  erratic, 147, 164
  flood plain, 145, 147
  glacial, 126, 132, 149
  glacial drift, 129
  glacial lake, 129, 135ff
  glacial moraine, 135ff
  glacial outwash, 129, 137
  glacial till, 43, 129, 131
  glaciofluvial, 129, 133ff
  heterogeneous, 293
  lacustrine, 135, 145
  laminated, 135
  marsh, 134, 147
  organic, 134, 135, 147
  permeability of stratified, 42
  river, 144ff
  shore, 147, 150
  shoreline, 136
  stream-channel, 144
  valley, 145
  varved, 135
  windblown, 140ff
Deposition, profile of, 126
Depth of boring, 164
  of embedment of sheeting, 452
  of footing, 378
  of foundation, 270
  of scour, 325
  of surcharge, 270, 311
Desiccation, drainage by, 50
Desiccation cracks, 297
Design, charts for footing on sand, 311ff
  of footing, 375ff
    on clay, 273ff
    on sand, 313ff
  of foundations, in swelling soil, 340ff
    on unwetted collapsible soil, 334ff
  loads on foundation, 264ff
  of raft foundation, 405ff
  reinforced concrete, 383
Detailed exploration, 164
Deterioration of concrete piles, 207
  of shale, 363
Development length, critical section (concrete design), 376
Dewatering, 237, 330, 367, 458
Diabase, 31, 100
Diagonal tension, 376
Diesel hammer, 209ff, 220
  damage to piles, 212
Differential equation of consolidation, 70
Differential expansion, 126
Differential settlement, 188, 193, 254, 265, 273, 286, 319, 320, 343
Differential steam hammer, 210
Dike, 153ff
Dilatancy, 5, 7, 86, 90, 114, 325
Dimension-stone footing, 185
Diorite, 31
Dip, 151
Direct shear test, 83
Direction of earth pressure, 423
Discharge velocity, 40
Dispersed structure, 19, 20
Dispersing agent, 8
Dispersion test, 6
Displacement due to filling, 194
  due to pile driving, 258
Dissipation of pore pressure, 96
Distribution of earth pressure, 423
Disturbance, 65, 97, 107, 108, 291
Ditches, 177ff
Dolerite, 31
Dolomite, 31, 100, 152ff
Double acting steam hammer, 210
Double oedometer test, 333ff
Double shaft pier, 245
Double tube core barrel, 111
Double wall sheet pile cofferdam, 232
Downdrag on pile, 226, 285

Drain, clay tile, 189
  floor, 189
  footing, 189ff
  intercepting, 189
  perforated pipe, 189
  sand, 180ff, 195
Drainage, 39, 169, 176, 177ff, 189ff, 248
  by consolidation, 50
  by desiccation, 50
  by electrical methods, 22
  gravity, 49
  permanent, 177
  by pumping, 348
  surface, 128
  time required by pumping, 326
  well, 180
Drainage methods, 181ff
Drainage pattern, see Airphoto pattern
Drained angle of internal friction, 90
Drained test, 88, 97, 101
Dredge bulkhead, 448
Dredge level, 448
Dredging, 256
  of caissons, 326ff
  wells, 234
Drift, glacial, 129
Drift plain, 130
Drilled pier, 230, 235ff, 243, 340, 366, 368
  belling, 239ff
  bottom cleaning, 241
  casing, 241
  concreting, 240
  conditions for success, 242
  danger of gas, 241
  excavation, 237ff
  inspection, 241
  lateral displacement, 341
  lining, 241
  pulling casing, 240ff
  in stiff clay, 292
  use of telltales in concreting, 242
  verticality, 241
Drill hole pattern for blasting, 369
Drilling, cable tool, 106
  percussion, 106, 109
  pneumatic, 153
  rotary, 106ff
Drilling bit, 106
Drilling fluid, 106, 110, 111
Drilling machine, 104, 231
Drilling mud, 106, 107, 173, 235, 237, 326
Drill rods, dimensions, 112
Drive head, 210
Driving of piles, see Pile Driving
Driving resistance, 212
Driving stress in precast concrete pile, 207
Drop hammer, 209
Drought, effect on shrinkage, 343
Dry density, 12, 18
  maximum, 15

Drying, effect on limits, 157
Dry strength, 5
Dry unit weight, 12, 15, 18
Dune sand, 141, 143
Dutch cone penetrometer, 113, 115, 198, 323
Dynamic analysis of pile driving, 216ff
Dynamic penetration test, 112, 121, 133, 136
  cone, 164
Dynamic pile formula, 227, 282
Dynamic pile resistance, 222, 223, 323
  correlation with static, 217

Earth pressure, 414, 417ff
  active, 385, 418
  against flexible support, 448
  at rest, 417
  charts for retaining wall, 424ff
  Coulomb's theory, 420
  direction, 423
  distribution, 423
  due to concentrated load, 423
  due to live load, 423
  locus, 422
  passive, 419
  point of application, 423
  Rankine theory, 417ff
  trial wedge method, 422
  wedge theories, 420ff
Earthquake, 97, 195
Eccentrically loaded footing, 274, 314, 385ff
Ecology, 129
Effective grain size, 9, 48
Effective stress, 44, 56, 88
  analysis, 101
  principle of, 44, 68
Efficiency equation, 292
Efficiency of pile hammer, 211, 220
Elasticity, modulus of, 81
Electrical methods of drainage, 22
Electrolyte, 19
Electron microscope, 9, 10
Electroosmosis, 182, 200
Elizabeth, N. J., 192
E-log p curve, 60, 333
Embedment of sheet piles, 452, 462
  of soldier piles, 462
Empress Hotel, 305
End bearing pile, load test, 215
End moraine, 129
Engineering-News formula, 217, 324
Engineering report, 165
Epoxy coating on pile, 208
Equilibrium moisture content, 337
Equipotential drop, 53
Equipotential line, 51
Erosion, 128
Erosion tunnel, 49
Erratic deposit, 147, 164
  of clay, 290
  of sand, 314

settlements on, 292ff
Error in N-value, 114ff, 314
Esker, 135
Evaporation, 339
Excavation, 169ff, 173, 174, 177, 194
  blow-up of bottom, 459
  bracing, 170ff
  in clay, 256, 297ff
  drilled pier, 237ff
  in stratified deposits, 256
  movements due to, 173ff
  open, 169ff
  in rock, 369ff
  in sand, 255ff, 325ff
  settlement, 255
  shallow, 169
  of slopes in non-uniform soil, 354ff
  stability, 177
  test, 93
  with unsupported slopes, 169ff
  below water table in sand, 325ff
  wide, 172
Excess pressure, 68
  hydrostatic, 40
  porewater, 53, 70, 96, 195, 269
Expansion, 126
Expansive soil, 79, 346
Exploration, detailed, 164
  geophysical, 122, 153
  preliminary, 163
  program, 163ff
  resistivity, 120
  seismic, 118
  soil, 103ff, 129, 163ff, 236
Exploratory boring, 103ff
Exploratory program, 103ff, 163ff
Explosive gases, drilled pier, 241
Extrasensitive clay, 20, 65, 66

Fabric of rock, 30, 32
Fact-finding survey, 163
Factor, bearing capacity, *see* Bearing capacity factor
  capacity (concrete design), 375, 378
  load (concrete design), 375, 379
  safety, 265, 271, 277, 301, 335
Failure, bearing capacity, *see* Bearing capacity failure
  of anchored bulkhead, 448
  modes of (concrete design), 376
  plane, 84
  by scour, 325
  of slope, 225
  wedge, 420
Falling head permeameter, 41
Fargo elevator, 304
Fault, 30, 99, 125, 149, 151, 366, 370
  gouge, 151, 366
Feather wedge, 370
Felsite, 31
Field density test, 16

Field exploration, records, 120
Field measurements, 122
Field observations, 174, 258ff, 446
Fill, 301
  compacted, 94, 198
  compaction, 195ff
  controlled, 193
  effect on piles in clay, 285
Fill bulkhead, 448
Fill-supported structure, 193ff
Filter, 49, 177, 178, 247
  particle-size requirement, 49
  ratio, 49
  skin, 41, 42, 181
Filter-protected sump, 178
Fine-grained soil, *see* Soil, fine-grained
Fissured clay, 98
Flapper valve, 108
Flat slab analogy, raft design, 406
Flexible earth-retaining structure, 447ff
Flexible support, earth pressure against, 448
Flexure, critical section (concrete design), 376
  minimum reinforcement, 379
Floating type open caisson, 234
Flocculated structure, 19
Flocculation, 8, 20, 136
Flooding, compaction by, 197, 198
  of foundation soil, 336
Flood-plain deposit, 145, 147
Floor drain, 189ff
Floor, soil supported, 199
Flow channel, 53
Flow line, 51ff
Flow net, 51ff
Flow slide, 138, 139
Fluid, drilling, *see* Drilling fluid
Fold, 125, 151
Foliated rock, 30
Footing, 368
  cantilever, 185, 186, 402, 405
  on clay, 270ff, 291
    design, 273ff
  column, 375ff
  column loads on, 403
  combined, 185ff, 401ff
  on compacted fill, 199
  continuous, 185
  depth, 378
  design of, 375ff
  dimension-stone, 185
  distribution of soil pressure beneath, 386
  eccentrically loaded, 274, 314, 385ff
  individual column, 185, 186
  isolated, 185
  moment about both axes, 391
  masonry, 185
  minimum depth curves (concrete design), 378
  on piles, 379
  proportioning, 187
  reinforced concrete, 185, 186

reinforcement, 377
rubble-stone, 185
on sand, 307ff
   above clay layer, 352
   bearing capacity, 309ff
   design, 311ff
   proportioning, 308
   settlement, 308ff
   water table influence, 312
settlement, 290ff
spread, 185
steel I-beam, 185
subject to moment, 384ff
trapezoidal, 401, 404
types, 185, 186ff
unsymmetrical, 392
wall, 185, 186, 375ff
Footing drain, 189ff
Footing foundation, 185ff; *see also* Foundations
Formula, pile, 202
Fossil, 149
Foundation, compensated, 291
   on clay, 269ff, 286ff
   on collapsible soil, 334ff
   on compacted fill, 193ff
   depth, 270
   footing, 185ff
   grillage, 185, 187
   load, 289
   mat, 185ff, 188, 400ff
   on nonplastic silt, 307ff
   on nonuniform soil, 349ff
   pier 188, 229ff
   pile 188, 203ff, 281ff
   on plastic silt, 269ff
   raft 185, 188, 400ff
   on rock, 361ff
   on sand, 307ff
   on swelling soil, 337ff
   subject to moment, 384ff
   types, 167ff, 259, 263ff
   uplift due to swelling, 345
   on weathered rock, 367ff
Foundation blasting, 369
Franki pile, 209, 324
Fraser River, 78
Free water surface, 47, 55, 177
Freeze of pile, 207, 223ff, 282ff, 323, 325
Freezing, 126, 153
   stabilization by, 182
Friction, wall, 423
Friction angle, 90
   drained, sand, 87ff
   silt, 87ff
Friction pile, 212ff, 281, 304
   in clay, settlement, 291ff
   group in clay, 284
   load test, 215
Frost action, 50, 247, 424
Frost heave, 56

Frost penetration, 186
Fuentes pile, 206
Fully compensated foundation, 277

Gabbro, 31
Gap grading, 5, 9, 66
Geologic classification of rock, 32
Geologic processes, 125
Geologic terminology, 126
Geology, 29, 120, 122, 127, 129, 159
Geophysical investigation, 118ff, 122, 153
George Washington Bridge, 228
Georgia, 153, 154
Glacial deposit, *see* Deposit, glacial
Glacial epoch, 125
Glacial ice, 64
Glacial lake, 139
Glaciation, 125, 129ff
Glacier, 126, 129
Glass, volcanic, 31
Gneiss, 31, 32, 100, 127
Gothenberg, 363
Gouge, *see* Fault gouge
Gradation, 5, 9
Grade beam, 203, 204, 340
Graded filter, 49
Gradient, hydraulic, 40, 45, 47, 68
   pressure, 40
Grain elevator, 335
Grain properties, 8, 10
Grain size, 8, 311
   distribution, 8, 9, 25
   effective, 9, 30, 48
   requirement for filter, 49
Grains, crushing of, 66, 83
Granite, 31, 32, 100, 127, 155, 158
Granite gneiss, 31
Grant Park Underground Garage, Chicago, 400
Gravel, 4, 9, 27, 53, 196, 247
   permeability, 43
   resistivity, 120
   river, 143
   seismic velocity, 120
   shear strength, 87, 97, 99
   stress-strain relations, 85
Gravity abutment, 248
Gravity retaining wall, 246, 415
Graywacke, 31
Great Lakes, 135, 140
Great Plains, 146
Granada, Miss., 9
Grillage foundation, 185, 187
   rail, 185
   timber, 185
Grooving tool, 24
Gross soil pressure, on clay, 277
   on sand, 310, 320
Ground moraine, 133
Groundwater, 152, 153, 235
   observations, 118

recharge, 330
Groundwater level, 47, 189
  lowering, 64
Groundwater table, 176
Group index, 26
Group of piles, 215ff
Grout, 366
  cement, 181
  chemical, 182
  slush, 365
Grouted anchor, 468
Grouting, 181, 200, 237, 242, 366
  cement, 182
  chemical, 182
  clay, 181, 182
Guadalupe Shrine, 58
Gypsum, 31, 32, 153

Hackensack, N. J., 100
Hammer, pile, see Pile hammer
Hammer cushion, 210
Hammerhead pier shaft, 245
Hand carved sample, 112, 117, 121
Hand excavated pit, 230
Hand-held pneumatic tamper, 197
Hard driving, piles, 208
Hawaii, 126, 154, 156, 157, 158
Head, hydraulic, 40, 47
  piezometric, 39, 47
  position, 39
  velocity, 45
Header pipe, 178
Heave, due to excavation, 256, 291, 348
  of desiccated clay, 346
  due to pile driving, 212, 258, 304
Height of capillary rise, 47ff
Helical auger, 104
Hennepin, Ill., 306, 384
Heterogeneous deposit, 293
Holland, 258, 296, 323
Hollow stem auger, 104, 105, 115
H pile, 171, 172
  damage to, 208
  point reinforcement, 208
  stresses, 207
Horizon, soil, 127
Horizontal load on pile, 225, 432
Hornfels, 31
Houston, 259
Humid tropics, 126
Hydrated lime, 199, 200
Hydration, 126
Hydraulic gradient, 40, 45, 47, 68, 180
  critical, 45
Hydraulic head, 40, 47
Hydraulic properties of soil, 39ff
Hydrometer, 8
Hydrostatic pressure, 170
  excess, 40
Hydrostatic uplift on raft, 320

Hydrothermal alteration, 30
Hydrous aluminosilicate, 10
Hydrous iron oxide, 155

I-beam grillage footing, 185
Ice lenses, 50
Idaho, 100
Igneous rock, 31, 32, 368
  seismic velocity, 120
Illinois, 9, 129, 131, 133, 134, 139, 141, 142, 160, 306, 384
Illite, 10, 11, 21
Impact pile hammer, 209
  characteristics, 210
Impedance of pile, 218ff, 224
Inclined bracing, 172, 463
Index, compression, 62
  density, 13, 14, 17, 18, 66, 198
  group, 26
  liquidity, 23
  plasticity, 21, 26, 98
Index properties, 4, 7, 21, 24, 163
  of glacial lake deposits, 137
  of glacial moraine deposits, 137
  of rock, 30
  of tills, 131
  of wind blown deposits, 142
Indiana, 100, 137, 142, 143
Individual column footings, 186, 375ff
Inevitable movement, 255
Influence chart for vertical pressure, 288ff
Initial consolidation pressure, 68
Initial tangent modulus, 86
Injection well, 348
Inorganic clay, 22, 27
Inorganic silt, 5, 22, 27
  permeability, 43
Injection methods of soil stabilization, 181, 182
Inspection of drilled pier, 241
Inspection pit, 116
Inspection shaft, 116
Inside clearance ratio, 108
Installation of piles, 209ff
Interlocking fabric, rock, 30
Intermontane basin, 146
Internal friction, 90
Investigation, subsurface, 103ff
Ion, adsorbed, 21
Ion exchange, 21, 35
Iowa, 143, 145, 146, 346
Iroquois moraine, 129
Irregular deposit, 353ff
Iron, 159
Isochrome, 68
Isolated footing, 185
Istanbul, 444
Iwan auger, 104

Jackson County, Ill., 139
Jack pile, 252

Jacking of piles, 252
Jet-eductor, 178, 179
Jetting of piles, 212
John Hancock Building, Chicago, 365
Joint, 19, 20, 30, 44, 98, 99, 149, 152, 153, 177, 361, 370
  in abutment at wing wall, 439
  clay-filled, 151
  in clay, 297
  columnar, 32
  in rock, 361, 362, 364
  relict, 154

Kame, 135
Kansas, 335
Kansas City, 230
Kaolinite, 10, 21
Karst, 152
Kern, 387
Kicker block, 463

Lacustrine deposit, 79, 126, 135, 138, 139, 145, 348
Lagging, 171, 172, 230, 471
Lake Agassiz, 135, 138
Lake Bonneville, 138
Lake Chicot, 146, 148
Lake Michigan, 147, 150
Lake, Oxbow, 145
Laminated deposit, 145
Laminated rock, 30
Land form, 128
Landslide, 341
Lateral displacement, in clay, 301ff
  of drilled pier, 341
  of pile, 324
Lateral load on pile, 213, 225ff, 227, 324
Lateral movement, 174, 258
  due to pile driving, 212, 304
  of excavation walls, 463
Laterite, 155
Laterization, 155, 157, 159
Lattice structure, 10
Layered deposit, 135
Lead, pile driving, 209
Leakage, cofferdam, 232
Leaching, 127, 137, 155
Leda clay, 138
Legal importance of field observations, 258
Lenticular deposit, 353
Levee, Natural, 145
Lift thickness, 369
Lime stabilization, 199, 200, 343, 346
Limestone, 31, 32, 100, 149, 153, 154, 155, 159, 363
  Bedford, 100
  cavities, 366
  seismic velocity, 120
  Solenhofen, 100
Limit, Atterberg, 20ff, 36, 94, 107, 270

  effect of drying, 157
  liquid, 21, 26, 27, 64, 66
  plastic, 21, 64
  shrinkage, 21, 24
Limnoria, 204
Line drilling, 370
Line load, earth pressure due to, 423
Line of seepage, 47, 55
Liner, 108
  for drilled pier, 241
Link Belt pile hammer, 210
Lithology, 32
Liquefaction, 97, 137, 195
Liquidity index, 23
Liquid limit, *see* Limit, Atterberg
Load, for settlement calculation, 289
  on abutment, 439
  on footing, 403
  on raft, 188
  repetitive, 95
  shock, 97
  in struts, 460
  transient, 97
Load factor (concrete design), 265, 375, 379
Load-settlement relation, 117, 144, 281, 282, 307, 334, 375, 379
Load test, 93, 117, 270, 283, 284, 352
  on fissured clay, 273
  on pile, 214, 215ff, 224, 227, 292, 323ff
  on rock, 367
  standard, 118, 143, 164, 198, 272
Loading, rate, 95
Loess, 66, 141, 142, 160, 164, 333, 334, 336, 346
London, 176, 305
Long-term stability, slope, 98
Longyear Co., 122
Loose sand, 66, 88, 214, 308, 324
  cemented by salt, 333
Loosest state of sand, 13, 14
Loss of ground, drilled pier, 242
Louisiana, 147
Lowering of water table, 64, 256

Magmatic water, 127
Mandrel, 181, 205, 207, 219
Manganese, 159
Manhattan schist, 360
Map, pedologic, 128
Marathon, Ont., 9, 135
Marble, 31, 100
Marina City, 304
Marine soil, 126
Marsh deposit, 134, 147
Masonry footing, 185
Maskinongé River, 138, 139
Massachusetts, 9, 146, 149
Mat foundation, 188
Maximum dry density, 15ff
Meander, 145, 148
Mechanical analysis, 8, 107

Meltwater, 129, 145
Membrane water proofing, **189**ff
Metamorphic rock, 31, 32, 368
   seismic velocity, 120
Metamorphism, 125
Method of, isolated piers, 184
   slices, **354**ff
      modifications, **358**
Mexico City, 58, 65, 79, 137, 138, 139, 160, 348
Mica, 66
Microrelief, 128
Millisecond delays, 369
Milwaukee River, 149, 150
Mineral, clay, **10**, 35, 127, 155
   three-layer, 10
   two-layer, 10
Mineralogical composition, 10
Mineralogy of rock, 32
Minimum depth curves, concrete footing, 378
Minnesota, 131
Mississippi, 9, 142, 143, 160
Mississippi River, 139, 147, 148
Missouri valley, 144, 146, 337, 368
MIT, 304
   classification system, 25, 30
Mixing time, concrete, 242
MKT pile hammer, 210
Modified AASHO compaction, 198
   test, **15**
Modified swelling pressure test, 343
Modulus of elasticity, 81
   initial tangent, **86**
   of intact rock, 100
   of subgrade reaction, 408
Mohr's circle of stress, **83**ff, 91, 417
Mohr's rupture diagram, **86**, 418
Moisture, contact, **48**
   capillary, 99, 118, 190
   optimum, **15**
Moisture-density relation, 15, 94, 128, 157, 158, 198
   test, **14**
Moment, on foundation, **384**ff, 391
   on pile footing, **392**ff
   in sheet pile bulkhead, 453
Moment of inertia, pile group, 393
Monolithic retaining wall, 247
Monotube pile, working loads, 207
Montana, 9
Montmorillonite, **10**, 11, 21, 151, 334
Montour, Iowa, 346
Montreal, 176
Moraine, 129, 133, 134
   Iroquois, 129
Movement, associated with excavation, **173**ff
   of braced excavation, 463
   lateral, 174
   vertical, 174
Muck, 7, 147
Mud, drilling, 106, 107

Mud line, 448
Mud wave, 192, 453
Multiple stage wellpoint, **178**ff
Muskeg, 78, 147
Mylonite, **152**

Natural levee, 145
Natural water content, 23
Nebraska, 142, 143
Needle beam, 251, 252
Negative pore pressure, 88, 96, 325
Negative skin friction, 226, 285, 305, 322, 323, 336, 462
Net soil pressure, 378
   allowable, 272, 320
   ultimate, 271, 278, 310
Netherlands, 115
Neutral stress, **44**
Nevada, 100
New Hampshire, 131
New Jersey, 100, 192, 360
New York, 100, 137, 151, 173, 330, 366, 367
   Thruway, 151
Nonplastic silt, foundation on, **307**ff
Nonuniform soil, slopes in, **354**ff
Normally loaded clay, 61, 66, 73, 90, 290, 292
North Carolina, 155, 158
Norway, 28, 161
Norwegian Geotechnical Institute, 268
Notched-wall shoring, 251
Nuclear density meter, 17
Nuclear moisture meter, 17
N-value, **113**ff, 198, 310, 311, 314
   correction for overburden pressure, 114
   errors, **114**ff, 314

Oahu, 154, 156
Oakland, Cal., 358
Observational method, 166, 196, 446
Observation well, 42, 237
Odor, 7
Oedometer test, **59**
   double, 337
Ohio, 131, 137, 138
Oklahoma, 337
Ontario, 9
Open caisson, **234**ff
   dredged, 243
   sand island, 234
Open cut, 172, 174, 178, **456**ff
   base failure in clay, 461
   embedment of piles, 462
   in clay, **461**ff
   lateral movement, 463
   settlement, 463
   sheeting design, 463
   strut loads, **459**ff, 462
   tieback system, **466**ff
Open-end diesel hammer, 209
Open excavation, **169**ff

Subject Index

Optimum moisture content, 15, 18, 195, 198
Oregon, 100
Organic matter, 127
Organic soil, 7, 27, 78, 164, 195
   clay, 7, 22
   deposit, 134, 145, 147
   silt, 7, 22, 27, 43, 179, 257, 323
Oslo, 363
Outwash, 129, 130
   plain, 134
Overburden pressure, correction of N-value, 114, 312
Overcompaction, 197
Overconsolidated clay, 61, 90, 93, 98
Overconsolidation, 60, 337
   ratio, **60**, 93
Overdriving of piles, 363
Oven drying, 23
   effect on liquid limit, 23
Overturning moment, 384
Oxbow lake, 145
Oxidation, 126

Pacific Coast Range, 147
Paired consolidation test, 334
Parapet wall, 441
Parent material, 127, 128
Partially saturated soil, strength, 99
Particle, angular, 66
   crushing of, 83
   plate-shaped, 66
   rounded, 66
   shape, 5
   size, **8**, *see also* Grain size
   size distribution, 8, 25
      for filter, 49
Passive earth pressure, 419
Paulding clay, 136, 137
Pea gravel, 9
Peat, 27, 78, 133, 134, 137, 147, 164, 192, 257, 323
Pedestal pile, 206
Pedologic map, 128
Pedologic profile, 127
Pedology, 29, 120, 122, 127ff, 159, 160
Pegmatite, 31
Penetration resistance, 114, 121, 133, 134, 136, 146
Penetration test, 112ff, 121
   cone, 113, 115, 133, 136
   dynamic, 112, 121, 131, 136, 164
   standard, 97, 108, 113ff, 121, 133, 136, 141, 142, 147, 150, 164, 198, 270, 308, 311ff
   corrections, 114, 312
   errors, 114ff, 314
   static, 112
Penetrometer, 112, 121, 163
   conical drive point, 113
   Dutch cone, 113, 115
   static, 115

Peoria, Ill., 9
Per cent compaction, 15, **16**, 18, 198
Percussion drilling, **106**, 109
Perforated drain, 178, **189**
Peridotite, 31
Permafrost, 56
Permeability, 39ff, 97, 146, 149, 180, 181
   coefficient, **40**, 43, 70, 177
   of bedrock, 149
   of clay, 43
   of delta deposits, 146
   of flood plain deposits, 145
   of glacial till, 43
   of gravel, 43
   of inorganic silt, 43
   of organic silt, 43
   of rock, 44, 151
   of sand, 43
   of silt, 43
   of stratified deposits, 43
   of till, 43
Permeability test, **41**, 56
   constant head, 41
   falling head, 41
Pervious blanket, 180
Philadelphia, Pa., 228
Photography, aerial, *see* Airphoto
Phyllite, 31
Physical weathering, 125
Physico-chemical changes, 137
Physico-chemical properties of soil, 35
Physiography, 159
Pier, 203, 229, 243, 256, 335, 368
   allowable load, 281
   belled, 239ff
   below water table, 326
   bridge, 244ff, 385, 395
   in clay, 277ff
   construction, 229ff, 244, 262
   dead load, 341
   defective, 241
   double-shafted, 245
   drilled, *see* Drilled pier
   large diameter, 231
   on rock, 363
   on sand, 321ff
   settlement, 292
   on swelling clay, 343
   tensile forces in, 340
Pier shaft, 229, 244ff
Pierre, S. D., 332
Pierre shale, 332
Piezometer, 39ff, 118, 119, 195, 196
Piezometric head, 39, 47
Piezometric level, 39, 47, 51, 174
Piezometric tube, 39, 47
Pike's Peak, 100
Pile, 203, 335, 336, 368
   adhesion, 214
   anchor, 214, 448

, 204, 225, 226, 302, 430, 432ff, 448
dense sand, 322ff
, 212ff
strength, 302
206
buckling, 213
capacity, 215, 218, 225, 323
cased, 205
cast in place, 205, 206, 213
  capacity, 282
  choice of type, 225
  in clay, 281ff
    progressive movement, 302
    reduction factor for adhesion, 286
    uplift resistance, 286
cluster, 215ff, 283, 291, 329
concrete, 205ff, 213, 219
compaction, 200, 323ff
composite, 209
creosoted, 204
critical section, 213
cushion, 210, 219ff
cushion block, 220
cylindrical pipe, 224
damage to point, 208
decay, 204
downdrag, 226, 285
driver, 209ff
dynamic analysis, 223ff, 323
efficiency, 292
end bearing, load test, 215
extractor, 212
footing, 379
  moment on, 392ff
formula, 202, 216, 217, 227, 282
  defects, 217
  Engineering News, 217
foundation, 188, 203ff, 216, 286
  lateral load, 225ff
freeze, 207, 223ff, 323, 325
Franki, 206, 209, 324
friction, 212ff, 281
  load test, 215
  increase of capacity with time, 283
  settlement in clay, 291ff
group, 215ff, 283, 291, 329
  moment of inertia, 393
  safe load, 285
H, 172
  steel points, 363
hard driving, 208
head assembly, 219
heave, 212, 304
horizontal load, 225, 432
impedance, 218ff
installation, 209ff
lateral displacement, 324
lateral support, 213, 227

leads, 209
load, allowable, 383
  ultimate, 215
load test, 214ff, 224, 227, 283, 284, 324, 325
  interpretation, 215
  long-time, 292
negative skin friction, 226, 285, 305, 323
overdriving, 363
parallel sided, 214
pipe, 208
point bearing, 212ff, 216, 281
  safe load in clay, 284
  settlement in clay, 292
precast concrete, 205ff
  cracking, 218
  driving stresses, 207
rate of settlement in clay, 294
Raymond step-taper, 205
redriving test, 225, 285ff, 292
reduction coefficient on adhesion, 282
relaxation, 223, 325
resistance to penetration, 224, 281ff, 285
roughness, 330
safe load, 281ff, 285ff
  in sand, 322ff
  uplift resistance, 325
scour prevention, 325
seismic velocity, 218
settlement, 226
sheet 171ff; *see also* Sheet pile
in silt, 324ff
skin friction, 213, 216, 223, 291
soldier, 172, 292
spacing, 284
spotter, 209
spudding, 212
steel, 207ff, 219
  corrosion, 208
  H, point reinforcement, 208
  H, stresses, 207
  pipe, 207ff
  sheet, 171ff
  working stresses, 213
stress, 223
stress-transmission characteristics, 219
subject to tension, 394
support for retaining wall, 428
taper, 213, 324, 330, 452
timber, 203ff, 219, 363
tip, precast concrete, 214
transmission of driving stress, 218
types, 203ff
ultimate capacity, 222
uncased concrete, 205
uplift, 225
velocity of wave propagation, 218
verticality, 209
Wakefield, 231, 448
wave analysis, 218ff, 227
wood, 203ff, 219, 363

sheet, 231
working load, 207
Z, 244
Pile bent, 225ff, 248
Pile driving, 195, 209ff, 257, 304
   cause of displacement, 258
   cause of vibration, 323
   dynamics, 216ff, 222
   equipment, 209
   lateral movement due to, 212
   resistance, 212, 223
   stress transmission, 218
   with vibatory equipment, 209, 211
   wave analysis, 218ff, 227
Pile hammer, 209ff, 219, 220
   closed-ended diesel, 209
   Delmag, 210
   diesel, 211
   differential steam, 210
   double acting steam, 210
   efficiency, 211, 220
   drop, 209
   impact, 209, 210
   Kobe, 210
   Link Belt, 210
   manufacturer's rating, 211
   MKT, 210
   open-ended diesel, 209
   racking, 211
   Raymond, 210
   single acting steam, 209
   Vulcan, 210
Pine plywood pile cushion, 221
Pinnacled limestone, 153, 154
Pipe, perforated, 178
Pipe pile, 207ff
Piston sampler, 109ff, 121
Pit, inspection, 116
Pittsburgh, Pa., 259
Plasticity, 5, 6
   index, 21, 26, 98, 337, 343
Plastic limit, see Limit, Atterberg
Plastic range, 21
Plastic silt, 5, 290
Plate, settlement, 196
   vibrating, 197
Plate-shaped particle, 66
Plucking, 126
Pneumatic caisson, 235, 262
Pneumatic drilling, 152
Pneumatic tamper, 197
Pneumatic-tired roller, see Roller, pneumatic-tired
Pneumatically placed concrete, 363, 369
Point-bearing pile, 212ff, 216, 281, 292
Point of application of earth pressure, 423
Point of contraflexure, sheet pile bulkhead, 453, 456
Point reinforcement of H pile, 208
Poisson's ratio, 81
Polymer stabilization, 181

Poorly graded soil, 5, 9, 27
Pore-air pressure, 94
Pore pressure, 88, 118, 217, 324, 325
   dissipation, 96
   during shear, 90
   excess, 96, 195, 269
   negative, 88, 96, 325
Porewater pressure, 44, 68, 70, 94
   excess, 53
Porosity, 11, 13, 17
Porous tube piezometer, 196
Port Alberni, 133, 135
Port Allen Lock, 330
Port Everglades, 330
Portland cement stabilization, 199
Portsmouth, 166
Portugal, 234
Position head, 39
Post hole auger, 104
Powder factor, 369
Power auger, 117, 230
Precast concrete pile, 205ff; see also Concrete pile
   reinforcement for handling, 207
   tips, 214
Precompression, 200
Preconsolidation load, see Preconsolidation pressure
Preconsolidation pressure, 60, 78
   graphical procedure, 60
Predrainage, 325
Predrilling, 212, 336
Preliminary exploration, 163
Preloaded clay, 61, 63, 67, 73, 290, 292
Preloading, 60, 200
Preshearing, 370
Presplitting, 370
Pressure, allowable soil, 186ff
   on base of retaining wall, 425
   confining, 86
   earth, see Earth pressure
   envelope for strut loads, 460
   excess, 53, 68, 96, 195
   excess hydrostatic, 40
   excess porewater, 53, 96, 195
   hydrostatic, 170
   initial consolidation, 68
   preconsolidation, see Preconsolidation pressure
   pore, see Pore pressure
   pore-air, 94
   porewater, see Porewater pressure
   seepage, see Seepage pressure
   total, 44
Prestressed pile, working stress, 213
Prestressing, 250, 252
   anchors, 468
   bracing, 463
   piles, 207
   struts, 174
Prewetting by flooding, 342
Primacord, 369
Primary consolidation, 72, 73

Primary structure, 18
Principal stress, 84
Principle of effective stress, 68
Proctor test, 14, 15, 94, 342
    standard, 15, 18, 198
Profile, soil, 126, 127
    of deposition, 126
    pedological, 127
    of weathering, 126, 153
Progressive movement in clay, 301
    bridge piers, 301
    piles, 302
Properties, index, *see* Index properties
Puerto Rico, 126
Pulling casing, drilled pier, 241
Pulling test, pile, 282
Pullman, Wash., 100
Pull-rise curve, pile, 282
Pumice, 31
Pump, deep well submersible, 180
Pumping, 64
    test, 42, 164, 326
    well, 42

Q-test, 92ff
Quarry blasting, 369
Quartzite, 31, 32, 100, 127
Quebec, 131, 137, 138, 139
Quick clay, 20, 137, 334
Quick condition, 46, 54, 212, 326
Quicksand, 46, 458

R-test, 88ff, 91, 97
Racking of pile hammer, 211
Raft, allowable soil pressure, 319
    on clay, 276ff
    compensated, 188
    flat-slab analogy, 406
    flexible, 412
    loads, 188
    on sand, 318ff
        hydrostatic uplift, 320
    stiff, 412
    stiffened, 188
Raft foundation, 152, 185ff, 188, 400ff
    compensated, 277
    design, 405ff
    differential settlement, 320
    rigid frame, 188
    settlement, 291, 293
    stiffened, 188
Rail grillage, 185, 187
Raker, 170, 172, 174, 251, 463, 464
Rankine earth pressure theory, *see* Earth Pressure
Rankine state, active, 418, 459
Rate of consolidation, 68ff
    of loading, 95
    of secondary consolidation, 73, 296
    of settlement, 68ff, 81, 195
        piles in clay, 294

Ratio, area, 108
    c/p, 93, 98
    filter, 49
    inside clearance, 108
    overconsolidation, 60, 93
    Poisson, 81
    recovery, 111, 151
Raymond pile hammer, 210
Raymond step-taper pile, 207
Rebound curve, 60
Recharge of groundwater, 330
Recovery ratio, 111, 151
Redriving of pile, 292
    test, 225, 285ff
Reduction coefficient, piers in clay, 280
    piles in clay, 282, 283, 286
Reduction factor, strength of undrained clay, 96
Refraction, seismic, 118ff
Reinforced concrete design, 375ff, 383
Reinforced concrete footing, 185, 186
Reinforcement, placement, 377
    spacing, 377
Relative density, 11, 13, 14, 117, 118, 198, 307, 308, 311, 325, 327
    correct measurement, 112ff
Relaxation of piles, 223, 224, 325
Relict joint, 154
Reloading curve, 60
Remolded clay, 59, 162
Remolded soil, 20, 23
Removal of casing, drilled pier, 240ff
Repetitive loading, 95
Report, engineering, 165
Representative sample, 107, 108, 163
Residual boulder, 153
Residual clay, 155
Residual soil, 153ff, 333
    collapsible, 335
    consolidation characteristics, 65
Resistance, cone, 115
    penetration, *see* Penetration resistance
    shearing, 214
    uplift, 384
Resistance diagram of pile, 224, 285
Resistivity survey, 120
Resonance, 327
Resonant pile driver, 212
Retaining wall, 246ff, 385, 414ff
    backfill, 247
    base width, 415
    cantilever, 246, 247, 415, 427
    continuous back drain, 247
    counterfort, 246
    crib, 246ff
    damage due to overcompaction, 197
    earth-pressure charts, 424, 425
    gravity, 246, 415
    monolithic, 247
    pile-supported, 428
    pressure against base, 425

rock-supported, 426
semi-gravity, 246, 415
sliding resistance, 426
soil-supported, 426
stability analysis, 416, 424
weep hole, 247
Retarder, 181
Rhyolite, 31
Richmond, Va., 305
River deposit, 144ff
River gravel, 143
Robert College, 444
Rock, 3
  allowable pressure, 361, 362
  anchor, 360
  bedding plane joint, 367
  blasting, 369
  bolts, 370
  cemented, 30
  classification, 30
  compressibility, 151
    load test, 367
  core, 36, 109ff, 121
  defects, 149
    treatment, 364ff
  deformation, 99
  engineering characteristics, 31
  excavation, 369ff
  fabric, 31, 32
  foliated, 30
  foundation, 361ff
    on weathered, 367ff
  igneous, 31, 32, 368
  index properties, 30
  joint, *see* Joint
  laminated, 30
  metamorphic, 31, 32, 368
  mineralogy, 32
  permeability, 44, 151
  pier, design, 363
  quality, 112, 150
  resistivity, 120
  sampling, 122
  sedimentary, 31, 32, 151, 367
  seismic velocity, 120
  settlement on, 361
  solubility, 30
  solution features, 152; *see also* Solution
  splitting, 370
  strength, 99
    of intact, 30
  structural features, 370
  tangent modulus, 100
  texture, 32
  unconfined compressive strength, 30
  volcanic, 154
  weathered, 153ff
  weathering, 368
Rock flour, 5, 9, 153
Rock mechanics, 36

Rock quality designation, 112
Rock salt, 31, 32, 153
Rocky Mountains, 146
Roller, pneumatic-tired, 195, 197, 200
  sheepsfoot, 196, 197, 200
  vibratory, 196, 200, 330
Roothole structure, 142
Rotary drilling, 106ff, 212
Rounded particles, 66
RQD, 112, 150, 151, 361, 370
Rubber balloon method, 16
Rubble stone footing, 185
Rubber tired roller, *see* Roller, pneumatic-tired
Rupture circle, 86ff, 92, 418, 419
Rupture diagram, 86ff
Rupture line, 417

S-test, 88, 97
Safe load, on friction pile, 281ff
  on group of piles, 285, 286
  on piles in clay of increasing stiffness with depth, 285ff
  on point-bearing piles, 284, 322ff
Safe soil pressure, 271, 276
Safety factor, 266
St. Lawrence River, 78, 138
St. Lawrence Valley, 334
St. Thuribe, Que., 137
Salt, 153
Salt Lake, Utah, 138
Sample, 102
  auger, 106
  continuous, 164
  hand carved, 117, 121
  representative, 107, 108, 163
  spoon, 107
  tube, 147, 164
  undisturbed, 107, 108, 163, 164
Sampler, piston, 109, 121
  split barrel, 108, 164
  thin walled, 109
Sampling, 106ff, 122
  block, 113
  spoon, 107, 121
San Antonio, 343
San Francisco-Oakland Bay Bridge, 228, 243
Sand, 5, 27, 95, 145, 153, 170, 177, 178, 247
  bearing capacity, 307ff, 318
  braced cut in, 456
  cemented, 236
  cohesion in, 257
  cohesionless, 176, 196
  compaction, 200
    by vibration, 327
    by vibratory roller, 330
  cone resistance, 115
  consolidation characteristics, 66
  dense, 66, 88
  deposits, significant characteristics, 307
  design charts, footings, 311ff

design of raft foundation, 318ff
drained friction angle, 87
dune, 141ff
effect of, lowering water table, 257
    vibrations, 327ff
erratic deposit, 314
excavation, 255ff, 325ff
flowing, 242
footing on, 307ff, 314ff
gap graded, 66
load-settlement diagram, 117
loose, 66, 88, 214, 308
penetration resistance, 114, 146
permeability, 43
piers on, 321ff
piles in, 322ff
quick condition, 326
resistivity, 120
seismic velocity, 120
settlement, 265, 308
    due to vibration, 327ff
shear strength, 87, 97
slope, 170, 325
strength, 97, 99
stress-strain relations, 85
strut load, open cut, 459ff
undisturbed sample, 109
uniform, 66
well graded, 66
wind blown, 141ff
Sand cone method, 16
Sand drain, 180ff, 195
Sand island caisson, 234
Sandstone, 31, 100
    seismic velocity, 120
Saprolite, 153, 369
Saturated unit weight, 13, 17, 18
Saturation, degree, 12, 17, 18, 47
Scandinavia, 65, 138, 334
Schist, 31, 32
Schistosity, 153, 370
Scour, 245
    of bridge pier, 330
    depth, 325
    use of piles to prevent, 325
Seattle First National Bank, 472
Secondary consolidation, 73, 195
    coefficient, 73
    rate, 73, 296
Secondary structure, 19, 116, 269, 270
Sectional precast concrete pile, 206
Sedimentary rock, 31, 32, 151, 367
Sedimentation, 8
Seepage, 51ff, 235
    force, 177
    line of, 47, 55
    pressure, 39, 46, 49, 51, 177, 180, 182, 451
Segregation, 41, 50
    of concrete, 240
Seismic survey, 118

Seismic velocity, 119, 120
    in pile, 218
Semi-arid climate, 65, 334
Semi-gravity retaining wall, 246, 247, 415
Sensitive clay, 20, 290
    consolidation characteristics, 64
    penetration test, 115
Sensitivity, 20, 137, 139
Serpentine, 31, 151
Settlement, 58, 59, 62, 67, 166, 173, 174, 187, 195, 252, 256, 258, 264, 269, 270, 284, 307, 334, 335, 350, 351
    adjacent to open cut, 463
    of anchor rods, 455
    on clay, 265
    computation, 62, 290
    deep seated, 194ff, 199
    differential, 188, 193, 254, 319
    due to, defects in rock, 361
        excavation, 255
        lowering water table, 257
        vibration, 257, 327ff
    effect, of water table fluctuation, 257
        on structures, 266
    on erratic deposit, 292ff
    of footing, 290ff
        on clay, 291
        on sand, 308ff
    forecast, reliability, 290
    of foundation, 290ff
        underlain by clay, 286ff
    of friction piles in clay, 291ff
    of pier in clay, 292
    of pier on sand, 322
    of piles, 216, 226, 292
    of raft foundation, 291, 293
    on sand, 265
    rate, 68ff, 81, 195
        piles in clay, 294
Settlement plate, 195, 196
Settlement-time curve, 71
Shaft, inspection, 116
Shaking test, 5
Shale, 31, 136, 363, 368, 369
    deterioration, 363
    seismic velocity, 120
    swelling, 67
Shallow excavation, 169
Shape of particle, 5
Shear, critical section (concrete design), 376
    granular mass, 81ff
    pore pressure during, 90
    test, direct, 83
    zone, 30, 99, 149, 151, 366, 370
Shear apparatus, vane, 116
Shearing resistance, 170, 214
    clay, 97, 102
    direct measurement, 116
    dry sand and gravel, 87
    gravel, 87, 97

mylonite, 152
sand, 97
silt, 97
undrained, 80, 269
unsaturated soil, 94
Shearing strength, *see* Shearing resistance
Sheepsfoot roller, 196, 197, 200
Sheet pile, 173
  arch web, 171
  cutoff, 51
  design for, bulkhead, 452
    open cut, 463
  flat web, 171
  splined, 231
  steel, types, 171
  tongue and groove, 231
  Wakefield, 231
  wood, 231
  Z, 171
Sheeted pit, 229
Sheeting, 170ff
  depth of embedment, 452
  design for, bulkhead, 452
    open cut, 463
Shock loading, 97
Shore deposit, 125, 147, 150
Shoring, 251
Shoreline deposit, 136, 138
Shot core barrel, 117
Shrinkage, 159, 186, 247, 337, 343
  crack, 247
  limit, 21, 24
  steel, 390
  stress, 64, 390
Sierra-Cascades, 146
Sieve, 8
Silica tetrahedron, 10
Silt, 5, 145, 164, 179, 196, 242, 247
  cohesionless, 307
    piles in, 324ff
  deposits, significant characteristics, 307
  drained friction angle, 87
  foundations on plastic, 269ff
  inorganic, 5, 22, 27
  nonplastic, foundations on, 307ff
  organic, 7, 22, 27, 192, 257
  penetration resistance, 146
  permeability, 43
  plastic, 5, 290
    foundations on, 269ff
  resistivity, 120
  seismic velocity, 120
  shear strength, 97
  skin friction on pier, 281
  stress-deformation characteristics, 90
  wind blown, 141ff
Silty clay, strength, 94
Siltstone, 31, 100
Single acting steam hammer, 209
Single grained structure, 19

Single tube core barrel, 111
Single wall cofferdam, 232, 244
Sinkhole, 30, 32, 152, 153, 257, 366
Skin friction, 216, 279, 284, 291, 292
  pier, 280, 281
  pile, 213, 223
Slaking, 50
Slate, 31, 32
Slickenside, 19, 20, 30, 93, 98, 152, 272, 297, 339, 340, 343
Slide, 169
  flow, 138, 139
Sliding of retaining wall, 426
Slope, 128, 169, 177
  clay, 170, 299
  construction, 169
  critical height in clay, 299
  in homogeneous soft to medium clay, 298ff
  in stiff clay, 98, 299ff
Slope circle, 298
Slope stability, 298, 325
  in clay, 300
  in nonuniform soil, 358
Sloping rock surface, 363
Slump of concrete, 240
  cast in place pile, 213
Slurry, 173, 181, 237, 238, 256
Slurry trench, 173
Slurry wall, 360, 472
Slush grout, 365
Smead, Mont., 9
Smectite, 10, 11
Soapstone, 31
Soil, 3, 126, 127
  aeolian, 126
  aggregate, consistency, 18
    structure, 18
  aggregate properties, relations, 17
  alluvial, 126
  black cotton, 159
  chemical stabilization, 22
  classification, 24ff
    shortcomings, 28
  coarse grained, 4, 8, 13, 27
    strength, 99
  creep, 341
  collapsible, 65, 164, 334ff
  colluvial, 126
  color, 128
  constituents, specific gravity, 12
  deposits, *see* Deposit
  disturbance of structure, 180
  exploration, 103ff; *see also* Exploration, soil
  fine grained, 4, 5, 8, 13, 27
    strength, 90ff, 99
  gap graded, 9
  glacial, 126
  heave, 348
  horizon, 127
  lacustrine, 126

lime treated, 200, 346
marine, 126
normally loaded, 61, 73
organic, 7, 78, 164, 195
partially saturated strength, 99
physico-chemical properties, 35
poorly graded, 9, 27
preloaded, 61
profile, see Profile, Soil
remolded, 23
residual, 65, 126, 153ff
   collapsible, 335
sampling, 122
saturated strength, 97
series, 127
shrinkage, 343
solidification, 181
stabilization, 182
strength, see Shearing resistance
stress-strain relations, 59, 81
swamp, 27
transported, 126
tropical weathering, 156
types, 4
uniform, 9
well graded, 9, 27
Soil pressure, allowable, 186ff
  distribution beneath footing, 396
  net, 378
    allowable, 272
  raft on sand, 318ff
  safe, 271
Soil survey, 128, 164
Soldier pile, 172
  design, 463
Solidification of soil, 181
Solubility of rock, 30
Solum, 127
Solution, 126, 153, 165
  cavity, 44
  channel, 125
  features, 149, 367
    of rock, 152
    of rock salt, 32
Sounding, 102
South Africa, 346
South Dakota, 332
Spacing of piles, 285
Specific gravity, 13
  of solid constituents, 12
Spheroidal weathering, 154, 157
Spill-through abutments, 248
Splined sheet pile, 231
Split barrel sampler, 108, 164
Spread footing, 185
Spring, 257
Spoon, sampling, 107, 108, 164
Spotter, pile, 209
Sprague and Henwood, 121
Spudding of pile, 212

Stability, cuts in clay, 297ff
  excavations, 177
  long term, 98
  number, 298
  retaining wall, 416, 424
  slope, 298, 305, 354ff, 358ff
Stabilization, 177ff, 337
  chemical, 22
  freezing, 182
  lime, 199, 200
  Portland cement, 199
Static cone test, 121
Standard load test, 118, 143, 164, 198
Standard penetration resistance, see Penetration test, standard
Standard penetration test, see Penetration test, standard
Standard Proctor test, see Proctor test
Static cone test, see Cone test
Static penetration test, 112, 115
Statistical relationships, 66
Steam hammer, 209, 220
Steel, sheeting, 171, 172
  shoe for pile, 204
  shrinkage reinforcement, 390
Steel deterioration in soil, 147
  I-beam grillage footing, 185, 187
  pile, 207ff, 219
    concrete encasement of, 208
    H-, 207
    pipe, 207, 208, 223
    point for H-pile, 363
Stiff clay, 20, 236
  crust, 350, 351
  crust underlain by soft clay, 352
  slope, 98, 299ff
Stiffener in wale, 463
Stiffening of raft, 188
Stokes' law, 8
Straits of Mackinac, 262
Stratified deposit, excavation, 256
  permeability, 42
Stream, braided, 146
  channel deposit, 144
Strength, clay, 94, 97, 102
  coarse-grained soil, 99
  compressive, 85
  creep, 95
  direct measurement, 116
  dry, 5
  dry sand, 87
  fine sand, 90ff
  fine-grained soil, 99
  gravel, 87, 99
  rock, 99, **100**
  sand, 99
  soil, 81ff
  unconfined, rock, 20
  undrained, 80, 116, 269
Strength design (concrete), 375

Subject Index

Stress, circle of, 83ff, 417
 in anchor rods, 455
 effective, 44, 56, 88
 lateral, 97
 neutral, 44
 principal, 83
 principle of effective, 44
 shrinkage, 64
  concrete, 390
 tensile, 82
 total, 44, 92, 98
Stress-deformation characteristics, normally loaded clay, 90ff
 overconsolidated clay, 90
 silt, 90
Stress difference, 84
Stress-strain characteristics, soil, 59ff, 81
 dry sand and gravel, 85
Stress transmission characteristics of piles, 219
Structure, dispersed, 19, 20
 fill supported, 194
 flocculated, 19
 lattice, 10, 11
 primary, 18
 root hole, 142
 secondary, 19, 116
 single grained, 19
Strut, 170ff, 231
 buckling, 456, 461
 distance between, 463
 prestressing, 174
Strut load, open cut, 460
 clay, 462
 sand, 459ff
Subgrade reaction, coefficient, 408, 413
 modulus, 408, 413
Submerged unit weight, 44
Submersible pump, 180
Subsidence, 257
Subsurface exploration, 103ff, 125ff, 236
 program, 163ff
Suction pump, 178
Suffolk, 166
Sump, 177ff, 326
 filter-protected, 178
Surcharge, 192, 195, 196
 depth of, 270
Surface drainage, 128
Surface tension, 47ff, 50
Survey, fact finding, 164
 geological, 164
 geophysical, 118ff
 resistivity, 120
 seismic, 118ff
 soil, 128, 164
Swamp soil, 27
Swedish Geotechnical Commission, 162
Swedish State Railways, 162
Swelling, 60, 63, 92, 98, 159, 186, 247, 333
 prevention, by prewetting, 342
 of damage by, 340
 unsatisfactory performance due to, 339
Swelling capacity, 337
Swelling clay, 67, 195, 199
 pier on, 343
Swelling potential, 198, 199, 337, 343, 346
Swelling pressure test, 338, 343
Swelling shale, 67
Swelling soil, 337
 damage caused by, 340
 design of foundation on, 340ff
 foundation on, 337ff
 identification, 337
Swelling test, 337ff
 modified, 343
 unrestrained, 337
Syenite, 31

Tagus River Bridge, 234, 243
Tampa, Fla., 359
Tamper, hand-held compacting, 197
Tangent modulus, initial, 86
Taper, pile, 213, 214, 324, 330
T-beam stiffened raft, 188
Telltale, 292
 in drilled pier, 242
Tensile force in pier, 340
Tensile wave in pile, 218
Tension, diagonal (concrete design), 376
Tension pile, 394, 452
Tension in soil, 82
Teredo, 204
Terminal moraine, 129ff
Termite, 204
Terra rossa, 154
Test, AASHO modified compaction, 15
 anchor, 468
 Atterberg limits, see Limit, Atterberg
 classification, 4, 7
 compaction, 14, 15, 36
 cone penetration, see Cone penetration test
 confined compression, 59; see also Consolidation test
 consolidated undrained, 88
 consolidation, 59; see also Consolidation test
 direct shear, 83
 dispersion, 6
 double oedometer, 333ff
 drained, 88, 97, 101
 Dutch cone, see Cone test, Dutch
 dynamic penetration, 112, see Penetration test, dynamic
 field density, 16
 liquid limit, 16
 load, 93, 117, 270, 282; see also Standard load test
  on pile, 292, 323
  on rock, 367
 modified swelling pressure, 343
 moisture-density, 14

oedometer, 59
permeability, *see* Permeability test
penetration, 121, *see also* Standard penetration test
pile load, 214ff, 224, 283, 284
   pulling, 282
   redriving, 285ff
plastic limit, 23
Proctor, *see* Proctor test
pumping, *see* Pumping test
Q, 92
R, 88, 91, 97
S, 88, 97
shaking, 5
standard load, 118, 143, 164, 198
standard penetration, *see* Penetration test, standard
standard Proctor, *see* Proctor test
static cone, 121
static penetration, 112
swelling, 337ff
swelling pressure, 338
torvane, 92, 116
triaxial, *see* Triaxial compression test
unconfined compression, *see* Unconfined compression test
undrained, 101
unrestrained swelling, 338
vane, 92, 96, 164, 270, *see also* Vane
Test excavation, 93
Test pit, 116
Texas, 79, 337
Textural classification, 25, 28
Texture of rock, 32
Thawing, 153
Theory, of consolidation, 68ff, 81
   of elasticity, 288
   of pile-driving dynamics, 221
Thermal stabilization, 200
Thin-wall sampler, 109
Thixotropy, 20, 282
Three-layer mineral, 10
Tieback, 172
   anchorage, 467
   cut, 446
   inclination, 468
   support system, 173, 360, 466ff, 472
Till, 9, 129, 130, 132
   index properties, 131
   permeability, 43
   seismic velocity, 120
   Wisconsinan, 130
Till plain, 130, 134
Timber grillage, 185, 187
Timber pile, 203ff, 363
   working load, 207
Time lag, 118
Time factor (consolidation), 71
Time of consolidation, 179
Time-settlement curve, 71

Titanium, 159
Tongue and groove sheet pile, 231
Torvane, 92, 116
Total pressure, 44
Total stress, 44, 92, 98
   analysis, 101
Tower Latino Americana, 304, 348
Transcona elevator, 304
Transient load, 97
Transmission of driving stress in pile, 218
Transported soil, 126
Trapezoidal footing, 401, 404
Trees, cause of shrinkage, 343, 346
Tremie, 173, 235, 240, 242
Trench, braced, 173
   slurry, 173
Trench brace, 170
Trial wedge method, 422
Triangular classification chart, 25, 127
Triaxial compression test, 83ff, 88, 101
Triple tube core barrel, 111
Tropical weathering, 155
Tropics, 126
Truck-mounted drilling machine, 104
Tube sample, 109, 147, 164
Tuff, 31, 100
Two-layer mineral, 10
Two-stage wellpoint system, 306

U-abutment, 248
Ultimate bearing capacity, 276, 351
   clay, 270ff
   footing on sand, 310
   friction pile, 283
   net, 271
   pile, 215, 222, 282
   skin friction, pier, 280
   see also Bearing capacity
Uncased concrete pile, 205
   Franki, 205
Unconfined compression test, 19ff, 92
Unconfined compressive strength, 19, 272, 361
   rock, 30, 100
Underpinning, 58, 251, 252, 295
Undisturbed clay, 162
Undisturbed sample, 107, 108, 163, 164, 366
   sand, 109
Undrained condition, 92, 96, 461
Undrained strength, 80, 116, 269
   reduction factor, clay, 96
Undrained test, 101
Unified classification system, 25, 27ff, 36, 128
Uniform soil, 5, 9, 66
Uniformity coefficient, 9, 28, 30
Union Metal monotube pile, 205
Unit weight, 12, 13
   dry, 12, 15, 18
   zero air void, 12
   of solid constituents, 12
   saturated, 12, 13, 17, 18

submerged, 44
Unsaturated soil, shear strength, 94
Unsymmetrical footing, 392
Unweathered bedrock, 149ff
Uplift force on pile, 225
Uplift of swelling soil, 345
Uplift resistance, 385
 piles, 226
  in clay, 286
  in sand, 325
Utilities, damage due to swelling soil, 339
USDA, 128
US Forest Service, 129
USGS, 129
Utah, 138

Vacuum well point, 179
Valley, buried, 139
 deposit, 145
Vancouver, B. C., 78, 305
Vancouver Island, 131
Vane, 98, 113, 116, 122, 164
 test, 92, 96, 164, 270
Varved clay, 138
Varved deposit, 135
Vegetative cover, 128
Velocity, discharge, 40
 of longitudinal wave in pile, 218
 seismic, 119, 120
Velocity head, 45
Venezuela, 358
Vermiculite, 341
Vermont, 100
Vertical cleavage, 142
Verticality, of drilled pier, 241
 of pile, 209
Vertical joint, 364, 365
Vertical pressure, computation, 287ff
 influence chart, 288ff
Vertical movements, 174
Vibration, due to pile driving, 323
 of concrete, 240
 effect on, settlement, 257
 sand, 327ff
Vibrating plate compactor, 197
Vibratory pile driver, 209, 211
Vibratory roller, 196, 200, 330
Vibroflotation, 195, 200, 327
Victoria, B. C., 305
Virgin consolidation line, 60
Virginia, 100, 305
Visual classification, 107
Void ratio, 11, 13, 17, 18, 60
 critical, 86
 in densest state, 13
 in loosest state, 13
Volcanic deposit, 139
Volcanic glass, 91, 156
Volcanic rock, 154, 156
Volume change, 85, 94, 333

 damage due to, 332
Volume compressibility, 70
 coefficient, 73
Vulcan pile hammer, 210

Wagoshance Lighthouse, 262
Wakefield sheeting, 231, 448
Wale, 170, 171, 174, 231, 467
 crippling, 456, 461
 design, 463
 stiffeners, 463
Wall, anchor, 448
 parapet, 441
 retaining, 385
Wall footing, 185, 186, 375ff
 reinforcement, 377
Wall friction, angle, 423
Wash boring, 104
Washington, 100, 142, 143
Water, in drilled pier, 240
Water content, 11, 17, 64, 198
 natural, 23
 optimum, 15, 18, 195, 198
Water level, 180
Water surface, free, 47
Water table, 42, 47, 118, 120, 169, 307, 308, 311, 313
 correction factor for foundation on sand, 313
 effect on design in sand, 316
 fluctuation, 257
 lowering 47, 226, 256
  effect on settlement, 257
Waterproofing, 189ff
Wave analysis of pile, 216ff, 224
Wave equation for pile driving, 202, 207, 208, 221, 227, 282ff, 323ff
Weathered rock, 153ff
Weathering, 30, 99, 126, 127, 151, 158, 368
 chemical, 125, 126, 153, 155
 mechanical, 153
 physical, 125, 126
 profile, 153
 spheroidal, 154, 157
 tropical, 155, 156
Wedge theory, earth pressure, 420ff
Weep hole, 247
Weight-volume relationship, 11
Well, drainage, 180
 dredging, 234
 observation, 42
 pumping, 42
Well graded soil, 5, 9, 27, 66
Well point, 54, 178ff, 326
 multiple stage, 179, 306
 vacuum, 179
West Nyack, N. Y., 100
Wetting, effect on loess, 144
Wheel load on abutment, 440
Will County, Ill., 160
Winchester, Mass., 9

Wing wall, 438, 439
Winnipeg, Man., 79
Windblown deposit, 140ff
Wire-line drilling, 111
Wisconsin, 100, 131, 149, 150
Wood pile, 219
   cushion, 220
   resistance to driving, 212
   sheet piles, 231
Working chamber, 235
Working hours, pneumatic caisson, 235
Working load on pile, 207
Working stress, concrete pile, 212
   steel pile, 213
Working stress method (concrete design), 375
World Trade Center, 151, 173, 360, 366, 367

X-ray analysis, diffraction, 10

Yugoslavia, 161

Z-pile, 244
Zero air voids, 12, 18
Zone, capillary, 118
   of depletion, 127
   of seasonal variation, 339

CPSIA information can be obtained
at www.ICGtesting.com
Printed in the USA
BVHW090749191218
535830BV00007B/51